Proceedings in Life Sciences

Blue Light Effects in Biological Systems

Edited by H. Senger

With 298 Figures

Springer-Verlag
Berlin Heidelberg New York Tokyo 1984

Prof. Dr. HORST SENGER
Fachbereich Biologie-Botanik
der Philipps-Universität
Marburg Lahnberge
3550 Marburg, FRG

ISBN-13:978-3-642-69769-2 e-ISBN-13:978-3-642-69767-8
DOI: 10.1007/978-3-642-69767-8

Library of Congress Cataloging in Publication Data Main entry under title: Blue light effects in biological systems. (Proceedings in life sciences) 1. Blue light-Physiological effect. 2. Plants, Effect of blue light on. 3. Ultraviolet radiation-Physiological effect. I. Senger, H. (Horst), 1931 – II. Series. QH515.B566 1984 574.19′154 84-5449

Softcover reprint of the hardcover 1st edition 1984

2131/3130-543210

This book is dedicated to the memory of Warren L. Butler,
a pioneer in photobiology

Preface

Four years ago *The Blue Light Syndrome* was published as the Proceedings of the *1st International Conference on the Effect of Blue Light in Plants and Microorganisms.*

Subsequently the interest in this fascinating and growing field of research has further increased, as is reflected by numerous publications. Blue light effects cover such a wide spectrum of organisms, responses and methods that communication among scientists with backgrounds in biology, biochemistry, and biophysics is particularly necessary. These facts not only justified, but demanded calling the "Blue Light Family" together again. In spite of many financial problems, the second conference attracted 113 active members from 19 countries.

The *2nd International Conference on The Effect of Blue Light in Plants and Microorganisms* was held in July 1984, like the first at the University of Marburg. The organizer could again rely on the help of the International Advisory Committee (W. Briggs, Stanford; M. Furuya, Tokyo; J. Gressel, Rehovot; S. Miyachi, Tokyo; W. Rau, München; J. Schiff, Waltham; P.-S. Song, Lubbock). The very generous financial assistance from the DFG and the support of the Philipps-Universität Marburg and its Sonderforschungsbereich "Zellenergetik and Zelldifferenzierung" were the prerequisites to organizing the conference.

The present book consists of 56 original papers. The partitioning into eight chapters is always a problem. The grouping of different aspects of the papers into these chapters has not always been obvious, so that one or the other contribution could possibly fit in another chapter.

Compared to the *Blue Light Syndrome*, the Proceedings of the first conference, the review articles are considerably reduced in the present book; the network of the various blue light effects of some organism has been summarized. Most room has been given to original papers reflecting the advances of the last 4 years of blue light research.

It is my great pleasure to thank, on behalf of all participants, the above-mentioned organizations, my secretaries Mrs. Ilse Krieger and Mrs. Petra Schreiber, and my unnamed co-workers and students whose indefatigable efforts made this conference and the publication of these Proceedings possible. Thanks are also due to the Springer-Verlag and its editorial staff who again devoted their expertise, efforts, and patience to publishing this book.

I owe my sincere thanks also to all contributors for their prompt and willing cooperation. It was a pleasure to edit the book for them with the Springer-Verlag. It is the editor's hope that this book will help to further stimulate all researchers in the field of blue light effects, and that it will serve as a source of information to those interested in photobiology.

Marburg, September 1984 Horst Senger

Contents

Development

Movement and Growth

Oecological Aspects

Contributors

You will find the addresses at the beginning of the respective contributions

Diversity of Photoreceptors

The Blue Light Responses in Evolutionary Studies

M. S. KRITSKY[1]

The ability to respond to environmental changes by adaptive corrections to physiological processes is a very important characteristic of any living system. This circumstance explains the interest in the origin and evolution of receptors for environmental stimuli providing regulation of metabolic reactions and physiological functions.

For widely differing organisms a change of illumination conditions supplies most of the essential information about the surrounding medium. This paper is an attempt to comprehend the evolutionary position of some photoregulation mechanisms, the so-called blue light responses. This term comprises processes differing in ultimate physiological response but with similar action spectra, with maximal activity in the short wave visible and near ultraviolet regions [13].

The results obtained by using different experimental approaches lead to the conclusion that the key element of the photoreception mechanism is the excitation by light of some flavin molecules. A more detailed chemical nature of the photoreceptor is still unclear. Unsuccessful attempts have been made to identify it with some defined flavoprotein components such as nitrate reductase [2]. The operation in the photoreception of free, i.e., covalently unbound flavin molecules, FMN or FAD, cannot be ruled out.

In the purely chemical, nonenzymic systems, excitation of free flavins by light induces effective photocatalysis of redox reactions. There are also some experimental grounds to suggest functioning of the flavin-dependent photocatalysis of redox reactions within the framework of the blue light response systems. Some experimental evidence is in favor of the relationship between the function of the photoreceptor and flavin-dependent photoreactions: oxidation of NAD.H [6] and reduction of cytochrome b-type [9].

It should be emphasized that the photoregulation processes involving flavin photocatalysis are inherent not only in aerobic, i.e., evolutionarily advanced organisms. Recent results have shown that photoregulation of the electron transport chain of the extremely primitive organism, methanogenic bacteria, may involve a flavin component named factor 430. This factor contains deazaflavine, a substance similar to "common" flavins except for the presence of a carbon instead of a nitrogen atom in the 5-position of the isoalloxasine ring, and with lower redox potential (–0.38 mV instead of –0.30 mV for FMN) [1, 10].

1 A.N. Bach Institute of Biochemistry, Moscow, USSR

Blue Light Effects in Biological Systems
ed. by H. Senger

The mechanism of photoregulation in light activation of flavin molecules may be regarded as an evolutionary rudiment of ancient photobiological processes, whose roots lie in the early development of life. This hypothesis is based mainly upon the following experimental observations.

1. Flavins could be accessible for chemical processes at the early stages of biopoesis. It has been established in the laboratory of K. Dose that these substances could appear in abiotic reactions under primitive Earth conditions [4].

2. There are grounds to believe that flavins joined the metabolic system of the actualistic cell at the early stage of evolution. This suggestion is supported by the presence of flavins in all living organisms, as well as by the evolutionary antiquity of some flavoproteins. For instance, the evolutionary age of flavodoxins is estimated at 3.5 billion years [12].

3. Flavins absorb light and thus manifest the photocatalytic activity in the relatively short wave spectral zone (not exceeding 500 nm) [14], i.e., the photochemical systems are especially sensitive to the spectral zone of solar radiation which reached the Earth surface during early evolution periods. It should be mentioned again that the covalent binding to a protein molecule is not a necessary prerequisite for flavin photocatalysis.

4. In contrast to other photobiological processes (the exception seems to be photoreparation of DNA) the blue light responses are not inherent in a limited taxonomic group, but characteristic of very different organisms: prokaryotes and eukaryotes, heterotrophs and photoautotrophs, plants and animals. The blue light response systems in these organisms control phototropic growth, realization of developmental programs, and biological motility [13].

These features may distinguish the blue light responses from other photobiological phenomena, including photosynthesis, phytochrome regulation or visual reception, as possible contenders for the role of a primordial photobiological mechanism. There is some ground to consider them as a rudiment of the most ancient photobiological processes, where light activated the molecules of coenzymes such as nucleotides and flavins. The significance of such processes for the early steps of evolution were proven in theoretical and experimental works (Krasnovskii [5], Gaffron [3]) as early as the late 1950's and early 1960's.

It is interesting that in the process of biological evolution the flavin photoactivation at the base of the blue light responses is associated with the universal mechanisms of cell metabolic control. When considering the regulatory effects on the metabolism, the shift of redox balance should be mentioned first of all. Other important properties under the control of the blue light photoreceptor are the key parameters of cell homeostasis – selective permeability of the cell membrane for ions and the level of cyclic nucleotides (Table 1). The ability of the blue light response systems to control these parameters has recently been demonstrated in our laboratory for the photoreceptor system of the fungus *Neurospora crassa* [7, 11]. It is interesting that in this respect some similarity can be observed between two photoregulation systems: the blue light response and the visual one. Despite the fundamental difference in photoreceptor pigments (rhodopsin with 11′-cis-retinal as a chromophore in the rod cells of retina and flavin chromophore in the "blue light response" systems), their activation by light pro-

Table 1. Blue light control of the key parameters of cell homeostasis

	Mycelium cells of a fungus *Neurospora crassa*	Retinal cells of vertebrate animals
Chromophore of photoreceptor	Flavin	11′-cis-retinal
Cyclic nucleotide level after illumination	cAMP decreases	cGMP decreases
Membrane potential after illumination	Hyperpolarization	Hyperpolarization
Time scale of maximal response	10^3 s	10^{-1} s

duces similar results: the activation of phosphodiesterase, decrease in cyclic nucleotide content, and hyperpolarization of the cell membrane [8].

At the present time we are not in a position to answer the question of how deep the apparent similarity is between the transduction mechanisms of these two photoregulation systems. Nevertheless, the comparison of the response time reveals that the blue light system acts more slowly: the development of the response to light stimulus develops much more slowly than in the photoreceptor cells of visual system. It can be suggested in a very hypothetical manner that this kinetic imperfection may predetermine the functioning of blue light response systems mainly in the ecological niche of slow processes such as growth and development.

References

1. Balch WE, Fox GE, Magrum LJ, Woise CR, Wolfe RS (1979) Microbiol Rev 43:N 2, 260
2. Belozerskaya TA, Burikhanov SS, Chernysheva EK, Kritsky MS (1982) Neurospora Newslett 29:14
3. Gaffron H (1962) In: Horizons of biochemistry. Academic Press, London New York
4. Henz B, Ried W, Dose K (1979) Angew Chem 91: N 6, 510
5. Krasnovskii AA (1959) In: The origin of life on the Earth. Proc Ist Int Symp. Pergamon Press, London New York Paris Los Angeles, p 606
6. Kritsky MS, Chernysheva EK, Soboleva IS (1977) Prikladnaya Biokhim Mikrobiol (Appl Biochem Microbiol) 13: N 6, 901 (in Russian)
7. Kritsky MS, Sokolovsky VYu, Belozerskaya TA, Chernysheva EK (1982) Arch Microbiol 133: 206
8. Maeda A, Yoshizava T (1982) Photochem Photobiol 35:891
9. Munoz V, Butler WL (1975) Plant Physiol 55: N 2, 421
10. Pantskhava ES (1977) Biokhimia 42: N3, 549 (in Russian)
11. Potapova TV, Levina NN, Belozerskaya TA, Kritsky MS, Chailakhjan LM (1984) in press
12. Rossmann MG, Moras D, Olsen KW (1974) Nature (London) 250:194
13. Schmidt W (1980) In: Struct Bonding, (Berlin) 41:1
14. Singer TP (ed) (1976) Flavins and flavoproteins. Elsevier, Amsterdam

Coaction of Phytochrome and Blue/UV Light Photoreceptors

H. MOHR, H. DRUMM-HERREL and R. OELMÜLLER [1]

Scotomorphogenesis Versus Photomorphogenesis

Seedlings or sprouts of higher plants follow different strategies of development depending on the ambient light conditions, namely scotomorphogenesis (etiolation) or photomorphogenesis (Fig. 1). The adaptive value of having different strategies is

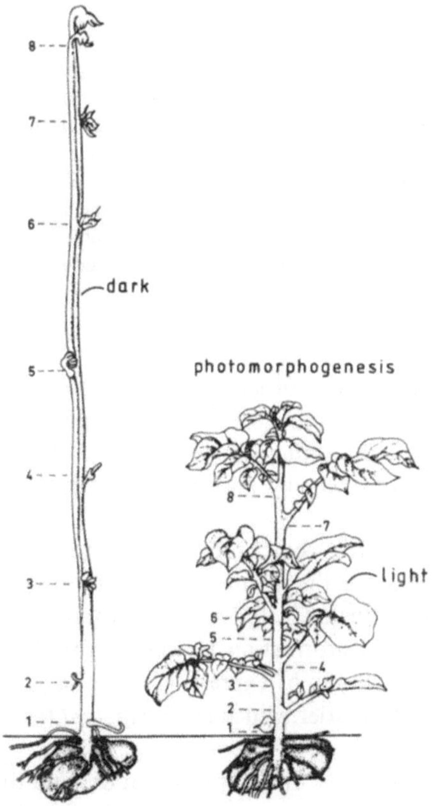

Fig. 1. Illustrations of alternative developmental strategies. Genetically identical potato plants (*Solanum tuberosum*) were grown in the dark – scotomorphogenesis – or in natural daylight – photomorphogenesis. (After [20]). *Numbers* indicate the position of the corresponding leaves along the main axis to document the constancy of the phyllotactic pattern in light and dark: pattern specification was found to be independent of light even in those cases where pattern realization, i.e., actual growth of leaves, depends on light. [13]

1 Biological Institute II, University of Freiburg, Schänzlestr. 1, D–78 Freiburg i.Br., FRG

Blue Light Effects in Biological Systems
ed. by H. Senger

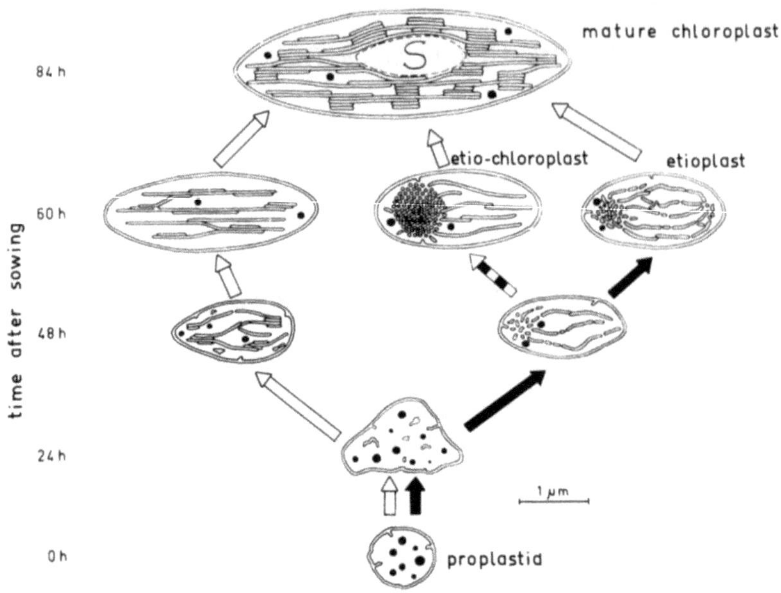

Fig. 2. A scheme showing the time course of plastid development in mustard (*Sinapis alba*) cotyledons. *Dark arrows* development in complete darkness; *white arrows* development in continuous white light (7000 lx); *white/dark arrow* treatment of the seedling with 4 red light pulses given 36, 40, 44 and 48 h after sowing to form P_{fr} (and thus to activate the phytochrome system). It is indicated that formation of P_{fr} does not suffice to transform the ultrastructure of an etioplast into that of a chloroplast. Chlorophyll formation, i.e., protochlorophyllide → chlorophyllide-a photoconversion is required as well. However, it is known (e.g., [1, 8]) that synthesis of soluble stroma enzymes, including ribulosebisphosphate carboxylase and NADP-dependent glyceraldehyde-3-phosphate dehydrogenase are controlled exclusively by phytochrome. (After [18])

obvious. Scotomorphogenesis is the appropriate strategy of survival under conditions where light is lacking or limiting, whereas photomorphogenesis is the appropriate strategy of development under conditions of light affluence. In fact, the double strategy of development can be considered one of the major evolutionary adaptations of higher plants to terrestrial life.

For the developmental biologist the *transition* from scotomorphogenesis to photomorphogenesis (called de-etiolation) is fascinating for the reason that in this process a single, well-defined environmental factor – light – causes a plant to change expression of genes. Naturally, light does not carry morphogenetic information. However, light can be considered an elective factor which controls the degree of expression of those genes which are pre-programmed to be affected by light, in brief, which are *competent* to the light factor.

Photomorphogenesis vs. scotomorphogenesis can be distinguished not only on the level of the organism but also in the case of intracellular morphogenesis of organelles such as mitochondria or plastids. As an example (Fig. 2), in the presence of light a proplastid develops into a green mature chloroplast, while development in darkness follows a different strategy, leading to an etioplast (prolamellar bodies, small amount of protochlorophyll, no chlorophyll, only small amounts of Calvin cycle enzymes, etc.). Upon irradiation with white light the transformation of the etioplast into a mature chloroplast occurs. This rapid phototransformation of the organelle is connected with extensive growth in length and volume of the plastid and with dramatic changes at the level of molecular composition, fine structure and plastid function.

The Essential Point

The transition from scotomorphogenesis to photomorphogenesis implies a strong and largely irreversible investment of matter and energy. As a result of Darwinian evolution, a seedling or a sprout may be expected to perform this transition only once the ambient light conditions have become favorable for net photosynthesis (economy principle). This expectation is derived from the fact that during seedling or sprout development a particularly strong selection pressure exists in favor of economic usage of limited storage material. In order to comply with the economy principle, i.e., in order to make the right decision with regard to developmental strategy, a plant has to sense the light conditions in its environment continuously and accurately. This implies that a plant must be capable of sensing the light conditions throughout the sun's spectrum as far as sunlight leads to electronic excitations (290–800 nm). For reasons of molecular physics it is improbable that a single photoreceptor can fulfill this task. Rather, we might expect that a higher plant will use several sensor pigments to check the whole spectral range with the sensitivity and accuracy required.

Sensor Pigments Related to De-Etiolation in Higher Plants

As far as we know today, three different sensor pigments occur in higher plants [15]. These are phytochrome (operating predominantly in the red-far-red spectral range), cryptochrome (operating in the blue/UV-A spectral range), and a UV-B photoreceptor.

The action spectrum related to the latter photoreceptor shows a single intense peak at 290 nm and no action at wavelengths longer than 350 nm [26]. It could be that the UV-B photoreceptor occurs widely but has been overlooked so far owing to experimental difficulties in working with UV-B. In any case, at least some recently investigated plants (e.g., the milo seedling, *Sorghum vulgare*) are obviously capable of sensing the light conditions throughout the sun's spectrum from UV-B to the near infrared [5].

The Problem Proper of this Article

What ist the mode of coaction between the three sensor pigments? Do they operate independently of each other, eventually causing the same photoresponses and thus the same kind of photomorphogenesis? Or, as an alternative, is there a hierarchy in coaction?

Comparative studies indicate that *phytochrome* is the dominating sensor pigment in higher plants, not only for the detection of photosignals but also, and in particular, for signal transduction (see [16], for a more comprehensive argument). However, in many instances phytochrome action depends on coaction with cryptochrome and/or UV-B photoreceptor. The physiologically active phytochrome (P_{fr}) is crucially involved as the "effector" in the transition from skoto- to photomorphogenesis. However, to establish or to increase and maintain responsivity in a plant towards P_{fr}, concomitant light absorption in cryptochrome or, sometimes, UV-B photoreceptor is required. It seems that a coaction of this kind between phytochrome and the blue/UV light sensor pigments is characterisitic for most photomorphogenesis. This coaction is highly economic since a single "effector" (namely P_{fr}) suffices to bring about the molecular events leading to photomorphogenesis, and yet information about the whole solar spectrum – as far as it is relevant for the plant – can contribute to the *rate and extent* of the photomorphogenetic responses.

Remark on the Mode of Signal Expression in Phytochrome-Mediated Responses

The mode of action of phytochrome in bringing about de-etiolation has been investigated with biochemical model systems of de-etiolation, namely light-mediated synthesis of flavone glucosides in cell suspension cultures of parsley and light-mediated synthesis of juvenile anthocyanin in the epidermal cells of mustard (*Sinapis alba*) cotyledons. Regarding the "molecular" mechanism of phytochrome action in these cases there is hardly any doubt that signal expression, i.e., appearance of the flavonoid pigments, is due to a stimulation of transcription of competent genes (Fig. 3). Experimental evidence in favor of this scheme has been summarized repeatedly (see, for example, [9, 10, 14, 18, 24, 25]), and need not be repeated here.

Regarding de-etiolation of the plastid, i.e., etioplast-chloroplast photoconversion (see Fig. 2) the mechanism of signal expression seems to be settled as well: phytochrome causes, in some way or other, activation of competent genes (review: [17]).

P_{fr} as "Effector"

In crucial experiments on phytochrome-mediated anthocyanin synthesis in mustard cotyledons, evidence was obtained that a mustard seedling responds to the amount of P_{fr} and not to the P_{fr}/P_{tot} ratio. In brief, P_{fr} can be considered as the "effector" in the photoresponse [22]. Thus, the unsolved problem is signal transduction, which means, at least in the above cases, the transduction of the P_{fr} signal to the competent genes (see Fig. 3). In an effort to elucidate the mechanism of signal transduction the phenomena of responsivity amplification were investigated.

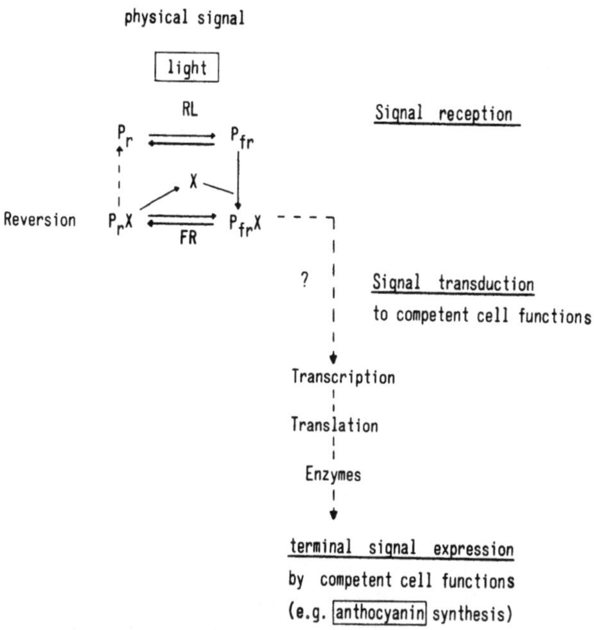

Fig. 3. A scheme to illustrate the sequence of events occurring in the case of light-mediated synthesis of flavonoid compounds. The basic observation is that the environmental (physical) signal light leads to synthesis of particular pigment molecules, e.g., anthocyanin, in particular cells. The core of the explanatory argument is that terminal signal expression (i.e., the appearance of the final response, in the present case cyanidin- or flavon glucosides) is due to stimulation of transcription of competent genes. The premises that signal reception is taking place through phytochrome and that competent cells in a de-etiolating plant respond to the amount of P_{fr} (rather than to the P_{fr}/P_{tot} or P_{fr}/P_r ratio) are well substantiated [22]. The P_{fr}-receptor X is largely hypothetical, and the mechanism of signal transduction from P_{fr} to the competent genes is still an enigma. On the other hand, the sequence of events between induced appearance of RNA and the appearance of flavonoid compounds is well documented. (After [17])

Light Control of Responsivity to P_{fr}: Some Case Studies Involving Anthocyanin Synthesis

1. Mustard seedling (*Sinapis alba*). In the mustard seedling synthesis of anthocyanin can be induced by single light pulses that operate through phytochrome [11]. Transduction of the P_{fr} signal takes place at a considerable rate and with high precision [3] as soon as P_{fr} is available. In spite of some effort we have not been able to find significant specific effects of blue or UV light on anthocyanin synthesis in mustard cotyledons [16]. However, it was noticed that the effectiveness of phytochrome (P_{fr}) in mediating anthocyanin synthesis in the epidermal cells of mustard cotyledons can be increased strongly by a light pretreatment of the seedling prior to competence (Fig. 4). The effect can best be described as "responsivity amplification". The effective light is absorbed by phytochrome. Thus, a light pretreatment, operating through phytochrome, leads to a strong, albeit transient, amplification of responsivity in phytochrome-(P_{fr}-) mediated anthocyanin synthesis. It is obvious that responsivity amplification plays a crucial role in phytochrome action under long-term or continuous light

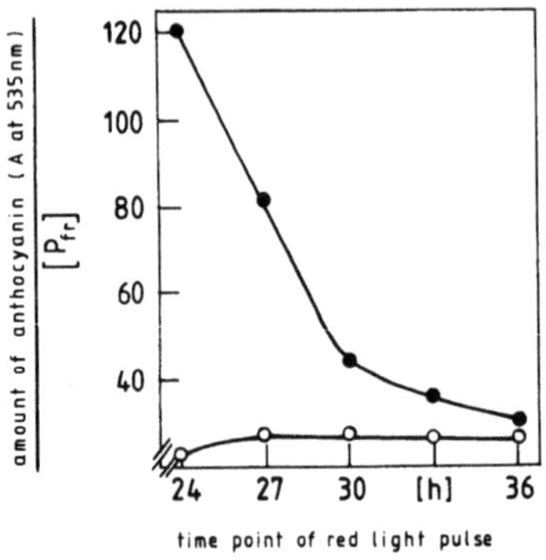

Fig. 4. Time courses of responsivity to a saturating red light pulse (5 min, 6.8 Wm^{-2}). Light-induced anthocyanin synthesis in mustard (*Sinapis alba*) cotyledons was measured. The light pulses were applied at the points indicated on the abscissa (time after sowing). Anthocyanin of the cotyledons was assayed after completion of synthesis. Pretreatments: •, 24 h red light, from sowing onwards, terminated by a saturating 756 nm light pulse to return virtually all P_{fr} back to P_r prior to competence[1]; ○, no light pretreatments. The extent of the response (amount of anthocyanin) is referred to the amount of P_{fr} established by the red light pulse. (After [22])

(e.g., under natural light conditions). Phytochrome has a double function: with its P_{fr} form it operates as the effector of anthocyanin synthesis and simultaneously it functions as a sensor pigment which determines responsivity to P_{fr}.

2. Milo seedling (*Sorghum vulgare*, cv. Weider-hybrid). The mesocotyl of the milo seedling does not produce anthocyanin in complete darkness. As described originally by Downs and Siegelman [2] even long-term red or far-red light does not lead to any anthocyanin synthesis while white light causes strong and rapid pigmentation (Table 1). The data show that phytochrome can only act once a blue/UV light effect has occurred. On the other hand, the expression of the blue/UV light effect is controlled by phytochrome. In experiments with dichromatic irradiations, i.e., simultaneous irradiation with two kinds of light to strongly modulate the level of P_{fr} on a constant background of blue/UV light, it was found that the blue/UV light photoreaction as such is not affected by the presence or virtual absence of P_{fr} [4].

The interpretation of the data obtained with milo is that phytochrome (P_{fr}) is the effector which causes anthocyanin synthesis through activation of competent genes, while the blue/UV light effect must be considered as establishing responsivity towards P_{fr}. In the case of anthocyanin synthesis in the milo mesocotyl there is no responsivity without the operation of a blue/UV photoreceptor.

3. Wheat seedling (*Triticum aestivum*, cv. Schirokko). Anthocyanin synthesis in the wheat coleoptile takes place readily in natural sunlight and in fluorescent white light (Table 2). Short and long term treatments with red and far-red light proved ineffective. Surprisingly, blue and pure UV-A have very little effect, while UV light with a small content of UV-B is very effective in eliciting anthocyanin synthesis. Following a UV treatment, the response was found to be controlled by phytochrome (Fig. 5). In fact, at least the major effect of the UV treatment is to establish responsivity to phytochrome. In contrast to milo, UV cannot be substituted for by blue light in its function. Rather, the function of the UV-B photoreceptor instead of cryptochrome is indicated. The responsivity amplification for P_{fr} brought about by UV is only short-lived. It is

---

1 In developmental biology the term competence means that a cell or a tissue is able to respond to a specific (inductive) stimulus with a specific response. In the present example, epidermal cells of a mustard cotyledon acquire competence towards P_{fr} with respect to anthocyanin synthesis approximately 27 h after sowing. Competence disappears approximately 36 h later

Table 1. Induction (or lack of induction) of anthocyanin synthesis in the mesocotyl of milo seedlings (*Sorghum vulgare*) by light of different qualities (white light: Xenon arc light, similar to sunlight, 250 Wm^{-2}). In the case of a 3 h light treatment the seedlings were kept in the dark for 24 h before extraction of anthocyanin. (After [4])

Treatment (onset 60 h after sowing)	Amount of anthocyanin (measurement: 87 h after sowing) (A at 510 nm)
27 h dark	0
27 h white light	1.85
27 h red light	0
27 h far-red light	0
3 h white light	0.19
3 h blue/UV	0.19
3 h white light + 5 min red light[a]	0.19
3 h white light + 5 min 756 nm light[a]	0.06
3 h white light + 5 min 756 nm light + 5 min red light	0.20
3 h blue/UV + 5 min red light	0.19
3 h blue/UV + 5 min 756 nm light	0.05
3 h blue/UV + 5 min 756 nm light + 5 min red light	0.19

[a] Photoequilibria of the phytochrome system are of the order of $\varphi_{red} = 0.8$ and $\varphi_{756} < 0.01$ [21]. A 5 min light pulse suffices under the present circumstances to virtually establish the photoequilibrium

$$\varphi_\lambda = \frac{[Pfr]_\lambda}{[Ptot]}$$

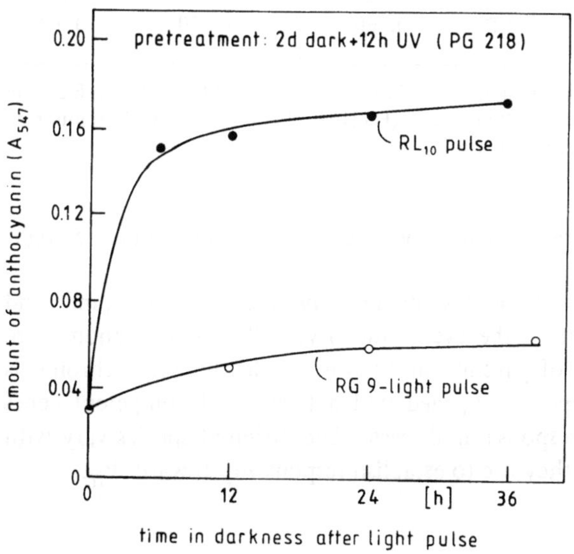

Fig. 5. Anthocyanin accumulation in the wheat (*Triticum aestivum*) coleoptile in darkness following a UV pretreatment. The pretreatment was terminated either with a saturating red light (φ_{RL} = 0.8, see Table 1) or with a saturating RG 9-light pulse ($\varphi_{RG\,9} < 0.01$, see Table 1). d, day; RL_{10}, 6.8 Wm^{-2}. (After [16])

Table 2. Induction (or lack of induction) and reversion of anthocyanin synthesis in the coleoptiles of *Triticum aestivum*, cv. Schirokko with light of different qualities (for details see [5]) and different lengths of time. At the time indicated by a point (•) seedlings received either a saturating RL or RG 9-light pulse. Anthocyanin contents were measured at the age of 98 h. RL, red light, 6.8 Wm^{-2}, $\varphi = 0.8$; RG 9-light, long wavelength far-red light, 10 Wm^{-2}, $\varphi < 0.01$.

As a gauge for responsivity to phytochrome we consider the extent of the reversible response, ΔR. Definition: ΔR = $response_1 - response_2$; $response_1$, response obtained if the light treatment is terminated with a saturating red light pulse; $response_2$, response obtained if the light treatment is terminated with a saturating long wavelength far-red (RG 9-) light pulse to return almost all P_{fr} to P_r (for definition of φ_λ-values, see legend to Table 1). (After [16])

Treatment[a]		Amount of anthocyanin (A at 546 nm) response$_1$	response$_2$	ΔR
5	d D (dark)	0.004		
2	d D + 10 h SL• + 40 h D	0.169	0.043	0.126
2	d D + 12 h WL• + 38 h D	0.092	0.030	0.062
3	d D + 2 h D• + 24 h D	0.004	0.004	0.000
3	d D + 2 h RL• + 24 h D	0.004	0.004	0.000
3	d D + 2 h BL• + 24 h D	0.004	0.004	0.000
3	d D + 2 h UV (WG345)• + 24 h D	0.008	0.004	0.004
3	d D + 2 h UV (PG218)• + 24 h D	0.012	0.004	0.008
3	d D + 2 h UV-B• + 24 h D	0.033	0.012	0.021
2	d D + 12 h FR• + 38 h D	0.005	0.005	0.000
2	d D + 12 h RL• + 38 h D	0.004	0.004	0.000
2	d D + 12 h BL• + 38 h D	0.010	0.004	0.006
2	d D + 12 h UV (WG345)• + 38 h D	0.021	0.005	0.016
2	d D + 12 h UV (PG218)• + 38 h D	0.160	0.071	0.089
2.5	d D + 12 h UV (PG218)• + 26 h D	0.195	0.141	0.044
2	d D + 12 h UV (PG218) + 12 h D• + 26 h D	0.160	0.160	0.000

[a] SL = sunlight (changing fluxes due to clouds); WL = fluorescent white light, 12,000 lx; BL, blue light, 7 Wm^{-2}; UV (WG345), UV-A, 9.3 Wm^{-2}; UV (PG218), UV-A with a small amount of UV-B, 12.6 Wm^{-2}; UV-B, λ_{max} at 310 nm, 3 Wm^{-2}

completely lost if 12 h darkness are inserted between the UV treatment and a red light pulse (see Table 2, last two lines).

The three species whose behavior was described briefly, namely mustard, milo and wheat, differ greatly with respect to the events underlying the same phenomenon, namely light-mediated synthesis of juvenile anthocyanin. Clearly, phytochrome is always involved in the photoresponse. P_{fr}-mediated activation of competent genes seems to be at the core of the response in all cases. The different species vary with regard to the photoreceptors that they use to establish responsivity towards P_{fr}.

Light Control of Responsivity to P_{fr} in Plastid-Related Photoresponses

As pointed out in the foregoing chapter a prolonged light treatment strongly increases responsivity to P_{fr} in many instances of phytochrome-controlled biogenesis of flavone

Table 3. Responsivity amplification in phytochrome-mediated GPD and Carboxylase accumulation. The seedlings were either kept in red light (RL, 6.8 Wm^{-2}) or blue/UV-light (BL/UV, 7 Wm^{-2}) or in complete darkness (D) for 72 h after sowing. The RL and RG 9-light pulses (given to obtain ΔR-values at 96 h after sowing, see Table 2) were given 72 h after sowing. Until harvest at 96 h seedling were kept in darkness

Enzyme	Light Treatment	s [μkat / Δ (ΔA)]	r_a [a]
GPD	D	0.07	–
	RL	2.80	40
	BL/UV	3.42	49
Carboxylase	D	0.25	–
	RL	9.67	39
	BL/UV	12.79	51

[a] Responsivity amplification (r_a) is defined by the equation

$$r_a = \frac{s_L}{s_D} = \frac{\left[\frac{\Delta R}{P_{tot}}\right]_L}{\left[\frac{\Delta R}{P_{tot}}\right]_D}$$

whereby s_L, responsivity in light-grown material; s_D, responsivity in dark-grown material $\Delta(\Delta R)$ is the gauge for P_{tot}

or cyanidin glycosides. Now we turn to the question of whether light also leads to a corresponding increase of responsivity towards P_{fr} in such photoresponses as are not related to synthesis of protective flavonoid pigments in outer tissue layers of seedlings.

Phytochrome-mediated accumulation of chloroplast GPD (glyceraldehyde-3-phosphate dehydrogenase, EC. 1.2.1.13) and Carboxylase (ribulosebisphosphate carboxylase, EC. 4.1.1.39) were chosen as responses and the milo shoot (*Sorghum vulgare*, cv. Weider-hybrid) as an appropriate subject. Since both enzymes respond identically with regard to responsivity amplification (Table 3) kinetic data will be given only for GPD.

Table 4. Induction (or lack of induction) of anthocyanin synthesis in the hypocotyl of tomato seedlings (*Lycopersicon esculentum*). Emphasis is on the extent of the reversible response, ΔR, as a function of light pretreatment [6, 7].

Programme: 3 days dark (D) – 12 h light (or dark) – 5 min light pulse (saturating red or RG 9-light) – 24 h dark – assay of anthocyanin. For definition of ΔR, see Table 2

Pretreatment (after 3 days D)	Amount of anthocyanin (A at 535 nm)		
	response$_1$	response$_2$	ΔR
12 h D	0.001	0.001	0.000
12 h RL (0.67 Wm^{-2})	0.020	0.009	0.011
12 h RL (6.7 Wm^{-2})	0.032	0.011	0.021
12 h FR	0.025	0.010	0.015
12 h BL	0.119	0.051	0.068
12 h UV (PG 218)	0.137	0.066	0.071

Control without any irradiation (108 h D) has A_{535} = 0.001

Fig. 6. Time course of responsivity amplification in continuous red (*RL*) and ultraviolet (*UV*) light for phytochrome-mediated GPD induction in the milo (*Sorghum vulgare*) shoot. The light treatment commenced 48 h after sowing. s_D (dark control) at 60 and 72 h refers to s_D at 48 h = 1. For definition of s, see Table 3. (After [19])

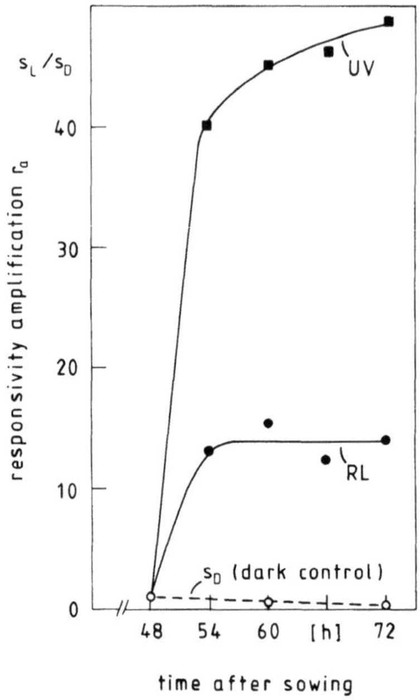

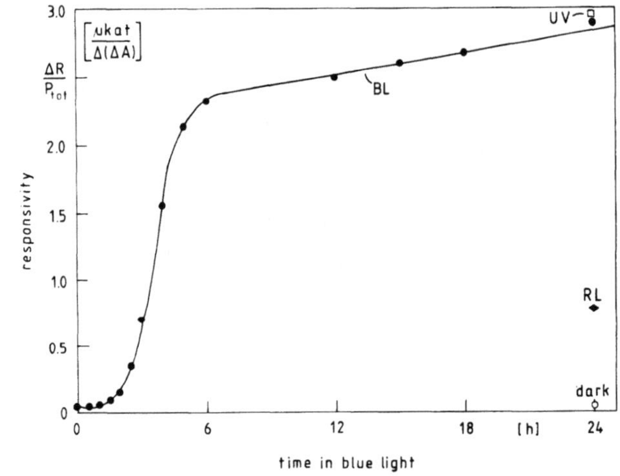

Fig. 7. Light-mediated responsivity increase in the milo (*Sorghum vulgare*) shoot with regard to phytochrome-mediated accumulation of GPD. As a gauge for responsivity to phytochrome we consider the extent of the reversible response, ΔR (for definition see Table 2). The ΔR value is referred to the amount of total phytochrome present in the milo shoot at the time of the red and RG 9-light pulse treatment. We consider the ratio $\Delta R/P_{tot}$ the most precise gauge for the effectiveness of P_{fr} (or, expressed another way, for the responsivity towards P_{fr}).

Dark-grown seedlings were pretreated with blue light (BL, 7 Wm^{-2}) of different durations. At 72 h after sowing the pretreatment was terminated with a red or RG 9-light pulse. Until enzyme assay (96 h after sowing) seedlings were kept in darkness. UV-A (9.3 Wm^{-2}) is as effective as BL while red light (RL, 6.8 Wm^{-2}) is much less effective. *Dark* level of responsivity in dark-grown seedlings. (After [19])

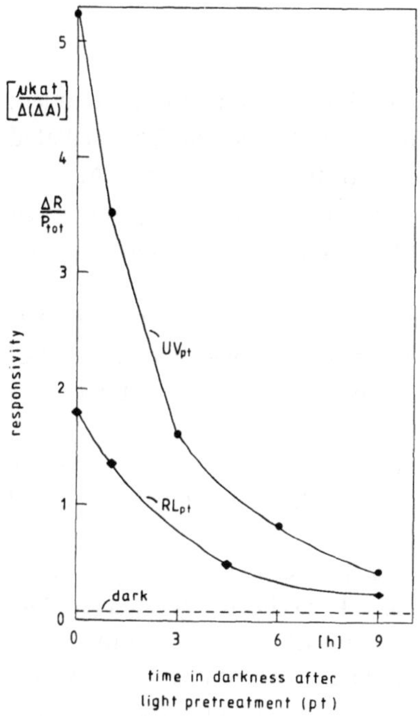

Fig. 8. Decrease of responsivity in darkness following a 24-h treatment with UV-A, terminated with a RG 9-light pulse to return almost all P_{fr} back to P_r (see Table 2). The experimental system was as in Fig. 7. UV-A (9.3 Wm^{-2}) and blue light (7 Wm^{-2}) were adjusted so as to give approximately the same effect with regard to responsivity amplification (see Fig. 7). Red light (RL, 6.8 Wm^{-2}) of corresponding fluence rate is far less effective. Red or RG 9-light pulses were given after different durations of darkness but always at 72 h after sowing. Enzyme assay was performed 24 h after the light pulse, i.e., always 96 h after sowing. *UV-A$_{pt}$* plants pretreated with 24 h of UV-A; *RL$_{pt}$* plants pretreated with 24 h of red light. The low endogenous level of responsivity (*dark*) does not change during the period of examination. (After [19])

It was found that responsivity towards P_{fr} established by single light pulses is extremely weak in dark-grown milo shoots, while prolonged light treatments lead to a dramatic increase of responsivity (degree of response per unit P_{fr}, $\frac{\Delta R}{P_{tot}}$) (Table 3).

Figure 6 indicates that continuous light causes a rapid and strong responsivity amplification which tends to saturate after approximately 6 h. Blue and UV light are equally effective, and far more effective than RL. Since light-mediated changes of P_{tot} levels are the same in RL, BL and UV, it is clear that UV and BL cause a several times higher responsivity than RL. Detailed kinetics of the increase of responsivity after the onset of BL (or UV) are given in Fig. 7. There is a small residual responsivity to P_{fr} which can be increased by BL and UV approximately 60-fold within 6 h. RL is less effective. High responsivity is only maintained in the light. When the seedlings are placed in darkness the level of responsivity drops rapidly with a half-life of the order of 2 h (Fig. 8). In view of these data, the level of responsivity to P_{fr} in the light must be considered a quasi-steady state which results from responsivity amplification by light and from a light-independent decay reaction which decreases responsivity [17]. We conclude that responsivity of a plant towards P_{fr} strongly depends on the quality and quantity of the ambient light. A plant measures light throughout the spectrum and this information determines the efficiency of P_{fr} action or, in other words, the actual responsivity towards P_{fr}, the effector molecule proper of photomorphogenesis.

Site of Responsivity Amplification?

An amplification at the site of signal expression (see Fig. 3) seems to be excluded. Recent data [10] have confirmed the previous conclusion that the rate of light-mediated enzyme/protein synthesis is proportional to the amount of the pertinent mRNA.

An amplification on the level of P_{fr} receptors seems to be excluded by the observation [3, 19, 22] that the response depends linearily on P_{fr} up to maximum P_{fr} level irrespective of degree of amplification.

The time course of loss of reversibility was found not to be affected by light pretreatments which amplify responsivity [23]. The interpretation was that the rate of signal transduction from P_{fr} to competent genes is not affected by light treatments which lead to strong amplification of responsivity.

The starting point of a photoresponse is not changed by light pretreatments (see [13], for review). Moreover, it seems that the temporal pattern of competence is not affected by light treatments.

At present we adhere to the working hypothesis that responsivity amplification occurs in connection with the process of transcription. It has been suggested (see Discussion between Ellis and Mohr, in [17]) that light via phytochrome "causes the amplification of processes already occurring in absolute darkness", albeit at a very low rate. In our view responsivity amplification means that the endogenous amplification factor (extent of response elicited by a given amount of P_{fr}) is increased by light treatment. This light, which modulates the effectiveness of P_{fr}, can be absorbed by phytochrome, cryptochrome and the UV-B photoreceptor.

Exclusion of Carotenoids as Photoreceptor Chromophores of Cryptochrome

The chemical nature of the blue light/UV-A photoreceptor (cryptochrome) was studied in connection with light-mediated anthocyanin synthesis in the tomato seedling [6, 7].

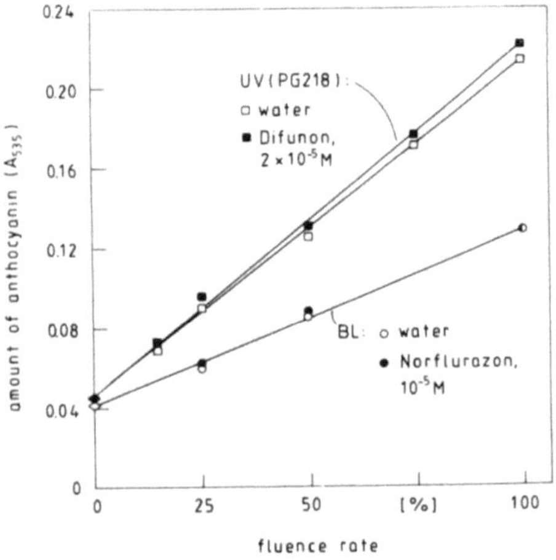

Fig. 9. Fluence-rate-response-curves for blue light (●, 7 Wm^{-2}) and UV (PG 218, ■, 12.6 Wm^{-2}). (For technical details see [5]). Seedlings were germinated with distilled water (*open symbols*) or with herbicide solution (*closed symbols*). Anthocyanin extraction was performed according to the programme 3 days dark + 12 h light + 5 min RL + 24 h dark. The zero values (◇,◆) represent the amount of anthocyanin formed as a response to phytochrome only, i.e., with the programme: 3 days dark + 12 h RL (or 12 h FR) + 24 h dark, on water (◇) or on herbicide solution (◆). RL, high fluence rate red light (6.7 Wm^{-2}). (After [16])

Anthocyanin synthesis in the hypocotyl of tomato seedlings responds strongly and specifically to blue/UV light (Table 3). The herbicides Norflurazon (SAN 9789) and Difunon (EMD-IT-5914) were used to inhibit synthesis of colored carotenoids almost completely without significantly affecting growth and development or phytochrome-mediated anthocyanin synthesis. Even though carotenoid content was reduced to less than 2% of the normal and the fluence rate response function for blue and UV light was linear within the experimental range, Norflurazon or Difunon treatments did not reduce seedlings sensitivity towards blue/UV light (Fig. 9). It was concluded that at least bulk carotenoids are not the photoreceptor chromophore of the blue/UV photoreceptor pigment. In positive terms, flavins rather than carotenoids are strongly suggested to be the photoreceptor pigments for blue/UV-A light.

Acknowledgment. Research supported by the Deutsche Forschungsgemeinschaft (SFB 206). We are grateful to Dr. E. Schäfer for valuable discussions.

References

1. Brüning K, Drumm H, Mohr H (1975) On the role of phytochrome in controlling enzyme levels in plastids. Biochem Biophys Pflanz 168:141–156
2. Downs RJ, Siegelman HW (1963) Photocontrol of anthocyanin synthesis in milo seedlings. Plant Physiol 38:25–30
3. Drumm H, Mohr H (1974) The dose response curve in phytochrome-mediated anthocyanin synthesis in the mustard seedling. Photochem Photobiol 20:151–157
4. Drumm H, Mohr H (1978) The mode of interaction between blue (UV) light photoreceptor and phytochrome in anthocyanin formation of the Sorghum seedling. Photochem Photobiol 27:241–248
5. Drumm-Herrel H, Mohr H (1981) A novel effect of UV-B in a higher plant (Sorghum vulgare). Photochem Photobiol 33:391–398
6. Drumm-Herrel H, Mohr H (1982) The effect of prolonged light exposure on the effectiveness of phytochrome in anthocyanin synthesis in tomato seedlings. Photochem Photobiol 35:233–236
7. Drumm-Herrel H, Mohr H (1982) Effect of blue/UV light on anthocyanin synthesis in the absence of bulk carotenoids. Photochem Photobiol 36:229–233
8. Frosch S, Bergfeld R, Mohr H (1975) Light control of plastogenesis and ribulose-bisphosphate carboxylase levels in mustard seedling cotyledons. Planta 133:53–56
9. Hahlbrock K, Knobloch KH, Kreuzaler F, Potts JRM, Wellmann E (1976) Coordinated induction and subsequent activity changes of two groups of metabolically interrelated enzymes. Eur J Biochem 61:199–206
10. Kreuzaler R, Ragg H, Fautz E, Kuhn DN, Hahlbrock K (1983) UV-induction of chalcone synthase mRNA in cell suspension cultures of Petroselinum hortense. Proc Natl Acad Sci USA 80: 2591–2593
11. Lange H, Shropshire W, Mohr H (1971) An analysis of phytochrome-mediated anthocyanin synthesis. Plant Physiol 47:649–655
12. Mohr H (1972) Lectures on photomorphogenesis. Springer, Berlin Heidelberg New York
13. Mohr H (1978) Pattern specification and realization in photomorphogenesis. Bot Mag Tokyo (special issue) 1:199–217
14. Mohr H (1982) Phytochrome and gene expression. In: Hélène C, Charlier M, Montenay-Garestier T, Laustriat G (eds) Trends in photobiology. Plenum Press, New York, pp 515–530
15. Mohr H (1983) Criteria for photoreceptor involvement. In: Smith H, Holmes MG (eds) Techniques in photomorphogenesis. Academic Press, London New York, in press

16. Mohr H, Drumm-Herrel H (1983) Coaction between phytochrome and blue/UV light in anthocyanin synthesis in seedlings. Physiol Plant 58:408–414
17. Mohr H, Schäfer H (1983) Photoperception and de-etiolation. Philos Trans R Soc London Ser B 303:489–501
18. Mohr H, Schopfer P (1977) The effect of light on RNA and protein synthesis in plants. In: Bogorad L, Weil JH (eds) Nucleic acids and protein synthesis in plants. Plenum Press, New York, pp 239–260
19. Oelmüller R, Mohr H (1983) Responsivity amplification by light in phytochrome-mediated induction of chloroplast glyceraldehyde-3-phosphate dehydrogenase (NADP-dependent, EC. 1.2.1.13) in the shoot of milo (Sorghum vulgare Pers.). Plant Cell Environ 7:29–37
20. Pfeffer W (1904) Pflanzenphysiologie. Engelmann, Leipzig
21. Schäfer E, Lassig TU, Schopfer P (1975) Photocontrol of phytochrome destruction in grass seedlings. The influence of wavelength and irradiance. Photochem Photobiol 22:193–202
22. Schmidt R, Mohr H (1982) Evidence that a mustard seedling responds to the amount of P_{fr} and not to the P_{fr}/P_{tot} ratio. Plant Cell Environ 5:495–499
23. Schmidt R, Mohr H (1983) Time course of signal transduction in phytochrome-mediated anthocyanin synthesis in mustard cotyledons. Plant Cell Environ 6:235–238
24. Schröder J, Kreuzaler F, Schäfer E, Hahlbrock H (1979) Concomitant induction of phenylalanine ammonia-lyase and flavanone synthase mRNAs in irradiated plant cells. J Biol Chem 254: 57–65
25. Wellmann E, Schopfer P (1975) Phytochrome-mediated de novo synthesis of phenylalanine ammonia-lyase in cell suspension cultures of parsley. Plant Physiol 55:822–827
26. Yatsuhashi H, Hashimoto T, Shimizu S (1982) Ultraviolet action spectrum for anthocyanin formation in Broom Sorghum first internode. Plant Physiol 70:735–741

Effects of UV Radiation on Plants in the Transition Region to Blue Light

M. M. CALDWELL [1]

Generalizations concerning the effects of UV radiation on plants are easy targets of criticism and certainly subject to considerable qualification. Nevertheless, in this brief review I will generalize rather than present a detailed compilation of available literature, as has been admirably undertaken by Klein [16]. Other reviews on this subject are also available [2, 3, 11, 27]. Wavebands within the UV spectrum have been defined as UV-A, UV-B, and UV-C, as shown in Fig. 1. Additionally, the terms near-UV (300–

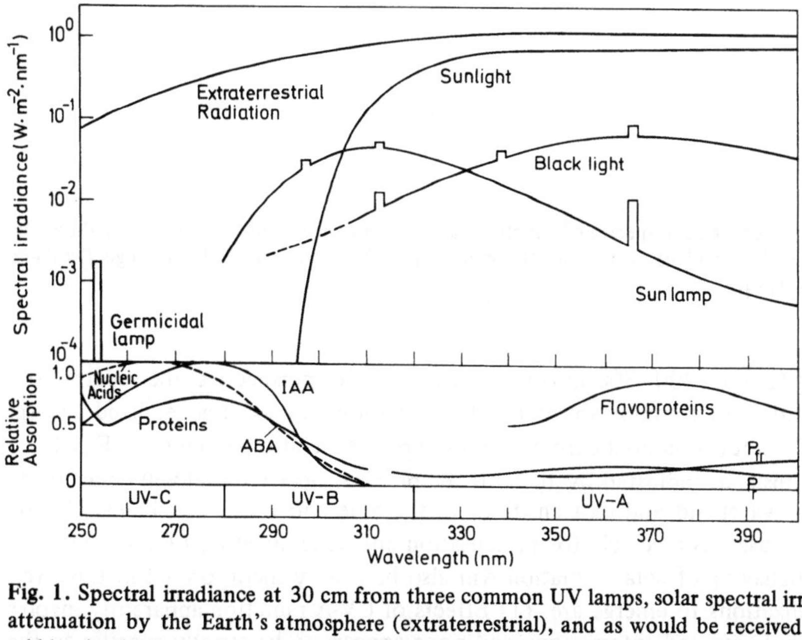

Fig. 1. Spectral irradiance at 30 cm from three common UV lamps, solar spectral irradiance before attenuation by the Earth's atmosphere (extraterrestrial), and as would be received at sea level at midday in the summer at temperate latitudes. The absorption spectra of some common potential UV chromophores in plants are also represented: abscisic acid, *ABA*, is represented by the same curve as for nucleic acids; indoleacetic acid, *IAA;* the two forms of phytochrome, P_{fr} and P_r; and general absorption spectra for flavoproteins and proteins. Major subdivisions of the ultraviolet spectrum are indicated. As originally defined *UV-C* is UV radiation less than 280 nm, *UV-B* is 280 to 315 nm, and *UV-A* is 315 to 400 nm; however, 320 nm is now more commonly used as a division between UV-B and UV-A. The near-UV is commonly defined as 300–400 nm [3]

1 Utah State University, Logan, Utah 84322, USA

Blue Light Effects in Biological Systems
ed. by H. Senger

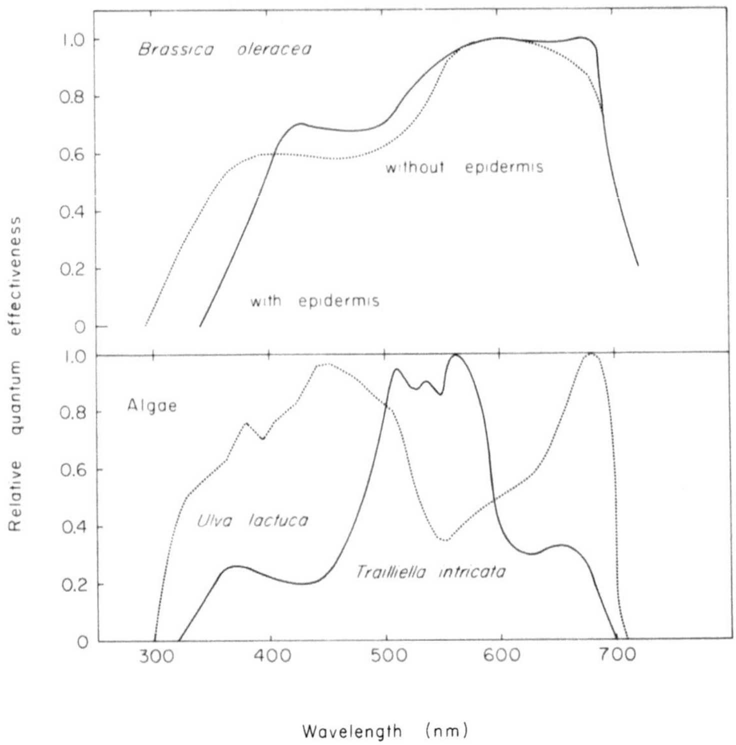

Fig. 2. Photosynthetic action spectra of a higher plant leaf, *Brassica oleracea,* with and without the epidermis (after [15]) and of two algae, the green alga *Ulva lactuca* and the red alga *Trailliella intricata.* (After [18])

400 nm) and far-UV (200–300 nm) are often used. For perspective, the spectral irradiance received from common sources of UV radiation and some generalized absorption spectra for a few potential chromophores in plants are also portrayed in Fig. 1.

This synopsis, as indicated by the title, emphasizes near-UV and will concentrate on the UV-A waveband and its transitions to the blue and UV-B wavebands. Action spectra will be used as a vehicle for presentation and generalization of UV effects on plants. The behavior of solar radiation will also be briefly incorporated in this overview. Generalizations to emerge are: (1) Effects of UV-A radiation apparently involve many chromophores and interactions and none appears to be strictly specific to the UV-A; (2) qualitative differences in radiation responses are more apparent in the transition from UV-A to UV-B rather than in the transition between UV-A and blue light; (3) solar UV-A behaves in a manner very similar to total solar radiation including the blue waveband, whereas solar UV-B is much more variable in nature; (4) with decreasing wavelength in the near UV, damaging reactions occur with increasing efficacy – these appear to operate by several different mechanisms; (5) protective effects mediated by UV-A radiation also occur.

The transition from the blue waveband to UV-A is generally considered to be marked by two major biological phenomena, i.e., the lower wavelength limits of both human vision and photosynthesis of higher plants. However, there is no basic photo-

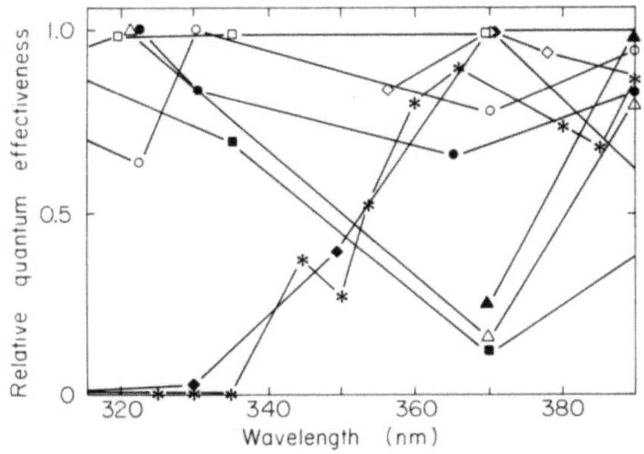

Fig. 3. Plant action spectra in the UV-A waveband involving regulatory phenomena. *Asterisks* indicate photoinduction of carotenoid biosynthesis in *Neurospora crassa* [6]; *open diamonds* chlorophyl formation in *Chlorella protothecoides* [21]; *solid diamonds* phototropic bending in *Phycomyces* [7]; *open circles* induction of seed germination in *Lactua sativa* [26]; *filled circles* unfolding of leaves for *Triticum* [26]; *open triangles* alterations of crown gall tissue culture growth of *Parthenocissus tricuspidatus* [15]; *solid triangles* induction of protoplasmic streaming in *Avena* [1]; *solid squares* stimulation of respiration in chlorophyll-free tissues of *Solanum* and *Acer* [20]; *open squares* induction of chloroplast movement in *Mougeotia* [9]

chemical lower limit at 400 nm for these phenomena. Just as the inability of the human eye to see UV is limited by UV absorption of the lens, the ability of higher plants to carry on photosynthesis in the UV appears to be limited by absorption of UV in the epidermis. Without the epidermis, higher plant photosynthesis can be driven by UV just as photosynthesis of algae responds to UV (Fig. 2).

Action spectra of many plant responses which might be termed regulatory in nature are presented for the UV-A waveband in Fig. 3. These have been normalized to the peak response within the UV-A waveband. While this is hardly a comprehensive summary, it does convey the diversity of spectral response and behavior in this waveband. None of these responses is specific to UV-A, since all of these spectra continue into the blue or UV-B. Of course, the classic "blue light" responses, represented in this figure by spectra for chlorophyll formation in *Chlorella* and carotenoid synthesis in *Neurospora,* include a significant response to UV-A.

Damaging photoreactions in the near-UV can involve several chromophores and mechanisms. Yet all have the common property that the quantum effectiveness of these reactions increases with decreasing wavelength (Fig. 4). These spectra shown in Fig. 4 have been selected to represent different modes of damage such as protein carrier inhibition, inactivation of ATPase, nucleic acid damage and loss of membrane selectivity. (However, none of these photoreactions involves exogenous sensitizing agents.) In the far-UV, most of these damaging photoreactions exhibit still greater quantum effectiveness. Demonstration of damaging photoreactions in the UV-A is comparatively recent. Earlier work had largely discarded UV-A as effective in damaging photoreactions, especially for nucleic acids [27]. Although the quantum effectiveness declines by orders of magnitude, UV-A is effective in producing both pyrimidine dimers and single-strand chain breaks of DNA. However, the ratio of these lesions

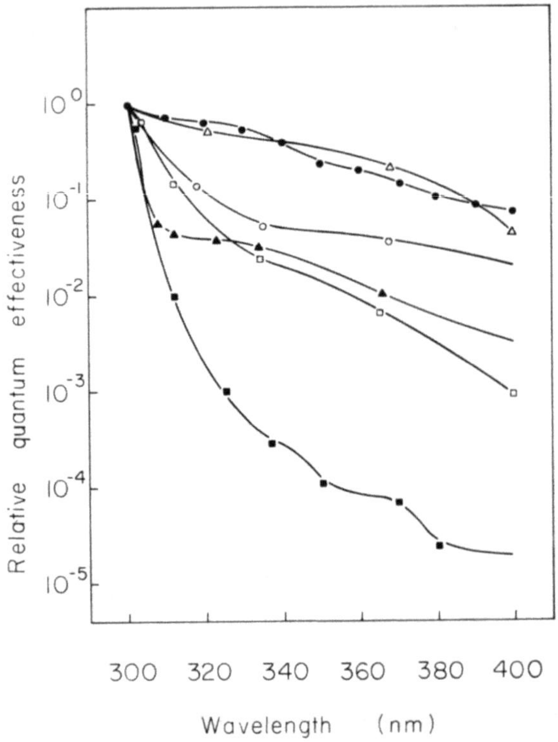

Fig. 4. Near UV action spectra of plants representing basic modes of damage to plant tissues. *Filled circles* inhibition of Hill reaction activity in isolated chloroplasts of spinach [13]; *open triangles* inhibition of adenosine triphosphatase from plasma membranes of *Rosa damascena* [10]; *open circles* induction of single strand breaks of *Bacillus subtilis* DNA [22]; *solid triangles* lethality of stationary phase cells of *E. coli* associated with cell membrane damage [14]; *open squares* inhibition of alanine uptake in *E. coli* linked to inactivation of carrier proteins [24]; *closed squares* represent inactivation of stationary-phase cells of *E. coli* associated with nucleic acid damage [27]

shifts with increasing wavelength so that the proportion of dimers formed decreases much more abruptly than the incidence of chain breaks [27].

The differences between UV-A- and UV-B-induced photoreactions are not merely quantitative, there are also several qualitative differences: (1) The action spectra for lethality, mutagenic events and DNA breaks always exhibit a discontinuity or shoulder in the curves at approximately 330 nm in a variety of biological systems. This has been interpreted as indicating radical changes in the mechanisms of photoreaction between UV-B and UV-A although these mechanisms have not yet been identified [17]. (2) Lethality caused by UV-A has been reported to be strongly enhanced by oxygen for a number of different microorganisms whereas at shorter wavelengths in the UV-B and UV-C, lethality is largely independent of oxygen [27]. (3) Molecular repair systems exist which can correct damage to nucleic acids caused by either far or near UV. Some of these repair systems are themselves induced by UV radiation; however, this induction of repair systems appears to be caused only by UV-B and UV-C and not by UV-A radiation [19]. (4) Membrane damage induced by UV radiation at wave-

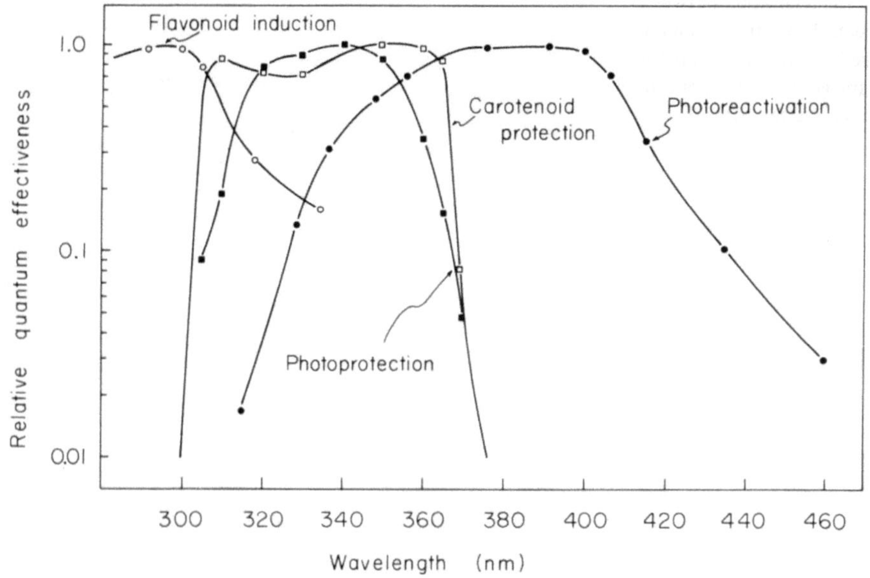

Fig. 5. Action spectra for phenomena that ameliorate UV radiation damage in plants. *Filled circles* photoreactivation of stationary-phase *E. coli* [12]; *solid squares* growth delay (photoprotection) of *E. coli* [17]; *open circles* induction of flavonoid pigments in cell cultures of parsley [28]; *open squares* carotenoid protection of *Sarcina lutea.* (After [27])

lengths greater than 310 nm contributes significantly to lethality of *E. coli;* however, at wavelengths shorter than 310 nm UV-induced membrane damage appears not to play a role in lethality of *E. coli* [14].

Although UV-A radiation may not be effective in inducing certain nucleic acid damage repair systems, it is important in driving a specific process termed indirect photoreactivation. This is an enzyme-mediated system that repairs specific lesions, namely pyrimidine dimers, which can be caused by either near- or far-UV [12]. There are several other nucleic acid repair systems which do not require radiation and are thus termed dark repair systems [25]. Another process induced by UV-A is termed photoprotection, a phenomenon by which near-UV administered before inactivating UV will reduce the effect of the damaging UV. This operates primarily by delay of growth which allows normal dark repair systems, which are comparatively slow, to have more time to repair UV damage. A recent action spectrum for photoprotection and an action spectrum for photoreactivation are shown in Fig. 5.

Flavonoid pigments are selective filters for UV and they can attenuate damaging UV before it reaches sensitive targets within plant tissues [5]. These pigments can be induced by UV radiation in some plants as indicated by the action spectrum in Fig. 5 [28]. Although carotenoids absorb principally in the visible portion of the spectrum, their presence in plant tissues can reduce oxygen-dependent UV-A damage, as indicated by an action spectrum for carotenoid protection in Fig. 5. Quenching of singlet oxygen has been proposed as a mechanism for this protection [27].

Thus, near-UV can cause damage, it can induce certain repair systems to reverse UV-induced damage, it can drive the photoreactivation process, delay growth to allow

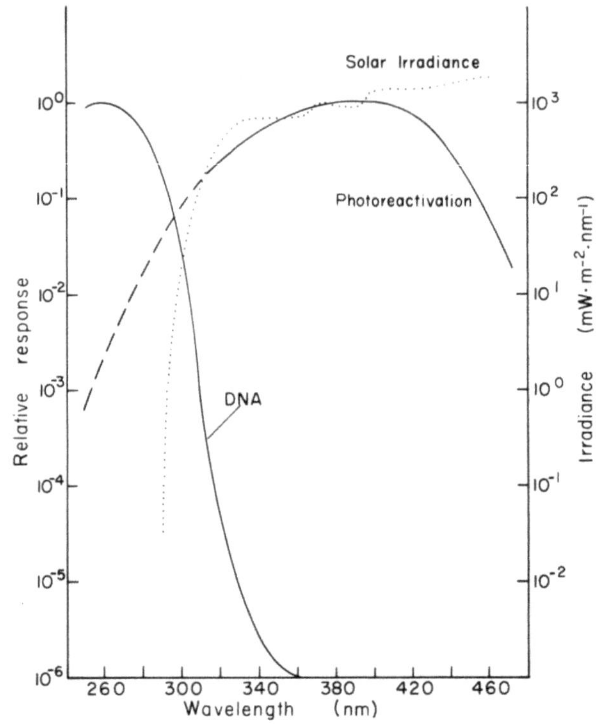

Fig. 6. Relationship between solar spectral irradiance and a generalized action spectrum for DNA damage [23] and photoreactivation. (After [12])

dark repair systems to operate more effectively, and it can also cause direct damage to the repair systems themselves. Damage by UV can operate by several basic mechanisms, as can the protective processes. Regulatory phenomena in the near-UV also involve several chromophores and mechanisms.

Most of the photobiology of near-UV is the result of laboratory study and often with monochromatic UV. To extend the knowledge of photobiological and physiological processes to a consideration of solar UV radiation effects on plants in nature involves another order of complexity. Not only is one dealing with polychromatic radiation with the attendant interactions between UV-induced phenomena, but also the solar UV spectrum undergoes considerable change, especially in the UV-B component of sunlight that reaches the Earth's surface. Figure 6 portrays the change in solar spectral irradiance along with the action spectrum for biological damage associated with DNA damage (a generalized spectrum from [23] and a photoreactivation spectrum. In the waveband where solar spectral irradiance is declining by orders of magnitude, the DNA damage response is increasing by orders of magnitude. Even if one assumes that there are no interacting effects of different wavelengths, a quantitative evaluation of the potential of solar radiation to cause damage depends on the reliability of the damage action spectrum. Even though all action spectra associated with different modes of UV damage (Fig. 4) decline with increasing wavelengths, the rates of decline differ considerably and this greatly influences the assessment of the damaging effects of solar radiation (see [3] for further discussion).

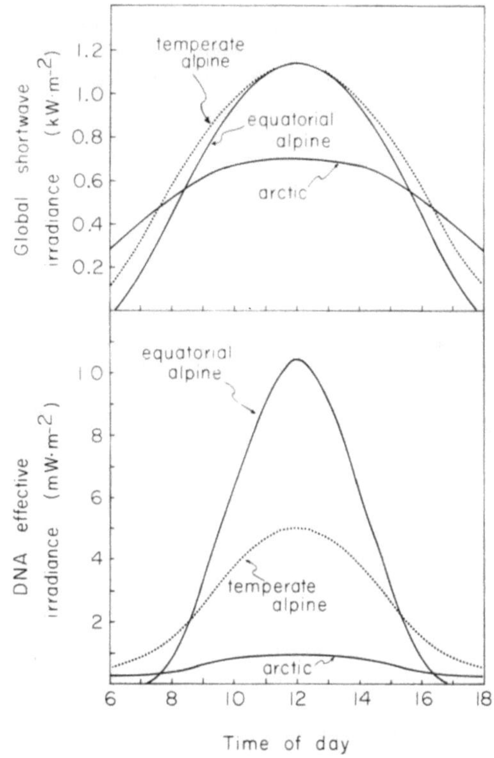

Fig. 7. Total solar shortwave irradiance at the time of maximum solar radiation at three locations in a latitudinal gradient in the arctic-alpine life zone from northern Alaska to the Andes of South America and solar UV-B radiation weighted for DNA damaging effectiveness based on the action spectrum of Setlow [23] for the same locations. (After [4])

Flux densities of solar UV-A radiation per unit wavelength are of about the same magnitude as in the blue waveband (Fig. 6). Furthermore, UV-A is not influenced by atmospheric ozone absorption and behaves in a manner similar to the blue waveband and to total solar shortwave radiation. The UV-B waveband, however, is much more variable since it is influenced by atmospheric ozone concentration and also undergoes greater changes as a function of solar angle [2, 3]. This is illustrated in a comparison of solar radiation measured over a latitudinal gradient ranging from northern Alaska at sea level to the Andes at elevations of 3000 to 4000 m. Over this latitudinal gradient prevailing solar angles, atmospheric ozone concentrations, and total atmospheric thickness vary considerably. Yet at the time of year of maximum solar radiation, total solar shortwave radiation is similar at different locations along this gradient. In contrast, a comparison of the solar UV-B flux, weighted according to its effectiveness to cause DNA damage, undergoes very large changes – a full order of magnitude (Fig. 7). Also it is only the UV-B portion of the spectrum that would be influenced if the Earth's atmospheric ozone layer were to be reduced. Thus, the transition from UV-A to UV-B for solar spectral irradiance in nature is represented by large quantitative differences which do not occur in the transition between the UV-A and blue wavebands. It is also in the transition between the UV-B and UV-A where qualitative, and often quantitative, changes in the nature of photobiological effects also occur rather than in the transition from UV-A to blue light.

Acknowledgments. Research supported by the U.S. Environmental Protection Agency (Contract CR 8081677010) led to many of the concepts presented here; however, this review has not been subjected to the Agency's required peer and policy review and therefore does not necessarily reflect the view of the Agency and no official endorsement should be inferred. Charles Warner drafted the figures and Stephan Flint and Peter Jordan provided helpful review of the paper.

References

1. Bottelier HP (1934) Über den Einfluß äußerer Faktoren auf die Protoplasmaströmung in der *Avena*-Koleoptile. Rec Trav Bot Neerl 31:474–582
2. Caldwell MM (1971) Solar UV irradiation and the growth and development of higher plants. In: Giese AC (ed) Photophysiology, vol 6. Academic Press, New York, pp 131–177
3. Caldwell MM (1981) Plant response to solar ultraviolet radiation. In: Lange OL, Nobel PS, Osmond CB, Ziegler H (eds) Encyclopedia of plant physiology, vol 12A. Physiological plant ecology I. Responses to the physical environment. Springer, Berlin Heidelberg New York, pp 169–198
4. Caldwell MM, Robberecht R, Billings WD (1980) A steep latitudinal gradient of solar ultraviolet-B radiation in the arctic-alpine life zone. Ecology 61:600–611
5. Caldwell MM, Robberecht R, Flint SD (1983) Internal filters: Prospects for UV-acclimation in higher plants. Physiol Plant 58:445–450
6. DeFabo EC, Harding RW, Shropshire Jr W (1976) Action spectrum between 260 and 800 nanometers for the photoinduction of carotenoid biosynthesis in *Neurospora crassa.* Plant Physiol 57:440–445
7. Galland P (1983) Action spectra of photogeotropic equilibrium in *Phycomyces* wild type and three behavioral mutants. Photochem Photobiol 37:221–228
8. Halldal P (1967) Ultraviolet action spectra in algology. A review. Photochem Photobiol 6: 445–460
9. Haupt W (1959) Die Chloroplastendrehung bei *Mougeotia* L. über den quantitativen und qualitativen Lichtbedarf der Schwachlichtbewegung. Planta 53:484–501
10. Imbrie CW, Murphy TM (1982) UV-action spectrum (254–405 nm) for inhibition of K^+-stimulated adenosine triphosphatase from the plasma membrane of *Rosa damascena.* Photochem Photobiol 36:537–542
11. Jagger J (1981) Near-UV radiation effects on microorganisms. Photochem Photobiol 34:761–768
12. Jagger J, Stafford RS, Snow JM (1969) Thymine-dimer and action-spectrum evidence for indirect photoreactivation in *Escherichia coli.* Photochem Photobiol 10:383–395
13. Jones LW, Kok B (1966) Photoinhibition of chloroplast reactions. I. Kinetics and action spectra. Plant Physiol 41:1037–1043
14. Kelland LR, Moss SH, Davies DJG (1983) An action spectrum for ultraviolet radiation-induced membrane damage in *Escherichia coli* K-12. Photochem Photobiol 37:301–306
15. Klein RM (1964) Repression of tissue culture growth by visible and near visible radiation. Plant Physiol 39:536–539
16. Klein RM (1978) Plants and near-ultraviolet radiation. Bot Rev 44:1–127
17. Kubitschek HE, Peak MJ (1980) Action spectrum for growth delay induced by near-ultraviolet light in *E. coli* B/r K. Photochem Photobiol 31:55–58
18. McCree KJ (1981) Photosynthetically active radiation. In: Lange OL, Nobel PS, Osmond CB, Ziegler H (eds) Encyclopedia of plant physiology, vol 12A. Physiological plant ecology I. Responses to the physical environment. Springer, Berlin Heidelberg New York, pp 41–55
19. Menezes S, Tyrrell RM (1982) Damage by solar radiation at defined wavelengths: involvement of inducible repair systems. Photochem Photobiol 36:313–318
20. Montfort C, Rosenstock G (1950) Die Lichtatmungsreaktion des Protoplasmas und ihre Beziehung zur Qualität der Strahlung. Z Naturforsch 5b:171–173

21. Oh-hama T, Senger H (1978) Spectral effectiveness in chlorophyll and 5-aminolevulinic acid formation during regreening of glucose bleached cells of *Chlorella protothecoides*. Plant Cell Physiol 19:1295–1299
22. Peak MJ, Peak JG (1982) Single-strand breaks induced in *Bacillus subtilis* DNA by ultraviolet light: action spectrum and properties. Photochem Photobiol 35:675–680
23. Setlow RB (1974) The wavelengths in sunlight effective in producing skin cancer: a theoretical analysis. Proc Nat Acad Sci USA 71:3363–3366
24. Sharma RC, Jagger J (1981) Ultraviolet (254–405 nm) action spectrum and kinetic studies of alanine uptake in *Escherichia coli* B/r. Photochem Photobiol 33:173–177
25. Smith KC (1978) Multiple pathways of DNA repair in bacteria and their roles in mutagenesis. Photochem Photobiol 28:121–130
26. Wagne C (1966) Effect of UV light on lettuce seed germination and on the unfolding of grass leaves. Physiol Plant 19:128–133
27. Webb RB (1977) Lethal and mutagenic effects of near-ultraviolet radiation. Photochem Photobiol Rev 2:169–261
28. Wellmann E (1983) UV radiation in photomorphogenesis. In: Shropshire W, Mohr H (eds) Encyclopedia of plant physiology, vol 16. Springer, Berlin Heidelberg New York, pp 745–756

Mycochrome System in the Induction of Fungal Conidiation

T. KUMAGAI [1]

Introduction

Since the end of the 19th century, light has been known to influence the formation of various reproductive structures such as conidia, coremia, perithecia and basidiocarps. As previously reviewed by Kumagai [6, 7], many fungi are classified into three groups according to the light-dark cycle necessary for conidiation. (1) Light is required for the induction of conidiophores but conidial development is suppressed by light. (2) Light is not required for the induction of conidiophores and conidial development is suppressed by light. (3) Light is required for the induction of conidiophores and conidial development is not suppressed by light. In other words, based on whether or not light is needed to induce conidiophores, the first and third types can be called photoinduced sporulators and the second nonphotoinduced sporulators.

Induction of conidiophore formation in photoinduced sporulators is promoted either by near-UV and blue light shorter than 530 nm or by near-UV alone. Conidial development after induction of conidiophore formation in *Alternaria tomato* [1], *A. solani* [14], *Stemphylium botryosum* [12], *A. tenuis* [2] and *S. solani* [17] is suppressed by blue light applied at the conidiophore maturation stage. Suppression of conidial development by blue light in *A. solani* [15] and *Botrytis cinerea* [19] has been found to be cancelled by red light applied immediately after blue light. An action spectrum in *A. solani* shows a peak around 600 nm for a wild species, and peaks at 575 and 650–675 nm for a mutant; blue light inhibition can be cancelled by added FMN. This suggests that semiquinones of flavin molecule may be the receptor for the response effected by light of wavelengths 550–600 nm. It has also been proposed that the response in the 625–650 nm region may be mediated by phytochrome; however, these red light effects for reversing inhibition by blue light have not been further reconfirmed so far. It has been found that a photoreceptor system, mycochrome, is involved in a blue- and near-UV-reversible photoreaction which in turn plays an important role in the photocontrol of conidial development in *A. tomato* [4], *B. cinerea* [18, 19], a photoinduced [3] and a nonphotoinduced [22] strain of *Helminthosporium oryzae* and *A. cichorii* [20]. That is, conidial development in these fungi is controlled by alternating exposures to blue and near-UV light applied at a definite stage of conidiophore development, and the final response of whether conidiation occurs can be determined by the final quality of light received. When the final light is near-UV, a conidium develops;

1 Institute for Agricultural Research, Tohoku University, Sendai 980, Japan

Blue Light Effects in Biological Systems
ed. by H. Senger

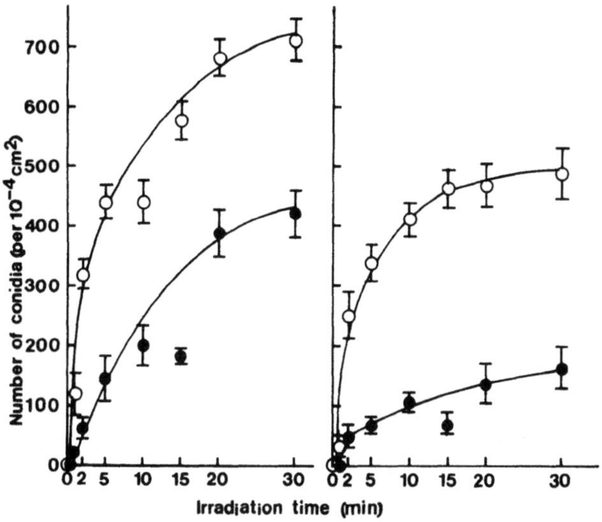

Fig. 1. Conidiation in *A. tomato* induced by near-UV radiation with and without simultaneous blue radiation. The fluence rate of blue light (415 nm) was 7000 mW m^{-2}. That of near-UV light (315 nm) was 590 mW m^{-2} (*left*) and 120 mW m^{-2} (*right*). (○–○) Conidiation induced by exposure to near UV radiation alone; (●–●) simultaneous irradiation with near UV and blue light [8]

when it is blue, conidiation is inhibited and conidiophores dedifferentiate into "sterile" ones. These two light effects can be repeatedly reversed. This kind of antagonistic action in conidial development was not found in any stage with the exception of conidiophore maturation until 1981. This article deals with the role of the mycochrome system in the induction of conidiation.

Antagonistic Action of Blue and Near-UV Radiation in the Induction of Conidiation

Antagonistic action of blue and near-UV radiation in the induction of conidiation has been examined by two methods: one is by simultaneous irradiation with these two lights, the other by alternating exposure to each light [8]. The two lights were introduced vertically (near-UV light) or obliquely (blue light) from the top using surface mirrors, and were superimposed on the culture plate. *A. tomato,* a photoinducible strain of *H. oryzae, B. cinerea* and *A. cichorii* do not form conidiophores or conidia in total darkness when grown on a potato dextrose agar medium. However, when dark-grown colonies are exposed to near-UV radiation, conidiophore formation is induced and conidia are then produced in subsequent darkness. Figure 1 shows conidiation in *A. tomato* induced by near-UV light with and without blue light. There was a 15-s lag period and then the number of conidia increased with irradiation time with the higher fluence rate of near UV radiation (490 mW m^{-2}). Results were similar with the lower fluence rate (120 mW m^{-2}) except the lag time was about 30 s. When colonies were exposed to both near-UV and blue light simultaneously instead of near-UV alone, the number of conidia increased with a duration of irradiation, but the amount of conidiation was less as compared with that when only near-UV radiation was used. The lower the fluence rate of near-UV radiation against a given fluence rate of blue radiation (7000 mW m^{-2}), the less conidiation. Furthermore, with an increase in the amount of

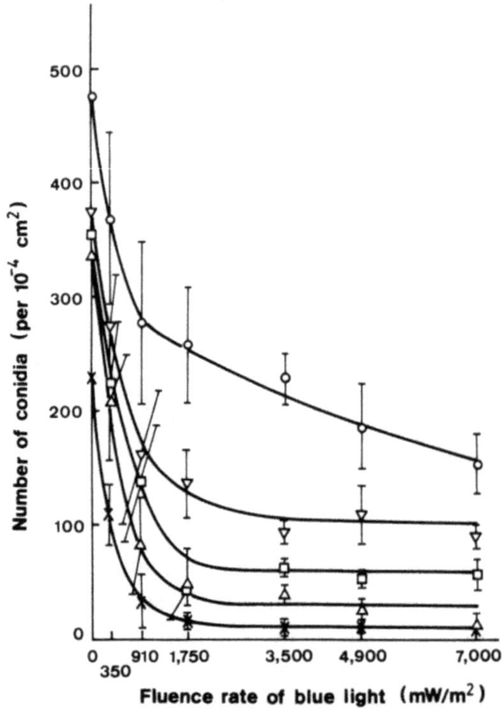

Fig. 2. Conidiation in *A. tomato* induced by light composed by near-UV and blue light. *Abscissa* shows the fluence rate of blue light at 415 nm. The fluence rate of near-UV light at 315 nm was 490 (○), 370 ( ), 250 (□), 120 (△) and 50 (X) mW m^{-2}. Irradiation time for inducing conidiation was 5 min [8]

near-UV radiation from 120 to 490 mW m^{-2} against a given fluence rate of blue light (7000 mW m^{-2}), the time lag for inducing conidiation shortened from 60 s to 15 s. Figure 2 shows the conidiation caused by simultaneous exposure for 5 min to near-UV and blue radiation of different fluence rates. The degree of conidiation decreased as the ratio of blue to near-UV increased and ultimately reached a steady state. In summary, when conidiation is induced by simultaneous irradiation with near-UV and blue light, both the time lag and the amount of conidia formed are regulated by the fluence rates of both lights.

It has recently been found that conidiophore maturation in *A. cichorii* is under control of a blue- and near-UV-reversible photoreaction, but not the induction of conidiation itself [20]. In their works, black light lamps (310–420 nm, mainly 360 nm) were used as a light source for near-UV radiation. Black light lamps do not seem to be suitable for investigating light effect on conidial development because they emit very little in the wavelength region effective in induction of conidiation. When dark-grown colonies of *A. cichorii* was irradiated with near-UV (290–400 nm, centered at 313 nm), conidiophore formation could be induced by exposures less than 15 s at 1.6 W m^{-2} and conidiation occurred during a subsequent dark period. Here, this near-UV light was provided by a C-63 Mineralight Trans-Illuminator, in which the radiation below 290 nm was removed by a layer of cellulose acetate film. Induction of conidiation could be cancelled by blue light at 450 nm applied just after irradiation with near-UV. Figure 3 shows dose-response curves for reversal of induction of conidiation by blue light in *A. cichorii.* Various fluence rates of blue light at 450 nm were applied for 30 s

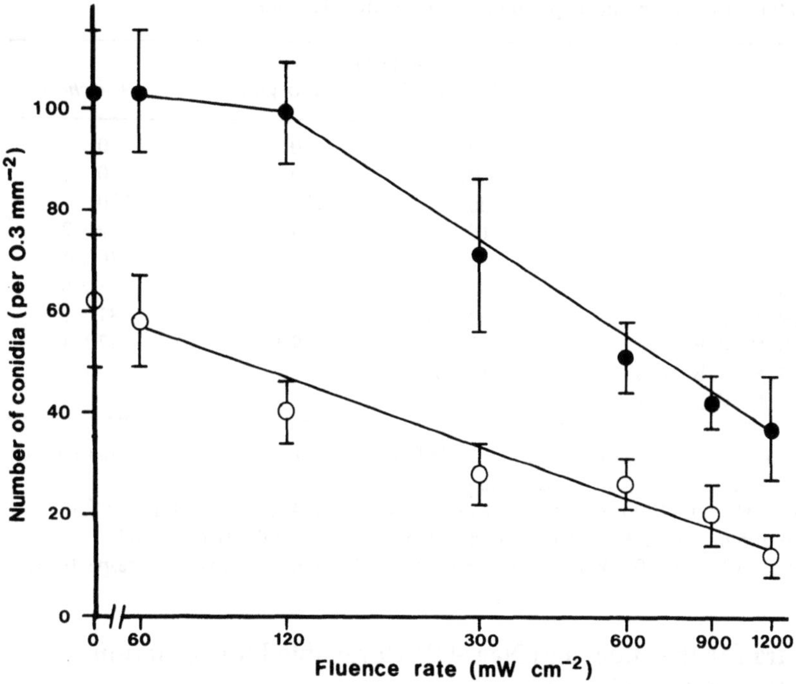

Fig. 3. Fluence-response curves for reversal of induction of conidiation in *A. cichorii* by blue light. Various fluence rates of blue light at 450 nm were applied for 30 s immediately after 2 min of near-UV (1600 mW m^{-2} : •–•) and (800 mW m^{-2} : ○–○) [10]

immediately after exposure to either a higher fluence rate of near UV (1.6 W m^{-2}, 2 min) or a lower fluence rate (0.8 W m^{-2}, 2 min). When conidiation was induced by a lower fluence rate of near-UV, the reversion by blue light was linear with the logarithm of fluence rate of blue light. However, in the case of a higher fluence rate of near-UV, reversion occurred linearly after a lag period.

Table 1 shows the effect of alternating exposure to near-UV and blue radiation on the induction of conidiation in *A. tomato, H. oryzae* and *A. cichorii.* With an increase in the number of light cycles, there was an increased percentage of conidiation that escaped reversion. However, it is clear that the induction of conidiation in these fungi could be repeatedly controlled by alternating doses of near-UV and blue radiation, with the final response depending on the last light received. When the final light was near-UV, conidiation was induced; when it was blue, induction of conidiation was reversed.

Thus, it has been shown that conidial development in these fungi can be regulated by a blue- and near-UV-reversible photoreaction at two developmental stages; one is at the inductive stage of conidiophore formation, the other at the maturation stage of conidiophore development.

Table 1. Conidiation after alternating exposure to near UV and blue radiation

Light treatment	Conidiation (%) *A. tomato*[a]	*H. oryzae*[b]	*A. cichorii*[c]
Dark control	0	0	0
B	0	0	0
NUV	100	100	100
NUV, B	16 ± 8	9 ± 5	22 ± 2
NUV, B, NUV	78 ± 10	59 ± 11	70 ± 4
NUV, B, NUV, B	33 ± 4	13 ± 8	28 ± 6
NUV, B, NUV, B, NUV	93 ± 14	70 ± 13	81 ± 5
NUV, B, NUV, B, NUV, B	41 ± 8	39 ± 13	37 ± 4
NUV, B, NUV, B, NUV, B, NUV	110 ± 12		
NUV, B, NUV, B, NUV, B, NUV, B	63 ± 9		

Dark-grown colonies of *A. tomato* [8], *H. oryzae* [9] and *A. cichorii* [10] were alternatingly exposed to near-UV and blue radiation. Doses of light were
[a] 2 min of NUV at 315 nm (490 mW m^{-2}) and 1 min of blue (B) at 415 nm (7000 mW m^{-2})
[b] 30 s of NUV at 305 nm (13 mW m^{-2}) and 1 min of B at 459 nm (9400 mW m^{-2}), and
[c] 2 min of NUV at 313 nm (800 mW m^{-2}) and 2 min of B at 455 nm (600 mW m^{-2}), respectively

Action Spectra for the Blue- and Near-UV-Reversible Photoreaction in the Induction of Conidiation

When *A. tomato* was irradiated with monochromatic light, number of conidia increased with photon fluence. An action spectrum was calculated by plotting the reciplocal value of incident quanta needed to produce 5 × 10^5 conidia per cm^2. Action spectra for induction of conidiation in *A. tomato* and *H. oryzae* show a peak near 300 nm with a shoulder between 310 nm and 330 nm, while radiation above 340 nm is ineffective (Fig. 4). Above 270 nm, the spectrum for *A. tomato* is similar to that for *H. oryzae.* These action spectra are similar to action spectra for induction of conidiation

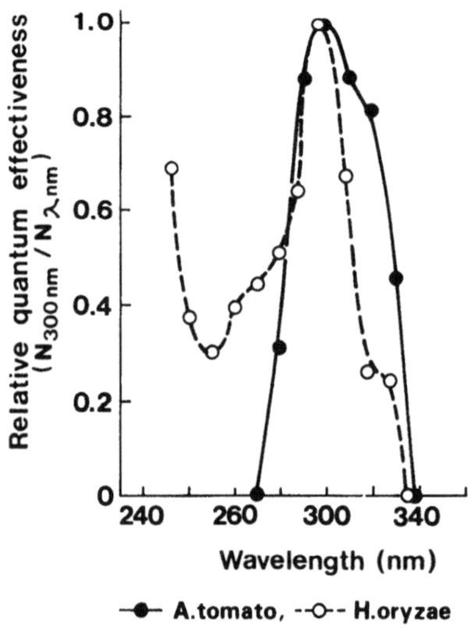

Fig. 4. Action spectra for induction of conidiation in *A. tomato* (●–●: [9]) and *H. oryzae* (○–○: [21])

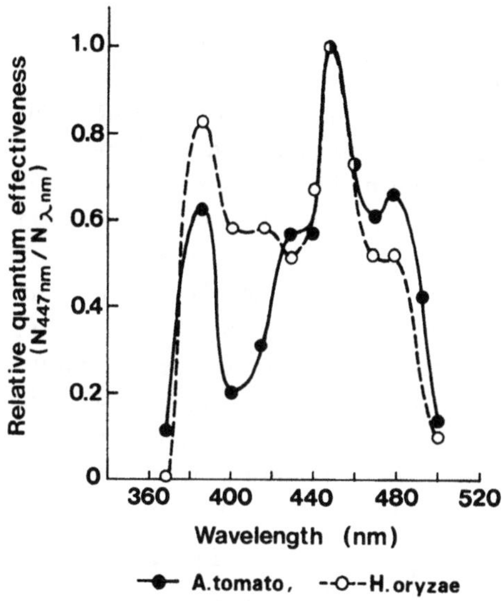

Fig. 5. Action spectra for reversion of induction of conidiation in *A. tomato* (•–•) and *H. oryzae* (○–○) [9]

in some fungi which are sensitive to only near-UV radiation [13]. The reversion of induction of conidiation occurred linearly with the logarithm of photon fluence (Fig. 3). Hence, action spectra for *A. tomato* and *H. oryzae* were calculated by plotting the reciprocal value of incident quanta needed to produce 50% reversal of induction by near-UV radiation. The action spectra have an action maximum near 447 nm with a shoulder between 400 nm and 440 nm, one minor peak around 478 nm, and another one near 385 nm (Fig. 5). Wavelengths below 334 nm or above 520 nm have no effect. This is very similar to the spectrum for the ubiquitous blue light receptor(s) [16].

Tentative Explanation of the Mechanism of Action of the Mycochrome System in the Induction of Conidiation

As stated above, a blue- and near-ultraviolet-reversible photoreaction is involved in conidial development in *A. tomato, H. oryzae* and *A. cichorii.* From in vitro experiments with *A. tomato,* at least two pigments, P_B and P_{NUV}, have been assumed to be involved in this photoreaction (Kumagai et al. 1976). P_B is a redox pigment and the reduced form of P_B [P_B(red)] may be physiologically active in conidial development [6, 7]. P_{NUV} (near-UV absorbing type) is a low weight molecular compound containing iron (unpublished) and the cooperative action of P_{NUV} and near-UV radiation can reduce P_B(ox) (the oxidized form of P_B). It has been further suggested that P_{NUV} may function as a photoreceptor for near-UV radiation in the induction of conidiation, because the absorption peak coincides with the action spectrum. On the other hand, the absorption spectrum of P_B is different from the action spectrum for reversal of the near-UV radiation effect (this action spectrum is of the usual, flavin type). Hence, some flavin molecule may be involved as a blue light receptor. Based on these results, the mechanism of the mycochrome system in the induction of conidiation in these fungi can be tentatively explained as follows. The cooperative action of P_{NUV} and near-UV radiation causes a reduction of P_B(ox) [$P_B(ox) \xrightarrow{NUV} P_B(red) \longrightarrow$ conidiation]. Accumulation of P_B(red) over a given level by irradiation with near-UV is necessary for the induction of conidiation, which could explain the time lag or threshold observed in Fig. 1. The duration of the time lag was determined either by the fluence rate of near-UV radiation or by the ratio of the fluence rates of near-UV to blue radiation. The lower the fluence rate of near-UV radiation, the longer the time lag. Furthermore, the time lag lengthened with a lowering of the fluence rate of near-UV given against the fluence rate of blue radiation, which may be due to the lower level of P_B(red) caused by the action of blue radiation. When conidiation was induced by simultaneous exposure to near-UV and blue light, the level of conidiation was decreased by an increase in the fluence rate of blue radiation against the given fluence rate of near UV radiation (Fig. 2). This seems to be due to a decrease in the level of P_B(red) caused by the difference between the reduction rate of P_B(ox) caused by near-UV radiation (V_1) and the oxidation rate of P_B(red) caused by blue radiation (V_2) [$P_B(ox) \underset{Blue\ (V_2)}{\overset{NUV\ (V_1)}{\rightleftharpoons}} P_B(red) \longrightarrow$ conidiation]. As the fluence rate of blue radiation was increased, its effect drew near to saturation level and the degree of conidiation reached steady state. The number of conidia increased as the duration of simultaneous irradiation with near-UV and blue radiation was lengthened (Fig. 1); photo-equilibrium appeared to be established at irradiation times longer than 5 min (Fig. 2). It appears that the maintenance of a constant level of P_B(red) produced continuously by the difference between V_1-V_2 may cause an additional effect on the production of conidia; hence, the amount of conidia gradually increased, though slowly. Some threshold appeared in the reversal of induction of conidiation in *A. cichorii* by high fluence rates of near-UV radiation (Fig. 3). In this case, an excess amount of P_B(red) over the amount necessary for inducing the maximum number of conidia per given area of hyphae may be accumulated. Therefore, blue radiation must first eliminate this pool before it can effect the number of conidia formed by reducing the level of P_B(red) further.

For further characterization of such photo-oxidation reduction reaction, the effect of molecular oxygen on photoinduced conidiation in *A. cichorii* was investigated [10]. Photoinduced conidiation was promoted strikingly by O_2 deficiency in which the cultures were flushed with N_2 gas for 1 h or 30 min both before and after irradiation with near-UV radiation (Table 2). Conidiation was not induced by this oxygen-deficient atmosphere without near-UV radiation. Conidiation was stimulated only slightly by O_2 deficiency for 1 h either before or after irradiation with near-UV radiation. Exposure to blue radiation immediately after near-UV radiation suppressed induction of conidiation irregardless of oxygen tension. Conidiation was also suppressed by anaerobiosis for 24 h irrespective of light irradiation. In this connection, it has been found that photoreduction of P_B(ox) by the cooperative action of P_{NUV} and near-UV radiation under anaerobic conditions was much larger than that under aerobic conditions and, moreover, P_B(red) could be oxidized by molecular oxygen [6, 7]. Hence, we can suggest that accumulation of P_B(red) in vivo is also larger under anaerobic conditions. In other words, molecular oxygen decreases the amount of P_B(red). The number of conidia produced under anaerobic conditions exceeded the maximum level under

Table 2. Photoinduced conidiation in *A. cichorii* under anaerobic conditions

Experimental conditions			Number of conidia (0.3 mm^{-2})
Air	NUV	Air	64 ± 7
	NUV, B		9 ± 4
N_2 (60 min)	NUV	N_2 (60 min)	208 ± 25
	NUV, B		8 ± 4
N_2 (30 min)	NUV	N_2 (30 min)	129 ± 19
	NUV, B		15 ± 8
N_2 (60 min)	NUV	Air	80 ± 13
	NUV, B		14 ± 9
Air	NUV	N_2 (60 min)	78 ± 9
	NUV, B		16 ± 7

Colonies of *A. cichorii* grown for 4 days in the dark were put into the chamber without lids. Each chamber was continuously flushed with N_2 gas for a given time as shown before or after exposure to 90 s of near-ultraviolet radiation (NUV: 1750 mW m^{-2}) and 2 min of blue radiation (B) at 450 nm (1000 mW m^{-2}), followed by 24 h darkness

aerobic conditions. This may also be due to the amount of P_B(red) being kept lower under aerobic conditions. Another possibility is that some aspect of the hyphae may be changed by anaerobic conditions, but it must be emphasized that the reversing effect by blue radiation was not influenced by molecular oxygen. Therefore, that oxidation of P_B(red) by blue light may be mediated through a possible flavin molecule which may not directly connect with molecular oxygen.

Although there are many obscure points, the blue and near ultraviolet reversible photoreaction in the induction of conidiation may be summerized as follows:

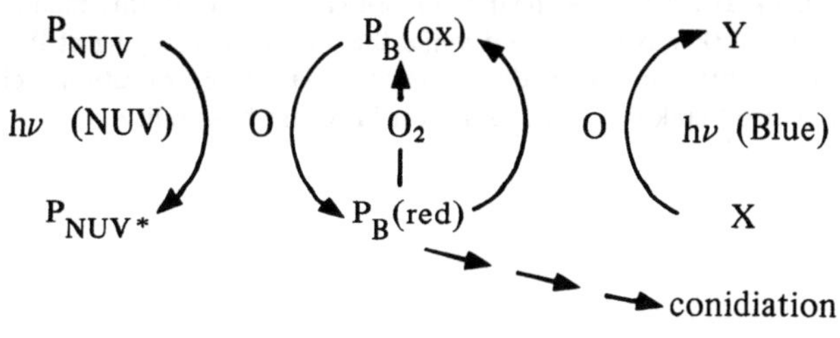

Here, P_{NUV*} is an unknown photoproduct and O shows an unknown coupling facter or site where electron transfer occurs. In this connection, it has been shown with experiments on action dichroism [11] that photoreceptors for near-UV and blue radiation are bound to some structure. Furthermore, when P_B(red) is oxidized by blue light, unknown X-Y reaction may be also involved, or it may be replaced by flavin molecule. When blue radiation was applied within 15 min following exposure to near-UV radiation, reversal of the induction of conidiation occurred almost completely. This suggests that the life time of P_B(red) is about 15 min (unpublished). This may support the possibility that P_B(red) is the physiologically active form for conidial development rather than the flavin molecule which is considered to be a photoreceptor for blue radiation.

Concluding Remarks

It has been found that conidiophore maturation which follows the induction of conidiophore formation in *Alternaria tomato, Helminthosporium oryzae, Botrytis cinerea* and *Alternaria cichorii* is under control of a blue and near-UV reversible photoreaction, namely, the mycochrome system. This article deals with the role of the mycochrome system in the induction of conidiation in *Alternaria tomato, Helminthosporium oryzae* and *Alternaria cichorii.* The discussion is based on the effects of simultaneous irradiation with near-UV and blue light, the effects of alternating exposure to each light, and action spectra for induction and reversal of induction of conidiation. Induction of conidiation in these fungi can be repeatedly controlled by alternating doses of near-ultraviolet and blue light. When the final light is near-ultraviolet, conidiation is induced and conidia then develop; when it is blue, induction of conidiation is suppressed. When conidiation is induced by simultaneous irradiation with both near-ultraviolet and blue light, both the time lag for inducing conidiation and the amount of conidia formed are regulated by the fluence rates of both lights. Action spectra for induction of conidiation show a peak near 300 nm, which is very common in photoconidiation of various fungi. Furthermore, P_{NUV}, which has been proposed as a candidate of the photoreceptor for near-ultraviolet radiation in vitro, may function as a receptor for conidial development. Action spectra for reversing the near-ultraviolet effect suggest participation of the flavin molecule. The inductive effect of near-ultraviolet radiation is promoted by anaerobiosis, but the reversing effect by blue radiation is not influenced by molecular oxygen. Thus, oxidation-reduction processes are thought to be involved in the "mycochrome" system.

Acknowledgments. I wish to thank Dr. T.C. Vogelmann for critical reading of this manuscript and linguistic revision. Part of this work was carried out while the author was staying at the University of Lund and was supported by grant from the ministry of Japanese Education, Science and Culture. In this connection, I also thank Prof. L.O. Björn for his warm hospitality.

References

1. Aragaki M (1962) Quality of radiation inhibitory to sporulation of *Alternaria tomato.* Phytopathology 52:1227–1228
2. Aragaki M (1969) Inhibitory and stimulatory radiation effects on sporulation of *Alternaria tenuis.* Phytopathology 59:1016
3. Honda Y, Sakamoto K, Oda Y (1968) Blue and near ultraviolet reversible photoreaction on the sporulation of *Helminthosporium oryzae.* Plant Cell Physiol 9:603–607
4. Kumagai T, Oda Y (1969) Blue and near ultraviolet reversible photoreaction in conidial development of the fungus, *Alternaria tomato.* Dev Growth Differ 11:130–142
5. Kumagai T, Yoshioka N, Oda Y (1976) Further studies on the blue and near ultraviolet reversible photoreaction with an intracellular particulate fraction of the fungus, *Alternaria tomato.* Biochim Biophys Acta 421:133–140
6. Kumagai T (1978) Mycochrome system and conidial development in certain fungi imperfecti. Photochem Photobiol 27:371–379

7. Kumagai T (1980) Blue and near ultraviolet reversible photoreaction in conidial development of certain fungi. In: Senger H (ed) The blue light syndrome. Springer, Berlin Heidelberg New York, pp 251–260
8. Kumagai T (1982) Blue and near ultraviolet reversible photoreaction in the induction of fungal conidiation. Photochem Photobiol 35:123–125
9. Kumagai T (1983) Action spectra for the blue and near ultraviolet reversible photoreaction in the induction of fungal conidiation. Physiol Plant 57:468–471
10. Kumagai T (1984) Mycochrome system and the effect of molecular oxygen in the induction of conidiation in the fungus, *Alternaria cichorii.* Physiol Plant 59:590–594
11. Kumagai T, Björn LO (1984) Light-induced linear dichroism in photoreversibly photochromic sensor pigments. V. Relation between the components of the mycochrome system in *Alternaria cichorii.* Submitted
12. Leach CM (1968) An action spectrum for light inhibition of the "terminal phase" of photosporogenesis in the fungus, *Stemphylium botryosum.* Mycologia 60:532–546
13. Leach CM (1971) A practical guide to the effects of visible and ultraviolet light on fungi. In: Boothe C (ed) Methods in microbiology, vol IV. Academic Press, London New York, pp 609–664
14. Lukens RJ (1963) Photoinhibition of sporulation in *Alternaria solani.* Am J Bot 50:720–724
15. Lukens RJ (1965) Reversal by red light of blue light inhibition of sporulation in *Alternaria solani.* Phytopathology 55:1032
16. Senger H (ed) (1980) The blue light syndrome. Springer, Berlin Heidelberg New York
17. Sproston T (1971) An action spectrum for ultraviolet-induced sporulation in the fungus *Stemphylium solani.* Photochem Photobiol 14:571–576
18. Suzuki Y, Kumagai T, Oda Y (1977) Locus of blue and near ultraviolet reversible photoreaction in the stages of conidial development in *Botrytis cinerea.* J Gen Microbiol 98:199–204
19. Tan KK (1974) Red-far-red reversible photoreaction in the recovery from blue light inhibition of sporulation in *Botrytis cinerea.* J Gen Microbiol 82:201–202
20. Vakalounakis DJ, Christias C (1981) Sporulation in *Alternaria cichorii* is controlled by a blue and near ultraviolet reversible photoreaction. Can J Bot 59:626–628
21. Yamamura S, Kumagai T, Oda Y (1977) An action spectrum for photoinduced conidiation in *Helminthosporium oryzae.* Plant Cell Physiol 18:1163–1166
22. Yamamura S, Kumagai T, Oda Y (1978) Mycochrome system and conidial development in a nonphotoinduced isolate of *Helminthosporium oryzae.* Can J Bot 56:206–208

Near-UV and Blue Light Effects in Cyanobacteria

T. HIROSAWA[1]

Red and green photoreversible phenomena have been extensively studied in cyanobacteria. In contrast to the above photobiological phenomena which have relatively longer wavelengths of light, very little has been reported on the effects of shorter wavelengths. In this review, I will focus on the following two subjects:

1. The effects of blue light on carbon metabolism.
2. Near-UV effects and photoreactivation.

Carbon Metabolism

The enhancement of respiration and concomitant carbohydrate metabolism by blue light have been extensively studied with green algae such as *Chlorella* and *Scenedesmus* and their various colorless mutants. It has been shown that PEP carboxylase [14] and pyruvate kinase [15] are activated by blue light. Since enzymes can be activated by the addition of glucose, primary blue light-enhanced reaction has been considered to be a step in the provision of substrates for these enzymes. Whether this step is degradation of reserve polysaccharide or transport across the plastid membrane remains to be determined (see review by Senger and Briggs [36]). However, experiments to test these possible mechanisms are hindered by the difficulty of isolating active chloroplasts or other organella from these green algae. It is therefore of interest to look at blue light effects on carbon metabolism in cyanobacteria which do not have discrete intracellular organella. However, the number of reports on this subject is limited.

Table 1 is a list of cyanobacteria used for the respiratory enhancement study (Hirosawa and Miyachi unpublished). None of these species tested responded to blue light in a manner similar to that reported with green algae. Neither prolonged dark incubation (24–36 h) nor heterotrophic growth condition (for *A. variabilis*) affected this type of response. Instead, blue light was often found to be inhibitory to respiration. *Choccochloris elabens* grown under blue light showed 50% less respiratory activity than those grown under white light [32]. Possible reason(s) for being unable to observe the enhancement were considered.

The light-induced inhibition of respiration has been reported in *Anabaena* sp. [2, 3], in *Anacytis nidulans* [10, 12] and in *Anabaena variabilis* [11, 24]. The inhibition

1 Institute of Applied Microbiology, University of Tokyo, Tokyo 113, Japan

Blue Light Effects in Biological Systems
ed. by H. Senger

Table 1. The effects of blue light on respiration in certain cyanobacteria

Strains		Blue light effects	
	Enhancement	No effect	Inhibition
Chroococcaceae			
Anacystis nidulans M-6		*	*
Oscillatoriaceae			
Oscillatoria neglecta M-83		*	
Phormidium molle M-77		*	
Nostocaceae			
Anabaena cylindrica M-1		*	*
A. cylindrica ATCC 29414		*	
A. variabilis M-3		*	*
A. variabilis ATCC 29413		*	*
Scytonemataceae			
Plectonema boryanum M-101		*	
Tolypothrix tenuis M-29		*	*
Rivulariaceae			
Calothrix parientina M-121		*	

Cells were grown at 28C in eightfold diluted Allen and Arnon's medium (1955) with or without KNO_3

appears to be insensitive to DCMU and the light specifically responsible for photosynthetic photosystem I (PSI) is active. Hoch et al. [10] suggested that the light inhibition of respiration could be due to cyclic photophosphorylation around PSI. Such a dependency has been shown to exist in *Chlorella* [35]. However, Imafuku and Katoh [11] showed that the light inhibition of respiration could be observed with *A. variabilis* cells treated with CCCP. It was therefore concluded that the inhibition is not related to cyclic photophosphorylation, as suggested by Hoch et al. [10]. Jones and Myers [12] suggested that the photoinhibition could be a result of competition for electrons between Chl a in PSI and oxygen analogous to photosynthetic bacteria. Murai and Katoh [24] showed that certain respiratory substrates such as glycorate, malate, succinate and isocitrate could donate electrons to PSI through cyt f. These results suggest the direct interaction between photosynthetic and respiratory electron flows. Recently, Lockau [17] showed that cytochrome C-553 and plastocyanin participate in both photosynthetic and respiratory electron transports of membrane preparation from *A. variabilis.* It is likely that when PSI is activated, it accepts part of the respiratory electrons from NADPH at around cyt f which would otherwise flow to oxygen. In turn, respiratory O_2 uptake is inhibited. Since cyanobacteria do not have intracellular organella and it seems that both respiratory and photosynthetic electrom flow systems are embedded in thylakoid, interaction between the two systems could be direct. In contrast to cyanobacteria, the interaction in green algae is more indirect and possibly mediated by respiratory substrates, NAD(P), ATP etc.; thus permeability of the plastid membrane is important.

Cyanobacteria have also been shown to have an ability to perform anoxygenic photosynthesis with electron donors other than H_2O [28]. This anoxygenic photosynthesis is also insensitive to DCMU and appears to be dependent on PSI. It seems that cyanobacteria still retain a bacterial photosynthetic capacity in addition to enkaryotic oxygenic photosynthesis. Since (1) most (85–90%) of Chl a has been shown to be associated with PSI in cyanobacteria [20, 43, 46] and (2) most of phycobiliproteins do not have strong absorption at around 450 nm, several blue light-dependent phenomena seem to be mediated by PSI. Döhler and Koch [5] showed that glycorate excretion into the medium was inhibited in blue and far-red light in *Anacystis nidulans.* At the same time, CO_2 release was stimulated in these spectral regions. These results lead the author to conclude that the blue light stimulates photorespiration, and thus excretion of glycorate is suppressed. A similar action spectrum for glycorate excretion has been shown in the same organism [39], however, interpretation of the result in this case by the author is somewhat different [6, 40]. Photosynthetic enhancement studies with 619 and 446 nm showed that glycorate excretion was reduced by 31% while photosynthetic ^{14}C fixation was enhanced by 18% [6]. Similar experiments showed that the formation of serine and glycine was reduced by blue (446 nm) and far red (687 nm) light as well as glycorate excretion. Thus, it was suggested that glycorate was consumed by a PSI-mediated oxidation process as shown by Murai and Katoh [24].

Nitrate reduction of a nitrite reductase-less mutant of *A. quadruplicatum* was shwon to be dependent on PSI [41]. The reduction of nitrate proceeded for nearly 12 h in medium containing DCMU, implying a pool of electrom donors to PSI.

To our knowledge, there is only one report concerning blue light effects on photosynthetic CO_2 fixation in cyanobacteria. Hauschild et al. [7] reported that supplementing red light (650 nm) with blue light (470–480 nm) had little effect on the distribution of ^{14}C among photosynthetic products in *Microcystis aeruginosa.* After 24 h of dark starvation (incubation), photosynthetic ^{14}C incorporation into glutamate under red with blue light added was increased by 40% when compared to cells kept under red light alone. Only a slight increase was observed in ^{14}C-aspartic acid and malic acids.

Miyachi et al. [22, 23] reported that NH_4Cl can mimic the blue light effects on dark CO_2 fixation and respiration in *Chlorella.* Lawrie et al. [6] and Ohmori [26] showed that NH_4Cl enhanced dark CO_2 fixation in *A. cylindrica.* The distribution of fixed ^{14}C is quite similar to that reported for *Chlorella* by Miyachi et al. [22]. However, Ohmori [26] stated that the dark respiration (1 μmol O_2 mg^{-1} protein h^{-1}) was not significantly influenced by the addition of NH_4Cl. This indicates a possibility that carbon flow into the TCA cycle could be enhanced without affecting respiration in cyanobacteria.

A speculative scheme of what might be happening under the blue light is shown in Fig. 1. In the dark, respiration proceeds through an oxidative pentose phosphate pathway. NADPH produced by the first two steps donate electrons to the respiratory electron transport system. Under the standard condition for blue light enhancement study (with DCMU + blue light), some, if not all, of the electrons are bypassed to P-700 of PSI and then to NADP (first direct inhibition: Lockau [17]). NADPH produced by this process may further inhibit the first reaction of OPP pathway (G6P dehydrogenase: second indirect inhibition). There is a possibility that blue light enhances glycogen degradation without affecting respiration. To our knowledge, no report concerning this question has yet been published. The TCA cycle in cyanobacteria has been known to be interrupted due to the absence of α-oxoglutarate dehydrogenase and

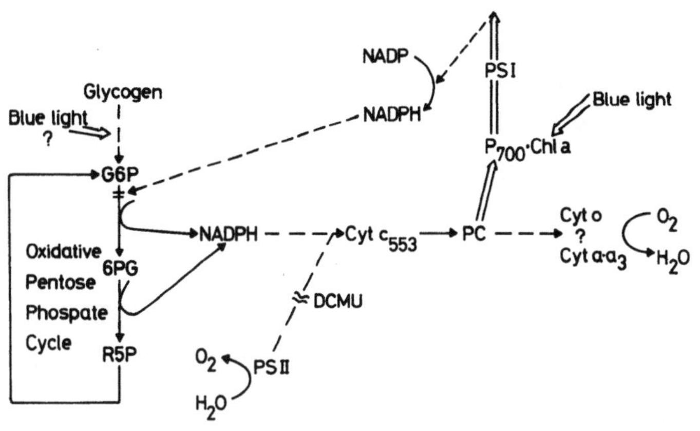

Fig. 1. A model indicating what might be happening in cyanobacterial cells under blue light. (Based on [11, 17, 24])

succinyl-CoA synthetase [29, 30, 31]. Therefore, the role of the TCA cycle in the organisms is oriented toward a synthetic rather than a respiratory oxidative purpose. The production of a respiratory reductant seems to be solely dependent on the oxidative pentose phosphate pathway (review: [38]). Since the carbon supply to the TCA cycle is limited in the dark due to the slow respiration rate, activation of PEP carboxylase and pyruvate kinase (if it occurs) may not be significant. In cyanobacteria, it is probable that the blue light effects on carbon metabolism are most important in the distribution of the photosynthetic CO_2 fixation product in which carbon flow to the TCA cycle is significant [7].

Near-UV Effects and Photoreactivation

Photoreactivation of UV-inactivated cyanobacteria and their viruses has long been reported. Blue light and near-UV (black) light were active in the reactivation of UV-inactivated *Agmenellum quadruplicatum* [45], *Plectonema boryanum* [37, 47], and *Fischerella muscicola* [37]. An action spectrum reported for the photoreactivation of *A. quadruplicatum* showed a sharp maximum at around 430 nm (Fig. 2a; [45], cf. [44]). In this organism, no photoreactivation was found at wavelengths above 500 nm. *Cylindrospermum* sp. is the only exception in which photoreactivation could be observed under red, yellow and green lights as well as blue and black lights [37]. A DNA-photoreactivating enzyme has been purified 3760-fold from *Anacystis nidulans* [34]. The purified enzyme with apparent molecular weight of 93,000 showed a sharp action maximum at 440 nm (Fig. 2b). Further purified by affinity chromatography, the enzyme showed a fluorescence exitation spectrum similar to the in vitro photoreactivation spectrum (Fig. 2c; [21]). Tang and Asato [42] showed that the partially photoreactivable mutant of *A. nidulans* exhibited partial photolysis of thimine dimer with black light while a wild type exhibited a higher level of photolysis. O'Brien and Houghton [25] also showed that in vivo photolysis of pyrimidine dimers by blue light

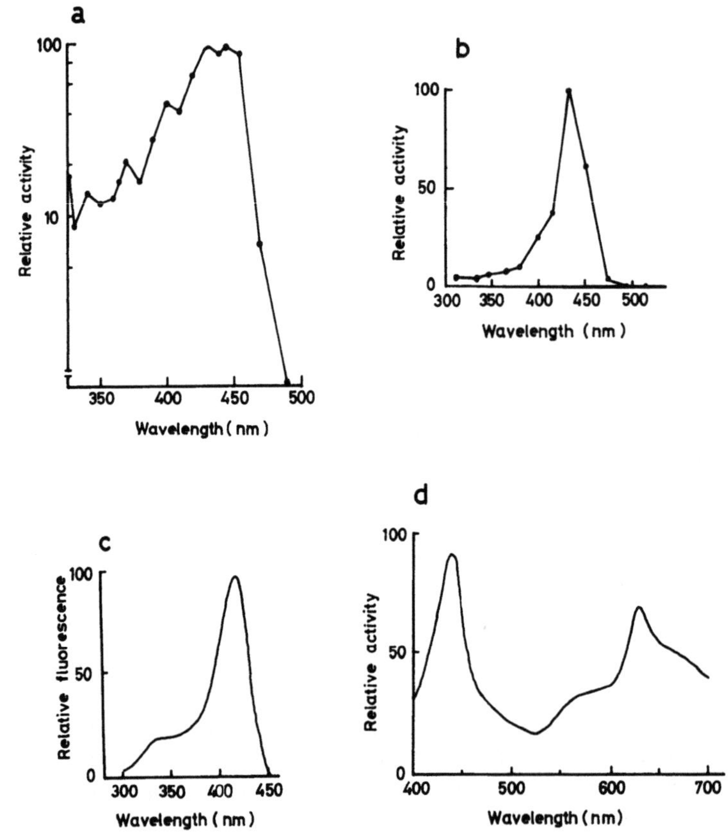

Fig. 2a–d. Action spectra for a Cellular photoreactivation of UV-irradiated *A. quadruplicatum.* (After [45]); b DNA-photoreactivation of *Haemophilis* by enzyme purified from *A. nidulans.* (After [34]); c Excitation of fluorescence from chromophore associated with enzyme purified from *A. nidulans.* (After [21]); d Photoreactivation of UV-A inactivated p-BQ Hill reaction in *A. nidulans.* (After [9])

was accompanied by cellular survival in *Gloeocapsa alpicola.* These data indicate that the major cause of cellular death by UV is due to damage to DNA and that blue and black lights reactivate the organisms through the DNA photoreactivating enzyme in cyanobacteria.

In photosynthetic organisms, it is well known that UV radiation inactivates photosynthesis [13, 18, 19, 27, 48]. Electron transport reaction between water and PSII seems to be the most sensitive reaction to UV. Van Baalen [44] showed that the UV-inactivated photosynthesis in *A. quadruplicatum* could be photoreactivated by subsequent illumination of blue light (430 nm). Since the kinetics of photosynthetic photoreactivation was quite similar to that of cellular photoreactivation, the author suggested that the cause of cellular death is the photosynthetic damage.

Near-UV light causes various deleterious effects on cellular metabolism in cyanobacteria at higher doses; it was found to be lethal to *Gloeocapsa alpicola* [4] and *A. nidulans* [1]. Unlike far-UV, near-UV is present in sunlight at the surface of the earth.

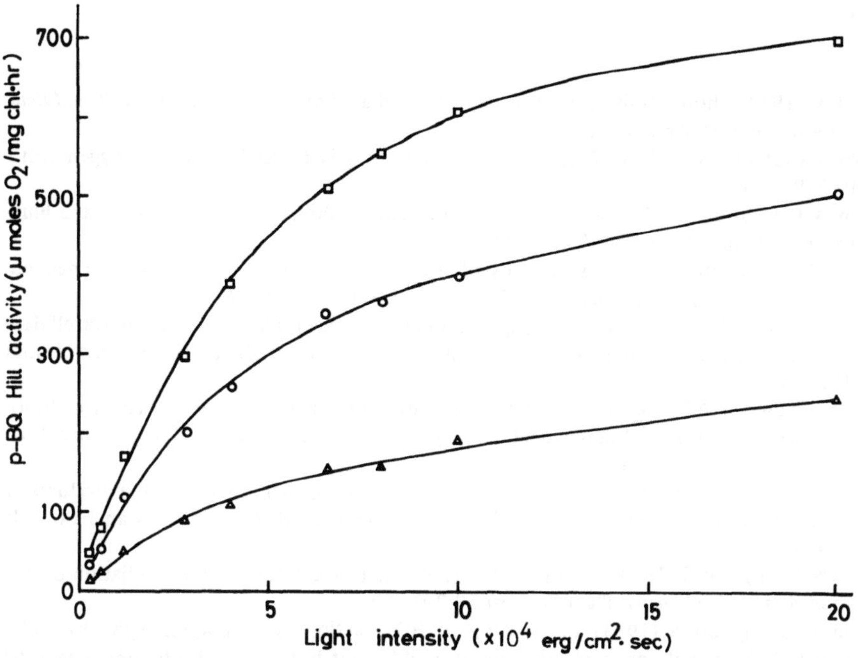

Fig. 3. Light saturation curves of control (□---□), UV-A treated (700 μW cm^{-2}, 30 min; △---△) and photoreactivated (800 μW cm^{-2}, 30 min; ○---○) *Anacystis nidulans* cells (Hirosawa and Miyachi, unpublished)

Using *A. nidulans*, Hirosawa and Miyachi [8] showed that UV-A (320–390 nm) radiation is inhibitory to growth and pigment synthesis. The primary cause of this effect appears to be the inhibition of the photosynthetic process which can be measured as the p-benzoquinone Hill reaction. The inactivated Hill reaction can be photoreactivated by the subsequent illumination of visible light. Light saturation curves of the Hill reaction by control, UV-A treated and photoreactivated cells are shown in Fig. 3. The ratio of the Hill activities of these three types of cells was constant throughout the intensities used. These results confirmed previous results that the primary target of UV-A inactivation is at around PSII rather than a component of electron transport chain between PSII and PSI. The quantum effectivity curve for photoreactivation showed maxima at 440 nm and 630 nm (Fig. 2d). The photoreactivation process was DCMU-sensitive. The peak at 440 nm and high efficiency at around 700 nm suggested that the process is dependent on the operation of PSI [9]. Photoreactivation of cellular death can only be observed when both DNA and photosynthetic damages are reactivated, and only blue light can play this dual role. Thus cellular photoreactivation can be observed only with blue or black light.

Acknowledgments. This review is due to the suggestion of Prof. S. Miyachi. The author thanks for his interest and support. Some of the research presented here was supported by Japanese Ministry of Education, Science and Culture and Nissan Science Foundation (Nissan Kagaku Zaidan). Thanks are also due to the Deutsche Forschungsgemeinschaft for generous support. I thank Miss S. Kurata for typing this manuscript.

References

1. Amla DV (1979) Photoreactivation of ultraviolet irradiated blue-green alga *Anacystis nidulans* and cyanophage AS-I. Arch Virol 59:173–179
2. Brown AH (1953) The effect of light on respiration using isotopically enriched oxygen. Am J Bot 40:719–729
3. Brown AH, Webster GC (1953) The influence of light on the rate of respiration of the blue-green alga *Anabaena*. Am J Bot 40:753–758
4. Buckley CE, Houghton JA (1976) A study of the effects of near UV radiation on the pigmentation of the blue-green alga *Gloeocapsa alpicola*. Arch Microbiol 107:93–97
5. Döhler G, Koch R (1972) Die Wirkung monochromatischen Lichts auf die extrazellulare Glykolsäure-Ausscheidung und die Lichtatmung bei der Blaualge *Anacystis nidulans*. Planta 105:352–359
6. Halldal P, Holmen AT (1979) The interrelationship between photosynthetic electron transport, glycolate excretion and amino acid metabolism in the blue-green alga *Anacystis nidulans*. Plant Cell Physiol 20:757–763
7. Hauschild AHW, Nelson CD, Krotkov G (1962) The effect of light quality on the products of photosynthesis in green and blue-green algae, and in photosynthetic bacteria. Can J Bot 40: 1619–1630
8. Hirosawa T, Miyachi S (1983) Effects of long-wavelength ultraviolet (UV-A) radiation on the growth of *Anacystis nidulans*. Plant Sci Lett 28:291–298
9. Hirosawa T, Miyachi S (1983) Inactivation of Hill reaction by long-wavelength ultraviolet radiation (UV-A) and its photoreactivation by visible light in the cyanobacterium, *Anacystis nidulans*. Arch Microbiol 135:98–102
10. Hoch G, Owens OH, Kok B (1963) Photosynthesis and respiration. Arch Biochem Biophys 101:171–180
11. Imafuku H, Katoh T (1976) Intracellular ATP level and light-induced inhibition of respiration in a blue-green alga, *Anabaena variabilis*. Plant Cell Physiol 17:515–524
12 Jones LW, Myers J (1963) A common link between photosynthesis and respiration in a blue-green alga. Nature (London) 199:670–672
13. Jones LW, Kok B (1966) Photoinhibition of chloroplast reactions. I. Kinetics and action spectra. Plant Physiol 41:1037–1043
14. Kamiya A, Miyachi S (1975) Blue light-induced formation of phosphoenolpyruvate carboxylase in colorless *Chlorella* mutant cells. Plant Cell Physiol 16:729–736
15. Kowallik W, Ruyters G (1976) Über Aktivitätssteigerungen der Pyruvatkinase durch Blaulicht oder Glucose bei einer Chlorophyll-freien *Chlorella*-Mutante. Planta 128:11–14
16. Lawrie AC, Codd GA, Stewart WDP (1976) The incorporation of nitrogen into products of recent photosynthesis in *Anabaena cylindrica* Lemm. Arch Microbiol 107:15–24
17. Lockau W (1981) Evidence for a dual role of cytochrome C-553 and plastocyanin in photosynthesis and respiration of the cyanobacterium *Anabaena variabilis*. Arch Microbiol 128:336–340
18. Malkin S, Jones LW (1968) Photinhibition and exitation quenching in photosystem II of photosynthesis, from fluorescence induction measurements. Biochim Biophys Acta 162:297–299
19. Mantai KE, Wong J, Bishop NI (1970) Comparison studies on the effects of ultraviolet irradiation on photosynthesis. Biochim Biophys Acta 197:257–266
20. Mimuro M, Fujita Y (1977) Estimation of chlorophyll a distribution in the photosynthetic pigment systems I and II of the blue-green alga *Anabaena variabilis*. Biochim Biophys Acta 459: 376–389
21. Minato S, Werbin H (1972) Excitation and fluorescence spectra of the chromophore associated with the DNA-photoreactivating enzyme from the blue-green alga *Anacystis nidulans*. Photochem Photobiol 15:97–100
22. Miyachi S, Kamiya A, Miyachi S (1977) Wavelength-effects of incident light on carbon metabolism in *Chlorella cells*. In: Mitsui A et al (eds) Biological solar energy conversion. Academic Press, London New York, pp 167–182

23. Miyachi S, Kamiya A, Miyachi S (1980) Effects of ammonia on carbon metabolism in photosynthesizing *Chlorella vulgaris* 11 h: the replacement of blue light by ammonium ion. In: Senger H (ed) The blue light syndrome. Springer, Berlin Heidelberg New York, pp 429–434
24. Murai T, Katoh T (1975) Photosystem I-dependent oxidation of organic acids in blue-green alga, *Anabaena variabilis.* Plant Cell Physiol 16:789–797
25. O'Brien PA, Houghton JA (1982) Photoreactivation and excision repair of UV induced pyrimidine dimers in the unicellular cyanobacterium *Gloeocapsa alpicola* (*Synechocystis* PCC 6308). Photochem Photobiol 35:359–364
26. Ohmori M (1981) Effect of ammonia on dark CO_2 fixation by *Anabaena* cells treated with methionine sulfoximine. Plant Cell Physiol 22:709–716
27. Okada M, Kitajima M, Butler WL (1976) Inhibition of photosystem I and photosystem II in chloroplasts by UV radiation. Plant Cell Physiol 17:35–43
28. Padan E, Cohen Y (1982) In: Carr NG, Whitton BA (eds) The biology of cyanobacteria. Blackwell, Oxford London Edinburgh Boston Melbourne, pp 215–235
29. Pearce J, Carr NG (1967) The metabolism of acetate by the blue-green algae *Anabaena variabilis* and *Anacystis nidulans.* J Gen Microbiol 49:301–313
30. Pearce J, Carr NG (1969) The incorporation and metabolism of glucose by *Anabaena variabilis.* J Gen Microbiol 54:451–462
31. Pearce J, Leach CK, Carr NG (1969) The incomplete tricarboxylic acid cycle in the blue-green alga *Anabaena variabilis.* J Gen Microbiol 55:371–378
32. Pulich WM, Van Baalen C (1974) Growth requirements of blue-green algae under blue light conditions. Arch Microbiol 97:303–312
33. Rippka R (1972) Photoheterotrophy and chemoheterotrophy among unicellular blue-green algae. Arch Microbiol 87:93–98
34. Saito N, Werbin H (1970) Purification of a blue-green algal deoxyribonucleic acid photoreactivation enzyme. An enzyme requiring light as a physical cofactor to perform its catalytic function. Biochemistry 9:2610–2620
35. Sargent DF, Taylor CPS (1971) Light-induced inhibition of respiration in DCMU-poisoned *Chlorella* caused by photosystem I activity. Can J Bot 50:13–21
36. Senger H, Briggs W (1981) The blue light receptor(s): primary reactions and subsequent metabolic changes. In: Smith KC (ed) Photochemical and photobiological reviews, vol VI. Plenum Press, New York London, pp 1–38
37. Singh PK (1975) Photoreactivation of UV-irradiated blue-green algae and algal virus LPP-1. Arch Microbiol 103:297–302
38. Smith AJ (1982) Modes of cyanobacterial carbon metabolism. In: Carr NG, Whitton BA (eds) The biology of cyanobacteria. Blackwell, Oxford London Edinburgh Boston Melbourne, pp 47–85
39. Sorensen L, Halldal P (1977) Comparative analyses of action spectra of glycolate excretion ("photorespiration") and photosynthesis in the blue-green alga *Anacystis nidulans.* Photochem Photobiol 26:511–518
40. Sorensen L, Holmen AT, Halldal P (1978) Excitation of system I and system II in photosynthesis as a possible control mechanism of glycolate metabolism in the blue-green alga *Anacystis nidulans.* Physiol Plant 42:425–427
41. Stevens SE, Van Baalen C (1973) Characteristics of nitrate reduction in a mutant of the blue-green alga *Agmenellum quadruplicatum.* Plant Physiol 51:350–356
42. Tang T, Asato Y (1978) Ultraviolet light-induction and photoreactivation of thymine dimers in a cyanobacterium, *Anacystis nidulans.* Arch Microbiol 118:193–197
43. Tel-Or E, Malkin R (1977) The photochemical and fluorescence properties of whole cells, spheroplasts particles from the blue-green alga *Phormidium luridum.* Biochim Biophys Acta 459:157–174
44. Van Baalen C (1968) The effects of ultraviolet irradiation on a coccoid blue-green alga: Survival, photosynthesis, and photoreactivation. Plant Physiol 43:1689–1695
45. Van Baalen C, O'Donnell R (1972) Action spectra for ultraviolet killing and photoreactivation in the blue-green alga *Agmenellum quadruplicatum.* Photochem Photobiol 15:269–274

46. Wang RT, Stevens CLR, Myers J (1977) Action spectra for photoreactions I and II of photosynthesis in the blue-green alga *Anacystis nidulans.* Photochem Photobiol 25:103–108
47. Wu JH, Lewin RA, Werbin H (1967) Photoreactivation of UV-irradiated blue-green algal virus LPP-1. Virology 31:657–664
48. Yamashita T, Butler WL (1968) Inhibition of chloroplasts by UV-irradiation and heat-treatment. Plant Physiol 43:2037–2040

Blue Light Effects in Halobacteria

G. WAGNER[1]

Rhodopsin-Like Photoreceptors in the Archaebacterium *Halobacterium halobium*

Halophiles, methanogenes and thermoacidophiles to the present knowledge comprise the phylogenetic kingdom of archaebacteria, equally distant from eubacteria and eukaryotes [16]. Compared to the relatively small number of archaebacteria discovered so far, a wide range of physiological capabilities seems realized in this ancient branch of evolution, including heterotrophism, chemoautotrophism and use of captured light [8]. Within the family Halobacteriaceae the species *Halobacterium halobium* is particularly well adapted to use sunlight in its brightly irradiated salty environment, i.e., the Dead Sea, other salt lakes and alterns close to the coast line. While bacteriocarotenoids function as screening pigments, photoactive retinal-proteins (rhodopsin-like photoreceptors) capture light and channel the photons into photoenergetic and/or photosensory processes. Two photoactive retinal proteins from the halobacterial membrane have been well characterized so far (Fig. 1): bacteriorhodpsin (BR)[2] as a light-driven electrogenic proton pump [11], and halorhodopsin (HR)[2] apparently to mediate electrogenic chloride influx [8a, 13].

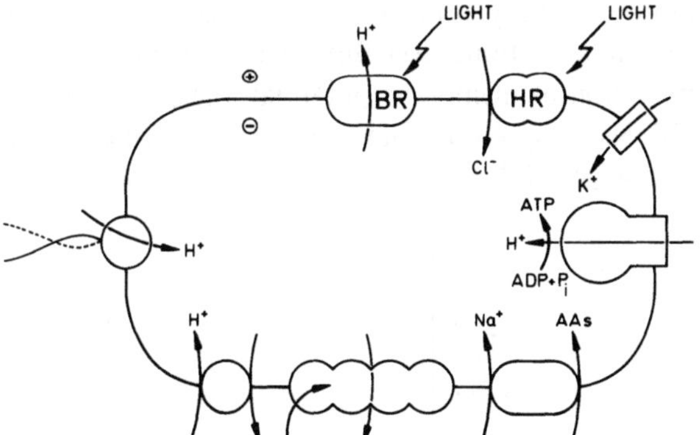

Fig. 1. Functions of energetics and behavior in the cell membrane of *Halobacterium halobium*, e.g., mutant strain R_1M_1, as shown in clockwise order: *BR* bacteriorhodopsin; *HR* halorhodopsin; K^+-uniport; H^+-driven ATP synthetase; Na^+-driven amino acids (*AAs*) symport; respiratory chain; H^+-driven Na^+ antiport; H^+-driven flagellar rotor. Sign of voltage across the membrane also indicated

1 Botanisches Institut I der Justus-Liebig-Universität, Senckenbergstraße 17–21, D–6300 Giessen, FRG

2 Abbreviations and halobacterial photoreceptor-mutants used: BR = bacteriorhodopsin, HR = halorhodopsin, SR = slow rhodopsin-like photoreceptor, F = Faraday constant
Mutant strain, R_1M_1 [22] : BR^+, HR^+, SR^+, Mutant strain, L-33 [23]: BR^-, HR^+, SR^+

Blue Light Effects in Biological Systems
ed. by H. Senger

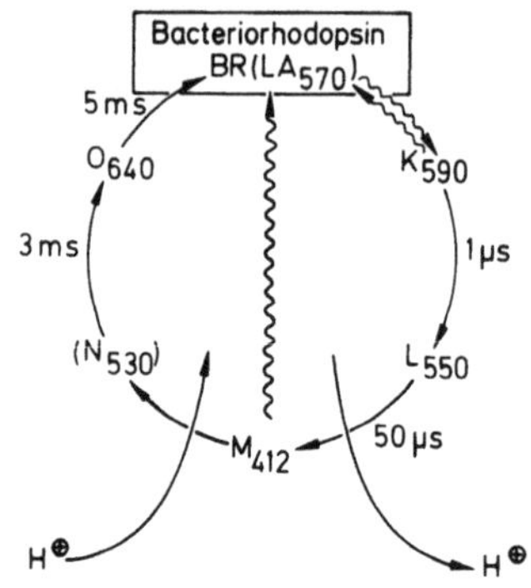

Fig. 2. Photoreaction cycle of light-adapted bacteriorhodopsin (LA_{570}). *Wavy arrows* light reactions. Additional light-driven as well as thermal backreactions also occur but are not shown. Half-lives are approximate values for a suspension of isolated purple membrane in water at room temperature and neutral pH. (After [14])

After photon absorption in the green to orange spectral range, BR (Fig. 2) and HR perform different photocycles which, however, by intermediate states span into the blue light region. BR transiently absorbs at λ_{max} = 412 nm (M_{412}) while the peak absorption of the respective intermediate state of HR was reported at λ = 380 nm but needs further substantiation. Besides dark reversion to the ground state, M_{412} is photoreversed by blue light [10]. Thus, BR (Fig. 2) and possibly HR perform photoreversible reaction cycles through photoactive intermediate states, and hence may be regarded as photoreversible photochromic pigments [4].

Halobacterial Photoenergetics in Blue-UV Versus Green-Orange Light

The balance sheet of electrogenic transport coupled to photoreversible photochromic BR and possibly HR has been studied in the wavelength range of λ = 370 to 565 nm. Rate of K^+-uniport and rate of photophosphorylation were chosen as the cell parameters proper (Figs. 3–5). Based on unidirectional ion flux measurements, atomic absorption spectroscopy and determination of membrane potential, the passive membrane permeability of K^+ (P_K) and Na^+ (P_{Na}) was calculated from constant field theory [22] and is shown for the pH range = 5.5 to 8.0 in Fig. 3. The data indicate a

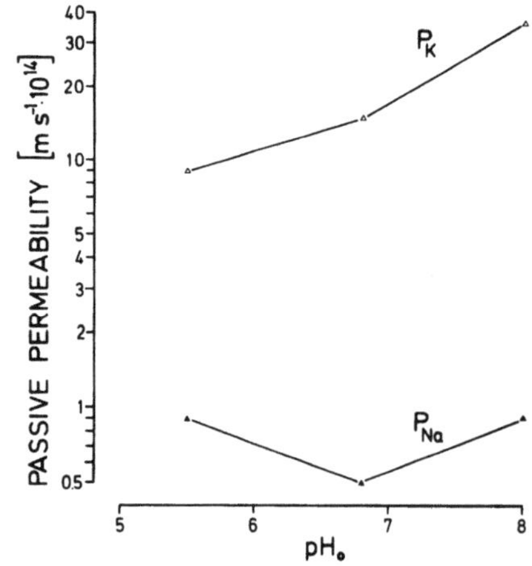

Fig. 3. Passive permeability of the halobacterial membrane, mutant strain R_1M_1, to K^+ (P_K) and Na^+ (P_{Na}) as a function of external pH

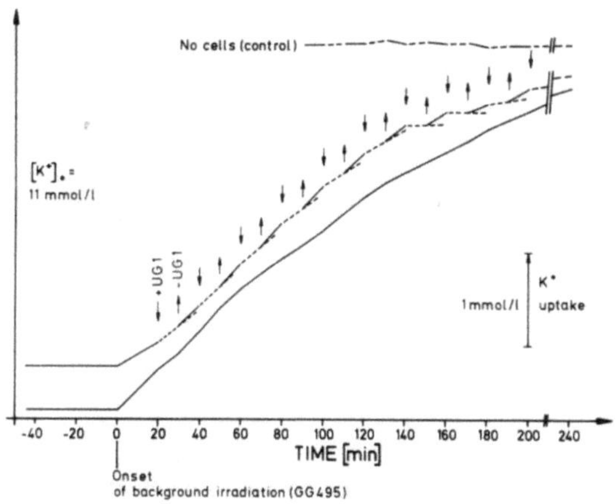

Fig. 4. Kinetics of halobacterial K^+ uptake in green-orange light (GG 495; $\lambda > 495$ nm, lower curve) and in green-orange light together with intermittent blue-UV light (UG 1; $\lambda = 360 \pm 25$ nm half band width) as indicated by *arrows*. Response of K^+-selective electrode in absence of cells (control) is also shown. Irradiance of green-orange light = 60 W m^{-2}; irradiance of blue-UV light = 39 W m^{-2}; cell density of mutant strain R_1M_1 = 2 mg protein ml^{-1}, $pH_o = 7.0$, T = 25 ± 0.1°C. (After [21])

significant selectivity of the halobacterial membrane to K^+ compared with Na^+ and possibly Cl^- (not shown). Also, rate of K^+ transport closely correlates with membrane potential (motive force [21]), and an electrophoretic mechanism of K^+ uptake (Fig. 1) through potassium selective pores (K^+-uniport) has been proposed [21, 22].

The effect of intermittent blue-UV light on a constant level of green-orange background light is shown in Fig. 4. When K^+ uptake in constant green-orange background light together with intermittent blue-UV light ceases after about 140 min, i.e., when K^+ uptake comes close to equilibrium with the potassium motive force ($\Delta\tilde{\mu}_{K^+}/F$), a significant decrease in K^+ uptake appears in blue-UV light; after blue-UV light is switched off, K^+ uptake recovers immediately (Fig. 4). The blue-UV light-mediated quench in energy transformation of absorbed green-orange light is evident also from the measured relative spectral quantum yield of photophosphorylation (Fig. 5), normalized to the absorption spectrum of BR [1]. In the wavelength range of green to orange light (λ = 440 to 565 nm), relative quantum yield of photophosphorylation turns out close to unity with a significant drop to about half at λ = 400 nm. Thus, blue-UV light absorbed concomitant with green-orange light in *H. halobium,* strain R_1M_1, leads to a quenched light energy transformation.

A cellular membrane process such as energy transformation, sensitive to the spectral distribution of light, should certainly allow the cell to notice spectral changes in light environment. In fact, spectral change in light is detected by *H. halobium* which

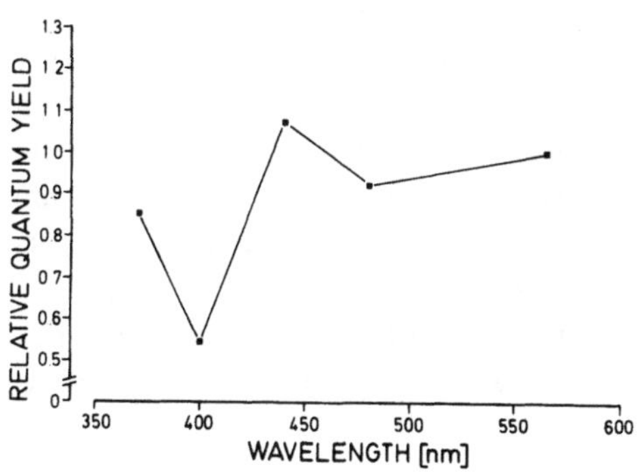

Fig. 5. Relative spectral quantum yield of photophosphorylation in halobacterial mutant strain R_1M_1, normalized to the absorption spectrum of light-adapted bacteriorhodopsin in intact cells

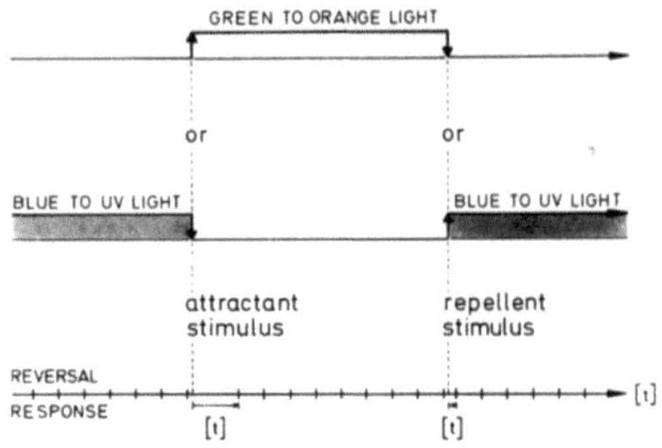

Fig. 6. Sequence in time (*bottom*) of halobacterial spontaneous reversal response as biased (*top*) by step-up or step-down photostimuli in the wavelength region of green-orange or blue-UV light. Attractant stimulus temporally suppresses spontaneous reversal, repellent stimulus induces reversal

behaves differently whether exposed to long wavelengths or to short wavelengths of light [5, 15]: e.g., step-up stimulus (i.e., increase in irradiance) by green-orange light temporally *suppresses* spontaneous reversal in bacterial swimming direction (attractant stimulus), while step-up stimulus by blue-UV light *induces* reversal response (repellent stimulus).

Halobacterial Photobehavior in Blue-UV Light Versus Green-Orange Light

Figure 6 gives a scheme on the spontaneous reversal response in *H. halobium* as biased by step-up or step-down photostimuli in the green-orange or the blue-UV wavelengths range. The spectral imbalance of photocoupled versus photouncoupled energy transformation as shown in Fig. 5 seems not to be the predominant signal in the halobacterial process of photosensing. This conclusion follows from several lines of evidence including reconstitution experiments by trans-retinol$_2$ of bacteriorhodopsin (BR_2) showing dismatch of the absorption spectrum of BR_2 and the action spectrum proper of photobehavior [12]. But as present in high amounts, BR and HR still seem to operate in photosensing of *H. halobium* [12, 18].

For the halobacterial mutant strain, L-33, deficient in BR but containing increased amounts of HR, an action spectrum of photobehavior is shown in Fig. 7. The action spectrum of blue-UV light (repellent light stimulus) shows a maximum at $\lambda = 370$ nm (Fig. 7a), which matches results from different mutant strains, irrespective of the deficiency of bacteriorhodopsin [6]. Also, no differences were found with respect to the main peak of sensitivity (Fig. 7b) to green-orange light at $\lambda = 565$ nm (attractant light stimulus). Contrary to earlier findings, however, the mutant strain, L-33, shows a secondary peak of sensitivity in the green-orange light range at $\lambda = 590$ nm in addition to the main peak at $\lambda = 565$ nm (Fig. 7b). The peak hights relative to each other are confirmed by fluence response curves (not shown). The secondary peak at $\lambda = 590$ nm corresponds to the absorption maximum of halorhodopsin but also to that of the spectroscopically detected slow rhodopsin-like photoreceptor at $\lambda = 580$ to 590 nm [2].

Work is in progress with double mutants deficient in BR and HR, to characterize further the predominant photosensory rhodopsin-like photoreceptor under consideration.

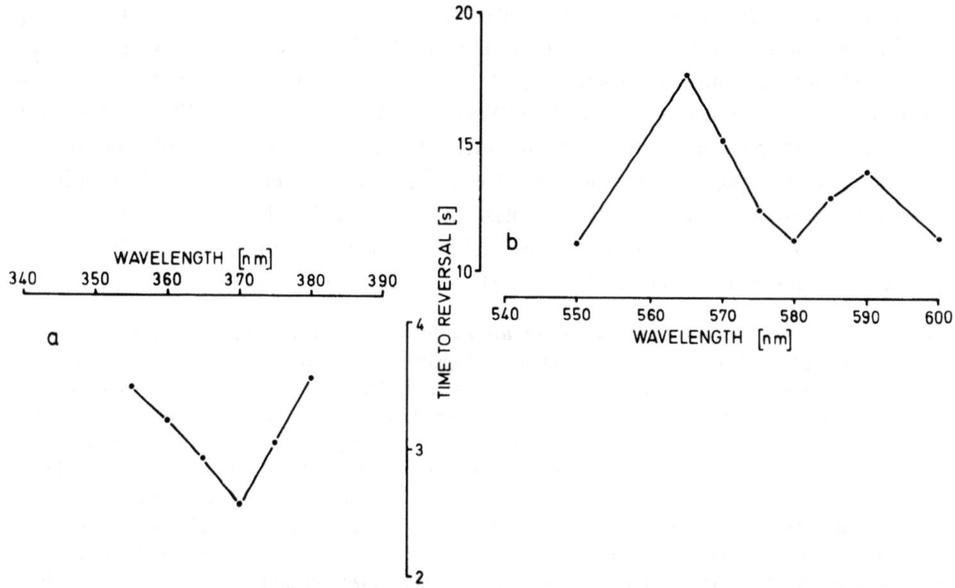

Fig. 7a,b. Action spectrum of photosensory behavior of *Halobacterium halobium,* mutant strain L-33, to blue-UV and green-orange light. Behavioral response is given as the time to reversal after the onset of a light stimulus. Light was applied immediately after a spontaneous reversal had been observed (see also Fig. 6). **a** Repellent stimulus by increase of ultraviolet light: 3×10^{11} photons mm^{-2} s^{-1}, applied as a pulse stimulus for 2 s at each wavelength; **b** Attractant stimulus by increase of visible light: 3×10^{14} photons mm^{-2} s^{-1}, applied as a step-up stimulus at each wavelength [19]

Fig. 8. *Right* Model of wavelength-dependent accumulation of a long-living intermediate state B relative to the ground state A of a photoreversible photochromic pigment. The absorption maximum of state A ($\lambda_{max, A}$) should be different from that of state B ($\lambda_{max, B}$); the absorption spectrum of state A, however, should cover $\lambda_{max, B}$ as realized, e.g., by rhodopsin-like photoreceptors in *Halobacterium halobium. Left* Action spectrum of photophobic response in *Halobacterium halobium,* mutant strain R_1M_1 [9]. The data possibly represent a difference spectrum: response to attractant light minus response to repellent light [15]

For the mechanism of spectral discrimination in halobacterial photosensing ("color" vision of blue-UV versus green-orange light), there are two opposing concepts: that of Hildebrand and coworkers [3, 7] based on the multipigment mechanism of color vision in vertebrates and that of Wagner and coworker [19, 20] based on the photoreversible photochromic pigment mechanism in plants (e.g., phytochrome). Recent spectroscopic data on slow rhodopsin-like photoreceptor show large photoreversible photochromic shifts [2] consistent with the halobacterial photobehavior observed in blue-UV versus green-orange light. The scheme in Fig. 8 thus allows for a probable mechanism of spectral discrimination in photosensing of *H. halobium*.

Epilogue. In the archaebacterium *Halobacterium halobium*, it seems intriguing to find rhodopsin-like photoreceptors closely related to animal rhodopsin [17], but to serve dual function, i.e., light energy conversion besides photostimulus discrimination. Thus, halobacterial rhodopsins cover a wide range of molecular capabilities to channel absorbed photons either into charge translocation across the membrane or into photoregulation of behavioral response. On the cellular level of integration, however, there seems no competition between these two light-induced processes: photobehavior allows the halobacterial cell to avoid environments of extensive blue to ultraviolet irradiances, harmful even to quantum yield of photophosphorylation, but enables it also to stay in photoenergetically effective green to orange light. The molecular mechanism of spectral discrimination and integration of absorbed light in *Halobacterium halobium* certainly represents a compelling problem in photobiology, to be resolved in future times by a joined venture of molecular biology, biophysics and biochemistry.

Acknowledgments. Work in the author's laboratory was financially supported by the Deutsche Forschungsgemeinschaft.

Skilful technical assistance of Mrs. A. Weisert as well as B. Traulich's support in preparing the manuscript is gratefully acknowledged.

References

1. Bogomolni RA, Baker RA, Lozier RH, Stoeckenius W (1980) Action spectrum and quantum efficiency for proton pumping in *Halobacterium halobium*. Biochemistry 19:2152–2159
2. Bogomolni RA, Spudich JL (1982) Identification of a third rhodopsin-like pigment in phototactic *Halobacterium halobium*. Proc Natl Acad Sci USA 79:6250–6254
3. Dencher NA, Hildebrand E (1979) Sensory transduction in *Halobacterium halobium*: Retinal protein pigment controls UV-induced behavioral response. Z Naturforsch 34c:841–847
4. Hartmann KM (1983) Action spectroscopy. In: Hoppe W, Lohmann W, Markl H, Ziegler H (eds) Biophysics. Springer, Berlin Heidelberg New York, pp 115–144
5. Hildebrand E, Dencher N (1975) Two photosystems controlling behavioural responses of *Halobacterium halobium*. Nature (London) 257:46–48
6. Hildebrand E, Schimz A (1983) Photosensory behavior of a bacteriorhodopsin-deficient mutant, ET-15, of *Halobacterium halobium*. Photochem Photobiol 37:581–584
7. Hildebrand E, Schimz A (1983) Consecutive formation of sensory photosystems in growing *Halobacterium halobium*. Photochem Photobiol 38:593–597
8. Kandler O (1981) Archaebakterien und Phylogenie der Organismen. Naturwissenschaften 68: 183–192

8a. Matsuno-Yagi A, Mukohata Y (1977) Two possible roles of bacteriorhodopsin: a comparative study of strains of *Halobacterium halobium* differing in pigmentation. Biochem Biophys Res Comm 78:237–243

9. Nultsch W, Häder M (1978) Photoakkumulation bei *Halobacterium halobium*. Ber Dtsch Bot Ges 91:441–453
10. Oesterhelt D, Hess B (1973) Reversible photolysis of the purple complex in the purple membrane of *Halobacterium halobium*. Eur J Biochem 37:316–326

11. Oesterhelt D, Stoeckenius W (1973) Functions of a new photoreceptor membrane. Proc Natl Acad Sci USA 70:2853–2857
12. Schimz A, Sperling W, Hildebrand E, Köhler-Hahn D (1982) Bacteriorhodopsin and the sensory pigment of the photosystem 565 in *Halobacterium halobium.* Photochem Photobiol 36: 193–196
13. Schobert B, Lanyi JK (1982) Halorhodopsin is a lightdriven chloride pump. J Biol Chem 257: 10306–10313
14. Schreckenbach Th (1978) The properties of bacteriorhodopsin and its incorporation into artificial systems. In: Barber J (ed) Photosynthesis in relation to model systems. Elsevier/North-Holland Biomedical Press, Amsterdam New York, pp 189–209
15. Spudich JL, Stoeckenius W (1979) Photosensory and chemosensory behavior of *Halobacterium halobium.* Photobiochem Photobiophys 1:43–53
16. Stackebrandt E, Woese CR (1981) The evolution of prokaryotes. In: Carlile MJ, Collins JF, Moseley BEB (eds) Molecular and cellular aspects of microbial evolution. Soc Gen Microbiol. Univ Press, Cambridge, pp 1–31
17. Stavenga DG, This volume
18. Traulich B, Hildebrand E, Schimz A, Wagner G, Lanyi JK (1983) Halorhodpsin and photosensory behavior in *Halobacterium halobium* mutant strain L-33. Photochem Photobiol 37:577–579
19. Traulich B, Wagner G (1983) Photosensory behavior of *Halobacterium halobium.* BioScience, 33:583–584
20. Wagner G (1981) Action spectra of ATP synthesis in the halobacterial mutants $R_1 M_1$ and L-33: Photoenergetic and photosensory implications. Discussion paper: EMBO Workshop Halophil Microorg, Ischia, Italy
21. Wagner G, Geissler G, Linhardt R, Mollwo A, Vonhof A (1980) Light dependent ion transport processes and photosensory transduction in *Halobacterium halobium.* In: Spanswick RM, Lucas WJ, Dainty J (eds) Plant membrane transport: Current conceptual issues. Elsevier/North-Holland Biomedical Press, Amsterdam New York, pp 641–644
22. Wagner G, Hartmann R, Oesterhelt D (1978) Potassium uniport and ATP synthesis in *Halobacterium halobium.* Eur J Biochem 89:169–179
23. Wagner G, Oesterhelt D, Krippahl G, Lanyi JK (1981) Bioenergetic role of halorhodopsin in *Halobacterium halobium* cells. FEBS Lett 131:341–345

Orange/Yellow Pigments in the Basidiomycete *Pleurotus ostreatus* (Jacq. ex. Fr.) Kummer[1]

W. MARGRAF[2]

Introduction

On the first blue light symposium held at Marburg in 1979 Eger-Hummel reported on several mutant types of *P. ostreatus* which are blocked at different stages of sporophore development and then reveal orange or yellow colorations [4]. A relation between color formation and photomorphogenesis in this mushroom was assumed. This paper describes investigations on the conditions for production and on the isolation of the respective pigments.

Materials and Methods

Yellow to orange stages of sporophores were obtained in the strains Y 102x42, 62_1B, 387, 117_1c, 439, 258a Myc C, 56_2B, 305cFk, 165 Mut 8/9 all from the culture collection of Eger-Hummel, F-3 from Gyurko (Sopron) and So 3004 from Somycel (ATCC no. 38546). The sterile dikaryon X77 × X53 was employed for the major part of this work. The mycelia were grown in plastic Petri dishes of 8.5 cm diameter on a deproteinated malt extract agar at 27.5°C as described previously [10, 11] or in glass Petri dishes of 8.5 cm diameter. Fruiting was induced with a filter sterilized solution of 133 mmol l^{-1} glutamine (Merck no. 289), 250 mmol l^{-1} glucose (Merck no. 8342) dissolved in 0.05 m Na phosphate buffer pH 6.5, called primordia-inducing solution [11]. Dependence of color production on N-sources was studied with Gln, Glu, Asn, Asp, $H_2N\text{-}CO\text{-}NH_2$, NH_4Cl, $(NH_4)_2H$-citrate in test tubes. From dark-grown mycelium of 8 cm diameter disks of 1 cm diameter were cut and transferred into test tubes (100 mm × 16 mm), which contained 2 ml of a sterile primordia-inducing solution, in which glutamine was replaced by other nitrogen compounds. The tubes were closed with Cap-o-test caps and incubated at 27.5°C for 20 days. Color production was controlled by measuring the extinctions of the solutions at 475 nm against water with a Zeiss PM 4 or a Beckman Acta M VI spectral photometer in 1-cm quartz cuvettes. For each solution 10 test tubes were examined and the significance of differences in extinction checked by t-test. Chromatography was performed with a Sephadex G 10 column (800 mm × 100 mm, Pharmacia) and a Lobar RP-8 reversed phase column

1 Part of Dr-dissertation, see: Bibliotheca Mycologica *93,* 1984, J. Cramer, Vaduz

2 Institut für Pharmazeutische Technologie der Universität, Ketzerbach 63, D-3550 Marburg, FRG

Blue Light Effects in Biological Systems
ed. By H. Senger

(Merck no. 11804). Sep-Pak C 18 cartridges (Millipore-Waters no. 51910) were used to remove impurities before passing the Lobar column. For freeze-drying of the solutions a lyophilizator from Kniese (Marburg) was used. The temperature did not exceed 26°C during the process.

Results

Depending on strains and temperature, the sporophores of *P. ostreatus* have dark to light gray, bluish, brown or cream pilei [10]. As yet orange or yellow colorations were only described for sporophores whose development was blocked by mutations [4]. In the course of observations of many strains (see Materials), commercial ones included, this coloration was also noticed when fruiting was arrrested by such unfavorable environmental conditions as too high temperature, too much light, poor nutrient medium or too low humidity. The color may appear on each stage of morphogenesis. It was not possible to standardize strains and growth conditions reliably in order to obtain reproduceable results and large enough quantities of yellow to orange stages of sporophores. It was noticed that sometimes mycelia excreted orange or yellow drops. These mycelia were unable to form primordia. Following these observations a large number of strains was screened. None of the dikaryotic mutants of Eger-Hummel's collection was suitable. Several monokaryotic mutants, however, excreted orange or yellow compounds into the nutrient medium each time after they were treated with primordia-inducing solution (see Materials and Methods). In this solution, glutamine, given simultaneously with light, is essential for optimal fruiting initiation [11]. Six selected monokaryons were paired to three dikaryons, which were sterile and produced the orange/yellow colors independend of light. One of these (X77 × X53) was superior and employed in the following investigations. In several experiments the dependence of N-sources on color production was studied in test tubes (see Materials and Methods). With 133 mmol l^{-1} Gln, Glu, Asn, Asp yellow/orange colors were formed, not with NH_4Cl, $H_2N\text{-}CO\text{-}NH_2$, $(NH_4)_2H$-citrate. In the positive cases an absorption peak was found around 475 nm (Fig. 1). With Gln the peak was significantly higher than with the other compounds. That glutamine is the most appropriate compound for primordia and color formation makes a relation between these two processes probable. Further experiments showed that a glucose concentration of 100 mmol l^{-1} and 50 mmol l^{-1} Na phosphate were sufficient for color production. The optimal pH value was 5.0.

In order to obtain large quantities of colored solution for isolation and identification of the contained pigments the test-tube method was inadequate. Therefore dark-grown mycelia of 8 cm diameter were underlaid with the solution for production (Fig. 2). After 5 to 6 days the yellow liquid was aseptically removed and replaced by fresh solution. This procedure was repeated every 2 to 3 days for 15 to 20 days. The production is optimal after 10 to 12 days and then decreases. Using 50 Petri dishes, 2 l of orange/yellow liquid could be obtained every 2 days. The liquid was freeze-dried and stored at −18°C. In this way sufficient quantities were produced to study the nature of the pigment(s). Ten g of the lyophilizate of the colored liquid were dissolved in 180 ml distilled water and chromatographed over a Sephadex G 10 column. Distinct

Fig. 1. Dependence of color production on the supplies of different amino acids. (△ Gln, □ Glu; ○ Asn; ○ Asp)

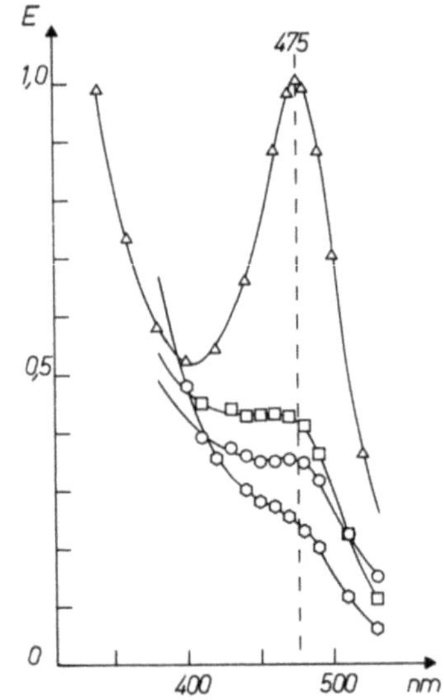

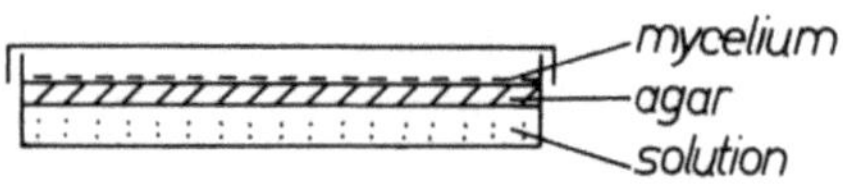

Fig. 2. The mycelium growing through the nutrient medium excretes orange/yellow pigments into the solution. (133 mmol l^{-1} Gln, 100 mmol l^{-1} glucose, 0.05 m Na phosphate buffer pH 5.0)

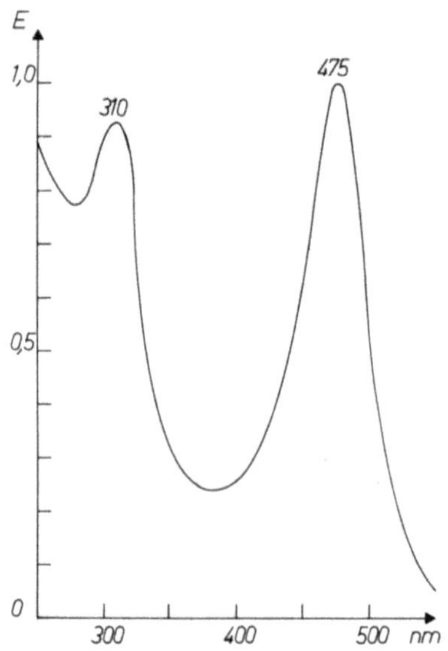

Fig. 3

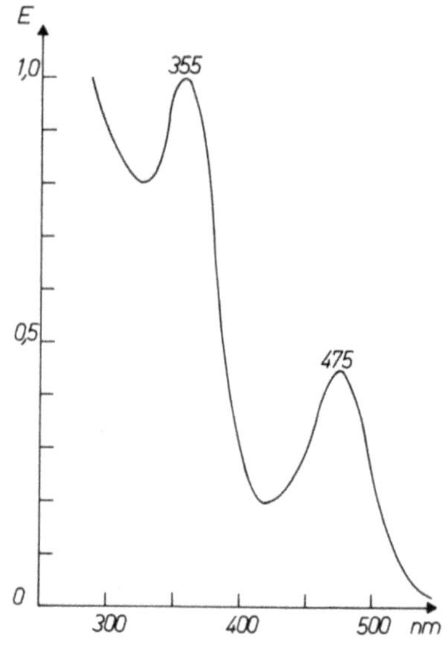

Fig. 4

Figs. 3 and 4. Absorption curves of two different yellow fractions after column chromatography ($E = \lg I_0/I$)

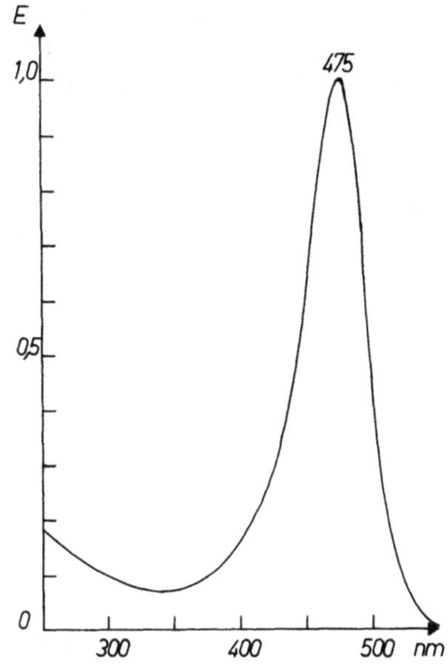

Fig. 5. Absorption curve of a purified pigment ($E = \lg I_0/I$)

bands of different colors could not be noticed. A widespread zone which was orange at the beginning and yellow at the end was collected in fractions. The last yellow fractions overlapped with ninhydrin-positive and Fehling-test-positive substances, which were probably glutamine and glucose. All colored fractions showed peaks around 475 nm. Two fractions showed additional peaks at 310 nm and 355 nm respectively (Figs. 3 and 4). The first orange and yellow ninhydrin- and Fehling-test-negative 640 ml were freeze-dried. The residue (about 200 mg) was dissolved in 3 ml of distilled water, passed through a Sep-Pak C 18 cartridge and chromatographed over a Lobar RP-8 reversed phase column. The eluent was water. An intense yellow band was separated and rechromatographed. The absorption spectrum of the so purified pigment has only one maximum at 475 nm (Fig. 5). The pigment is insoluble in ethanol, acetone, chloroform, ether and petrol ether, soluble in methanol and highly soluble in water, which is already colored intense yellow by traces of it.

Discussion

Carotenes and/or flavins have been suggested to be the absorbing part of the "characteristic" photoreceptor in fungi [12]. Their absorption spectra resemble the action spectra of many photoreactions, which often show two peaks, one in the region of 430 nm to 480 nm and another around 370 nm. Tan also described "atypical" photoresponses with one peak only. In *Penicillium isariiforme* [2] and *Stemphylium botryosum* [9] the maximum was found around 480 nm. A recent report cites an action spectrum with a single sharp peak of maximum effectiveness near 475 nm and a third class of photoreceptor substances besides flavins and carotenoids was assumed [3]. Action spectra of higher fungi sometimes correspond with the profile of the "characteristic" photoreceptor. The Basidiomycete *Favolus arcularius* was considered to possess two receptors [6, 7]. For complex photoreactions such as sporophore development, a combined photoreception system can be imagined in which several substances absorbing at different wavelengths are involved.

A substance absorbing at 310 nm was suggested to be correlated with certain photoreactions. It was first isolated by Leach [8], identified by Favre-Bonvin et al. [5] and called P 310 (mycosporine). It has been isolated from the Basidiomycete *Stereum hirsutum* by Arpin et al. [1]. Its role in the complex of photoreactions is not clear [2]. In *P. ostreatus* I also found a sharp peak at 310 nm, which probably originates from P 310. I do not know what substances may be responsible for the observed peaks at 355 nm and 475 nm. I am concentrating on the identification of the latter.

References

1. Arpin N, Thivend S, Favre-Bonvin J (1977) Substances photoabsorbantes de Stereum hirsutum; Isolement, purification et caracteristiques de P310 (mycosporine). Bull Soc Mycol Fr 93:no 1
2. Bennink GJH (1971) Fotomorfogenese bij Penicillium isariiforme. Ph D thesis, Univ Amsterdam, The Netherlands
3. Briggs WR, Iino M (1983) Blue light-absorbing photoreceptors in plants. Philos Trans R Soc London Ser B, pp 347–359
4. Eger-Hummel G (1980) Blue-light-photomorphogenesis in mushrooms. In: Senger H (ed) The blue light syndrome. Springer, Berlin Heidelberg New York, pp 555–562
5. Favre-Bonvin J, Arpin N, Brevard C (1976) Structure de la mycosporine (P310). Can J Chem 54:1105–1113
6. Kitamoto Y, Horikoshi T, Suzuki A (1974) An action spectrum for photoinduction of pileus formation in a Basidiomycete, Favolus arcularius. Planta 119:81–84
7. Kitamoto Y, Suzuki A, Furukawa S (1972) An action spectrum for light-induced primordium formation in a Basidiomycete, Favolus arcularius (Fr.) Ames. Plant Physiol 49:338–340
8. Leach CM (1965) Ultraviolet-absorbing substances associated with light-induced sporulation in fungi. Can J Bot 43:185–200
9. Leach CM (1968) An action spectrum for light inhibition of the "terminal phase" of photosporogenesis in the fungus Stemphylium botryosum. Mycologia 60:532–546
10. Li SF (1980) Studies on the tolerance to elevated temperatures in Pleurotus ostreatus. In: Bibliotheca mycologica, vol 76. Cramer, Vaduz
11. Netzer U von (1978) Induktion der Primordienbildung bei dem Basidiomyceten Pleurotus ostreatus. In: Bibliotheca mycologica, vol 62. Cramer, Vaduz
12. Tan KK (1978) Light-induced fungal development. In: Smith, Berry (eds) The filamentous fungi, vol III. Arnold, London, pp 334–357

Blue and Ultraviolet Light in Eyes: Primary Reactions and Light-Induced Metabolic Changes

D. G. STAVENGA[1]

Introduction

Animals have eyes for seeing, i.e., to discriminate variations in the environmental light distribution. The key role for detecting these light changes is played by the visual pigment molecules, because they trigger the phototransduction process: a chain of molecular reactions in the photoreceptor membrane resulting in a neural signal. The absorption spectra of an eye's visual pigments therefore determine the spectral sensitivity of the photoreceptors and thus the spectral range which is covered by the visual system.

In general, visible light extends from roughly 300–750 nm, i.e., from the ultraviolet up to the far-red. Many visual systems, however, have light sensitivity in a restricted spectral range only, the most eye-catching example being the normal human eye which is insensitive to ultraviolet light. In contrast, several insects have large eye parts which are exclusively sensitive to the ultraviolet, and most marine animals living in the deeper sea layers see blue light only. From the last example, it is clear that an understanding of the often bewildering variety of eye properties necessitates consideration of the structural elements in direct relation with the ecological factors which have shaped an animal's eye.

A brief description of the visual pigments and of the optical factors affecting light sensitivity is presented here in the perspective of the role of short-wavelength light. Furthermore, an optical method employing blue and ultraviolet light which reveals basic, metabolic processes is discussed in relation to the function of photoreceptor cells.

Visual Pigments and Retinochromes

Since the review presented at the first Blue Light Syndrome Meeting [51], major developments in the field of vision research have occurred. Most importantly, the complete peptide sequence of bovine visual pigment has been resolved [22, 34]. This substantiated the conception of a reliable three-dimensional model of the visual molecule embedded in the cell membrane [10]. The chromophore retinal is folded in an 11-cis configuration with the molecular dipole more or less parallel to the membrane surface.

1 Biophysics Department, Rijksuniversiteit Groningen, The Netherlands

Blue Light Effects in Biological Systems
ed. by H. Senger

Until recently two types of chromophore were known: retinal and 3-dehydroretinal, both absorbing in the ultraviolet (reviews: [9, 30]). They characterize two classes of visual pigments, the rhodopsins and porphyropsins respectively; this year Vogt [58] reported that flies have a third type of chromophore, presumably 3-hydroxyretinal. Since the chromophore can be synthetized from a hydroxy-xanthophyll lutein, the new class of visual pigment has been tentatively termed xanthopsin. Whether the newly discovered chromophore occurs in invertebrates other than flies has yet to be determined.

The interaction between chromophore and protein determines the visual pigment's absorption spectrum, which can be restricted to the ultraviolet or extend to longer wavelengths, eventually into the red region. It is of historical interest that before the discovery of the class of ultraviolet absorbing visual pigments [14], ultraviolet sensitivity of visual photoreceptors was presumed to originate from fluorescing ocular media [7]. Visual pigments absorbing in the blue region should then capture the fluorescence. A critical survey of this view has been presented by Stark and Tan [64].

Rather than this trivial, radiative transfer, energy transfer via a Förster resonance mechanism can enhance light sensitivity. Such a system which is well established in photosynthesis was recently concluded for phytochrome-mediated opening in the excised hook of bean [39, cf. 40].

In the area of vision research, energy transfer was first hypothesized to occur in fly photoreceptors by Kirschfeld et al. [24]. The existence of enhanced sensitivity in the ultraviolet region, mediated by sensitizing pigments, is now more firmly established due to meticulous studies performed by Hardie and Kirschfeld [21] and Vogt and Kirschfeld [59].

Another area where important new insights have been gained is that of visual pigment and membrane turnover. It should be recalled that after light absorption the chromophore of a visual molecule isomerizes from the 11-cis to the all-trans form, resulting in a series of thermally induced conformational changes in the molecule, demonstrable as changes in the absorption spectra. The subsequent stages of the visual molecule have been extensively studied in extracts or in excised retinae from cattle and squid by low-temperature spectroscopy, and later (at room temperature) by flash photolysis and laser methods (review: [63]). Figure 1 shows the photochemical cycle of fly visual pigment investigated in the first in vivo laser analysis of a visual pigment [26]. Photoconversion of the rhodopsin R leads via a few intermediates to a thermostable state metarhodopsin M which can be photoconverted, again via thermolabile intermediates, back into the native rhodopsin state (e.g., [17, 53]. The process is, in general, different in vertebrate eyes. After photoconversion of rhodopsin, retinal separates from the opsin part so that the absorption in the visible wavelength range vanishes: the visual pigment is bleached. Recent studies have shown, however, that invertebrate metarhodopsins can also decay rather rapidly, at least at elevated temperatures [2, 3]. Hence, the dogma that only the visual pigments of vertebrates bleach and those of invertebrates do not, is no longer tenable.

Although the photolysis of vertebrate rhodopsins has been a well-known fact for over a century (e.g., [37]) it is only in recent years that the complicated system for renewal of the visual molecules and their replacement in the photoreceptor membrane is beginning to become clarified (e.g., [6]). Renewal of visual molecules requires reiso-

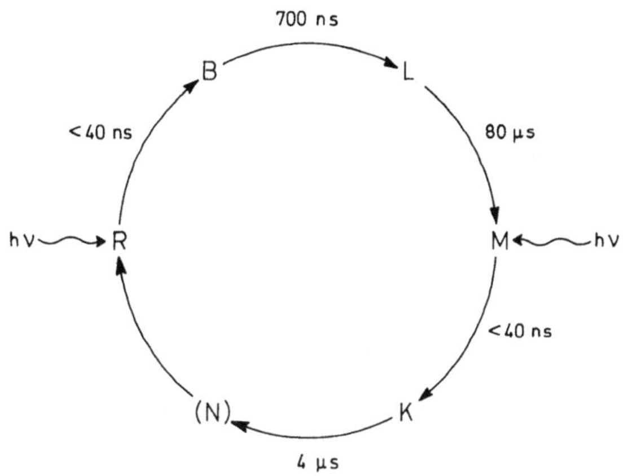

Fig. 1. Schematic presentation of the photochemical cycle of blowfly rhodopsin. Rhodopsin R excited by light absorption converts to bathorhodopsin B, which decays thermally, via lumirhodopsin L to metarhodopsin M. The light-induced reverse reaction proceeds via an intermediate K and possibly N. Time constants of the conversions are indicated [26]

merization of all-trans retinal into the 11-cis configuration. Whereas in the vertebrate eye this requires a multistage process involving transport to pigment cells, esterification, hydrolysis into 11-cis retinal and oxidation, with all these processes occurring in the dark [5], invertebrates have developed a more direct system in addition to membrane renewal [4, 60] which works inside the photoreceptor cells. This system is of specific interest in this review.

Hara and Hara (review: [19]) first noticed in octopus photoreceptors a protein binding all-trans retinal. This so-called retinochrome absorbs blue light. When doing so, the all-trans retinal converts into 11-cis retinal and, subsequently, the chromophore is liberated and becomes accessible for being built into a rhodopsin molecule (Fig. 2). From the viewpoint of retinal, retinochrome and rhodopsin are antagonistically working isomerizing enzymes using light as a cofactor.

Retinochromes have been mainly studied in cephalopods but accumulating data indicate that they are widespread in other phyla as well; e.g., bee [35, 36] and fly [42], see also [43, 53].

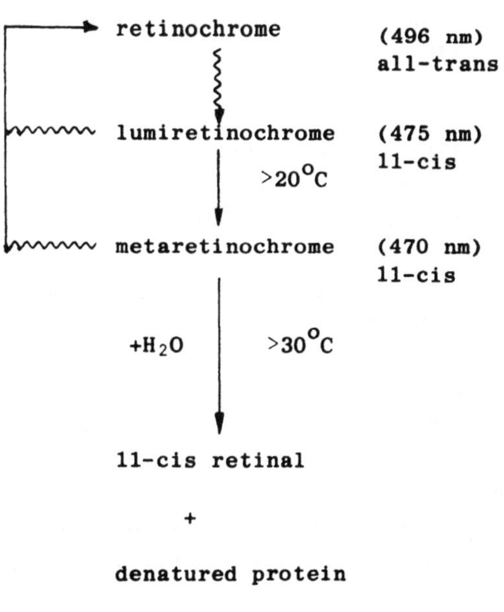

Fig. 2. Photochemical cycle of squid retinochrome. Photoconversions are indicated by *wavy lines*, thermal decays by *straight lines*. *Numbers* indicate the wavelength of maximal absorption [20]

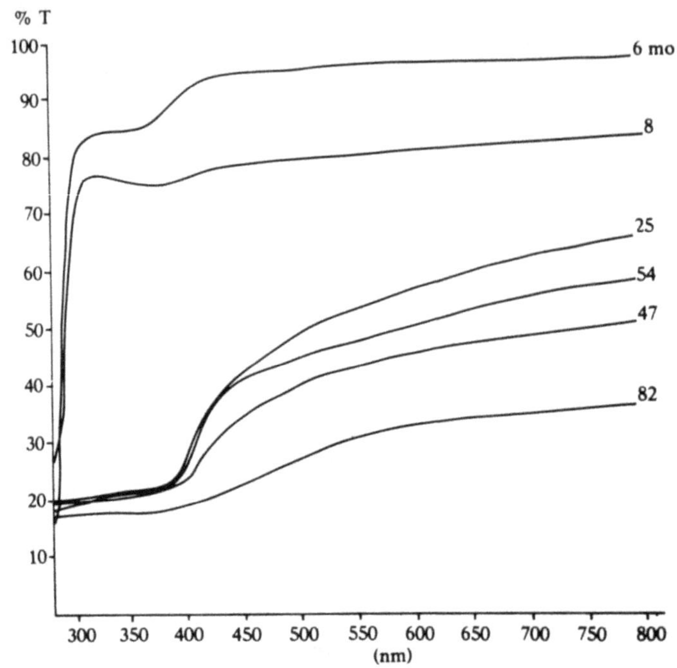

Fig. 3. Absorption by human eye lenses depends distinctly on age (given in years, except for the above curve, which is in months). A continuous increase in blue absorbing pigment causes the yellowing of the eye lens upon aging [28]

Optical Filters

Before light reaches the visual pigments in the photoreceptors its spectral content is modified by the optical properties of the dioptric system. This feature is a dominant one in the case of wavelength-selective absorption filters, which are found in the eyes of several animal species.

In human eyes absorption of short-wavelength light is quite substantial. The consequence is that normal eyes have negligible ultraviolet sensitivity. Relative to normal observers, aphakics (persons with the eye lens removed by operation) have a much enhanced ultraviolet sensitivity. This can be directly understood from the highly ultraviolet-absorbing eye lens (Fig. 3).

The optical density in the short-wavelength range progressively increases upon aging (Fig. 3; reviews: [28, 38, 62]), which is noticeable as the yellowing of the eye during life.

Yellow lenses are found in a variety of sea fish [30, 38, 47]. A curious case is the puffer fish: the cornea is transparent at low light levels and yellow upon illumination due to the migration of carotenoid pigment in the corneal chromatophores [1].

At the retinal level, yellow filtering is executed in the human eye by the macular pigment (e.g., [38]), in the eyes of birds by oil droplets situated in front of the visual pigment [30, 31], and in invertebrates by photostable pigments embedded in the visual membrane [23, 25]. Spectral filtering or enhancement is further realized by various optical devices: interference filters and reflectors, and scattering structures (review: [31]).

A single satisfactory interpretation of the function of the short-wavelength filters, which are widely encountered in animal eyes, probably does not exist. A number of consequences can be indicated, however. Filtering improves contrast transfer, because chromatic aberration and scattering, which progressively increases at shorter wavelengths, is reduced. Indeed, some correlation exists between enhanced short-wavelength absorption (i.e., more yellow eyes) and active life in brighter environments [33]. An interesting exception to this rule is a stomiatoid deep-sea fish which lives in an environment dimly lit by a band of blue light centred around 470 nm [47]. This animal possesses an orbital light organ with red filters, so that the blue absorbing eye lens appears to be tuned to the red bioluminescence. In this way then, vision is improved over that of its prey.

A quite different explanation for the existence of ultraviolet-blue ocular filters has been stressed by Ham et al. [15, 16, 48] and by Kirschfeld [23] since short-wavelength light can have a distinctly damaging effect on biological cells. Indeed, light-induced structural degradations have been demonstrated at all levels of the eye: that of the cornea, lens and retina (review: [28]). Notably the latter layer is prone to deleterious influences from intense light sources, due to the focusing effect of the dioptric apparatus. An obvious protection function is thus fulfilled by a yellow, short-wavelength absorbing, lens. The lens itself is apparently under continuous attack, because brown cataracts result from overexposure to ultraviolet light. (An enhanced sensitivity for cataract formation was found in patients who used photosensitizers during the treatment of skin diseases [28].) The increase in yellow pigmentation of the human eye lens with age may thus function as a continuously improving protection filter.

If protection from the devastating ultraviolet radiation is a necessity for prolonged well-functioning of an eye, are then all animals equipped with the same filters? This is not the case as numerous animals possess lenses which are transparent in the UV (e.g., [33]). The reason for clear optics in those eyes is that vision in the short-wavelength range is specifically valuable, particularly in birds and insects. The necessary protection from the UV has then presumably been sought in a way differing from that which utilizes an ultraviolet opaque lens. It has already been mentioned above that birds have strongly UV-absorbing oil droplets in front of the photoreceptor outer segments, and that flies have photostable pigments in the photoreceptive membrane. The pigments, moreover, can quench excited states and/or singlet oxygen (for extensive discussion, see [23]), and can thus provide effective defence against photic damage.

In addition to the static protection filters mentioned above, most eyes have moveable pigment filters as part of light control systems. The pupil in human eyes seems to have a minor function in obstructing dangerously intense light, but its primary function is to extend the working range of the eye [46, 61]. Optically dense pigments are definitely essential for eyes which cannot avoid gazing into the bright sun, which is the case for the compound eyes of insects. Apart from compound eyes, most insects have ocelli located dorsally on their heads. Considering that most dragonflies are active in bright light, it is comprehensible that brown pigment migrates in front of the ocellar photoreceptors upon bright illumination (Fig. 4). In addition, in the dragonfly *Anax*

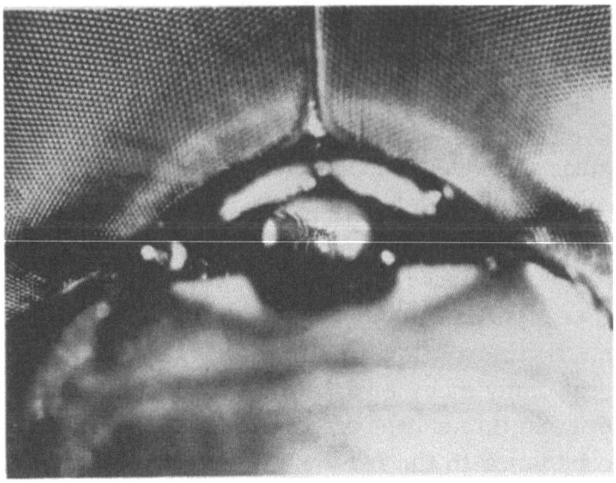

Fig. 4. The median ocellus of dragonflies is situated between the compound eyes, which touch each other dorsally. Inside the ocellus, predominantly in the lower left part, threads of pigment are seen. This pigment migrated due to bright illumination from the right. The pigment is retracted in the dark-adapted ocellus [55]

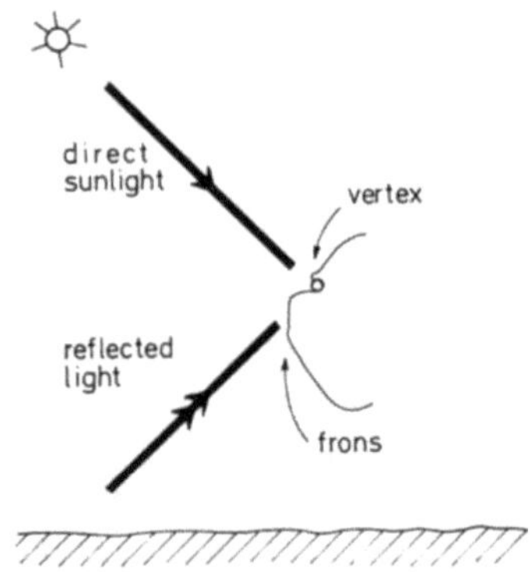

Fig. 5. Diagram demonstrating the light conditions of the dragonfly *Anax* when flying above a water surface. The dorsal vertex occludes the main area of the skies. The protruded frons shields the reflection from the water surface. Direct light from the sun impinging upon the ventral part of the ocellus is filtered by the flexible pigment layer (see Fig. 4) [55]

the median ocellus is protracted behind an extended frons, possible because in this way the light reflected from the water surface above which the dragonfly usually patrols is blocked (Fig. 5).

Mitochondrial Pigments Absorbing in the Ultraviolet and Blue

Anatomical data of retinal cells invariably show up a high concentration of mitochondria. It should be noted that these organelles contain certain pigments which absorb distinctly in the ultraviolet, i.e., the pyridine nucleotide NADH and flavoproteins. Spectroscopical studies on the mitochondrial pigments have played a central role in the elucidation of the mitochondrial molecular machinery. It has thus been shown that NADH and flavoprotein are elements of the so-called respiratory chain. This chain of molecules provides the energy for ATP production by electron transfer to oxygen (oxidative phosphorylation).

Both NADH and flavoproteins fluoresce. Notably NADH displays a high blue emission under ultraviolet excitation. The first optical study indicating the high concentration of NADH in visual sense cells was reported by Liebman and Leigh [29].

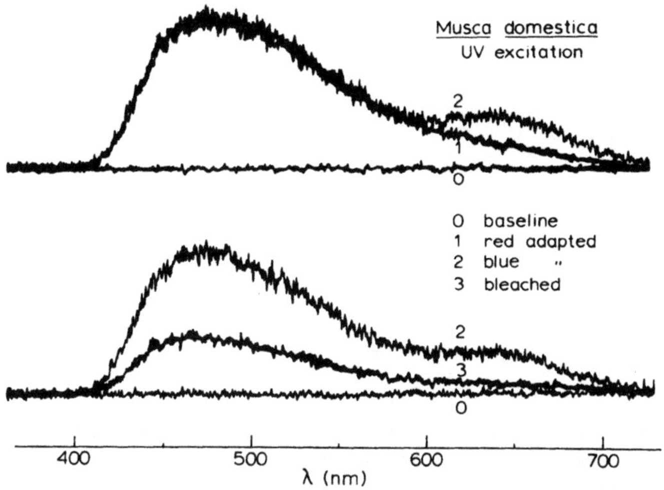

Fig. 6. Experimental emission spectra induced by ultraviolet light measured from the compound eye of a white-eyed mutant housefly *Musca.* A main emission in the blue is seen with a subsidiary peak in the red, which is dependent on the population of the visual pigment states. This population is varied by red (*1*) or blue (*2*) preillumination (for details, see [56]). Prolonged ultraviolet light destroys both the visual pigment as well as part of the main blue fluorescing pigment (*3*) [52]

Fluorescence microscopy of goldfish retinae demonstrated a high yield of blue emission (under UV-excitation) from the outer segments of the cone cells. A few years earlier Chance [7] hypothesized that the high UV-sensitivity of aphakics is due to the blue fluorescence from the pyridine nucleotide which is absorbed by the visual pigments. However, a quantitative evaluation of the idea is lacking.

Favorable subjects for fluorescence studies appear to be white-eyed mutant flies because the ocular screening pigments are missing and it is possible to study the fluorescence of the photoreceptors in vivo, i.e., in completely intact animals. Franceschini et al. [12, 13] characterized various classes of photoreceptor cells in the housefly *Musca* from the difference in ultraviolet- and blue-induced fluorescence displayed by the rhabdomeres, i.e., the cell structures where the visual pigments are located. The fluorescence of the visual pigments and the photoreceptor cell bodies is excellently illustrated in the beautiful colour photographs published by Franceschini et al. [12]. Emission spectra recorded from the eyes of white-eyed mutant housefly are presented in Fig. 6. The short-wavelength part of the spectra accord with those reported by Stratten and Ogden ([57], barnacle) and Stark et al. ([49], fruitfly), and point to the paramount presence of NADH and flavoproteins in visual sense cells.

Metabolic Processes

Oxidative phosphorylation involves oxidation of the mitochondrial pigments. Oxidation of NADH to NAD^+ results in a diminished UV-absorption. Measurement of a tissue's absorption in the ultraviolet hence has become a classical method for monitor-

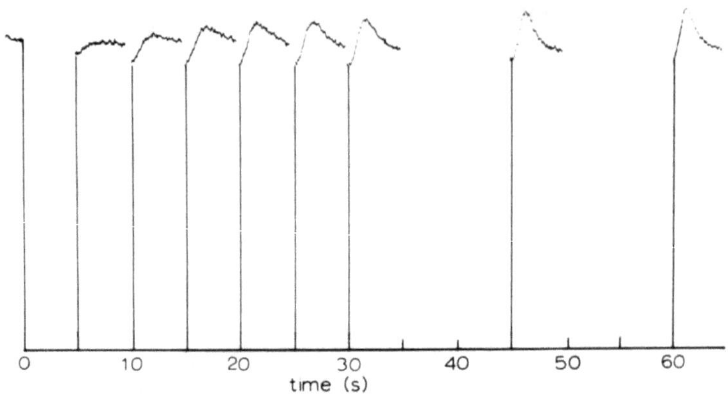

Fig. 7. Autofluorescence measured from the eye of a completely intact, living blowfly *Calliphora erythrocephala* mutant chalky. The excitation light, wavelength 456 nm and intensity 1.2×10^{17} quanta $cm^{-1} s^{-1}$, was applied for 5 s. Emission, measured in the wavelength range 510–600 nm, exhibits a biphasic time course, the shape of which depends on the dark period. The response is presented with time of onset at the preceding dark period (*abscissa*) [54]

ing mitochondrial activity. Nevertheless, although the retina has the highest rate of respiration, i.e., oxygen consumption per unit weight [28] of any tissue in the human body, and furthermore, the content of mitochondria in retinal cells in general is very high, optical measurements on metabolic processes have been virtually neglected in vision research. Only Sickel [44, 45] simultaneously measured the ultraviolet absorption of frog retinae and light-induced electrical activities. Of course, an obvious problem with the vertebrate eye is the lability of the visual pigments to intense or prolonged illumination, and thus also for test lights. This difficulty is not encountered in most invertebrate eyes because the visual pigments do not photolyse. Thus long-term absorption measurements can be executed without serious depletion of the visual molecules.

An alternative approach for studying metabolic processes is offered by measurable changes in fluorescence accompanying the oxidation-reduction reactions of the respiratory chain [8, 41]. Visual systems had not been investigated along this line until recently. The only published report is that by Laing et al. [27], who measured the fluorescence from the human cornea in relation to oxygen consumption. Photoreceptor processes in human eyes cannot be followed in vivo, due to the fluorescent media being distal from the retina. An attractive alternative is found in the eyes of invertebrates. Using white-eyed mutant flies, Stavenga and Tinbergen [54] monitored blue-induced green emission. Illumination of dark-adapted fly eyes causes a rapid emission increase which can be interpreted as due to enhanced oxidation of flavoprotein (Fig. 7). The time course accords with that found for light-induced oxygen consumption measured in isolated fly retinae with a microrespirometer [18]. The green emission readily drops upon hypoxia, as is expected from a decrease in oxidized flavoprotein, and the light-induced dynamic effect vanishes (Fig. 8). These experiments demonstrate the high metabolic activities of fly photoreceptors in the light as well as in the dark. Radiation of fly eyes with intense ultraviolet and blue light has dramatic effects [11, 32]. The visual pigment is destroyed by radiation with bright ultraviolet and blue lights

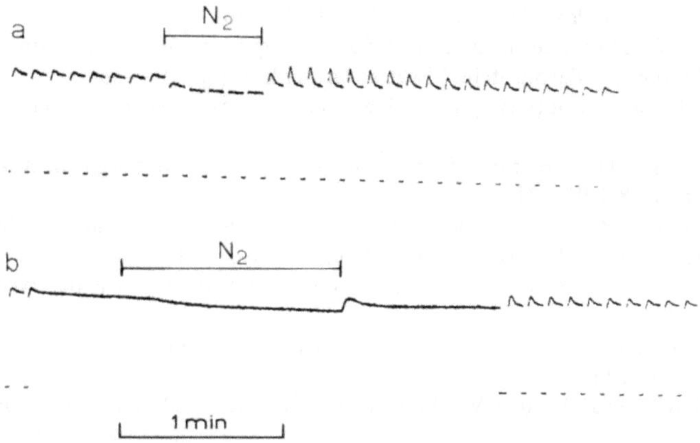

Fig. 8a,b. Hypoxia severely affects the fly's eye autofluorescence. The conditions were as in Fig. 7, except that the 5-s flashes were interrupted by a 30-s dark period (during which time the recorder stopped). Application of nitrogen rapidly lowered the fluorescence level, and the biphasic curve is flattened; after hypoxia an enhanced light-induced fluorescence change occurs which gradually settles down (**a**). Essentially the same effects are seen during continuous light (**b**) [54]

which have energies in the order of 10 J cm^{-2} [16, 23, 50]. Interestingly, Ham et al. [15] found substantial retinal damage in monkey eyes when radiation of similar intensities was applied (see also [46]). Prolonged illumination of fly eyes with bright UV induces a substantial drop in the (presumed) NADH fluorescence (Fig. 6), and furthermore the light-induced dynamic effects of Figs. 7 and 8 can no longer be elicited. Short-wavelength radiation presumably destructs mitochondria in addition to the visual pigments, due to the absorption properties of the pigments, NADH and the flavoproteins. Shielding by ocular filters which absorb in the short-wavelength range is therefore an important means for maintaining vital cellular functions.

Acknowledgments. The critical reading of the manuscript by Dr. M.G. Holmes is greatly appreciated.

References

1. Appleby SJ, Muntz WRA (1979) Occlusable yellow corneas in Tetraodontidae. J Exp Biol 83: 249–259
2. Bernard GD (1983a) Bleaching of rhabdoms in eyes of intact butterflies. Science 219:69–71
3. Bernard GD (1983b) Dark processes following photoconversion of butterfly rhodopsins. Biophys Struct Mech 9:277–286
4. Blest AD (1980) Photoreceptor membrane turnover in arthropods: comparative studies of breakdown processes and their implications. In: Williams TP, Baker BN (eds) The effect of constant light on visual processes. Plenum Press, New York, pp 217–245
5. Bridges CD (1976) Vitamin A and the role of pigment epithelium during bleaching and regeneration of rhodopsin in the frog eye. Exp Eye Res 22:435–455
6. Chader GJ (1982) Retinoids in ocular tissues: Binding proteins, transport, and mechanism of action. In: McDevitt D (ed) Cell biology of the eye. Academic Press, London New York

7. Chance B (1964) Fluorescence emission of the mitochondrial DPNH as a factor in the ultraviolet sensitivity of visual receptors. Proc Natl Acad Sci USA 51:359–361
8. Chance B, Schoener B (1966) Fluorometric studies of flavin component of the respiratory chain. In: Slater EC (ed) Flavins and flavoprotein. Elsevier, Amsterdam, pp 510–528
9. Dartnall HJA (ed) (1972) Handbook of sensory physiology, vol VII/1. Springer, Berlin Heidelberg New York
10. Dratz EA, Hargrave PA (1983) The structure of rhodopsin and the rod outer segment disk membrane. Trends Biochem Sci 8:128–131
11. Franceschini N, Stavenga DG (1981) The ultraviolet sensitizing pigment of flies studied by in vivo microspectrofluorometry. Invest Ophthalmol Vis Sci Suppl 20:no 3, 111
12. Franceschini N, Kirschfeld K, Minke B (1981a) Fluorescence of photoreceptor cells observed in vivo. Science 213:1264–1267
13. Franceschini N, Hardie R, Ribi W, Kirschfeld K (1981b) Sexual dimorphism in a photoreceptor. Nature (London) 291:241–244
14. Gogala M, Hamdorf K, Schwemer J (1970) UV-Sehfarbstoff bei Insekten. Z Vergl Physiol 70: 410–413
15. Ham WT Jr, Mueller HA, Sliney DH (1976) Retinal sensitivity to damage from short wavelength light. Nature (London) 153–155
16. Ham WT Jr, Mueller HA, Ruffolo JJ Jr (1980) Solar retinopathy as a function of wavelength: its significance for protective eyewear. In: Williams TP, Baker BN (eds) The effects of constant light on visual processes. Plenum Press, New York, pp 319–346
17. Hamdorf K (1979) The physiology of invertebrate visual pigments. In: Autrum H (ed) Handbook of sensory physiology, vol VII/6A. Springer, Berlin Heidelberg New York, pp 145–224
18. Hamdorf K, Schwemer J (1975) Photoregeneration and the adaptation process in insect photoreceptors. In: Snyder AW, Menzel R (eds) Photoreceptor optics. Springer, Berlin Heidelberg New York, pp 720–746
19. Hara T, Hara R (1972) Cephalopod retinochrome. In: Dartnall HJA (ed) Handbook of sensory physiology, vol VII/1. Springer, Berlin Heidelberg New York, pp 720–746
20. Hara T, Hara R (1982) Cephalopod retinochrome. In: Packer L (ed) Methods in enzymology, vol 81. Biomembranes, part H1, pp 827–834
21. Hardie RC, Kirschfeld K (1983) Ultraviolet sensitivity of fly photoreceptors R7 and R8: Evidence for a sensitising function. Biophys Struct Mech 9:171–180
22. Hargrave PA, McDowell JH, Curtis DR, Wang JK, Juszczak E, Fong S-L, Rao JKM, Argos P (1983) The structure of bovine rhodopsin. Biophys Struct Mech 9:235–244
23. Kirschfeld K (1981) Carotenoid pigments: Their possible role in protecting against photooxidation in eyes and photoreceptor cells. Proc R Soc London Ser B 216:71–85
24. Kirschfeld K, Franceschini N, Minke B (1977) Evidence for a sensitising pigment in fly photoreceptors. Nature (London) 269:386–390
25. Kirschfeld K, Feiler R, Franceschini N (1978) A photostable pigment within the rhabdomere of fly photoreceptors no R7. J Comp Physiol 125:275–284
26. Kruizinga B, Kamman R, Stavenga DG (1982) Laser induced visual pigment conversions in fly photoreceptors measured in vivo. Biophys Struct Mech 9:299–307
27. Laing RA, Fischbarg J, Chance B (1980) Non-invasive measurements of pyridinenucleotide fluorescence from the cornea. Invest Ophthalmol Vis Sci 19:96–102
28. Lerman S (1980) Radiant energy and the eye. Macmillan, New York
29. Liebman PA, Leigh RA (1969) Autofluorescence of visual receptors. Nature (London) 221: 1249–1251
30. Lythgoe JN (1979) The ecology of vision. Clarendon, Oxford
31. Miller WH (1979) Ocular optical filtering. In: Autrum H (ed) Handbook of sensory physiology, vol VII/6A. Springer, Berlin Heidelberg New York, pp 69–143
32. Miller GV, Itoku KA, Fleischer AB, Stark WS (1982) Damaging effects of UV treatment in normal and vitamin A deprived *Drosophila* eyes. Invest Ophthalmol Vis Sci Suppl 22:66
33. Muntz WRA (1972) Inert absorbing and reflecting pigments. In: Dartnall HJA (ed) Handbook of sensory physiology, vol VII/1. Springer, Berlin Heidelberg New York, pp 529–565

34. Ovchinnikov YA, Abdulaev NG, Feigina MY, Artamov ID, Zolotarev AS, Kostina MB, Bogachuk AS, Miroshnikov AI, Martinov VI, Kudelin AB (1982) The complete amino acid sequence of visual rhodopsin. Bioorg Khim 8:1011–1014
35. Pepe IM, Cugnoli C (1980) Isolation and characterization of a water soluble photopigment from honeybee compound eye. Vison Res 20:97–102
36. Pepe IM, Schwemer J, Paulsen R (1982) Characteristics of retinal-binding proteins from the honeybee retina. Vision Res 22:775–781
37. Rodieck RW (1973) The vertebrate retina. Freeman, San Francisco
38. Ruddock KH (1972) Light transmission through the media and macular pigment and its significance for psychophysical investigation. In: Jameson D, Hurvich LM (eds) Handbook of sensory physiology, vol VII/4. Springer, Berlin Heidelberg New York, pp 455–469
39. Sarkar HK, Song P-S (1982) Blue light induced phototransformation of phytochrome in the presence of flavin. Photochem Photobiol 35:234–246
40. Schmidt W (1980) Physiological blue light reception. In: Hemmerich P (ed) Structure and bonding, vol 41. Molecular structure and sensory physiology. Springer, Berlin Heidelberg New York, pp 1–44
41. Scholz R, Thurman RG, Williamson JR, Chance B, Bücher T (1969) Flavin and pyridine nucleotide oxidation – reduction changes in perfused rat liver. I. Anoxia and subcellular localization of fluorescent flavoproteins. J Biol Chem 9:2317–2324
42. Schwemer J (1979) Molekulare Grundlagen der Photorezeption bei der Schmeißfliege *Calliphora erythrocephala* Meig. Habilitationsschr, Ruhr Univ, Bochum
43. Schwemer J (1983) Pathways of visual pigment regeneration in fly photoreceptor cells. Biophys Struct Mech 9:287–298
44. Sickel W (1972) Retinal metabolism in dark and light. In: Fuortes MGF (ed) Handbook of sensory physiology, vol VII/2. Springer, Berlin Heidelberg New York, pp 667–727
45. Sickel W (1973) Energy in vertebrate photoreceptor function. In: Langer H (ed) Biochemistry and physiology of visual pigments. Springer, Berlin Heidelberg New York, pp 195–203
46. Sliney DH, Freasier BC (1973) Evaluation of optical radiation hazards. Appl Opt 12:1–24
47. Somiya H (1982) "Yellow lens" eyes of a stomiatoid deep-sea fish *Malacosteus niger*. Proc R Soc London Ser B 215:481–489
48. Sperling HG (1980) Prolonged intense spectral light effects on rhesus retina. In: Williams TP, Baker BN (eds) The effects of constant light on visual processes. Plenum Press, New York, pp 195–241
49. Stark WS, Stavenga DG, Kruizinga B (1979) Fly photoreceptor fluorescence is related to UV sensitivity. Nature (London) 280:581–583
50. Stark WS, Itoku KA, Srivastava K, Carlson SD (1983) Retinal degeneration induced by intense UV and blue stimuli in *Drosophila*. Photochem Photobiol Suppl 37:S 84
51. Stavenga DG (1980) Short wavelength light in invertebrate visual sense cells – Pigments, potentials and problems. In: Senger H (ed) The blue light syndrome. Springer, Berlin Heidelberg New York, pp 5–24
52. Stavenga DG, Franceschini N (1981) Fly visual pigment states, rhodopsin R490, metarhodopsins M and M′, studied by transmission and fluorescence microspectrophotometry in vivo. Invest Ophthalmol Vis Sci Suppl 20: no 3, 111
53. Stavenga DG, Schwemer J (1984) Visual pigments of invertebrates. In: Ali MA (ed) Photoreception and vision in invertebrates. Plenum Press, New York, pp 5–61
54. Stavenga DG, Tinbergen J (1983) Light dependence of oxidative metabolism in fly compound eyes studied in vivo by microspectrofluorometry. Naturwissenschaften 70:618–620
55. Stavenga DG, Bernard GD, Chappell RL, Wilson M (1979) Insect pupil mechanisms III. On the pigment migration in dragonfly ocelli. J Comp Physiol 129:199–205
56. Stavenga DG, Franceschini N, Kirschfeld K (1984) Fluorescence of housefly visual pigment . Photochem Photobiol, in press
57. Stratten WP, Ogden TE (1971) Spectral sensitivity of the barnacle *Balanus*. J Gen Physiol 57: 435–447
58. Vogt K (1983) Is the fly visual pigment a rhodopsin? Z Naturforsch 38c:329–333

59. Vogt K, Kirschfeld K (1982) Sensitising pigment in the fly. Biophys Struct Mech 9:319–328
60. White RH, Gifford D, Michaud NA (1980) Turnover of photoreceptor membrane in the larval mosquito ocellus: rhabdomeric coated vesicles and organelles of the vacuolar system. In: Williams TP, Baker BN (eds) The effects of constant light on visual processes. Plenum Press, New York, pp 271–296
61. Woodhouse JM, Campbell FW (1975) The role of the pupil light reflex in aiding adaptation to the dark. Vision Res 15:649–653
62. Wyszecki G, Stiles WS (1976) Color science: concepts and methods, quantitative data and formulas. Wiley, New York
63. Yoshizawa T, Shichida Y (1982) Low temperature spectrophotometry of intermediates of rhodopsin. In: Packer L (ed) Methods in enzymology, vol 81. Biomembranes, part H1. Academic Press, London New York, pp 333–353
64. Stark WS, Tan EWP (1982) Ultraviolet light: photosensitivity and other effects on the visual system. Photochem Photobiol 36:371–380

Cryptochrome, Some Terminological Thoughts

H. SENGER[1]

From the beginning of the investigation of responses of plants to blue light there was the dilemma that a proper expression for the photoreceptor did not exist. The term "blue-UV light photoreceptor" is rather long-winded, particularly when contrasted with the attractive term "phytochrome" for red light-triggered photomorphogenic responses.

Some 15 years ago Jonathan Gressel and Esra Galun from the Weizmann Institute of Science in Rehovot/Israel coined the term "cryptochrome" as a laboratory nickname for the "blue-UV light photoreceptor". Others picked up the term and used it in their publications without the necessary definition. Researchers outside the field were misled into assuming that there is only *one* "cryptochrome" in analogy to the single photoreceptor "phytochrome". Under the impression that there are several photoreceptors for blue-UV light and that the term cryptochrome was not properly defined, it was decided at the *1st International Conference of Blue Light Effects in Plants and Microorganisms* to avoid the use of the term.

Nevertheless, the term cryptochrome continued to appear in publications, textbooks, and in the Encyclopedia of Plant Physiology. Belatedly, Jonathan Gressel gave an explanation of how the term cryptochrome should be used and defined (*Photochemistry and Photobiology* 30, p. 749, 1979).

Under these circumstances it was decided at the *2nd International Conference on the Effect of Blue Light in Plants and Microorganisms* to adopt the expression cryptochrome as an operational term for a group of blue light photoreceptors until their true chemical nature is discovered, and to give a closer definition.

It is recommended that equivalent to terms such as photosynthetic pigments or phytochromes, the term blue light photoreceptors should be used for the group of photoreceptors absorbing below 520 nm extending into the UV region. Only for those photoreceptors having action spectra with peaks or shoulders at or around (± 5 nm) (370), 420, 450, and 480 nm, should "cryptomere(s)" be used as an operational term. Thus the term cryptochrome(s) would apply to a subdivision of blue light photoreceptors, equivalent to other subdivisions like UV-B photoreceptors absorbing between 280 and 320 nm or mycochrome. There are, as well, other blue light photoreceptors absorbing below 520 nm, for which none of the above subdivisions are adequate.

Some of the above-listed subdivisions of blue light photoreceptors might consist of flavoproteins or carotenoproteins; until this is proven, however, the names of these pigment groups should not be used as subdivisions in the operational terminology of blue light photoreceptors.

1 Fachbereich Biologie-Botanik der Philipps-Universität, D–3550 Marburg/Lahn, FRG

Blue Light Effects in Biological Systems
ed. by H. Senger

Nature and Properties of the Photoreceptors

Photophysical Aspects of Blue Light Receptors: The Old Question (Flavins Versus Carotenoids) Re-Examined

P. S. SONG[1]

Introduction

About 10 years ago, the question as to the possible identity of the chromophore(s) of blue light receptors was first examined [18, 19] in terms of the photophysical and photochemical reactivities of flavins and carotenoids. The main conclusions were that flavins were intrinsically more likely than carotenoids to be the blue light photoreceptor. At the first meeting of the *Blue Light Syndrome* [16], it was argued that falvins could act as the primary photoreceptor for a number of responses of organisms to blue light and that the short fluorescence lifetime of flavins bound to the corn coleoptile plasmalemma could serve as a photophysical assay for the photoreceptor flavins. On the other hand, the photobiological functions of carotenoids primarily involve the absorption of blue light as the light-harvesting antenna. The antenna function of carotenoids is facilitated by the binding of carotenoids to proteins and/or membranes and by exciton interactions between and/or among the bound carotenoid molecules.

In the present review, the photophysical properties of carotenoids will be examined to ascertain whether or not carotenoids should be re-evaluated as primary blue light receptors on the basis of the new assignment of electronic states in polyenes in general [9] and carotenoids in particular [17].

The New Electronic State Ordering in Carotenoids

The new electronic state ordering is different from the traditional picture. Thus, in the new assignment, the lowest electronically excited state (S_1) is an $^1A_g^-$ state (forbidden and too weak to be clearly observable in symmetric polyenes and carotenoids), in contrast to the allowed ^1B_u state (at 450 nm for β-carotene) as the lowest excited state. Two different orderings of the excited states for β-carotene as a typical carotenoid are illustrated in Fig. 1. The location of the forbidden $^1A_g^-$ state (S_1 in the new ordering) and its absorption maximum and intensity are not known at this time. It is predicted that the $A_g \longrightarrow {}^1A_g^-$ ($S_0 \longrightarrow S_1$) absorption occurs in the near IR region ($\sim$ 800 nm;

1 Department of Chemistry, Texas Tech Universtiy, Lubbock, TX 79409, USA

Blue Light Effects in Biological Systems
ed. by H. Senger

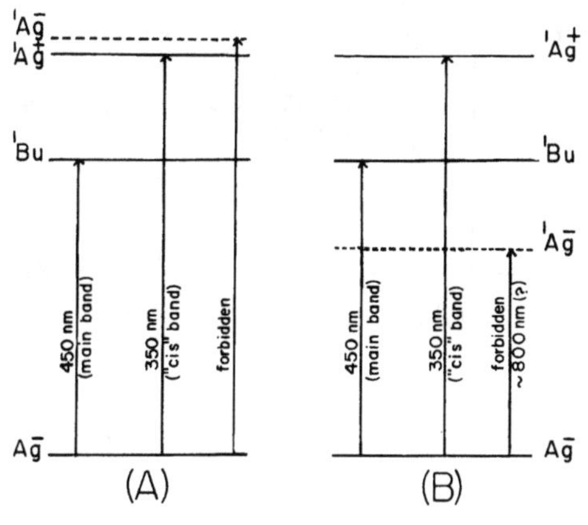

Fig. 1A,B. The electronic state diagrams and $\pi \rightarrow \pi^*$ transitions in carotenoids. **A** Traditional electronic state ordering; **B** New electronic state ordering

[10], private communication also). This long wavelength transition is symmetry-forbidden and is not readily observable under the condition of mono-photon excitation. However, the loss of molecular symmetry enhances the transition probability and it is then possible to observe such an absorption band (for example, weak absorption bands appear at λ longer than the main B_u absorption band in carotenenals [4]). The $^1A_g^-$ band is also strongly allowed in two-photon absorption.

In the traditional picture for β-carotene, it is well known that the first transition, $A_g \longrightarrow {}^1B_u$ (or $A \longrightarrow {}^1B$ in Platt notation), is strongly electric dipole allowed with its polarization direction along the long molecular axis. Both oscillator strength (f ∿ 3.4 for all-trans β-carotene) and λ_{max} increase with conjugation. The latter converges to an asymptotic value of 610 nm for "infinitely" long polyenes as a result of C-C and C=C bond-order alternations which restrict the electron motion in a one-dimensional periodic potential and/or π-electron correlation effects.

Both $A_g \longrightarrow {}^1A_g^+$ and $A_g \longrightarrow {}^1A_g^-$ transitions (Fig. 1) are symmetry-forbidden, but the former can be readily observed even in trans-polyenes, and it becomes strongly allowed in cis-polyenes ("cis peak") due to the loss of the center of symmetry.

The assignment of $^1A_g^-$ to the lowest excited $^1\pi,\pi^*$ state has a profound implication in re-interpreting the spectroscopic and photochemical properties of carotenoids. Recent results further confirm that the $^1A_g^-$ state is the lowest excited singlet state for polyenes of up to four conjugated double bonds [1, 2, 3, 6]. As mentioned earlier, however, the location of the $^1A_g^-$ state in longer polyenes, including carotenoids, has not been resolved unambiguously, and consequently, its assignment remains to be established [4, 8, 15].

A new band has been observed for β-carotene film at 536 and 537 nm for all-trans and 15,15′-cis isomers, respectively. On the basis of the resonance Raman excitation profile spectrum, Thrash et al. [21] proposed that the $S_0 \longrightarrow S_1$ transition in β-carotene is due to a transition to the $^1A_g^-$ state as the lowest state. However, the $^1A_g^-$ assignment for β-carotene cannot be viewed as definitively established until the possibility of impurities present and produced during spectral measurements has been rigorously dealt with. It is more likely that the $^1A_g^-$ band occurs at λ much longer than 536–537 nm, possibly at ∿ 800 nm (vide supra).

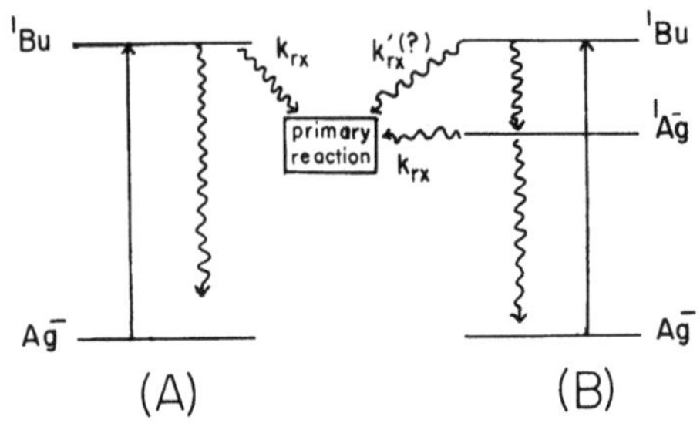

Fig. 2A,B. Electronic excitations, relaxations and primary photoreactions (k_{rx} and k'_{rx}). **A** Traditional electronic state ordering; **B** New electronic state ordering

Implications in Blue Light Receptors

The two different electronic state assignments for the lowest S_1 state of carotenoids imply that primary photoprocesses in blue light responses proceed from the ^1B_u or $^1A_g^-$ state in the traditional or new electronic state ordering, respectively, provided that carotenoid is indeed the photoreceptor.

With ^1B_u as the S_1 state, the primary photoreaction (k_{rx}) must compete with highly efficient internal conversion (Fig. 2). The radiationless relaxation (internal conversion) is so efficient that the mean lifetime (τ) of the ^1B_u state is predicted to be $\sim 10^{-15}$ s from the integration of the absorption band [$\epsilon(\bar{\nu})$ with maximum $\bar{\nu}_{max}$ in wavenumber; n is refractive index]

$$\tau_{(^1B_u)} = \tau_F^0 \times \Phi_F = [2.9 \times 10^{-9}\, \bar{\nu}^2_{max}\, n^2 \int_0^\infty \epsilon(\bar{\nu})\, d\nu]^{-1} \times \Phi_F$$
$$\leqslant 1 \times 10^{-9} \times 10^{-6}$$
$$\leqslant 10^{-15}\ \mathrm{s}$$

where τ_F^0 and Φ_F are the radiative fluorescence lifetime and quantum yield, respectively. Clearly, the ultrashort lifetime is not conducive to the carotenoid to act as a primary photoreceptor with rate constant k_{rx}, as has been pointed out previously [18].

On the other hand, with $^1A_g^-$ as the lowest excited state (S_1), the blue absorption band is attributable to the allowed $A_g \longrightarrow {}^1B_u$ ($S_0 \longrightarrow S_2$) transition. The primary photoreaction may then proceed from either the ^1B_u (S_2) or $^1A_g^-$ (S_1) state, if the radiationless transition between these two states is not efficient. This does not seem to be a likely possibility, since the vibronic coupling between the ^1B_u and $^1A_g^-$ states can be strong [9], suggesting that the relaxation from ^1B_u to the hypothetical lower state ($^1A_g^-$) is so fast that an efficient primary photoprocess for blue light responses is unlikely. If the $^1B_u \longrightarrow A_g^-$ relaxation is slow, then the assignment of $^1A_g^-$ to the S_1 state in carotenoids requires that the blue light photoreceptor with the carotenoid as the chromophore also acts as a red light receptor, especially if the $A_g \longrightarrow {}^1A_g^-$ transition is at least partially allowed as a result of the carotenoid molecule being nonsymmetric (e.g., carotenals, carotenones and their Schiff's bases). Furthermore, the radiative lifetime of the $^1A_g^-$ state is expected to be much longer than that of the ^1B_u state. These factors (nonsymmetric carotenoids and long lifetimes) would favor carotenoids to act as both blue and red receptors. However, there is no experimental evidence for the dual spectral nature of blue light receptors.

It has been pointed out that the primary photoreceptors for sensory transduction in both prokaryotes and eukaryotes possess covalently-linked chromophores [14]. For retinylic photoreceptors, such as rhodopsin and bacteriorhodopsin, the covalent linking between the chromophore and protein is provided by the protonated Schiff's base formation.

With carotenals and carotenones, the protonated Schiff's base linkage shifts the blue absorption maximum of these carotenoids toward red, thus eliminating protonated carotenoid Schiff's bases as primary photoreceptors [14]. Neutral Schiff's bases show very little spectral shift with respect to the absorption maxima of the parent carotenoids and can spectrally qualify as blue light receptors [14]. However, if the ${}^1A_g^-$ assignment for the lowest S_1 state is assumed, the neutral carotenoid Schiff's bases should also act as red light receptors, in addition to being blue light receptors. Since the spectral duality is not experimentally supported, it seems unlikely that carotenoids can be considered as the favored photoreceptor molecules as opposed to flavins.

Chromophores Other Than Flavins and Carotenoids

It was casually mentioned in the symposia on *Blue Light Syndromes* that chromophores which are different from flavins and carotenoids may be involved in blue light responses. In this section, we will consider several blue light absorbers as such possible candidates.

Bilirubin serves as a blue light receptor in the phototherapy of infant jaundice, in which the concentration of toxic bilirubin isomer in circulation is decreased [11, 13]. Bilirubin is unusual in that only one major absorption band in the blue region constitutes its UV-visible absorption spectrum. The blue absorption maximum and its intensity (ϵ) are sensitive to solvent polarity [7]. The natural isomer (4 *Z*, 15 *Z*) shows a λ_{max} at 438 nm (ϵ = 47,200) in phosphate buffer, pH 7.4 [12], whereas the absorption maximum shifts to 450 nm in chloroform with an increase in the molar extinction coefficient. Upon binding to human serum albumin, bilirubin absorbs maximally at 460 nm (48,500 M^{-1} cm^{-1}; pH 7.4; [12]). The photoisomerization of the *Z,Z* isomer to "photobilirubin" isomers (4 *Z*, 15 *E*; 4 *E*, 15 *Z*) is accompanied by some changes in λ_{max} and ϵ_{max}, but the most significant change occurs in the CD spectra of these isomers [11].

Some of the other naturally occurring compounds with the absorption spectra similar to the action spectra for blue light responses are listed below for future examination: chrysazine, coleone A, isocoleone A, dihydro-flavylium salts and naphthazarins.

Concluding Remarks

Flavins exhibit diverse photoreactivities. For example, the ${}^{1,3}(\pi,\pi^*)$ state flavins can mediate (a) proton and/or electron transfer, (b) photosensitization of a number of reactions including redox reactions, and (c) they are involved in chemiluminescent and bioluminescent processes. Some of the well-resolved action spectra of blue light responses [5, 20, 22] can thus be understood in terms of the intrinsic photoreactivity of flavins.

Can carotenoids be ruled out as primary photoreceptors? It has been pointed out that the excited state of carotene as a blue light receptor is too short-lived to initiate sufficiently efficient primary photoprocesses [16]. Recent studies [10] in the photophysics of carotene suggest that carotenoids may in fact photoisomerize with a high efficiency, especially in the presence of an electron transfer catalyst (T.Y. Lee, private communication, 1983). Nonetheless, carotenoids such as β-carotene without a functional group that can covalently link to the apoprotein moiety are not likely to be viable candidates as blue light receptors, since the chromophore relaxation resulting from photoisomerization and/or polarizability changes cannot lead to specific coupling with the transducing membrane/effector molecules. Furthermore, if the lowest electronic state is of $^1A_g^-$ genealogy (see [15] for review), carotenoids are also expected to act as far-red light receptors. There is no evidence among the reported action spectra for blue light responses that suggests the involvement of carotenoids acting in the far-red region.

Are there carotenoids with functional groups that can link to proteins (e.g., via Schiff's base) and that can possibly function as primary blue light receptors? The argument that rhodopsin- and bacteriorhodopsin-like carotenoprotein Schiff's bases (protonated) cannot be reconciled with the blue light action spectra have been presented. However, rhodopsin and carotenoprotein neutral Schiff's bases cannot be ruled out, provided that the lowest electronic state is of ^1B_u genealogy. In this connection, photophysical processes of the Schiff's bases should be re-examined as to their potential receptor roles.

There are a number of naturally occurring chromophores that also absorb blue light, with spectral shapes similar to blue light action spectra. These chromophores, including tetrapyrrolics such as bilirubin derivatives, should also be considered as possible photoreceptor candidates, if it is assumed that there is more than one type of blue light receptors in the diversity of blue light responses and their action spectra in photobiological systems.

Acknowledgments. This work was supported by grants from the Robert A. Welch Foundation (D-182) and the National Science Foundation (PCM 81-19907).

References

1. Andrews JR, Hudson B (1979) Chem Phys Lett 60:380–384
2. Birge RR, Pierce BM (1979) J Chem Phys 70:165–178
3. Birks JB, Tripathi GNR, Lumb MD (1978) Chem Phys 33:185–194
4. Chae Q, Song P-S, Johansen JE, Liaaen-Jensen S (1977) J Am Chem Soc 99:5609–5614
5. Diehn B (1969) Biochim Biophys Acta 177:136–143
6. Fang HLB, Thrash RJ, Leroi GE (1978) Chem Phys Lett 57:59–63
7. Holzwarth AR, Langer E, Lehner H, Schaffner K (1980) Photochem Photobiol 32:17–26
8. Hotchandani S, Paquin P, Leblanc RM (1978) Can J Chem 56:1985–1988

9. Hudson B, Kohler BE (1974) Annu Rev Phys Chem 25:437–460
10. Kohler B (1983) In: Zewail A (ed) Photochemistry and photobiology. Harwood Academic Publishers, Chur, Switzerland, in press
11. Lamola AA, Blumberg WE, McCleod R, Fanoroff A (1981) Proc Natl Acad Sci USA 78:1882–1886
12. Lamola AA, Eisinger J, Blumberg WE, Patel SC, Flores J (1979) Anal Biochem 100:25–42
13. McDonagh AFM, Palma LA, Lightner DA (1980) Science 208:145–151
14. Song P-S (1977) J Korean Agric Chem Soc 20:10–25
15. Song P-S (1977) Annu Rep (B) Chem Soc London 1977:18–40
16. Song P-S (1980) In: Senger H (ed) The blue light syndrome. Springer, Berlin Heidelberg New York, pp 157–171
17. Song P-S (1983) Photochem Photobiol Rev 7:77–139
18. Song P-S, Moore TA (1974) Photochem Photobiol 19:435–441
19. Song P-S, Moore TA, Sun M (1972) In: Chichester CO (ed) The chemistry of plant pigments. Academic Press, London New York, pp 33–74
20. Thimann KV, Curry GM (1960) In: Florkin M, Mason HS (eds) Comparative biochemistry: A comprehensive treatise. Academic Press, London New York, pp 243–304
21. Thrash RJ, Fang HLB, Leroi GE (1977) J Chem Phys 67:5930–5933
22. Zurzycki J (1980) In: Senger H (ed) The blue light syndrome. Springer, Berlin Heidelberg New York, pp 50–68

The Study of Basic Photochemical and Photophysical Properties of Membrane-Bound Flavins: The Indispensible Prerequisite for the Elucidation of Primary Physiological Blue Light Action

W. SCHMIDT [1]

Introduction

Some 100 different physiological blue light responses are known today, and the list is growing steadily. It is safe to assume that all blue light photoreceptors are of ancient lineage and have been probably derived from molecules and structures previously otherwise employed. Even if there appears to be a great diversity of the secondary and subsequent steps following the photoreceptor relaxation of the various light response signal chains, for many blue light responses, even if not for all, a uniform picture appears to crystallize concerning the *primary* reactions: (1) The photoreceptor pigment is a flavin, probably a flavoprotein, (2) the photoreceptor is bound to a membrane, often the plasmamembrane, in a highly dichroic manner, and (3) the primary reaction is a redox reaction. For details refer to several recent reviews which have been exclusively devoted to physiological blue light responses [12, 24, 27, 33, 40, 41].

The primary events of most other light-initiated reaction pathways are better understood than the blue light-receptor problem. Their common feature is the intimate relation to membraneous portions of the cells. Summarizing, the conversion of light energy as mediated by a specific "photoenzyme" into an electrical potential, a pH gradient or an ionic flow across a membrane appears to be a biophysically highly efficient transduction step, generally adopted by nature.

So far most photoreceptor molecules have been purified and subsequently reconstituted with artificial membranes in order to study their properties under more idealized conditions. These systems include phytochrome [9, 17], rhodopsin [21, 46], bacteriorhodpsin cf. [4, 46], stentorin [42, 43] and chlorophyll [6, 7, 10, 16, 45]. More artificial pigment/membrane systems have also been described [2, 3, 5, 15, 16]. Most recently a survey of this issue was given by Pepe and Gliozzi [23]. However, due to the ubiquitous appearance of flavins, the blue light receptor has so far escaped clear-cut identification, purification, and as a result, reconstitution. Therefore the most valuable approach to attack the mechanisms underlying the primary reactions in blue light physiology was to use completely artificial systems.

1 Universität Konstanz, Fakultät für Biologie, Postfach 5560, D–7750 Konstanz 1, FRG

Blue Light Effects in Biological Systems
ed. by H. Senger

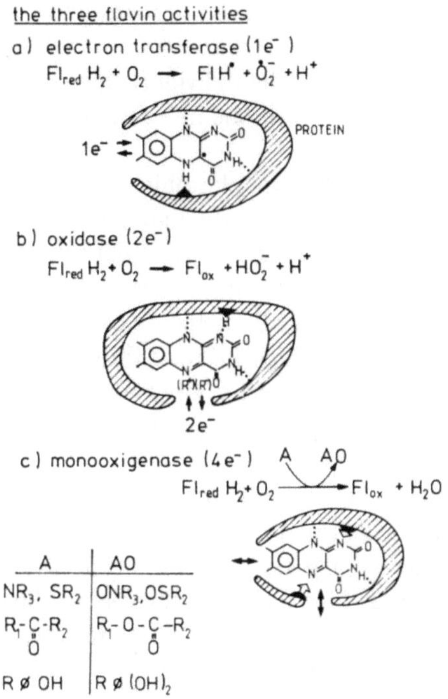

Fig. 1a–c. Visualization of the regulation of flavin reaction modes by the apoprotein as described by Hemmerich et al. [14]. We have to differentiate between three principal activities of the flacin chromophore which are mechanistically independent: **a** Single electron transfer, which requires a thermodynamic stabilization of the biologically active(blue) radical state. This is accomplished by blocking the N5-lone pair by stable hydrogen bridges. Electrons are supposed to enter the alloxazin ring via the positions 7/8; **b** Oxidase activity is characterized by $2e^-$ transfer to molecular oxygen, yielding hydrogen peroxide. This is accomplished by blocking the position $1/2\alpha$ which destabilizes the neutral radical in favor of a mixture of oxidized and fully reduced flavin; **c** Oxygenases exhibit so-called $4e^-$-activation, yielding water and oxene (oxygen atom) to be incorporated into various substrates A, as shown. All four oxidation equivalents contained in the oxygen molecule are used. Flavin is unique in making up the borderline between $2e^-$ and $1e^-$ transport. This implies continuous shuttle between $1/2\alpha$ and 5-blocked conformations, as sketched

Flavin-(Photo-) Chemistry. A Review

Flavin chemistry and flavin photo-chemistry are not distinct features inherent in the alloxazin chromophore. They are rather different aspects of the same phenomena, which are kinetically, thermodynamically, stochiometrically and sterically controlled or modified by light, the apoproteins and a membrane to which they are possibly bound. The various mechanisms have been extensively reviewed by Bruice [1] and Hemmerich [13]. Following Hemmerich, the central role of flavin enzymes is based on their ability to act as $1e^-$ or $2e^-$ redox reagent. It converts $1e^-$ to $2e^-$ packages, and vice versa. Flavin is unique in making up the borderline between $2e^-$ and $1e^-$ transport. Consequently flavin exists in three different forms: (1) the fully oxidized form, (2) the radical, i.e., semiquinone form and (3) the fully reduced species, i.e., the dihydroflavin. Based on spectral characteristics these are easily distinguished, according to the three main activities of flavin (Fig. 1):

(a) electrontransferase, (b) oxidase and (c) oxygenase. The individual flavin-apoprotein interaction defines which of these activities inherent in the free chromophore will be expressed in a specific flavoprotein.

Flavin photochemistry is an obligatory forerunner of flavin photobiology. Photosensitivity of flavin is inherent in the free form rather than the protein-bound entity. This is reflected by the low or absent fluorescence of most flavoproteins both in the purified form and in situ. However, taking into account a quantum efficiency of blue light responses close to unity, we would not expect significant fluorescence of the blue light photoreceptor flavin either, in spite of an expected high intrinsic fluorescence quantum yield. This excludes adoption of fluorescence as feasible assay for the blue light receptor flavin in situ, but not for the purified complex, as suggested previously [44].

In the presence of minute amounts of free flavin, which are hardly experimentally excluded (k_d about 10^{-4} M), all flavoproteins are amenable to the action of light, i.e., they will be photochemically active [18]. Therefore any flavoprotein, especially if membrane-bound, has to be considered a potential blue light photoreceptor. For example, the membrane-bound flavoproteins glycine oxidase [50] and nitratereductase [22] have been proposed as physiological blue light receptor.

It has long been recognized that the ground state chemistry of flavoenzymes has much in common with the excited state chemistry of free flavin (cf. [13]). This is due to the change in the acidic pattern of the flavin molecule upon excitation, elevating the N5-pK of 0 in the ground state to 5 in the first triplet state [38].

Flavin photochemistry is divided into following categories: (1) photoelectron transfer, (2) photodehydrogenation, (3) photoaddition, and (4) photodealkylation. Each of these activities might be involved in the primary physiological blue light action, respectively in the specific inhibition of blue light responses (cf. [35]). In addition, such an activity could be mimicked by some photoactive dye replacing the natural photoreceptor (cf. [27]). "5-deazaflavin" [11] is a typical $2e^-$-acceptor, but when photoreduced to the radical state, it is a most powerful reductant, capable of donating single electrons to any flavoprotein. On this basis the latter is photoactivated in a highly artificial way, giving rise to subsequent reactions which are difficult to interpret.

A specific flavin, originally detected by Zenk [48] as the bulk flavin in oat coleoptiles and termed FX because of its hitherto unknown structure, was suspected to be the photoreceptor for phototropism. This assumption was based on its capability of decarboxylating auxin, thereby causing destruction of the hormone on the lighted side (Blaauw's theory, cf. [27]). From our present viewpoint several arguments contradict this hypothesis: (1) the photoreceptor is known to be present only in very low concentrations, (2) photodecarboxylation is a rather unspecific flavin reaction and, most important, (3) phototropism does not imply photochemical interaction of flavin and auxin [47], as often implied (cf. [42]). Nevertheless, most recently S. Ghisla in this laboratory (personal communication) succeeded in unravelling the chemical structure of FX. FX is identical with 5′ riboflavin-malonic acid ester.

Materials and Methods

The preparation of single-shelled vesicles carrying amphiphilic flavins in their membraneous portions has been described previously [31]. In order to trap additional cyt c in the aqueous lumen of these vesicles, the procedure was slightly modified: Prior to sonication, the phosphate buffer (pH 8), which was added to the dry lipid/flavin film, was saturated with cyt c. Subsequent purification steps were performed as described [31]. Cytochrome c was purchased from Merck, Darmstadt (24804). In addition, highly sensitive spectrophotometric apparatuses were constructed [26, 30, 32] capable of handling the small effective pigment concentrations of the model systems (order 10^{-7} M), in spite of the scattering properties of the samples. For further experimental details refer to [25, 29, 31, 36].

Artificial Membrane/Flavin Systems

In 1977 we characterized plasmamembrane-enriched fractions obtained from various blue light-sensitive organisms [37]. All contained a specific b-type cytochrome amenable to photoreduction in vivo, and flavin as potential basic components of the physiological blue light photoreceptors systems. However, the concentrations available were not sufficient, and the preparation of the samples too laborious to allow systematic physicochemical experiments. This led us to study completely artificial flavin/membrane systems. Since this project will be closed this year (1983), the present article is intended to present a crude survey of the numerous data obtained with these systems during the past 10 years. Some latest results will be included which demonstrate a flavin-mediated, light-driven transport of redox equivalents across artificial membranes. Drastic differences between the classical "isotropic" and the "anisotropic" flavin chemistry described here have been established.

As a minimum model system for the blue light receptor three different amphiphilic flavins bearing C_{18}-hydrocarbon chains at positions 3, 7 and 10 have been synthesized (AFl 3, 7 and 10) [19] and anchored within artificial, single-shelled vesicles made from various natural and synthetic phospholipids [31]. On this basis we hoped to mimic the steric and mechanistic restrictions imposed by the apoprotein and/or the membrane in flavoproteins, including the flavin photoreceptor ("photoenzyme").

The fluorescence quantum efficiency of flavins bound to membranes of vesicles made from synthetic lipids which exhibit well-defined phasetransitions at T_c strongly depends on temperature, i.e., the phase of the membrane [25]. It is higher in the *liquid crystalline* state than in the *gel* state, in contrast to isotropic flavins which exhibit *lower* fluorescence at elevated temperatures. This is interpreted as a phase-dependent localization of the flavin nucleus, which is confined to the more hydrophilic interface region of the membrane below T_c, but to the more hydrophobic region above T_c [36]. This interpretation is strongly supported by characteristic spectral changes of the flavin upon phase transition of the membrane [29], and also by the photoreduction properties of the membrane-bound flavin: In the *gel* state of the membrane some *intrinsic* electron donor (the secondary hydrogen of the glycerol backbone) is available, however not in the *liquid crystalline* state [36]. Based on fluorescence polarization spectroscopy, the following data were obtained: (1) The rotational relaxation time of the flavin nucleus strongly depends on the specific flavin/membrane system used if the membrane is in the gel state ($\rho \geqslant 50$ ns), but decreases to about 15 ns in the liquid crystalline state in all systems [25]. This has been verified by electron spin resonance spectroscopy [20]. (2) The "flavin-selfcontact" as the dominant and highly disturbing effect in isotropic flavin chemistry, is suppressed to about 2% under membrane-bound conditions [29]. (3) The flavin/membrane interaction is capable of inducing drastic changes in the transition dipoles of the alloxazin ring [29], and consequently, in its quantum mechanical properties.

It has been further established that the accessability of membrane-bound flavin, i.e., essentially the permeability of the membrane/water interface, is strongly dependent on the phase of the lipid: For glycerol [25], molecular oxygen, iodide, azide and protons [36] it is greater above than below T_c.

pK Values, as most decisive parameters in photobiology, are also strongly modified by the membrane: (1) The pK of the N3-proton of aqueous riboflavin of 9.5 is increased by roughly three units upon membrane binding [36]. Similarly, the N1-pK of dihydroflavin of 6.3 in aqueous solution is increased by three units, if the flavin is membrane-bound [28]. The pK of the N5-proton of triplet flavin is 5 compared to the ground state pK of 0 [38]. When the flavin is bound to a membrane, the pH-dependence of photoreduction efficiency indicates an even more elevated photo-pK of 7.5 [31]. To generalize, all flavin-pK's investigated are significantly shifted into the alkaline range upon membrane binding.

Most photochemical and photobiological reactions are initiated by triplet states of the responsible photoreceptor chromophores. Whether this holds true for the phototropism in corn coleoptiles is still a matter of debate. Schmidt et al. [35] reported specific inhibition of phototropism by the flavin triplet quenchers iodide and azide. In contrast to these findings, Song [42] failed to establish any inhibiting influence of xenon, known to be a typical triplet quencher, as well (isoelectronic with iodide). From this he concluded that the flavin singlet is probably the active species. This is consistent with our recent observation that the photoreduction of flavins bound to vesicles made from natural phospholipids appears to be largely mediated by the flavin singlet state [31]. It is also consistent with our observation that at saturating aqueous concentration (4.8 mM) xenon inhibits flavin-mediated redox-transport in the system described below (lecithin vesicles, AFl 3/cyt c) by more than 50%. However, to our surprise, neither the flavin singlet (fluorescence) nor the flavin triplet state (photoreduction) is significantly inhibited by xenon under *isotropic* aqueous conditions. This is probably due to different solubilities of xenon and iodide in the membrane. Therefore, a discrimination of triplet and singlet states on the basis of xenon appears to be futile.

As pointed out by Hemmerich et al. [14], the transfer stochiometry of flavin is regulated by hydrogen bridges extending from the apoprotein to either position $1/2\alpha$ or 5 (Fig. 1). The hydrogen bridge or positive charge blocking off the $1/2\alpha$-lone pair switches the flavin for selective $2e^-$-transfer, blocking of the N5-lone pair for selective $1e^-$-transfer. On this basis the strong dependence of photoreduction efficiency on the type of electron donor and on the specific flavin/membrane system used is easily interpreted (Fig. 2). If the flavin is bound in position 3 to the membrane (AFl 3), the position $1/2\alpha$ is exposed. Consequently EDTA as a $1e^-$-donor is highly efficient. However,

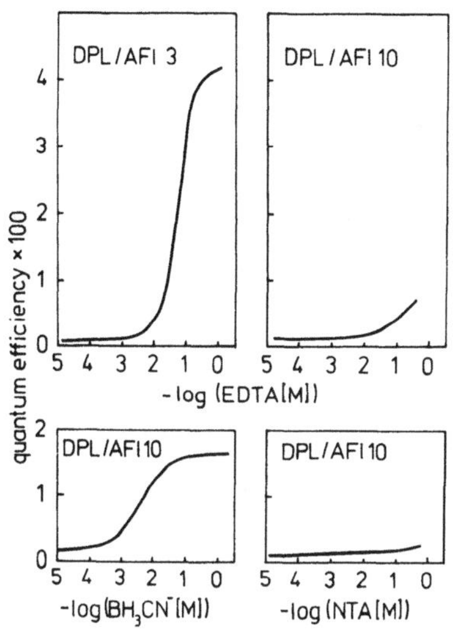

Fig. 2. Quantum efficiency of photoreduction of various amphiphilic flavins bound to DPL vesicles above T_c as a function of exogeneous concentrations of potential $1e^-$ (EDTA, NTA) and $2e^-$ (BH_3 CN^-) donors. (Data from [36])

if the flavin is anchored in position 10 within the membrane, the $2e^-$-donor BH_3 CN^- is more efficient, due to its better access to position N5. This is further supported by the fact that EDTA or NTA, which are both typical $1e^-$-donors, are not capable to photoreduce membrane-bound AFl 10.

A completely artificial, isotropic electron transport chain, driven by blue light via a flavin moiety, has been explored previously [34]. It includes EDTA, flavin, oxygen (optionally) and cyt c (cf. Table 1 for possible reaction modes). On this basis an advanced model system was designed, in order to study blue light-mediated transport of redox equivalents across vesicle bilayers. Single-shelled vesicles were prepared with cyt c trapped in their aqueous lumen, and amphiphilic flavin bound to the vesicle membrane. EDTA as potential electron donor for the flavin was added to the bulk medium. Figure 3 demonstrates that redox equivalents are indeed transferred across the vesicle membrane upon blue light irradiation, both under aerobic and anaerobic conditions. Membrane transport of EDTA itself had been excluded previously [36].

Table 1. Potential photoredox reactions of systems containing flavin, molecular oxygen and cytochrome c. The "marked" species are probably involved in transport of redoxequivalents across membranes

$Fl_{ox} + 2H^+ + 2e^-$	→	$Fl_{red}\,H_2$
$\mathit{Fl_{red}\,H_2}$ + cyt c (3+)	→	$FlH^{\cdot}$ + cyt c (2+) + H^+
2 $FlH^{\cdot}$	→	$Fl_{red}\,H_2 + Fl_{ox}$
$Fl_{red}\,H_2 + O_2$	→	$FlH^{\cdot} + H^+ + O_2^{\cdot-}$
cyt c (3+) + $\mathit{O_2^{\cdot-}}$	→	cyt c (2+) + O_2
$2\,O_2^{\cdot-} + 2H^+$	→	$H_2O_2 + O_2$
cyt c (2+) + 1/2 $H_2O_2 + H^+$	→	cyt c (3+) + H_2O
${}^3Fl_{ox}^{\cdot} + {}^3O_2$	→	${}^1Fl_{ox} + {}^1O_2^{\cdot}$
${}^1O_2^{\cdot}$ + cyt c (2+)	→	$O_2^{\cdot-}$ + cyt c (3+)
$\mathit{FlH^{\cdot}}$ + cyt c (3+)	→	cyt c (2+) + $Fl_{ox} + H^+$

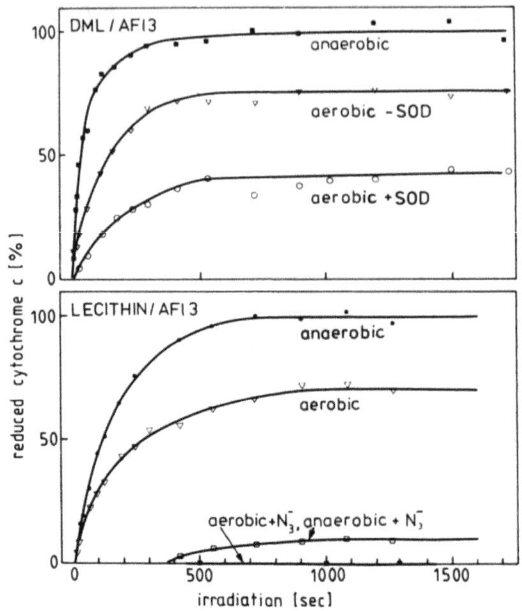

Fig. 3. Photoreduction kinetics of cyt c trapped in the lumen of single-shelled vesicles by oxygeneous EDTA. AFl 3 was bound to the vesicle membrane. *Top* Vesicles made from DML, measuring temperature 21°C (gel state of the membrane). [EDTA] = 66 mM; "anaerobic": 90 s de-aeration with argon (cf. Fig. 4, optimum). 450 units of SOD significantly decrease the rate of photoreduction in the presence of oxygen. *Bottom* Vesicles made from egg lecithin, measuring temperature 21°C. [EDTA] = 167 mM; "anaerobic": 90 s de-aeration with argon. 50 mM sodium azide significantly inhibit the photoreduction both under aerobic and anaerobic conditions

The efficiency of photoreduction under these anisotropic conditions is about 2–10% of that in the isotropic case, depending on the specific flavin/membrane system used and on the phase of the membrane.

In general, the photoreduction kinetics of the type shown in Fig. 3 do not follow a well-defined order, reflecting complex underlying mechanisms, again, depending on the specific system used. More information was obtained from inhibitor experiments. For example, the photoreduction of cyt c trapped in DML/AFl 3 vesicles is essentially diminished by SOD under aerobic conditions. This is indicative of the involvement of superoxide anions, as already established in the isotropic system [34]. In addition, azide is known as a highly efficient, even if not very specific quencher of radicals. Figure 3 clearly demonstrates that in this specific case the trans-membraneous redox transport is mainly mediated by some other radical species, in addition to $O_2^{-\cdot}$. Under anaerobic conditions a minute portion of the cyt c appears to be reduced by a non-radical species (insensitive to azide). The inhibitor efficiencies again depend on the specific system used. Exogeneous dithionite is capable of reducing the encapsuled cyt c indiscriminately and immediately. On this basis it could be demonstrated that even under anaerobic conditions only approximately 80% of the total cyt c was photoreduced via flavin. Therefore the assymptotic levels reached by the photoreduction kinetics (Fig. 3) reflect some redoxequilibrium, controlled by opposite tendencies: "photoreduction" of cyt c and its reoxidation by other oxidizing species as hydrogen peroxide or singlet oxygen, which are photochemically generated via flavin (cf. Table 1).

Two aspects of the involvement of molecular oxygen in membraneous redoxtransport processes appear to be relevant and are most probably involved in primary physiological blue light action, as well. This is exemplified in Fig. 4, top. The dependence of transmembraneous photoreduction of cyt c trapped in DML/AFl 3 vesicles upon

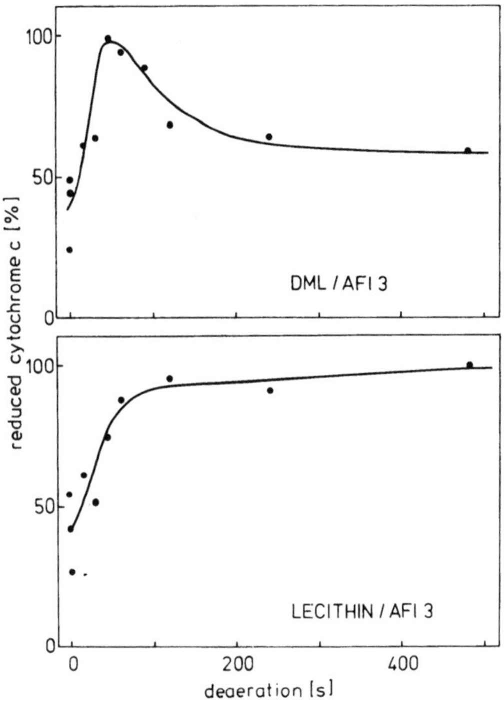

Fig. 4. Dependence of photoreduction efficiency of cyt c being trapped in the lumen of single-shelled vesicles on de-aeration. This was achieved by moderate bubbling of the sample with argon. *Top* [EDTA] = 12 mM (cf. Fig. 5), 300 s irradiation time (cf. Fig. 3). *Bottom* [EDTA] = 35 mM, 300 s irradiation time

progressive de-aeration reflects a clear-cut optimum curve. This is a feature generally found with vesicles made from synthetic phospholipids. Minimum amounts of oxygen promote the photochemical production of redox carriers, too large amounts of oxygen (a triplet in its ground state!) tend to quench the flavin triplet which, in turn, results in a decreased photoreduction rate of cyt c. This latter process obviously can be ignored in vesicles made from natural phospholipids as demonstrated with the lecithin/AFl 3 system (Fig. 4, bottom). In these cases the excited flavin appears to be protected from quenching by molecular oxygen. Similar arguments apply for the dependence of the rate of cyt c photoreduction as a function of the exogeneous EDTA (= electron donor) concentration (Fig. 5). Referring to the DML/AFl 3 system, EDTA concentrations exceeding 15 mM diminish the cyt c photoreduction rate. This is consistent with our finding that EDTA is an efficient quencher of excited flavin states (unpublished observations). Again, no optimum curve is observed in vesicles made from natural phospholipids as egg lecithin (Fig. 5) or cardiolipin.

Another crucial parameter for trans-membraneous, flavin-mediated and blue light-driven reduction of vesicle-trapped cyt c is the phase of the membrane. For DML-vesicles the temperature dependency reflects a strong increase in cyt c photoreduction rate with the transition gel → liquid crystalline near 23°C (Fig. 6).

Figure 7 summarizes the mechanisms which are probably involved in the flavin and blue light mediated transport of redox equivalents across artificial membranes, the single reactions being listed in Table 1. The participation of the various reactions appears to be strongly dependent on the specific flavin/membrane system used. For example, the cyt c photoreduction in the DML/AFl 10 system under aerobic conditions is completely abolished by SOD, but inhibited by only 50% in the DML/AFl 3

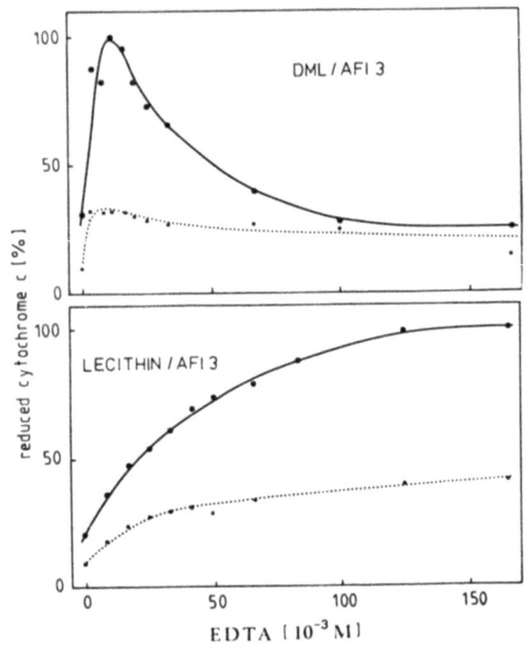

Fig. 5. Dependence of photoreduction efficiency of cyt c being trapped in the lumen of single-shelled vesicles on the concentration of exogeneous EDTA. *Dotted lines* aerobic conditions; *solid lines* 90 s de-aeration time (cf. Fig. 4). The reduction state of cyt c was measured after 300 s irradiation (cf. Fig. 3)

system (Fig. 3). Due to the N5-pK of the membrane-bound flavin triplet of 7.5 (vide supra), a flavin/blue light-mediated proton translocation across membranes under physiological pH's is predicted and at present under investigation. Similarly we expect a flavin/blue light-mediated charge separation in vesicles, as already observed in flavin-charged black lipid membranes [8].

In order to elucidate the number of quanta required for the reduction of one cyt c molecule, the photoreduction kinetics were replotted in Hill diagrams (Fig. 8). In the isotropic system containing EDTA, FMN and cyt c the regression line yields a Hill coefficient of 2., under anisotropic conditions involving membrane transport of redox equivalents, a Hill coefficient of 1. is obtained. This suggests that under isotropic conditions the flavin molecule is photoreduced by a single electron to the radical state, which in turn disproportionates into the fully oxidized and the fully reduced species

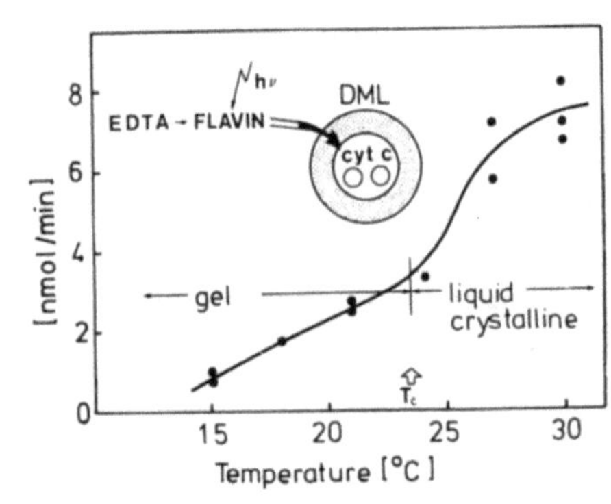

Fig. 6. Photoreduction rate of cyt c trapped in the aqueous lumen of single-shelled vesicles made from DML, as a function of temperature. DML-vesicle exhibit a broad phase transition gel-liquid crystalline close to 23°C (cf. [25]). Conditions: Exogeneous [EDTA] = 10 mM, exogeneous [3-methyl-lumiflavin] = 44 mM. De-aeration time: 180 s

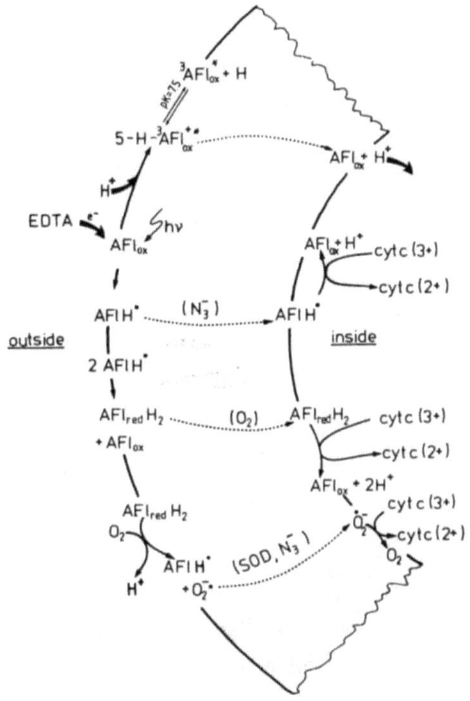

Fig. 7. Summarizing sketch of the mechanisms probably involved in amphiphilic flavin AFl-mediated, blue light-driven translocation of redox equivalents across membranes (cf. Table 1). Potential quenchers of the various species are given on the *dotted lines*. EDTA as a typical $1e^-$-donor is added to the outside medium, oxidized cyt c as efficient acceptor of redox equivalents [34] is trapped in the aqueous lumen of single-shelled vesicles, serving as a highly sensitive assay. Experiments were performed under various oxygen concentrations (cf. Fig. 4). A possible and physiologically highly significant proton-transfer across membranes might involve the flavin-triplet-pK of the position N5. This is at present under investigation

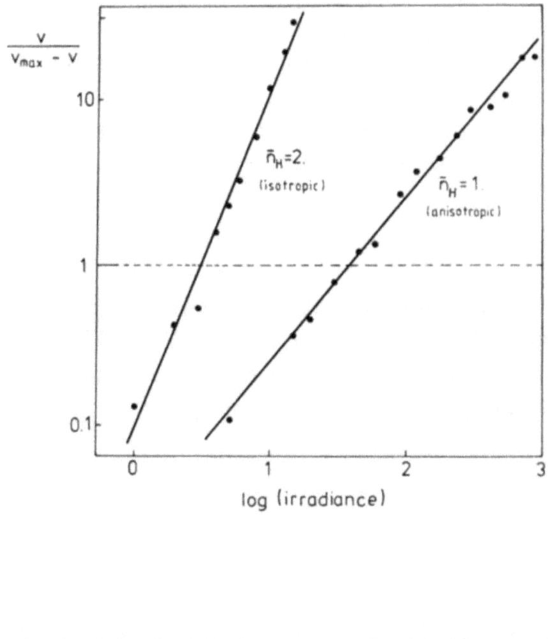

Fig. 8. Hill plots of the photoreduction kinetics of cyt c under isotropic and anisotropic conditions. Anaerobic conditions (180 s de-aeration with Ar). *Isotropic* [FMN] = 8×10^{-7} M, [cyt c] = 8×10^{-7} M, [EDTA] = 5×10^{-3} M. *Anisotropic:* AFl 3/DML-vesicles, cyt c trapped in the aqueous lumen, as used before (Figs. 3, 4, 5). Regression lines and Hill coefficients were obtained by standard procedures, using a HP 9825 microcomputer and HP 9862A digital plotter from Hewlett Packard

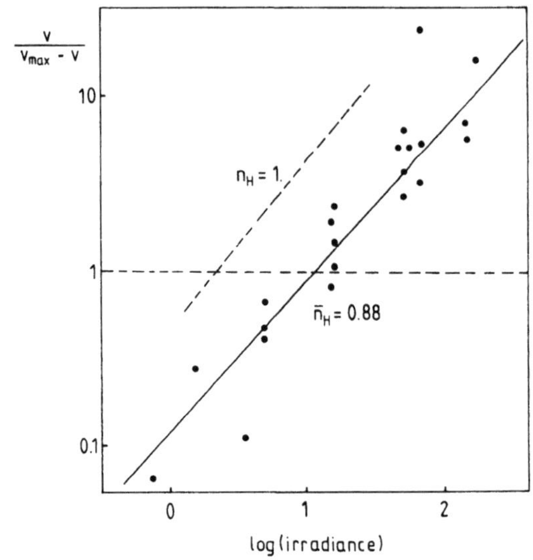

Fig. 9. Hill plot of the dosis-response curve for the first phototropic curvature of oat coleoptiles as measured by Zimmerman and Briggs [49]. No red light pretreatment, quantum flux = 1.4×10^{-13} cm^{-2} s^{-1} at 435 nm. A Hill coefficient of $\bar{n}_H = 0.88$ is obtained, which can experimentally not be distinguished from $n_H = 1$. (*dotted line*)

(Table 1, Fig. 7). The fully reduced species finally reduces a cyt c molecule by single electron transfer, effectively requiring 2 photons. In the membrane system the disproportionation of flavin is largely suppressed due to the efficiently reduced "selfcontact" [29], vide supra, and the radical species will stochiometrically reduce the cyt c by single electron transfer. This interpretation of (a) radical mediated mechanism(s) is consistent with the inhibitor experiments (Fig. 3) and a Hill coefficient of unity (Fig. 8).

Interestingly, replotting the dosis response curve for the first positive curvature of the avena coleoptile [49] in a Hill diagram reveals a Hill coefficient close to 1. (Fig. 9). By analogy, this can be taken to support a radical mechanism in vivo, as well. Unfortunately, however, we failed to detect specific blue light-induced EPR signals in vivo due to a high intrinsic sensitivity of EPR cavities to blue light. However, it is consistent with our previous observation that azide completely inhibits the photoreduction of cytochrome b in the mycelium of *Neurospora crassa* [34].

This work on completely artificial flavin/membrane systems certainly permitted no definite answers to the question of the nature of primary mechanisms involved in physiological blue light responses mediated by flavins. However, it provided valuable data not known before, which might serve as a molecular basis for future experiments.

Abbreviations

EDTA: ethylenediaminetetraacetic acid; NTA: nitrilotriacetic acid; DML: L-β, γ-dimyristoyl-α-lecithin; AFl 3: 7,8,10-trimethyl-3-octadecyl-isoalloxazin; AFl 10: 3,7,8-trimethyl-10-oxtadecyl-isoalloxazin; cyt c: horseheart cytochrome c; T_c: phase transition temperature; FMN: flavin mononucleotide; EPR: electron paramagnetic resonance; SOD: superoxide dismutase.

Acknowledgments. The author is indebted to Miss G. Wenzl for her excellent technical assistance. This work was financially supported by the Deutsche Forschungsgemeinschaft (SFB 138, Project A1, membrane-bound flavins).

References

1. Bruice RC (1976) Models and flavin catalysis. In: Kaiser ET, Kezdy FJ (eds) Progress in bioanorganic chemistry, vol IV. Wiley, New York London Sidney Toronto, pp 1–87
2. Cafiso DS, Hubbel WL (1980) Interfacial charge separation in photoreceptor membranes. Photochem Photobiol 32:461–468
3. Calvin M (1983) Artificial photosynthesis: quantum capture and energy storage. Photochem Photobiol 37:349–360
4. Darszon A, Strasser RJ, Montal M (1979) Rhodopsin-phospholipid complexes in apolar environments: photochemical characterization. Biochemistry 18:52-5–5213
5. Dodelet J-P, Lawrence MF, Ringuet M, Leblanc RM (1981) Electron transfer from chlorophyll a to quinone in mono- and multilayer arrays. Photochem Photobiol 33:713–720
6. Ford WE, Tollin G (1982) Direct observation of electron transfer across lipid bilayer: pulsed laser photolysis of an asymetric vesicle system containing chlorophyll, methyl viologen and EDTA. Photochem Photobiol 35:809–819
7. Ford WE, Tollin G (1982) Chlorphyll phtosensitized electron transfer in phospholipid vesicle bilayers: inside vs outside asymetry. Photochem Photobiol 36:647–655
8. Fröhlich O, Diehn B (1974) Photoeffects in flavin-containing lipid bilayer membrane and implication for algal phototaxis. Nature (London) 248:802–804
9. Georgevich G, Roux SJ (1982) Permeability and structural changes induced by phytochrome in lipid vesicles. Photochem Photobiol 36:663–671
10. Ginkel v G (1979) Photoreactivity of isolated photosystem I particles upon combination with artificial lipid membranes or triton X100 micelles. Photochem Photobiol 30:397–404
11. Goldberg M, Pecht I, Kramer HEA, Traber R, Hemmerich P (1981) Structure and properties of 5-deazaflavin radicals as compared to natural flavosemiquinones. Biochim Biophys Acta 673: 570–593
12. Gressel J (1979) Blue light photoreception. Photochem Photobiol 30:749–754
13. Hemmerich P (1976) The present status of flavin and flavoenzyme chemistry. Prog Chem Org Nat Prod 33:451–527
14. Hemmerich P, Massey V, Michel H, Schug Ch (1982) Scope and limitation of single electron transfer. In: Dunitz JD, Goodenoug JB, Hemmerich P, Ibers JA, Jørgensen CK, Neilands JB, Reinen D, Williams RJP (eds) Structure and bonding, vol 48. Springer, Berlin Heidelberg New York, pp 93–123
15. Hurley JK, Castelli F, Tollin G (1981) Chlorphyll-quinone photochemistry in liposames: mechanisms of radical formation and decay. Photochem Photobiol 34:623–631
16. Kano K, Tanaka Y, Ogawa T, Shimomura M, Kunitake T (1981) Photoresponse artificial membrane. Regulation of membrane permeability of liposomal membrane by photoreversible cistrans-isomerization of azobenzenes. Photochem Photobiol 34:323–329
17. Kim I-S, Song P-S (1981) Binding of phytochrome to liposomes and protoplasts. Biochemistry 20:5482–5489
18. Massey V, Hemmerich P (1977) A photochemical procedure for the reduction of oxidation-reduction proteins employing deazariboflavin as catalyst. J Biol Chem 252:5612–5614
19. Michel H, Hemmerich P (1980) Substitution of flavin chromophore with lipophilic sidechains: A novel membrane redox label. J Membr Biol 60:143–153
20. Michel H, Schmidt W, Hemmerich P (1982) On the environment and the rotational motion of amphiphilic flavins in artificial membrane vesicle as studied by electron paramagnetic resonance spectroscopy. Biophys Chem 2:121–130

21. Montal M, Darszon A, Strasser RJ (1978) Rhodopsin and bacteriorhodopsin in model membranes. In: Dutton PL, Leigh JS, Scarpa A (eds) Frontiers of biological energetics, vol II. Academic Press, London New York, pp 1109–1117
22. Ninnemann H, Klemm-Wolfgramm E (1980) Blue light-controlled conidiation and absorbance change in *Neurospora* are mediated by nitrate reductase. In: Senger H (ed) The blue light syndrome. Springer, Berlin Heidelberg New York, pp 238–243
23. Pepe IM, Gliozzi A (1983) Model photoresponsive membranes. In: Montegnoli G, Erlanger BF (eds) Molecular models of photoresponsiveness. Plenum Press, New York London, in press
24. Presti D, Delbrück M (1978) Photoreceptors for biosynthesis, energy storage and vision. Plant Cell Environ 1:81–100
25. Schmidt W (1979) On the environment and the rotational motion of amphiphilic flavins in artificial membrane vesicles as studied by fluorescence. J Membr Biol 47:1–25
26. Schmidt W (1980) A high performance dual wavelength spectrophotometer and fluorometer. J Biochem Biophys Methods 2:171–181
27. Schmidt W (1980) Physiological bluelight reception. In: Hemmerich P (ed) Structure and bonding, vol 41. Springer, Berlin Heidelberg New York, pp 1–44
28. Schmidt W (1980) Artificial flavin/membrane systems; a possible model for physiological blue light action. In: Senger H (ed) The blue light syndrome. Springer, Berlin Heidelberg New York, pp 212–220
29. Schmidt W (1981) Fluorescence properties of isotropically and anisotropically embedded flavins. Photochem Photobiol 34:7–16
30. Schmidt W (1982) A computerized single beam spectrophotometer: an easy set up. Anal Biochem 125:162–167
31. Schmidt W (1983) Further photophysical and photochemical characterization of flavins associated with single-shelled vesicles. J Membr Biol 76:73–82
32. Schmidt W (1983) The capability of the on-line computer in fluorescence spectroscopy. J Opt Eng 22:576–582
33. Schmidt W (1983) The physiology of blue-light systems. In: Cosens DJ, Vince-Prue D (eds) The biology of photoreception. Proc 16th Conf Br Soc Photobiol. Cambridge Univ Press, Cambridge, pp 305–330
34. Schmidt W, Butler WL (1976) Flavin-mediated photoreactions in artificial systems: a possible model for the blue-light photoreceptor pigment in living systems. Photochem Photobiol 24: 71–75
35. Schmidt W, Hart J, Filner P, Poff KL (1977) Specific inhibition of phototropism in corn seedlings. Plant Physiol 60:736–738
36. Schmidt W, Hemmerich P (1981) On the redoxreactions and accessibility of amphiphilic flavins in artificial membrane vesicles. J Membr Biol 60:129–141
37. Schmidt W, Thomson K-S, Butler WL (1977) Cytochrome b in plasma membrane enriched fractions from several photoresponsive organisms. Photochem Photobiol 36:407–411
38. Schreiner S, Steiner U, Kramer HEA (1975) Determination of the pK values of lumiflavin triplet state by flash photolysis. Photochem Photobiol 21:81–84
39. Senger H (ed) (1980) The blue light syndrome. Springer, Berlin Heidelberg New York
40. Senger H, Briggs WR (1981) The blue light receptor(s): Primary reactions and subsequent metabolic changes. In: Smith KC (ed) Photochemical and photobiological reviews, vol VI. Plenum Press, New Aork London, pp 1–38
41. Senger H (1982) The effect of blue light on plants and microorganisms. Photochem Photobiol 35:911–920
42. Song P-S (1980) Primary photophysical and photochemical reactions: Theoretical background and general introduction. In: Lenci F, Colombetti G (eds) Photoreception and sensory transduction in aneural organisms. Plenum Press, New York London, pp 189–210
43. Song P-S (1981) Photosensory transduction in *Stentor Coeruleus* and related organisms. Biochim Biophys Acta 639:1–29
44. Song P-S, Fugate RD, Briggs WR (1980) Flavin as a photoreceptor for phototropic transduction: fluorescence studies of model and corn coleoptile systems. In: Yagi K, Yamano T (eds) Flavin and flavoproteins. Jpn Sci Soc Press, Tokyo; Univ Park Press, Baltimore

45. Trissl H-W, Läuger P (1970) Photoelectric effects in thin chlorophyll films. Z Naturforsch 25: 1059–1060
46. Trissl H-W, Darszon A, Montal M (1977) Rhodopsin in model membranes: charge displacements in interfacial layers. Proc Natl Acad Sci USA 74:207–210
47. Webster WW, Schrank AR (1953) Electrical induction of lateral transport of 3-indoleacetic acid in the *Avena* coleoptile. In: Archives of biochemistry and biophysics, vol 47. Academic Press, London New York, pp 107–118
48. Zenk MH (1967) Untersuchungen zum Phototropismus der Avena Koleoptile: II Pigmente. Z Pflanzenphysiol 56:122–140
49. Zimmerman BK, Briggs WR (1963) Phototropic dosage-response curves for oat coleoptiles. Plant Physiol 38:248–253
50. Zurzycki J (1972) Primary reactions in the chloroplast rearrangements. Acta Protozool 23: 189–199

The Nitrate Reductase System

HELGA NINNEMANN[1]

Introduction

It has been proposed by Klemm and Ninnemann [29] that nitrate reductase (NR) may act as a photoreceptor in blue light-stimulated conidiation of *Neurospora crassa* (carotenoid-less mutant al-2 bd). Nitrate reduction and conidiation appear to be correlated in some specific ways: Hochberg and Sargent [19] and Weiss and Turian [61] reported that conidiation of *Neurospora* is maximally supported by nitrate as sole nitrogen source in the medium, whereas the mycelium grows optimally with ammonium nitrate or diammonium citrate. Weiss and Turian [61] also described suppression of conidiation on diammonium citrate (+ 2% sucrose). The same amino acids (e.g., try, ala, his) appear to be favorable for conidiation [47] and induction of NR [51]. We found [26] that hyphae grown on nitrate formed conidia under starvation conditions (nitrogen and carbon source depleted, possibly partial anaerobiosis) whereas conidia occurred much later if at all when the mycelium was starved after growth on medium containing NH_4NO_3 (Fig. 1A and B). Also light-induced absorbance changes at 423 and 557 nm – regarded as an assay system for a primary process in the vicinity of the photoreceptor – could be seen in vivo in *Neurospora* mycelium only at a time when light stimulated conidiation [28]. We have done preliminary experiments which show that the light requirement for promotion of conidiation decreased in somatic fusion products nit 1 + 3 made from protoplasts of nitrate reductase-deficient nit-1 and nit-3 mutants of *Neurospora* as compared with the parental strains [37].

Nitrate reduction and conidiation depend on complex parameters. Beside the nitrogen source, quality and concentration of the carbon source and partial pressure of O_2 (or low CO_2 tension) influence the process [47]. We therefore started to concentrate on those properties of the flavohemomolybdoenzyme NR from *Neurospora* which might be involved in mediating environmental stimuli for regulation, and which might be of importance for the possible photoreceptor role of this enzyme. I shall first survey some well-documented relevant facts on the nitrate reductase system and its regulation, and then proceed to new results from my laboratory.

1 Institut für Chemische Pflanzenphysiologie der Universität, Correnstraße 41, D–7400 Tübingen, FRG

Blue Light Effects in Biological Systems
ed. by H. Senger

Fig. 1A,B. A Conidiation in mycelium grown in liquid medium with $NaNO_3$ for 20 h, starved (no substrates) in darkness for 8 h; **B** No conidiation in mycelium grown in liquid medium with NH_4NO_3 for 20 h, starved (no substrate) in darkness for 8 h [26]

Structure and Localization of Nitrate and Nitrite Reductase

The seemingly simple processes of nitrate and nitrite reduction involve the functions of two complex enzymes, NR and nitrite reductase (NIR). Eukaryotic nitrate reductases of fungi, algae and higher plants have a high degree of similarity as to presence of subunits, kind of prosthetic groups and presence of a molybdenum cofactor (Moco). The enzyme of *Neurospora* will be especially considered here. Since NR and NIR show certain structural and functional resemblances, NIR will also be included in this survey.

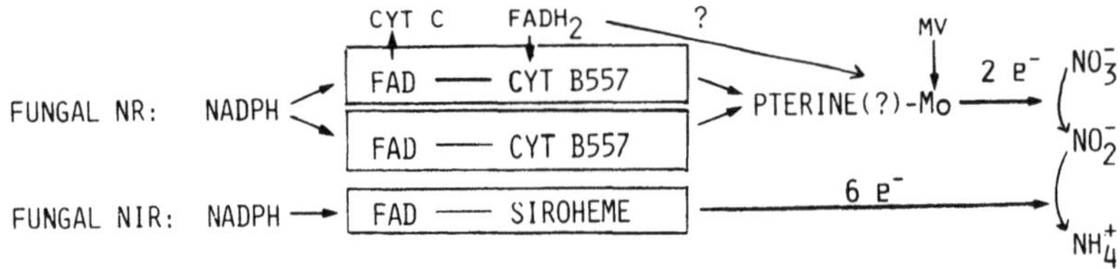

Fig. 2. Probable structure of NR ([40] modified) and NIR [58]

Pioneer contributions to the structural and functional characterization of *Neurospora* NR come from the groups of Nason, Garrett and their collaborators. They analyzed NR as a dimeric enzyme with MW 228,000 [16], using NADPH as electron donor and containing FAD, a b-type cytochrome (cyt b-557) which is highly autoxidizable [38] and molybdenum as prosthetic groups or cofactor. NR therefore is a catalytically active protein with multiple redox centers. NR (as well as NIR) constitutes a short electron transport system with a potential free energy between NADPH/NADP (–320 mV) and NO_3^-/NO_2^- (+420 mV) of –34.1 kcal/mole. Its presently most likely structure is pictured in Fig. 2.

Since at least for *Chlorella* NR [23] and also for *Neurospora* NR (Siefermann-Harms and Ninnemann, to be published) a pterine associated with the Moco has been reported, this moiety has also been indicated in the NR model. In Fig. 2 the functionally following NIR (MW 300,000) is included which also contains a flavin (FAD)-cytochrome (siroheme) couple [58] as prosthetic groups but no Moco, and reduces nitrite to ammonia in a 6-e^- step. It is remarkable that all prosthetic groups and the Moco of NR are not only capable of redox reactions but also have chromophoric properties.

Several activities are associated with the NR molecule: "total" activity is the electron transfer from NADPH to NO_3^-, so-called partial activities include electron transport from NADPH via the NR- (or NIR-)flavin moiety to cytochrome c (or DCPIP or Fecy), known as diaphorase activity, and electron transfer from exogenous substrates such as $FADH_2$ and reduced MV to nitrate, processes which require only functioning of the terminal part of NR ("terminal activity"). At least in *Neurospora* about 90% of the maximally active NR is found in the soluble fraction after homogenation and centrifugation at 20,000 g/20 min, around 10% remain in the membrane fraction. Thus if the enzyme was associated with membranes in situ it easily dissociates during the isolation process. Of the terminal MVH-NO_3^- activity, 80–85% are located in the soluble, 15–20% in the membrane fraction of mycelia with optimally active NR; in starving mycelia total and partial activities decrease with time, but terminal activity is better protected in the membrane fraction against destruction than in the soluble fraction. In this way, 50–70% of the remaining terminal activity could be found in the membrane fraction of mycelium grown for 50 h in nitrate medium (Ninnemann, unpublished). An association of NR or part of the NR molecule with *Neurospora* cell membranes was also demonstrated by Roldán et al. [46], who treated *Neurospora* mycelium sections with purified rabbit antinitrate reductase IgG and labeled it with ferritin-conjugated goat antirabbit IgG. They found in their electron micrographs preferential localization of ferritin indicator in the cell wall/plasmamembrane region and in the tonoplast. Dalling et al. [11] observed that NR of spinach and tobacco leaves was located in the cytosol but bound indiscriminately to membrane (protein) surfaces unless the extraction medium was fortified with exogenous protein.

Chemical Regulation of Nitrate Reductase

The enzymes of the nitrate reductase system are highly regulated proteins. Synthesis of NR and NIR is induced (derepressed in the case of NR) by nitrate and in many organisms by nitrite [14], it is repressed by ammonia [25]. NR activity declines rapidly after reaching a maximum whereas NIR activity persists in presence of NO_3^- or NO_2^- [14, 18]. Addition of ammonium salt to nitrate-grown cultures of *Chlorella, Chlamydomonas* or *Neurospora* inactivates their total NR and MVH-NO_3 activities (diaphorase activity is affected in *Neurospora* but not in *Chlorella*) and leads to decrease of extractable NR activity [18, 31, 50]. NR partial activities of NR-deficient *Neurospora* (mutants nit-1 and nit-3) and *Aspergillus* are not induced by nitrate, they are constitutive [10, 12], nor are they inactivaed in vivo by addition of ammonium salt [50]. Also in vitro, NR can be reversibly inactivated by chemicals. Incubation with NAD(P)H, $FMNH_2$ or reduced methylviologen in the absence of nitrate leads to inactive NR and terminal electron transport within several minutes in *Chlorella* [34]. This inactivation occurs more completely and more rapidly in presence of cyanide [44], which seems to bind to reduced molybdenum thus rendering the enzyme inactive. Studies done by us with *Neurospora* NR showed that one CN^- binds per active center in the enzyme-inhibitor- (or E-I-substrate-)complex; the inhibition has a K_i of 0.1 mM and is noncompetitive [13]. Cyanide is regarded as a physiological effector in *Chlorella* [30, 41].

Spontaneous reactivation proceeds only within hours [31] but can be enhanced by addition of ferricyanide [57]. Ferricyanide ($E^{\circ\prime}$ = 410 mV) is presumed to act by reoxidizing the Moco, thereby reactivating the electron transfer to nitrate. Redox titration for chemical reactivation done in my laboratory with various ratios of ferri-/ferrocyanide showed a midpoint potential at pH 7.2 of 371 mV with one electron transferred in the kinetically determining step [13]. Physiologically as important might be photoinactivation and photoreactivation of the nitrate reductase system.

Photoinduction and Photoactivation of Nitrate Reductase

Photoactivation of NR or NIR happens at two levels: by photoinduction of their syntheses or by photoactivation, photoinactivation or photoreactivation of already existing molecules. Photoinduction of NR has been reported by several authors: Hageman and Flesher [17] found an increase of maize NR levels after illumination of leaves and lower levels in shaded leaves or in those transferred into darkness. The NR level was restored after subsequent irradiation. Since then similar observations have been accumulated in the literature (review: [5]). The stimulation of NR synthesis by light appears to be due to an increased capacity for protein synthesis and polysome development [55] in illuminated leaves; decrease of NR activity in darkness is thought to be a consequence of NR decay [5].

In some higher plants light-stimulated de novo synthesis of NR, e.g., in etiolated pea and maize tissues, is caused by red light via the phytochrome system [24]. Johnson [21] described a rapid phytochrome-controlled increase in NR activity which was insensitive to inhibitors of protein synthesis. Stoy [48, 49] reported a blue light-induced increase of nitrate assimilation in wheat leaves and a riboflavin-catalyzed enzymic photoreduction of nitrate. Tischner and Hüttermann [52] showed photoactivation of NR in synchronized *Chlorella* which was not due to de novo synthesis. Light was supposed to act directly on NR or to stimulate nitrate uptake. Light-promoted availability or permeation of nitrate was proposed by many authors (e.g., [6, 54].

Apparently the nitrite reducing and NH_4^+ assimilating enzymes can also be photoregulated: stimulation of nitrite reduction by light has been described [4, 8, 53], as well as photoactivation of glutamine synthetase and glutamate synthase after a light/dark shift of synchronous *Chlorella* cultures. Therefore, besides a regulation by different nitrogen sources of NR, NIR, glutamine synthetase and glutamate synthase (e.g., in vivo in *Neurospora,* [20]) all steps in anorganic nitrogen metabolism can be controlled by blue or red light.

Photoreactivation and Photoinactivation

Aparicio and his coworkers [1, 3] reactivated *Chlorella fusca* and spinach NR – inactivated in vivo with ammonium or in vitro under reductive conditions with cyanide – with ferricyanide or with white and blue light, preferentially after addition of 20 μM FAD or FMN. Spinach and *Neurospora* NR can also be inactivated by acetylene ([32], Ninnemann, unpublished) and reactivated with blue light, but not with ferricyanide – an indication that the mechanisms of inactivation might be different with cyanide and acetylene where acetylene seems to form a stronger bond with the Moco than cyanide [33].

Photoinactivation and photoreactivation of NR from *Neurospora* and its kinetics were investigated extensively by Roldán and Butler [45] and in my laboratory [13]. For most of their experiments Roldán and Butler used the partially purified, ammonium sulfate-precipitated fraction P3 [16] inactivated chemically with cyanide in presence of NADPH. An irreversible, temperature-dependent partial photoinactivation of total NR and diaphorase activities was observed by them whereas MVH-NO_3^- activity via Moco remained stable in light under their conditions. At 24°C, 30 min of blue light (450 nm, 25 mW cm^{-2}) caused the loss of 30% of those two enzyme activities. Accordingly, maximally 70% of total NR activity could be reactivated in a chemically inactivated enzyme by light because of the superimposed photodestructive and photoreactivating processes. Terminal activity could be fully restored by light. The action spectrum for photoreactivation of *Neurospora* NR indicated a flavin-type photoreceptor. The authors claimed that exogenous flavin was not needed for the photoreactivation but enhanced the reaction and that oxygen was required. Neither SOD or catalase nor singlett oxygen quenchers of scavangers like β-carotene or histidine influenced photoreactivation.

We have been concerned with the kinetics of both photoprocesses affecting *Neurospora* NR and with the reactive molecule species involved ([13], Fritz and Ninnemann, submitted). It could be shown that photoinactivation and photoreactivation proceed via different mechanisms: photoinactivation is caused by singlet oxygen; in photoreactivation of CN^--inactivated NR, on the other hand, excited flavin interacts with the NR molecule. Evidence for these results came from the observation that *photoinactivation* of *Neurospora* NR could be inhibited by DABCO (50% with 0.3 mM), histidine (50% with 1 mM) and azide (50% with 0.2 mM); in D_2O where lifetime of $^1O_2$ increases, inactivation proceeded threefold faster than in H_2O. SOD and catalase showed little if any effect. We therefore deduced singlet oxygen as active species for photoinactivation of *Neurospora* NR.

These quenchers and scavengers (DABCO up to 5 mM, his up to 4 mM) were, however, ineffective in inhibiting *photoreactivation* of chemically inactivated enzyme; azide at 0.3 mM quenched the reactivation by 15%. SOD was without effect, catalase (160 $\mu g\ ml^{-1}$ or 10,400 units ml^{-1}) inhibited it, but this effect could be mimicked by comparable concentrations of hemin. In samples with low oxygen tension the reactivation rate increased by 100%. Thus we propose that excited flavin – probably in its triplet state – directly reactivates NR. Similar conclusions concerning the function of excited flavin in photoreactivation were reached recently by Mauriño et al. [33] for spinach NR. We could show that the rate of photoreactivation of *Neurospora* NR depended linearly on the concentration of free flavin [FAD tested up to 32 μM, FMN or riboflavin (RF) tested up to 4 μM]; bound flavin remaining after 15 h of dialysis of our enzyme preparation did not suffice for measurable photoreactivation. Free RF, FMN or FAD acted with relative efficiencies of 6:5:1. The ratio of apparent rate constants for photoreactivation k^+/photoinactivation k^- was approximately 3.

During the course of our studies, Vargas et al. [56] published a paper on photoinactivation of spinach NR and also proposed the involvement of singlet oxygen. They showed that blue light, especially when mediated by exogenous FMN, inhibited diaphorase and – to a lesser degree – terminal activities of spinach NR irreversibly. The photoinactivation of terminal NR activity by blue light had not been described before. It might be worthwhile to determine whether the proposed pterine of the Moco might act as phototarget.

Mechanisms of Regulation of the Nitrate Reductase System

It is widely accepted that NR plays a primary role in regulating inorganic nitrogen metabolism [59]. The data reviewed so far show that NR, as well as the other enzymes of inorganic nitrogen metabolism, are precisely regulated by physiological parameters such as kind and concentration of the nitrogen and carbon source (see [2]), by inhibitors like cyanide, by specific inhibitors like protease with a serine active center attacking the diaphorase activity of *Neurospora* NR [60] and last but not least by light. As far as the mechanisms involved are concerned, inhibition of the reduced Moco and reoxidation of the stable Mo(IV)-CN complex to the unstable Mo(VI)-CN complex from which cyanide would be dissociated has been proposed for chemical inactivation and reactivation of NR by ferricyanide or light [33]. This interpretation is supported by results of Nicholas et al. [36] with extracts from *Neurospora:* they partially purified NR and dialyzed it against KCN and glutathione which resulted in inactive NR. Incubation of the dialyzed enzymes with MoO_3 or Na_2MoO_4 restored the activity.

It has not yet been established whether similar redox changes of the cyt b-557 and/or FAD determine the activity of NR as well. But regulation of the enzyme by redox changes of the Moco is not the only means of influencing NR activity. Frequent observations in the literature are concerned with a time-dependent decline of NR activity or with an actively regulated decomposition of the NR molecule under nitrogen starvation: in *Chlorella,* NR and NIR were not detected in N-starved cells [53]. Garrett [14] and Cove [9] concluded that NR of *Neurospora* and *Aspergillus* was unstable in vivo in absence of nitrate. Purvis et al. [43] deduced from experiments on the protective effects of BSA on extracted NR from cotton cotyledons that it was due to enhancement of activity rather than extractability. BSA may prevent NR from dissociating into subunits. In our experiment on *Neurospora* NR under conditions of nitrogen limitation NR activity reached a maximum in late log phase in suspension cultures and declined afterwards even when nitrate was still present in the medium: starting with 56 mM $NaNO_3$ and 150 mM glucose in modified Vogel's medium we measured still 25 mM NO_3^- and 60 mM glucose after 24 h of growth; at this time total NR activity had already declined by about 65% or more as compared with maximal activity at 16–18 h of growth [18]. Garrett [14] transferred *Neurospora* mycelia with decreasing NR activity into fresh nitrate medium where activity increased again, suggesting exhaustion of an essential factor for NR synthesis in the original medium or accumulation of a toxic substance. Starvation conditions result in a similar decrease of total NR and of diaphorase activities whereas the MVH-NO_3 activity remains more stable (Ninnemann, unpublished). We have succeeded in photoreactivating MVH-NO_3^- and NR total activity in starved mycelia by irradiating them in vivo [29, Ninnemann, unpublished], a property which could increase the attraction NR may have as a photoreceptor candidate.

Photoreduction of Nitrate Reductase

The finding of Butler et al. [35, 42] of a flavin-mediated photoreduction of a b-type cytochrome occurring close to the primary act of photoreception has provided an assay system for some operating photoreceptors. Such light-induced absorbance changes have been reported since then in many systems (e.g., [7, 27, 28, 62], and others). If such flavin-cytochrome b couples participate in the act of photoreception, flavin and cytochrome should be located close to each other at or near the photoreceptor; hence the preference with which flavohemoenzymes are being considered as promising photoreceptor molecules. Since among the enzymes of the nitrogen metabolism the flavohemoprotein NR is photoregulated in such a refined and multiple way, and at the same time appears to be a determinante for conidiation, we chose to investigate this enzyme with respect to its involvement in light-stimulated conidiation. We do not advocate NR as THE key element of all photoreceptor systems in blue light-regulated processes but rather as a model representative for a principal mechanism Nature may have invented, namely using suitable enzymes for a photoreceptive activity at the same time.

Light-regulated changes of NR and light-stimulated fungal conidiation are complex processes; their causal relationship is hard to prove beyond doubt. It is preferentially the starved mycelium which responds best to blue light with a promotion of conidiation and ΔA423 nm absorbance changes in vivo. At this time NR activity in *Neurospora* is suboptimal and apparently ready for photoreactivation. But while total NR activity is low or absent, the presence of its prosthetic groups flavin and cyt b-557 can be measured in vitro in the inactive (inactivated) preparations of starved or NR-deficient cells ([38] and unpublished results). Since a NR with optimal activity requires presence of substrate which in turn results in an at least partially reduced NR electron transport chain, photoregulation at this state seems neither possible nor meaningful. Rather the enzyme oxidized and probably structurally altered, e.g., by starvation is proposed to be apt for the transduction of a photosignal. Before we can hope to understand the details of the in vivo responses of NR under various physiological conditions, we first started to investigate the prosthetic groups in the isolated enzyme.

Photoreduction of Cytochrome b-557 by Internal Free and Bound Flavin

It was shown recently that partially purified and gel-filtrated NR preparations can be photoreduced in presence of EDTA or certain amino acids without exogenously added flavin [38]. Consequently our interest centered on whether it is only free flavin which mediates photoreduction of NR and its cyt b-557 in a moderately specific photoreaction or whether also the bound FAD of NR can act as photoreceptor for cyt b-557 reduction. In contrast with the firmly bound cyt b-557 (see below), FAD of *Neurospora* NR has been reported as readily dissociable ([15, 45] who used mainly fraction P3; my own observations). Partially purified NR fractions (e.g., P3) contain appreciable amounts of free or loosely attached flavin beside bound flavin of NR and possibly of other flavoproteins. We therefore followed the photoreducibility of a NR preparation (P3) after various times of dialysis when flavins, especially free flavins, became depleted (Ninnemann and Siefermann-Harms, unpublished). Under aerobic conditions, photoreducibility ceased after 7 h of dialysis due to loss of sufficient concentration of free flavin for mediation of cyt b reduction (Fig. 3). Addition of external FAD restored the photoreducibility under aerobic conditions. Treatment of the preparations dialyzed up to 41.5 h with SDS in order to liberate bound flavin rendered them photoreducible again without exogenous flavin. At the same time these concentrations of bound flavins present after 21 h and 41.5 h of dialysis (no SDS treatment) sufficed to mediate a very pronounced photoreduction of the autoxidizable cyt b-557 under anaerobic conditions (Fig. 3 top curves on left side). Irradiated flavin bound to foreign enzymes (e.g., glucose oxidase, GOD) could not mediate this effect.

An analysis of the flavin content of preparations dialyzed for various periods of time was done by testing their fluorescence emission at 520 nm either directly or after a treatment for 4 min with 2% SDS: the former "class" of flavin was considered as "free" flavin, the latter as "total" flavin, the difference as SDS-liberated "bound"

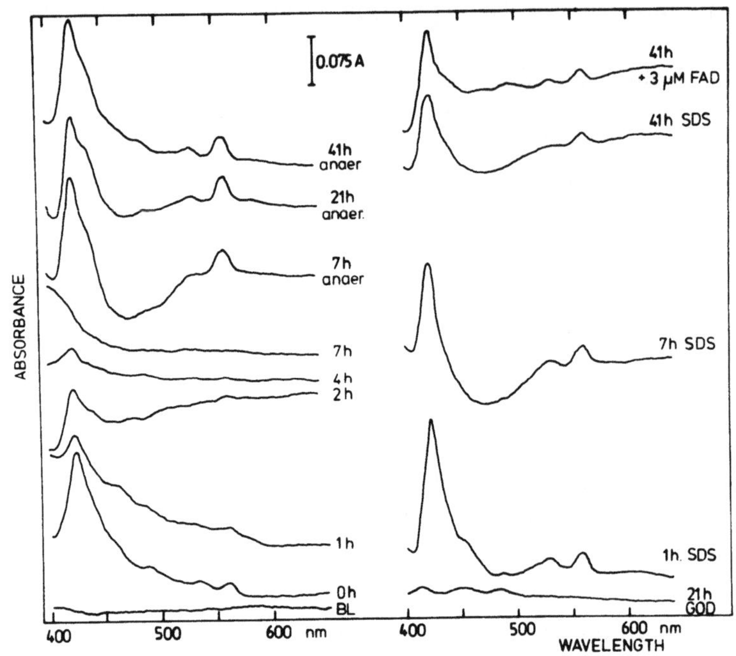

Fig. 3. Photoreduction with internal free or bound flavin of NR fraction P3 after various periods of dialysis. (Light-minus-dark difference spectra).
Sample: P3 (6.1 mg protein) + 50 mM EDTA in 250 μl. Anaerobic conditions: GOD (250 μg, 50 units) + glucose (15 μmol) + catalase (50 μg, 3250 units). SDS treatment: 2%, 4 min. Difference spectra at −196° C. Spectra recorded with a single beam spectrophotometer on line with a computer for calculation of difference spectra light (L)-dark (D = reoxidized in darkness by shaking for 30 s). Photoreduction: 3 min at 0° – 6° C, white light, 130 mW cm^{-2}. *BL* baseline
Left side from bottom to top
0 h–7 h: P3 dialyzed below 8° C, photoreduced aerobically
7 h anaer. – 41 h anaer.: P3 dialyzed for 7–41.5 h, photoreduced anaerobically
Right side from bottom to top
21 h, GOD: dialysis 21 h, P3 (+ EDTA) + GOD, no glucose
1 h SDS – 41 h SDS: 1–41.5 h dialyzed P3 + SDS treatment. Aerobic conditions.
41 h + 3 μM FAD: dialysis 41.5 h, 3 μM FAD added for aerobic photoreduction
Control experiment: 123 μM FAD dialyzed under comparable conditions ± BSA 50 mg ml^{-1}: after 4.5 h 98.5% (–BSA) or 92% (+BSA) FAD lost, after 20 h of dialysis ± BSA 99.4% FAD lost

flavin. SDS-treated samples lost all fluorescence during a subsequent dialysis. This shows that the majority of noncovalently bound flavins could be liberated by SDS. In Fig. 4 the changes within the different flavin classes are demonstrated with increasing dialysis time. "Total" and "free" flavin levels decreased substantially within the first 12 h of dialysis, their ratio free:total went down from 0.52 (undialyzed) to 0.27 (14–41.5 h dialyzed). Subsequent identification of the flavin types contained in the preparation revealed 46–51% FAD and 54–59% FMN plus RF before dialysis; after dialysis 50–52% FAD and 48–50% FMN + RF were found (Scheideler, unpublished). Because of the about tenfold higher fluorescence yield of FMN and RF as compared with FAD, the total FAD content can be calculated to be in the order of 2.0–2.4 μM before dialysis (beside 2.8–2.9 μM FMN + RF) and 0.36 μM FAD (plus 0.36 μM FMN + RF)

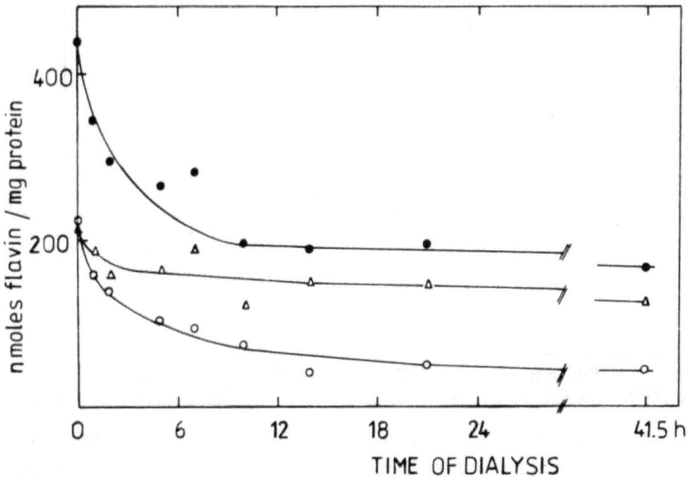

Fig. 4. Flavin content of P3 dialyzed for various periods of time. Flavin fluorescence measured at 520 nm (λ_{ex} = 420 nm). Flavin content calculated on the basis of 50% FAD + 50% (FMN + RF) contained in the sample (TLC analysis)
○–○ flavin content in untreated sample = "free" flavin, ●–● after SDS treatment (2%, 4 min) = "total" flavin, △–△ "total" minus "free" flavin = SDS-liberated "bound" flavin. Control: fluorescence of FAD unaffected by 2% SDS, 4 min; no enhancement by boiling

after dialysis. At the same time NADPH-NO_3 activity was unaffected in dialyzed preparations after addition of FAD as compared with non-dialyzed samples kept at 4°C for 41.5 h, and the ratio cyt b-557: protein remained constant. The stability of the cyt b-557 can be seen in all spectra of Fig. 3. We therefore believe to have good evidence that FAD bound to NR can act as photoreceptor for cytochrome b-557 reduction. This experiment has to be repeated as soon as we have sufficient amounts of highly purified NR.

A New Fluorescence Signal Increasing with Purity of *Neurospora* Nitrate Reductase

In 1980, Johnson et al. [23] reported on a pterine binding to molybdenum, both comprising the Moco of xanthine oxidase. They also found it present in the Moco's of sulfite oxidase and NR of *Chlorella.* In purified enzymes the active cofactor was nonfluorescent, whereas pterine fluorescence (excitation at 380 nm, maximal emission at 455 nm) was detected after denaturation in the presence of KJ and J_2 ("form A"). Another degradation procedure introduced by Johnson et al. [22] consisted of boiling for 20 min at pH 2.5 which yielded a fluorescent species "form B" (excitation at 400 nm, maximal emission at 465 nm). We investigated the fluorescence of homogenates and isolated NR of *Neurospora* (Siefermann-Harms and Ninnemann, unpublished) and obtained signals which we interpret as resulting from pterines [38a]. They will be characterized in more detail here: in crude homogenates of *Neurospora* mycelia

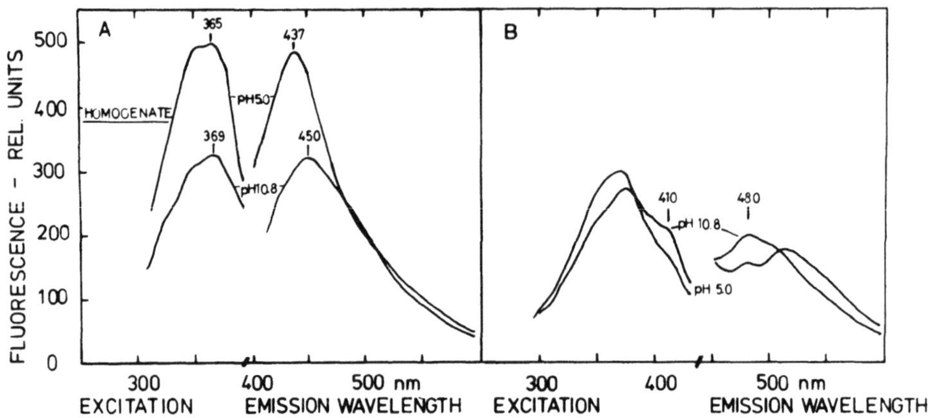

Fig. 5A,B. Fluorescence excitation and emission spectra of "unspecific" (**A**) and "specific" (**B**) pterine signals from *Neurospora* homogenate at various pH. Samples were boiled (20 min, pH 2.5) and their pH adjusted with NH_4OH; 0.44 mg protein ml^{-1}; NADPH-NO_3 activity: 190 nmol NO_2^- formed mg^{-1} × 10′

growing in log phase, fluorescence at 430–450 nm could be excited maximally by λ = 350–370 nm (Fig. 5A), a wavelength couple characteristic for many pterines. Beside this "unspecific" fluorescence, a smaller "specific" signal emitting at 480 nm and maximally excited at 410 nm was detected in the homogenates (Fig. 5B). This "specific" fluorescence increased with increasing purity of NR: in Figs. 6A and 7A the decreasing signal emitting at 430–450 nm and in Figs. 6B and 7B the increasing fluorescence at 480 nm in fraction P3 and in a purer fraction chromatographed on blue Sepharose are shown; the 480 nm signal increases still further after additional purification by Sephadex G-25 and DEAE 22 treatment (spectrum not shown). Both signals have conspicuously different properties: the "unspecific" fluorescence (360–370 nm/430–450 nm) can be found already in untreated samples but is enhanced by heat treatment; excitation and emission maxima of the underlying fluorescent species vary with pH as seen in Fig. 8. The "specific" signal (410 nm/480 nm) which we associate with NR is found only after heat treatment; optimal yields were obtained after boiling at acidic pH, i.e., the procedure of Johnson et al. [22] to create their "form B". The excitation and emission maxima of our "specific" signal are pH-independent but fluorescence disappears below pH 7 (Fig. 9A). The titration curve of this signal from purified *Neurospora* NR (Fig. 9B) shows a pK value of 8.1. Similar fluorescence spectra and pK values were obtained in homogenates of starved *Neurospora* where total NR activity had decreased but Moco-dependent terminal activity (and probably also inactive Moco) was present.

The wavelength characteristics of our "specific" component are unusually long; they are longer than for the pterine of xanthine oxidase Moco in its B-form as proposed by Johnson et al. [22]. But different origin, chemical substitutions and/or extraction procedure can result in varying fluorescence maxima.

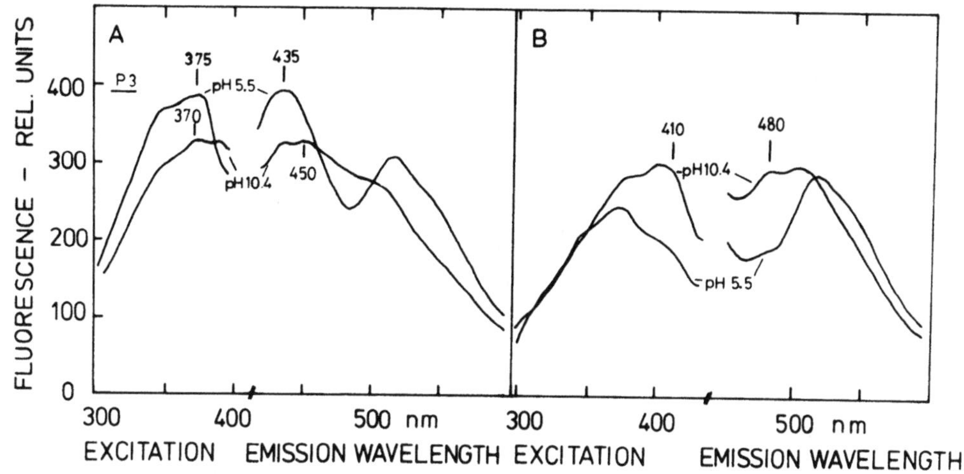

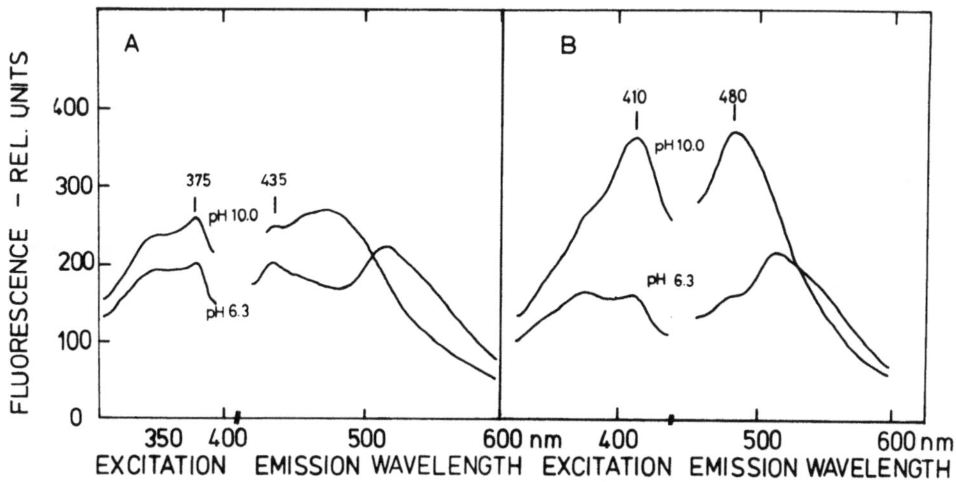

Figs. 6A,B and **7A,B**. Fluorescence excitation and emission spectra of "unspecific" (**A**) and "specific" (**B**) pterine signals in partially purified NR fraction P3 (Fig. 6) and NR chromatographed on blue sepharose (Fig. 7). Treatment of samples see Fig. 5

Fig. 6: 1.27 mg protein ml^{-1}, NR activity 980 nmol NO_2^- formed mg^{-1} × 10′

Fig. 7: 0.95 mg protein ml^{-1}, NR activity 6000 nmol NO_2 formed mg^{-1} × 10′

Spectra recorded at acidic pH show flavin content as indicated by emission maximum at 520 nm

Summary

I have emphasized the regulatory characteristics governing nitrate reductase and its related enzymes. Since we are mainly concerned with the transduction of blue light signals I concentrated on properties of the chromophoric redox groups flavin, cytochrome and pterine of NR. The chemical and photochemical reactivity of these groups in a system which has such a high molecular weight, which is dimeric and nearly as complex as a respiratory electron transport chain, open up the possibility of additional

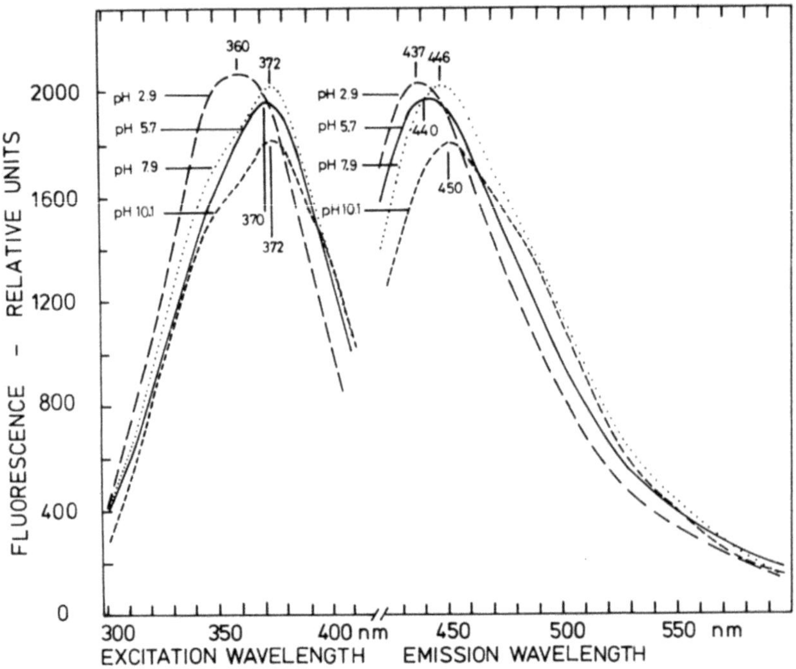

Fig. 8. Dependence of fluorescence excitation and emission maxima on the pH of "unspecific" signal. Supernatant of *Neurospora* homogenate (mycelium in log phase) centrifuged 20 min/20,000 *g*. Treatment see Fig. 5. 0.99 mg protein ml^{-1}, pH adjusted with NH_4OH. Fluorescence excitation and emission wavelengths as indicated

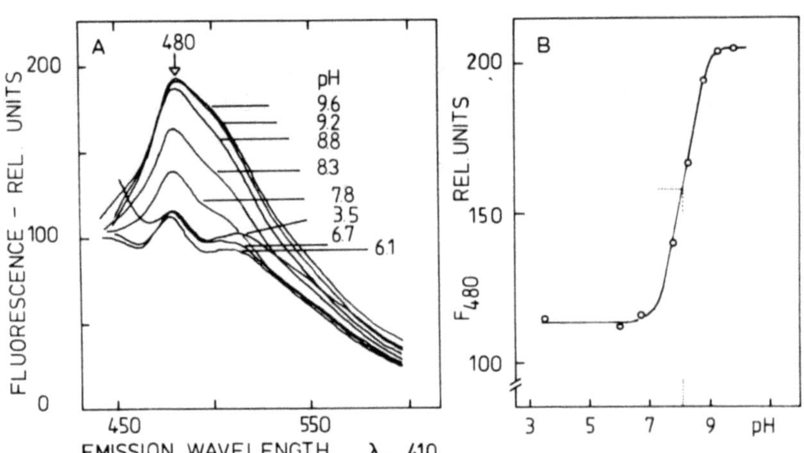

Fig. 9A,B. A Independence of emission maxima of the "specific" signal (λ_{ex} = 410 nm, λ_{em} = 480 nm) on the pH (pH 3.5 – pH 9.6). Sample treatment see Fig. 5. The emission maximum at 476 nm remaining at pH below 6.7 results from Raman scatter. **B** Fluorescence yield at various pH of the samples. Fluorescence data from **A** corrected for volume changes during titration with NH_4OH. NR sample purified by blue sepharose, 0.6 mg protein ml^{-1}, NR activity 3850 nmol NO_2^- formed mg^{-1} × 10′

functions of NR beside the two-electron transfer from NADPH to nitrate. This way mycelia might have the "option" to accommodate to special physiological conditions like growing under substrate shortage or starvation either by mobilizing additional substrates available only to excited flavin (like certain amino acids [39]) or by using the light signal for a conformational change, e.g., in order to reconstitute NADPH-NO_3^- or partial activities or to initiate a metabolic change which may alter vegetative growth into conidiation. Whether part or all of the photoreceptive electron transport chain could be used for transmitting a photosignal is still open to speculation.

Acknowledgments. The experimental work was supported by the Deutsche Forschungsgemeinschaft. I thank Frau Marianne Dussa for her competent and devoted technical assistance.

References

1. Aparicio PJ, Roldán JM, Calero F (1976) Biochem Biophys Res Commun 70:1071–1077
2. Aslam M, Oaks A (1975) Plant Physiol 56:634–639
3. Azuara MP, Aparicio PJ (1983) Plant Physiol 71:286–290
4. Beevers L, Hageman RH (1972) In: Giese A (ed) Photophysiology, vol VII. Academic Press, London New York, pp 85–113
5. Beevers L, Hageman RH (1980) In: Miflin BJ (ed) The biochemistry of plants, vol V. Academic Press, London New York, pp 115–158
6. Beevers L, Schrader LE, Flesher D, Hageman RH (1965) Plant Physiol 40:691–698
7. Brain RD, Freeberg JA, Weiss CV, Briggs WR (1977) Plant Physiol 59:948–952
8. Canvin DT, Atkins CA (1974) Planta 116:207–224
9. Cove DJ (1966) Biochim Biophys Acta 113:51–56
10. Cove DJ, Pateman JA (1969) J Bacteriol 97:1374–1378
11. Dalling MJ, Tolbert NE, Hageman RH (1972) Biochim Biophys Acta 283:505–512
12. Dantzig AH, Zurowski WK, Ball TM, Nason A (1978) J Bacteriol 133:671–679
13. Fritz B (1983) Master's thesis, Univ Tübingen
14. Garrett RH (1972) Biochim Biophys Acta 264:481–489
15. Garrett RH, Nason A (1967) Proc Natl Acad Sci USA 58:1603–1610
16. Garrett RH, Nason A (1969) J Biol Chem 244:2870–2882
17. Hageman RH, Flesher D (1960) Plant Physiol 35:700–708
18. Hahne G (1981) Master's thesis, Univ Tübingen
19. Hochberg ML, Sargent ML (1974) J Bacteriol 120:1164–1175
20. Hummelt G, Mora J (1980) Biochem Biophys Res Commun 96:1688–1694
21. Johnson CB (1976) Planta 128:127–131
22. Johnson JL, Rajagopalan KV (1982) Proc Natl Acad Sci USA 79:6856–6860
23. Johnson JL, Hainline BE, Rajagopalan KV (1980) J Biol Chem 255:1783–1786
24. Jones RW, Sheard RW (1972) Nature (London) New Biol 238:221–222
25. Kinsky SC (1961) J Bacteriol 82:898–904
26. Klemm E (1979) Doctoral thesis, Univ Tübingen
27. Klemm E, Ninnemann H (1976) Photochem Photobiol 24:369–371
28. Klemm E, Ninnemann H (1978) Photochem Photobiol 28:227–230
29. Klemm E, Ninnemann H (1979) Photochem Photobiol 29:629–632
30. Lorimer GH, Gewitz HS, Völkel W, Solomonson P, Vennesland B (1974) J Biol Chem 249: 6074–6079
31. Losada M, Paneque A, Aparicio PJ, Vega JM, Cardenas J, Herrera J (1970) Biochem Biophys Res Commun 38:1009–1015

32. Maldonado JM, Vargas MA, Mauriño SG, Aparicio PJ (1981) Biochim Biophys Acta 661: 112–119
33. Mauriño SG, Vargas MA, Aparicio PJ, Maldonado JM (1983) Physiol Plant 57:411–416
34. Moreno CG, Aparicio PJ, Palacian E, Losada M (1972) Febs Lett 26:11–14
35. Muñoz V, Brody S, Butler WL (1974) Biochem Biophys Res Commun 58:322–327
36. Nicholas DJD, Nason A, McElroy WD (1954) J Biol Chem 207:341–351
37. Ninnemann H (1981) Microsensory Newslett 3 no. 2
38. Ninnemann H (1982) Photochem Photobiol 35:391–398
38a. Ninnemann H (1983) In: Montagnoli G, Erlanger BF (eds) Molecular models of photoresponsiveness. Plenum Press, New York London, pp 133–152
39. Ninnemann H, Klemm-Wolfgramm E (1980) In: Senger H (ed) The blue light syndrome. Springer, Berlin Heidelberg New York, pp 238–243
40. Pan S-S, Nason A (1978) Biochim Biophys Acta 523:297–313
41. Pistorius EK, Jetschmann K, Voss H, Vennesland B (1979) Biochim Biophys Acta 585:630–642
42. Poff KL, Butler WL, Loomis WF Jr (1973) Proc Natl Acad Sci USA 70:813–816
43. Purvis AC, Tischler CR, Fites RC (1976) Plant Physiol 58:95–99
44. Relimpio AM, Aparicio PJ, Paneque A, Losada M (1971) Febs Lett 17:226–230
45. Roldán JM, Butler WL (1980) Photochem Photobiol 32:375–381
46. Roldán JM, Verbelen JP, Butler WL, Tokuyasu K (1982) Plant Physiol 70:872–874
47. Sargent ML, Kaltenborn SH (1972) Plant Physiol 50:171–175
48. Stoy V (1955) Physiol Plant 8:963–986
49. Stoy V (1956) Biochim Biophys Acta 21:395–396
50. Subramanian KN, Sorger GJ (1972) J Bacteriol 110:538–546
51. Subramanian KN, Padmanaban G, Sarma PS (1968) Biochim Biophys Acta 151:20–32
52. Tischner R, Hüttermann A (1978) Plant Physiol 62:284–286
53. Tischner R, Hüttermann A (1980) Plant Physiol 66:805–808
54. Tischner R, Lorenzen H (1979) Planta 146:287–292
55. Travis RL, Key JL (1971) Plant Physiol 48:617–620
56. Vargas MA, Mauriño SG, Maldonado JM, Aparicio PJ (1982) Photochem Photobiol 36:223–228
57. Vega JM, Herrera J, Relimpio AM, Aparicio PJ (1972) Physiol Vég 10:637–651
58. Vega JM, Garrett RH, Siegel LM (1975) J Biol Chem 250:7980–7989
59. Vennesland B, Guerrero MG (1979) In: Gibbs M, Latzko E (eds) Encyclopedia of plant physiology, New Ser, vol VI. Springer, Berlin Heidelberg New York, pp 425–444
60. Wallace W (1975) Biochim Biophys Acta 377:239–250
61. Weiss B, Turian G (1966) J Gen Microbiol 44:407–418
62. Widell S, Björn LO (1976) Physiol Plant 36:305–309

Re-Examination of Action Spectroscopy in Blue/Near-UV Light Effects

Y. Inoue [1]

Introduction

The nature of the primary photoreceptor(s) of blue light-induced reaction in plants and microorganisms is still an open question [23]. Flavoprotein(s) and carotenoprotein(s) are considered to be likely candidates of the photoreceptor(s). Precise examination of an action spectrum for blue light-induced reaction in the UV region should give some clue to elucidate whether the photoreceptor(s) of the blue light-induced reaction is flavoprotein(s) or carotenoprotein(s). As described below, absorption spectra of flavin and carotenoid show such clear differences in the UV region, and few action spectra determined in the UV region take into consideration the harmful effects of UV radiation and masking effect of samples in the UV region. This paper describes newly determined action spectra for blue light-induced reactions obtained in the UV region, and discusses some factors which distort the action spectrum in the UV region.

Absorption Spectrum Versus Action Spectrum

The action spectrum for blue light-induced reaction typified in phototropism of *Phycomyces* sporangiophore (Fig. 2,1) showes complex multiple peaks in the blue region and one broad peak in the near-UV region. Assignment of these peaks to the absorption spectra peaks of flavin or carotenoid was attempted [21]. The presence of a band in the near-UV region (maximum at about 370 nm) supported a flavin photoreceptor (see Fig. 1,1), the presence of multiple peaks in the blue region supported rather a carotenoid photoreceptor (see Fig. 1,3). But these arguments became meaningless, because cis isomer of carotenoid exhibited an absorption in the near-UV region (Fig. 1,4) and flavin in ethanol showed multiple peaks in the blue region (Fig. 1,2).

In several biological systems the action spectra for blue light-induced reaction in the shorter wavelength UV region have been determined (Fig. 2). The action spectra for phototropism (Fig. 2,1 [3]) and light growth response (Fig. 2,2 [5]) in *Phycomyces* sporangiophore, for induction of conidia formation in *Trichoderma* (Fig. 2,3 [16]) and for chloroplast rearrangement in *Funaria* (Fig. 2,4 [34]) show a discernible band

1 Department of Biology, Faculty of Science, University of Tokyo, Hongo, Tokyo 113, Japan

Blue Light Effects in Biological Systems
ed. by H. Senger

Fig. 1. Comparison of absorption spectra of flavin with that of carotenoid: *1* riboflavin in water [29]; *2* riboflavin in ethanol at 77 K [27]; *3* β-carotene in hexane [28]; *4* 15, 15′-cis-β-carotene in hexane [28]

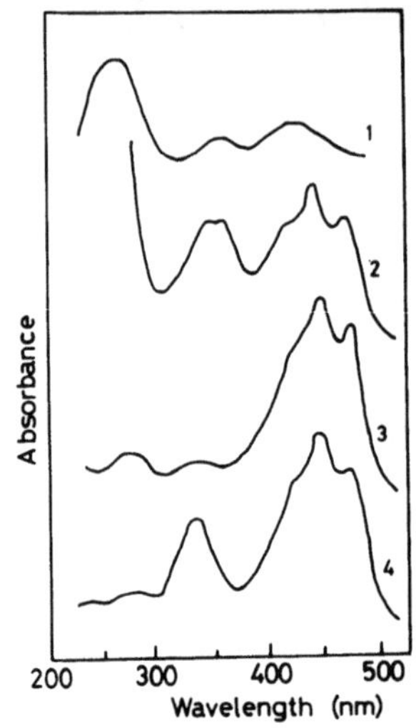

in the shorter wavelength UV region. The action spectrum for photoinduction of carotenoid synthesis in *Neurospora* (Fig. 2,5 [4]) shows only a very small action band in the shorter wavelength UV region. The action spectrum band in the former cases supports a flavin photoreceptor, and in the latter case the action spectrum suggests rather a carotenoid photoreceptor. Although the action spectra in the former cases have a band in the shorter wavelength UV region, the height of the UV band is still lower and the peak wavelength position of the UV band is longer than the UV absorption band of flavin except in the case of chloroplast rearrangement in *Funaria*.

It is well-known fact that UV radiation has some harmful effects on organisms [15]. Therefore, it is possible to suppose that the action spectrum in the shorter wavelength UV region is distorted to low by the harmful effects of UV radiation in some degrees. Indeed the harmful effects of UV radiation have rarely been considered in the previous action spectroscopy shown in Fig. 2. Thus, re-examination of action spectra in the shorter wavelength UV region is needed.

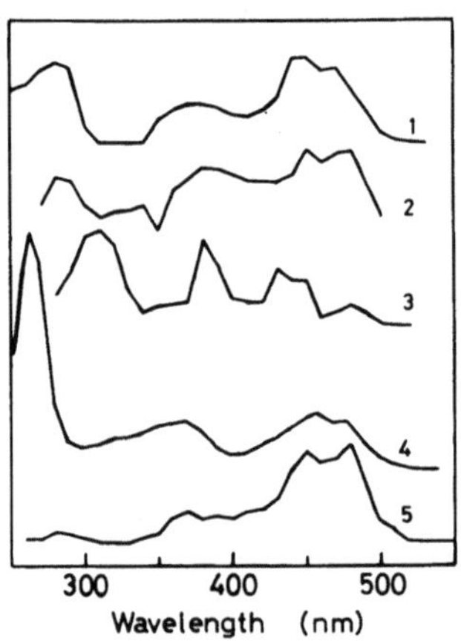

Fig. 2. Action spectra for blue light-induced reactions determined in the UV region: *1* phototropism in *Phycomyces* sporangiophore [3]; *2* light growth response in *Phycomyces* sporangiophore [5]; *3* photoinduction of conidial formation in *Trichoderma* [16]; *4* chloroplast rearrangement in *Funaria* [34]; *5* photoinduction of carotenoid biosynthesis in *Neurospora* [4]

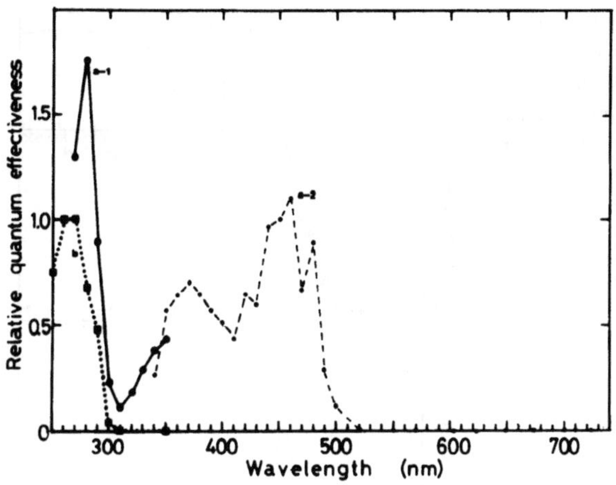

Fig. 3. Action spectra for photoinduction and photoinhibition of perithecial formation in *Gelasinospora reticulispora: a-1* photoinduction determined in the UV region [14]; *a-2* photoinduction determined in the near-UV and visible region [12]; *b* photoinhibition of blue light-induced perithecial formation [14]

Re-Examination of Action Spectra in the Shorter Wavelength UV Region

A new spectrograph for photobiological research has recently been constructed at the Natl. Inst. Basic Biol., Okazaki, Japan, and intense and purer monochromatic UV radiation ranging from 250 nm having fluence rate higher than 3×10^{15} photon s^{-1} cm^{-2} has become available [30]. Recently, action spectra for blue light-induced reaction were determined in the UV region on two different biological systems using this spectrograph.

1. Effect of UV Radiation on Perithecial Formation in Fungus Gelasinospora reticulispora [14]

Perithecial formation of *G. reticulispora* was fully light-dependent during linear mycelia growth [10]. Morphological differentiation of perithecium was induced by blue light given after an inductive dark period [11, 13] and an action spectrum for the induction of perithecia indicated that the "P_{NUV-B}" was responsible for photoperception (Fig. 3a-2 [12]). An action spectrum in the UV region has not yet been determined.

Fluence-response curve of UV radiation on perithecial formation suggests dual effects of UV radiation. Namely, one has an inductive and the other an inhibitory effect. Since, UV (at 280 nm) irradiation induces perithecia at low fluence, but UV irradiation exceeding the saturation fluence level (ca. 30 Jm^{-2}) reduces the number of induced perithecia.

Inhibitory effect of radiation of 260 and 270 nm which was irradiated just after inductive light irradiation became obvious at less than 10 Jm^{-2} and saturated at about 100 Jm^{-2}. An action spectrum for the photoinhibition of blue light-induced perithecial formation exhibited a peak between 260 and 270 nm and showed little effectiveness in the wavelength region longer than 310 nm (Fig. 3b).

Inductive effect of the shorter wavelength UV radiation given without inductive blue light irradiation became obvious at less than 10 Jm^{-2}. The slopes of the fluence-response curves of UV radiation having the wavelength longer than 260 nm were nearly parallel to that of blue light and the effect was saturated at about 30 Jm^{-2}. The slopes

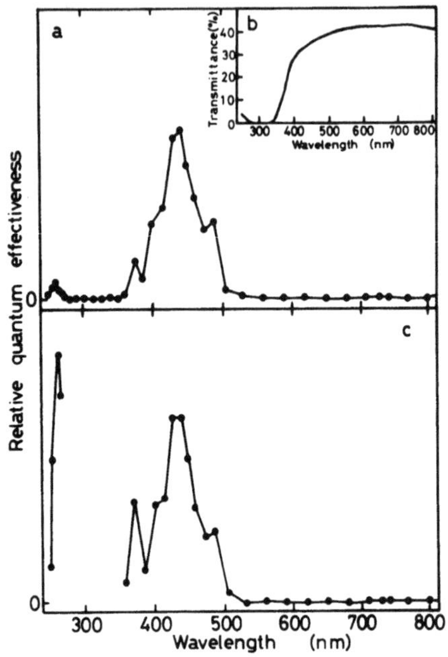

Fig. 4a–c. Action spectra for photoinhibition of red light-induced spore germination in *Pteris vittata:* **a** raw action spectrum; **b** transmission spectrum of a spore coat determined by microbeam spectrophotometer; **c** corrected action spectrum based on the transmission spectrum [26]

of the fluence-response curves at 250 and 260 nm were very shallow and the number of induced perithecia obtained saturated at a low level. These shallow slopes of fluence-response curves at 250 and 260 nm were probably brought about by the inhibitory effect of these radiations. An action spectrum in the UV range for the photoinduction of perithecial formation (Fig. 3a-1) indicated a sharp band with a peak at 280 nm and the lowest activity at 310 nm. Quantum effectiveness of 280 nm was ca. 1.7 times higher than that of 450 nm (cf. Fig. 3a-1 and 3a-2).

Comparison of the fluence-response curves for photoinduction and photoinhibition of UV radiation showed that inhibitory effects were observed in the fluence range where photoinductive effects became obvious. This observation suggests (a) that the height of the UV peak in the action spectrum for photoinduction of perithecia, when free from distortion by the inhibitory effect, should be higher than the peak showen in Fig. 3a-1 and, (b) that the photoreceptor pigment for photoinduction of perithecia in *G. reticulispora* is probably flavin.

2. Photoinhibition of Phytochrome-Controlled Germination of Spores in Pteris vittata [26]

Spore germination of *P. vittata* was induced by red light. Phytochrome mediated the photoperception of red light. This red light-induced spore germination was inhibited by brief blue light irradiation [25]. The action spectrum for this blue light effect suggested participation of "$P_{NUV\text{-}B}$" in the photoreaction [24]. An action spectrum in the UV region also has not yet been determined.

The newly determined raw action spectrum for photoinhibition of red light-induced spore germination showed a very low peak at 260 nm (Fig. 4a). Spore of *P. vittata* has a dark brown spore coat. It was reasonable to suppose that the raw action spectrum was distorted by the filter effect of the spore coat, since the photoperceptive site should be inside the spore coat. Transmission spectrum of the spore coat determined by a microbeam spectrophotometer showed very low transmittance value in the UV region (Fig. 4b). A corrected action spectrum based upon the transmission spectrum showed a sharp band having a peak at 260 nm, and quantum effectiveness at 260 nm was about 1.4 times higher than that at 450 nm (Fig. 4c).

This corrected action spectrum also supports a flavin photoreceptor.

General Discussion

The results obtained in *Gelasinospora* and *Pteris* may be taken as additional evidence in support of a flavin photoreceptor, and suggest that the action spectrum for blue light-induced reaction determined in the UV region is distorted to some extent by harmful effect of UV radiation and masking or filter effect of samples in the UV region.

The tendency, observed on the previous action spectra in Fig. 2, that the relative height of the action spectrum peak in the UV region was lower and the position of the peak longer than observed for the absorption spectrum of flavin in water [29] was probably also brought about by the same causes.

Theoretically, an action spectrum must be compared with a spectrum of the rate of light absorption expressed in terms of A, the fraction absorbed [2].

$$A = \frac{I_0 - I_x}{I_0}$$

where I_0 is the light intensity entered and I_x is the light intensity emerged. When the optical density ($OD = -\log\frac{I_x}{I_0}$) is low, A is practically in linear proportion to OD. But the relationship becomes nonlinear when OD exceeds to ca. 0.1 and finally A converges to 1. This fact means that the ratio of A_{265} to A_{450} of flavin depends highly on the concentration. This ratio of riboflavin calculated from the molecular extinction coefficients determined by Whitby [31] is 2.5, 2.2 or 1.1 when the concentration of riboflavin is 0.01, 0.1 or 1 in terms of OD at 450 nm, respectively. Therefore, the relative low peak of action spectrum in Fig. 2 seems to be explained by the difference of OD and A. But the concentration of flavin at 1 in terms of OD at 450 nm is abnormally high in organisms [8], and this supposition seems to be impossible.

Fluence of UV radiation which is required to kill about 50% of fungal cells is about several tens Jm^{-2} (Table 1, right column). This fluence is a little higher than the fluence of blue light which is required to induce about a half-maximum reaction of the blue light effect (Table 1, left column). It must, however, be kept in mind that killing effects were mainly determined on the dormant spores, and a dormant spore is more resistant to UV radiation than a growing cell [33]. Therefore, the data in Table 1 does not assure that the harmful effects of UV radiation can be neglected in action spectroscopy in the UV range, but rather suggests that in many biological systems the

Table 1. Fluence of blue light and UV radiation to induce the half-maximum blue light effects and injurious UV effects on fungi

Blue light reaction				UV radiation effect			
Organisms	Phenomena	Fluence (Jm^{-2})	Reference	Organisms	Phenomena	Fluence (Jm^{-2})	Reference
Penicillium	Induction of conidia	0.035	[1]	*Gelasinospora*	Inhibition of perithecia	20	[14]
Phycomyces	Induction of sporangiophore	0.4	[7]	*Trichophyton*	Killing of spore	26	[9]
Stemphyllium	Inhibition of conidiation	1.3	[17]	*Saccharomyces*	Killing of thallus	50 56	[19] [32]
Trichoderma	Induction of conidia	4.5	[16]	*Neurospora*	Killing of spore	76	[18]
Neurospora	Induction of carotenoid	10	[4]	*Aspergillus*	Killing of spore	130	[6]
Gelasinospora	Induction of perithecia	20	[12]	*Rhizopus*	Killing of spore	790	[6]
Penicillium	Induction of colemia	47	[22]	*Mucor*	Killing of spore	4200	[6]
Schizophyllum	Induction of basidiocarps	85	[20]				

sensitivity of cells to inductive blue light and harmful UV radiation are very close. In Table 1 the fluence values which were required for the phototropism or light growth response in *Phycomyces* sporangiophore and chloroplast rearrangement in *Funaria* in Fig. 2 unfortunately cannot be calculated, because these reactions are determined as fluence-rate-dependent phenomena.

In some systems, for example, conidia induction in *Penicillium* [1] and sporangiophore induction in *Phycomyces* [7], the effect of blue light can be observed at very low fluence level (Table 1). Precise action spectroscopy in the UV region for the blue light-induced reaction on these relatively photosensitive systems is desired to determine the nature of primary photoreceptor pigment(s) for blue light-induced reaction, as in these systems, distortion of action spectrum caused by the harmful effect of UV radiation should be lower than the other relatively light-insensitive systems.

Acknowledgments. The author would like to thank Dr. M. Watanabe and Mr. M. Kubota of Large Spectrograph Laboratory and Miss M. Hayashi of Division of Biological Reguration at Natl. Inst. Basic Biol., Okazaki, for excellent technical assistance. This study was supported in part by Grants (No. 81A10 and 82A08) for Research at Okazaki Large Spectrograph, Natl. Inst. Basic Biol., Okazaki.

References

1. Bennink GJH (1972) Photomorphogenesis in *Penicillium isariiforme.* I. The action spectrum for light-induced sporulation. Acta Bot Neerl 21:528–534
2. Clayton RK (1970) Light and living matter, vol I. The physical part. McGraw-Hill, New York
3. Curry G, Gruen HE (1959) Action spectra for the positive and negative phototropism of *Phycomyces* sporangiophores. Proc Natl Acad Sci USA 45:797–804
4. De Fabo EC, Harding RW, Shropshire W Jr (1976) Action spectrum between 260 and 800 nanometers for the photoinduction of carotenoid biosynthesis in *Neurospora crassa.* Plant Physiol 57:440–445
5. Delbrück M, Shropshire W Jr (1960) Action and transmission spectra of *Phycomyces.* Plant Physiol 35:194–204
6. Dimond AE, Duggar BM (1941) Some lethal effects of ultraviolet radiation on fungus spores. Proc Natl Acad Sci USA 25:459–468
7. Galland P, Russo VEA (1979) Photoinitiation of sporangiophores in *Phycomyces* mutants deficient in phototropism and in mutants lacking β-carotene. Photochem Photobiol 29:1009–1014
8. Goodwin TW (1963) The biosynthesis of vitamins and related compounds. Academic Press, London New York
9. Hollaender A, Emmons CW (1941) Wavelength dependence of mutation production in the ultraviolet with special emphasis on fungi. Cold Spring Harbor Symp Quant Biol 9:179–186
10. Inoue Y, Furuya M (1970) Perithecial formation in *Gelasinospora reticulispora* I. Effect of light at two different growth states. Dev Growth Differ 12:141–150
11. Inoue Y, Furuya M (1974) Perithecial formation in *Gelasinospora reticulispora* II. Promotive effects of near-ultraviolet and blue light after dark incubation. Plant Cell Physiol 15:195–204
12. Inoue Y, Furuya M (1975) Perithecial formation in *Gelasinospora reticulispora* IV. Action spectra for the photoinduction. Plant Physiol 55:1098–1101
13. Inoue Y, Furuya M (1975) Perithecial formation in *Gelasinospora reticulispora* V. Microscopically versus photobiologically recognizable differentiation of hyphae. Bot Mag Tokyo 88: 31–40

14. Inoue Y, Watanabe M (1984) Perithecial formation in *Gelasinospora reticulispora* VII. Action spectra in the UV region for the photoinduction and photoinhibition of photoinductive effect brought by blue light. Plant Cell Physiol 25:107–113
15. Jagger J (1967) Introduction to research in ultraviolet photobiology. Prentice-Hall, New Jersey
16. Kumagai T, Oda Y (1969) Action spectrum for photoinduced sporulation in the fungus *Trichoderma viride*. Plant Cell Physiol 10:387–392
17. Leach CM (1968) An action spectrum for light inhibition of the "terminal phase" of photosporogenesis in the fungus *Stemphylium botryosum*. Mycologia 60:532–546
18. Norman A (1951) Inactivation of *Neurospora* conidia by ultraviolet radiation. Exp Cell Res 2: 454–473
19. Oster RH (1934) Results of irradiating *Saccharomyces* with monochromatic ultra-violet light. I. Morphological and respiratory changes. J Gen Physiol 18:71–88
20. Perkins JH, Gordon SA (1969) Morphogenesis in *Schizophyllum commune*. II. Effects of monochromatic light. Plant Physiol 44:1712–1716
21. Presti D, Delbrück M (1978) Photoreceptors for biosynthesis, energy storage and vision. Plant Cell Environ 1:81–100
22. Salman AGF (1971) Das Wirkungsspektrum der lichtabhängigen Zonierung der Koremien von zwei Mutanten von *Penicillium claviforme* Bainier. Planta 101:117–121
23. Senger H (1982) The effect of blue light on plant and microorganisms. Photochem Photobiol 35:911–920
24. Sugai M (1971) Photomorphogenesis in *Pteris vittata* IV. Action spectra for inhibition of phytochrome-dependent spore germination. Plant Cell Physiol 12:103–109
25. Sugai M, Furuya M (1967) Photomorphogenesis in *Pteris vittata* I. Phytochrome-mediated spore germination and blue light interaction. Plant Cell Physiol 8:737–748
26. Sugai M, Tomizawa K, Watanabe M, Furuya M (1984) An action spectrum between 250 and 800 nm for the inhibition of phytochrome-controlled germination of spores in *Pteris vittata*. Plant Cell Physiol 25:205–212
27. Sun M, Moore TA, Song P-S (1972) Molecular luminescence studies of flavins. I. The excited states of flavins. J Am Chem Soc 94:1730–1740
28. Vetter W, Englert G, Rigassi N, Schwieter U (1971) Spectroscopic methods. In: Isler O (ed) Carotenoids. Birkhäuser, Basel, pp 189–266
29. Warburg O, Christian W (1938) Isolierung der prosthetischen Gruppen der d-Aminosäureoxydase. Biochem Z 298:150–168
30. Watanabe M, Furuya M, Miyoshi Y, Inoue Y, Iwahashi I, Matsumoto K (1982) Design and performance of the Okazaki large spectrograph for photobiological research. Photochem Photobiol 36:491–498
31. Whitby LG (1953) A new method for preparing flavin-adenine dinucleotide. Biochem J 54: 437–442
32. Wyckoff RWG, Luyet BJ (1931) The effects of X-rays, cathode, and ultraviolet rays on yeast. Radiology 17:1171–1175
33. Zelle MR, Hollaender A (1955) Effects of radiation on bacteria. In: Hollaender A (ed) Radiation biology, vol II. McGraw-Hill, New York, pp 365–430
34. Zurzycki J (1972) Primary reaction in the chloroplast rearrangements. Acta Protozool 11: 198–199

Photogeotropism of Phycomyces: Evidence for More Than One Photoreceptor

G. LÖSER and E. SCHÄFER [1]

Introduction

Recently action spectra for photogeotropism (PGT) of wild-type [7] mutant [3] sporangiophores of *Phycomyces* bl. were reported. In contrast to Delbrück et al. [1], an observed unexpected responsivity to orange-red light in these papers was interpreted as contribution of an unknown pigment and not as a direct triplet excitation. This is in agreement with the observation that 605 nm light given simultaneously with 450 nm light, can reduce the effect of blue light with respect to PGT [6].

In this paper we describe fluence-rate response curves for PGT of the sporangiophore of wild-type *Phycomyces* using monochromatic and pulsed dichromatic irradiation. Especially the time course of interaction between 450 and 605 nm light is investigated.

Materials and Methods

The experiments were carried out using wild-type *Phycomyces* NRRL 1555(–).

A modified threshold box (Fig. 1) as described in detail by Löser and Schäfer [6] was used, the light incident on the sporangiophores inclined by 20° from the horizontal.

For intermittent light treatment mechanical shutters and a timer as described by Heim and Schäfer [4] were used additionally. The PGT equilibrium angle was measured after 6 h. For more details see figure legends.

Results

Fluence-Rate Response Curves

Monochromatic fluence-rate response curves for wavelengths between 450 and 641 nm are shown in Fig. 2. The major observations concerning the fluence-rate response curves are the following:

1 Biologisches Institut II, Universität Freiburg, Schänzlestraße 1, D–7800 Freiburg, FRG

Blue Light Effects in Biological Systems
ed. by H. Senger

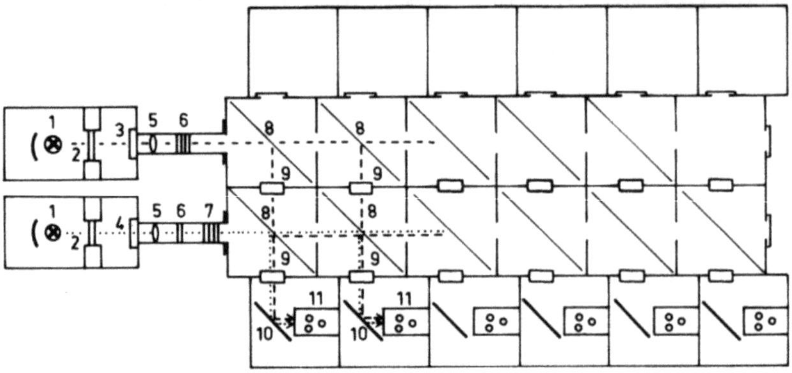

Fig. 1. Scheme of experimental arrangement (top view). *1* light sourse; *2* heat filter glass; *3, 4* interference filter; *5* lenses; *6, 9* neutral density glasses; *7* cut-off-glasses; *8* 50% transmission mirrors; *10* surface mirrors; *11* adjustable tables with specimens in glass vials

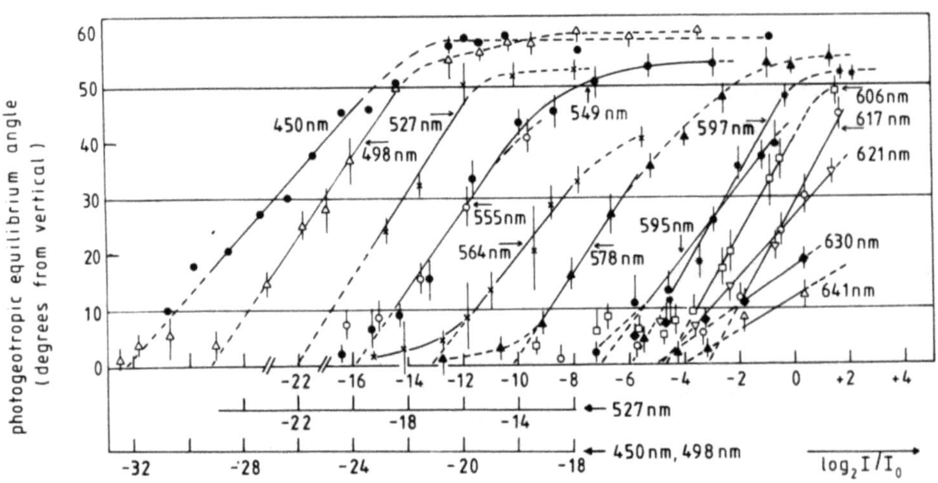

Fig. 2. Fluence-rate response curves for photogeotropism of the *Phycomyces* sporangiophores. The light incident on the sporangiophores is inclined by 20° down from the horizontal. Standard error of the mean indicated. Thresholds extrapolated from the intermediate part of the ascending fluence-rate response curves. $I_0 = 1\ Wm^{-2}$

a) a significant PGT can also be elicited using orange-red light (about 10^7 to 10^8 higher fluence rates than those for blue light are mandatory),
b) the slope of the fluence-rate response curves and
c) their apparent saturation behavior are wavelength-dependent. More data for 605 nm light are given by Löser and Schäfer [6].

Action Spectra

Up to about 570 nm the action spectra for the wild-type PGT equilibrium angle method (Fig. 3) can be approximately described by a gaussian approximation as done by

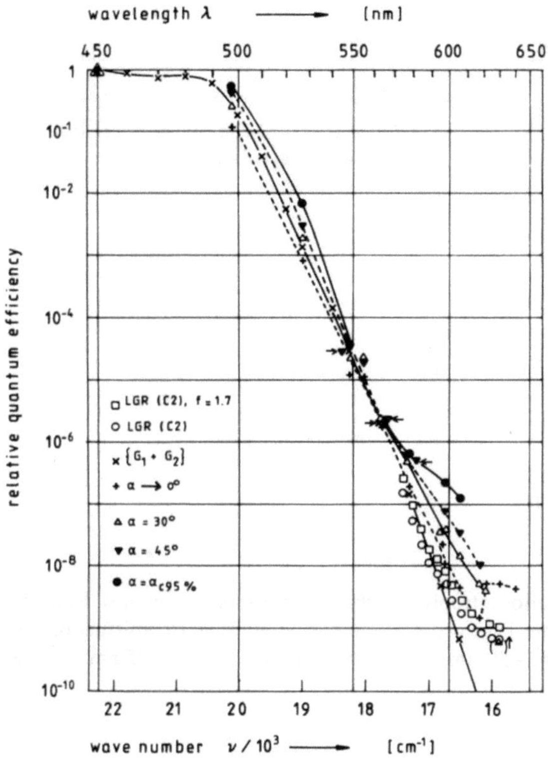

Fig. 3. Photogeotropic action spectra of *Phycomyces* sporangiophores for four different response levels as indicated (data from Fig. 2). Included for comparison: □ light growth response, mutant C2, corrected by factor 1.7, ○ uncorrected. (After [1]); x phototropic balance method, mutant C2, extrapolated by a sum of two Gaussians for $\lambda > 550$ nm ([2], after [1]). ⇄ point displaced because of graphic reasons; ↑ point represents lower limit; ( ) point at the limit of detectability

Foster [2] for the phototropic null method action spectrum of the albino mutant C2 and described by Delbrück et al. [1]. An additional unexpected action band could be observed between 580 and 640 nm. Because of the wavelength dependence of the slope of the fluence-rate response curves the shape of the action spectrum depends on the response level chosen, especially for $\lambda > 580$ nm (Fig. 3).

The action spectrum determined from photogeotropic fluence-rate response curves extrapolated to their thresholds may be considered as approximation of a phototropic action spectrum. It is shown to include a minimum near 615 nm and the appearance of a small band around 630 nm, which is not of riboflavin triplet type (cf. [1]).

Dichromatic Irradiation

To test the time course of a possible interaction between 450 and 605 nm light, intermittent irradiation conditions were chosen. Using 30 s dark/30 s light cycles a similar fluence-rate dependence of the inhibitory effect of 605 nm to PGT produced by light-dark cycles of 450 nm light is obtained as shown for continuous irradiation ([6] and Fig. 4).

For detailed analysis 2 s light/28 s dark cycles were applied using various time intervals between the start of the blue (450 nm) and the orange-red (605 nm) light pulses. Inhibitory effects were observed if the 605 nm light pulses were given 2–10 s

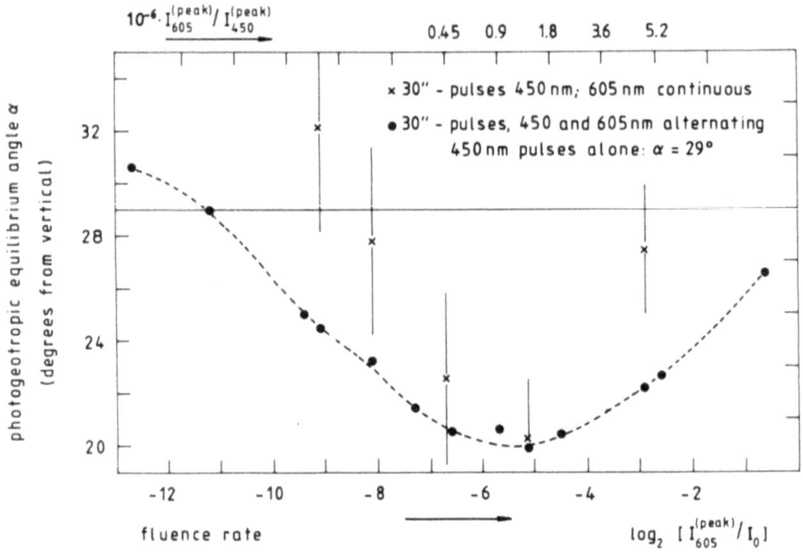

Fig. 4. Fluence-rate response curves for photogeotropism of *Phycomyces* sporangiophores. Response induced by continuous or pulsed 605 nm light and 450 nm background light consisting of 30 s light/30 s dark cycles. The fluence rate of 450 nm light is kept constant, that of 605 nm varied as indicated

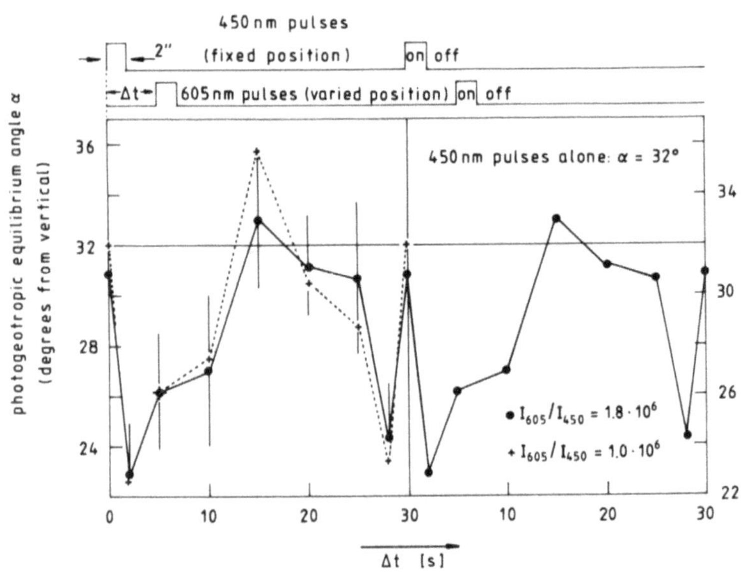

Fig. 5. Photogeotropic equilibrium angle of *Phycomyces* sporangiophores after 6 h of cyclic light pulses (2 s light/28 s dark) shown as a function of the time interval Δ t between the onset of the 450 nm and that of the 605 nm light pulses within one cycle.

The fluence rate of the blue light pulses was $I_{450} = 5.3 \cdot 10^{-4}$ nmolm^{-2} s^{-1}).

The fluence-rate ratios used were ——— ● $I_{605}/I_{450} = 1.8 \times 10^6$, - - - + $I_{605}/I_{450} = 1.1 \times 10^6$

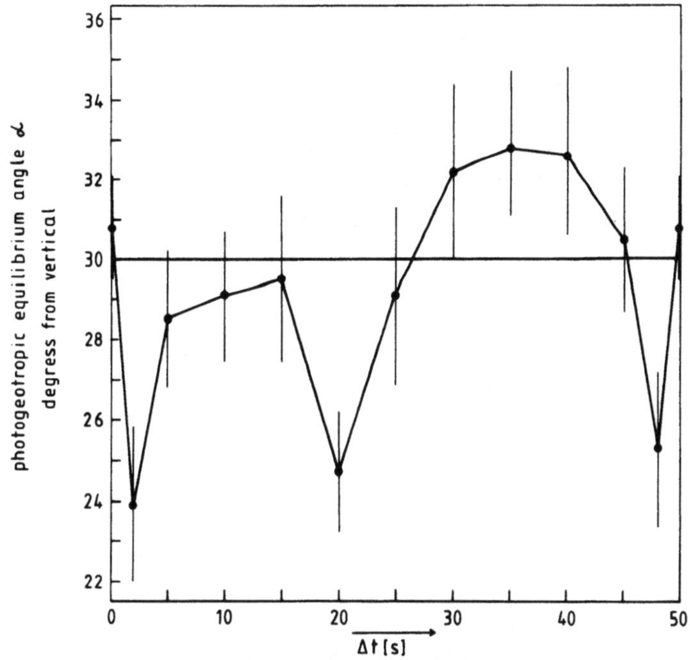

Fig. 6. As for Fig. 5, but using 2 s light/48 s dark cycles. $I_{605}/I_{450} = 1.2 \times 10^6$. Mean of three experimental series; standard error of the mean indicated. Each data point in the figure corresponds to 16 sporangiophores at least. Blue pulses alone: $\alpha = 30°$ as reference

after or immediately, i.e., 2 s before the 450 nm light pulses (Fig. 5). For 2 s light/48 s dark cycles inhibitory effects were found if the 605 nm light pulses were applied 2 and 20 s after and 2 s before the 450 nm light pulses (Fig. 6).

We also found indications of the inhibitory effect of 605 nm light for 5 s light/95 s and 45 s dark cycles; here response minima were obtained for 605 nm pulses applied 10 s after and 5 s before the onset of the blue pulses (data not shown).

It should be emphasized that for all the light cycles used (cf. Figs. 5, 6) no inhibitory effect of 605 nm pulses could be detected for simultaneous light pulses of 2 to 5 s duration using light/dark cycles; however, it was obtained for alternating 30 s light pulses or continuous 605 nm irradiation combined with 30 s pulses nm light (Fig. 4) as well as for continuous irradiation of both 450 nm and 605 nm light near thresholds, the latter as shown earlier [6].

Discussion

In agreement with the findings of Galland [3] and our previous suggestions [6], the fluence-rate response curves for PGT of the wild-type *Phycomyces* indicate that this so-called blue light response cannot be described by a single photoreceptor with only one chromophoric absorption. Wavelength-dependent saturation behavior and slopes of the fluence-rate response curves point to separate photoreceptors in agreement with the mutant analysis performed by Galland [3].

This conclusion is supported by our observations (action spectrum) that orange-red light given alone is more effective than expected for a single photoreceptor as described above, and on the other hand also supported by our previous results [6] that 605 nm light, if given simultaneously and continuously together with blue light, reduces the photogeotropic responsiveness to the background 450 nm light at fluence-rates near the threshold for PGT of both wavelengths used.

The present results for alternating light-dark cycles according to Fig. 4 still fit into the category of our dichromatic continuous light experiments, as there is a similar antagonistic effect of 605 nm light corresponding to a similar type of minimum of the dichromatic fluence-rate response curves situated at the same fluence ratio I_{605}/I_{450} about $1 \cdot 10^6$ to $2 \cdot 10^6$. This ratio is two orders of magnitude smaller than that of the ratio of quantum efficiencies inspected from the PGT-action spectrum for the same equilibrium angle (Fig. 3).

However, surprisingly, if the continuous dichromatic irradiation is substituted for by 2 s light/28 s dark cycles or by the other light cycle programs of this type mentioned in the results section, 605 nm light will not reduce the responsiveness to 450 nm light if given simultaneously. Another common feature of the dichromatic experiments of the latter type (cf. Figs. 5 and 6) is the antagonistic action of 605 nm light pulses given a few seconds after and a few seconds before the onset of the 450 nm light pulses.

A third type of minimum which appears at a delay of 20 s (Fig. 6) cannot be explained easily at present except by the assumption that an additional weakly absorbing band or long wavelength tail of an additional photoreceptor acts as or on an intermediate part of the transduction chain. The 20 s minimum might be hidden in Fig. 5 because of the shortness of the cycle used compared to some time constant of this unknown substance.

For the early and late minimum at 2 s after and 2 s before onset of the blue pulses, we suggest the following mechanisms as possible explanations of the antagonistic action of an orange-red light absorbing photoreceptor (or photoreceptors).

a) It is a transduction product formed by a 450 nm light pulse and possesses a limited "lifetime" of at most a few seconds thereafter. This would explain the first mininum.

b) It interacts with an "unstable" transduction product produced by the 450 nm light pulse or (cf. c) interacts with a preexisting element of the transduction chain. The former possibility again would explain the first minimum.

c) It is a preexisting element of the transduction chain, that if illuminated by 605 nm pulses at proper time intervals before or after the blue light pulses, acts antagonistically. Possible interpretations could be a flavin redox system [5] or a flavin-cytochrome redox system with the properties that the transduction element regarded returns to the state undisturbed by orange-red light within less than 1 s (late minimum) and does not significantly transduce before about 2 s after the onset of the blue pulse (Fig. 6).

d) It acts directly or indirectly antagonistically on the blue light absorbing pigment or system itself, thus possibly explaining the last minimum.

e) It forms, together with the blue light absorbing system, a photoreceptor system that interacts in a more complicated manner than (a) to (d), possibly comprising branched pathways and/or a complicated blue light photoreceptor itself. For the latter evidence is given by Galland [3].

f) Furthermore, the minimum appearing 2 s before onset of the blue pulses bould be described as an adaptational effect. However, our preliminary evidence shows that this minimum could not be produced using equivalent 450 nm pulses instead of 605 nm pulses.

At present, we prefer explanations (a) to (d). Case (c) could be of help with the paradox that orange-red light can elicit positive PGT action and act antagonistically. We would not attribute special ecological significance to the orange-red light effects because of the large difference between blue and orange-red light with regard to relative quantum efficiencies respectively dichromatic experiment fluence-rate ratios. However, we do not exclude that the time constants available from Figs. 5 and 6 may help in elucidating elements of the transduction chain as well as the nature of the photoreceptor(s) involved.

References

1. Delbrück M, Katzir A, Presti D (1976) Responses of Phycomyces indicating optical excitation of the lowest triplet state of riboflavin. Proc Natl Acad Sci USA 73:1969–1973
2. Foster KW (1972) The photoresponses of Phycomyces: analysis using manual techniques and an automated machine which precisely tracks and measures growth during programmed stimuli. Ph.D. thesis, Calif Inst Technol, Pasadena
3. Galland P (1983) Action spectra of photogeotropic equilibrium in Phycomyces wildtype and three behavioral mutants. Photochem Photobiol 37:221–228
4. Heim B, Schäfer E (1984) Hypocotyl growth in dark-grown Sinapis alba L.: the effect of red and far-red light in the high irradiance reaction of phytochrome. Plant Cell Environ 7:39–44
5. Hertel R (1980) Phototropism of lower plants. In: Lenci F, Colombetti G (eds) Photoreception and sensory transduction in aneural organisms. Plenum Press, New York London, pp 89–105
6. Löser G, Schäfer E (1980) Phototropism in Phycomyces: a photochromic sensor pigment? In: Senger H (ed) The blue light syndrome. Springer, Berlin Heidelberg New York, pp 244–250
7. Schäfer E, Heim B, Löser G (1983) Formalphysiologische Analyse der Signaltransduktion in der Photomorphogenese. Ber Dtsch Bot Ges in press

Ultraviolet Photoreceptors and Their Interaction in Broom Sorghum - Analysis of Action Spectra and Fluence-Response Curves

T. HASHIMOTO and H. YATSUHASHI[1]

Introduction

In recent years the importance of ultraviolet light (UV) as a normal photomorphogenic agent of plants has been noted [6, 11, 21, 22]. It is becoming clear that plants have several UV photoreceptors, and that UV exerts its action through interaction among the UV photoreceptors or between UV photoreceptors and photoreceptors of visible light such as the blue-absorbing photoreceptor or phytochrome [7, 8, 9, 23]. In this article we deal with the problems: what sorts of UV photoreceptors are involved in UV-induced anthocyanin synthesis, phototropic curvature and coiling in the first internode of broom sorghum?, what sorts of interaction take place between the photoreceptors and phytochrome?; and we propose a mechanism of interaction of UV-B photoreceptor and phytochrome.

Materials and Methods

Plants and Evaluation of Results. Seedlings of broom sorghum (*Sorghum bicolor* Moench, cv. Acme Broomcorn and Sekishokuzairaifukuyama) were grown in total darkness, irradiated and subjected to anthocyanin determination as described previously [23]. For determination of curvature and coiling, plants were harvested, unless otherwise stated, 24 h after irradiation and soaked in a 70% aqueous ethanol to fix the internodes, and the angle formed by the top and base parts of a first internode was measured with a protractor. For both measurements 20 to 25 plants served as one probe.

Irradiations. Monochromatic lights [20], UV-B (UV310) [11], red (R) and far red (FR) [23] lights were described previously. For dichromatic irradiation lights of two wavelengths from the spectrograph were reflected twice by means of mirrors, and combined. The ratio of intensities of the two beams to be combined was adjusted before mixing and the intensity of light incident to plants, after mixing with an inconel-coated quartz

1 Institute for Physical and Chemical Research, Wakoshi, Saitama, 351, Japan
Present address: Department of Biology, Tokyo Metropolitan University, Setagaya, Tokyo, 158, Japan

Blue Light Effects in Biological Systems
ed. by H. Senger

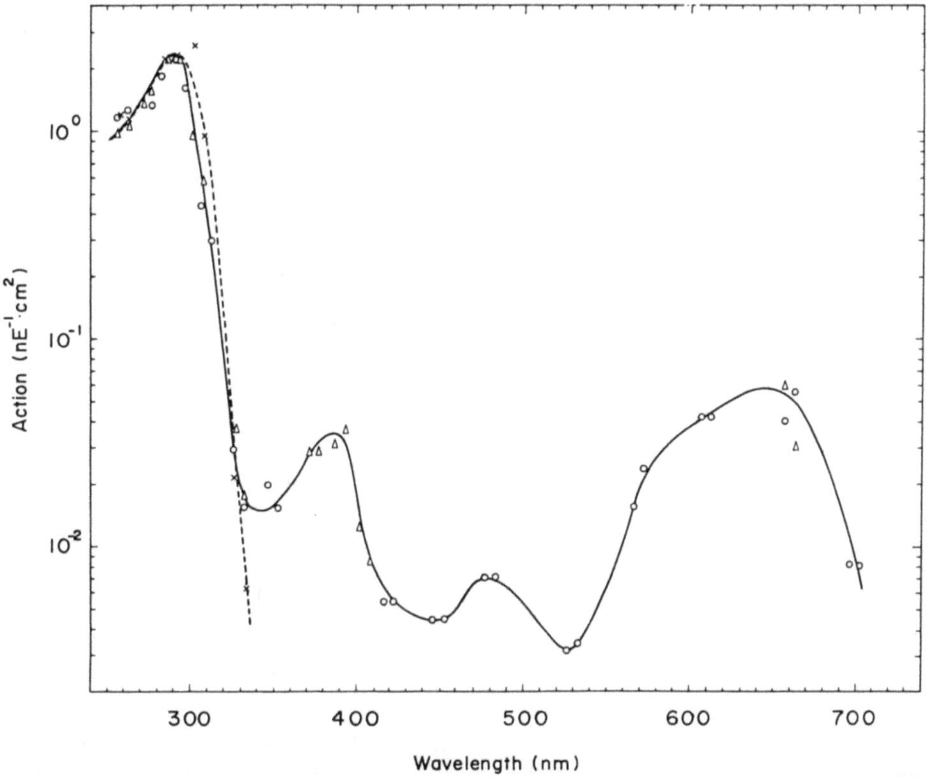

Fig. 1. Action spectrum for anthocyanin synthesis in Acme sorghum first internodes. An anthocyanin level as low as 0.02 absorbance unit was used to obtain the relative quantum effectiveness from fluence-response curves. Irradiation period, 1 or 2 min [23]

glass plate [4] having a negligible wavelength dependency of transmission. For sequential irradiation the unwanted wavelength of beam was shut alternately.

Results

Anthocyanin Synthesis. An action spectrum for anthocyanin synthesis obtained on the basis of a minimum detectable response showed action peaks at 385 and 290 nm in the UV region as well as the one at 660 nm due to phytochrome, and a very small unidentified one at 480 nm (Fig. 1). The 385 nm peak was identified as phytochrome based on its complete nullification with a saturating dose of FR and on its masking by a phytochrome-saturating dose of R. The 290 nm peak was concluded to be due to a new photoreceptor tentatively named P-290 because it was not affected by FR and was intensified by R [23].

The slope of fluence-response curves plotted on a logarithmic scale of axes is low at 290 and 385 nm, and is steep at 335 to 341 nm (Figs. 2, 3, the curves for UV alone).

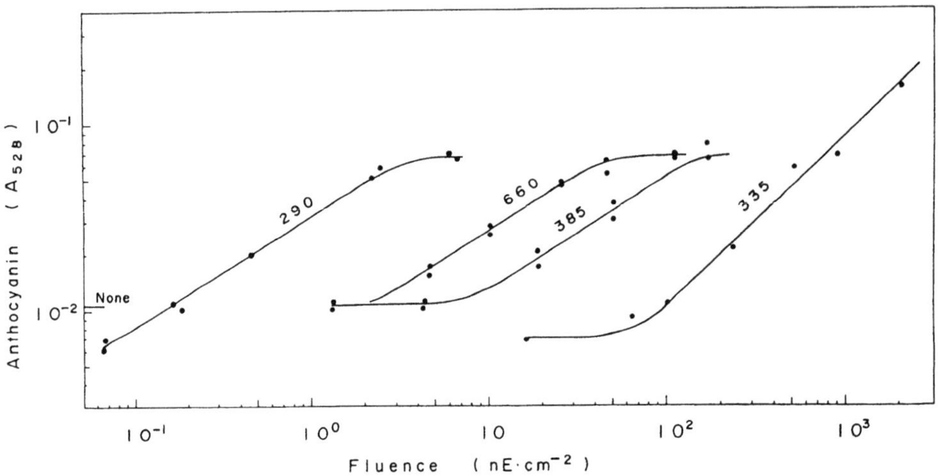

Fig. 2. Fluence-response curves at representative wavelengths including those of action peaks for anthocyanin synthesis in Acme sorghum first internodes. Irradiation periods; 60, 300, 60 and 100 s, respectively, for 290, 335, 385 and 660 nm

It can also be seen in Fig. 2 that the plateau of the curve is high at 335 nm compared with those at 290, 385 and 660 nm[1]. To examine the possibility of phytochrome being involved in these features of the fluence-response curves FR was given after the UV irradiation (Fig. 3). FR suppressed the action of UV's progressively more markedly at higher fluences, lowering the slope of the curves. This effect of FR was greatest at 341 nm of the tested wavelengths, and decreased with decreasing wavelengths. R also lowered the slope of the curves by raising the UV action. The slope lowered by FR or R was comparable to that of the curve for 290 nm (cf. Fig. 2). These results are well understood by assuming that both phytochrome and P-290 are excited with a single irradiation at 319 to 341 nm and act synergistically with each other. It is clear in Fig. 3 that the contribution of phytochrome in the UV action decreases and that of P-290 increases with decreasing wavelengths in this region. That the curves for +FR at 319 and 325 nm did not plateau in spite of enough height, as did those for 290 nm may be due to an incomplete reversion of P_{fr} to P_r [3].

To ascertain the above assumption dichromatic irradiations at 290 vs. 385 and 290 vs. 660 nm were made. To begin with, it was ascertained that sequential and its reverse, and simultaneous irradiations at the two wavelengths gave the same responses within the limit of accuracy of the experiment (Fig. 4). Figure 5 shows the result of the irradiations at 290 and 385 nm, which gave severalfold the sum of the actions of each irradiation, and a steep slope of the curve. Figure 6 shows a similar result with 290 and 660 nm lights. The steep slopes thus obtained are comparable with those for 335 to 341 nm (Fig. 3, UV alone). In Fig. 6 there seems to be an optimum ratio of the intensities of the two lights combined for giving steepest slope as in the curve –4.

1 The curves for short wavelengths such as 290, 335 and 341 often declined at low fluences below the nonirradiated control in which no anthocyanin could be detected. This was ascertained by a scanning spectroscopy to be due to a decrease in background absorption

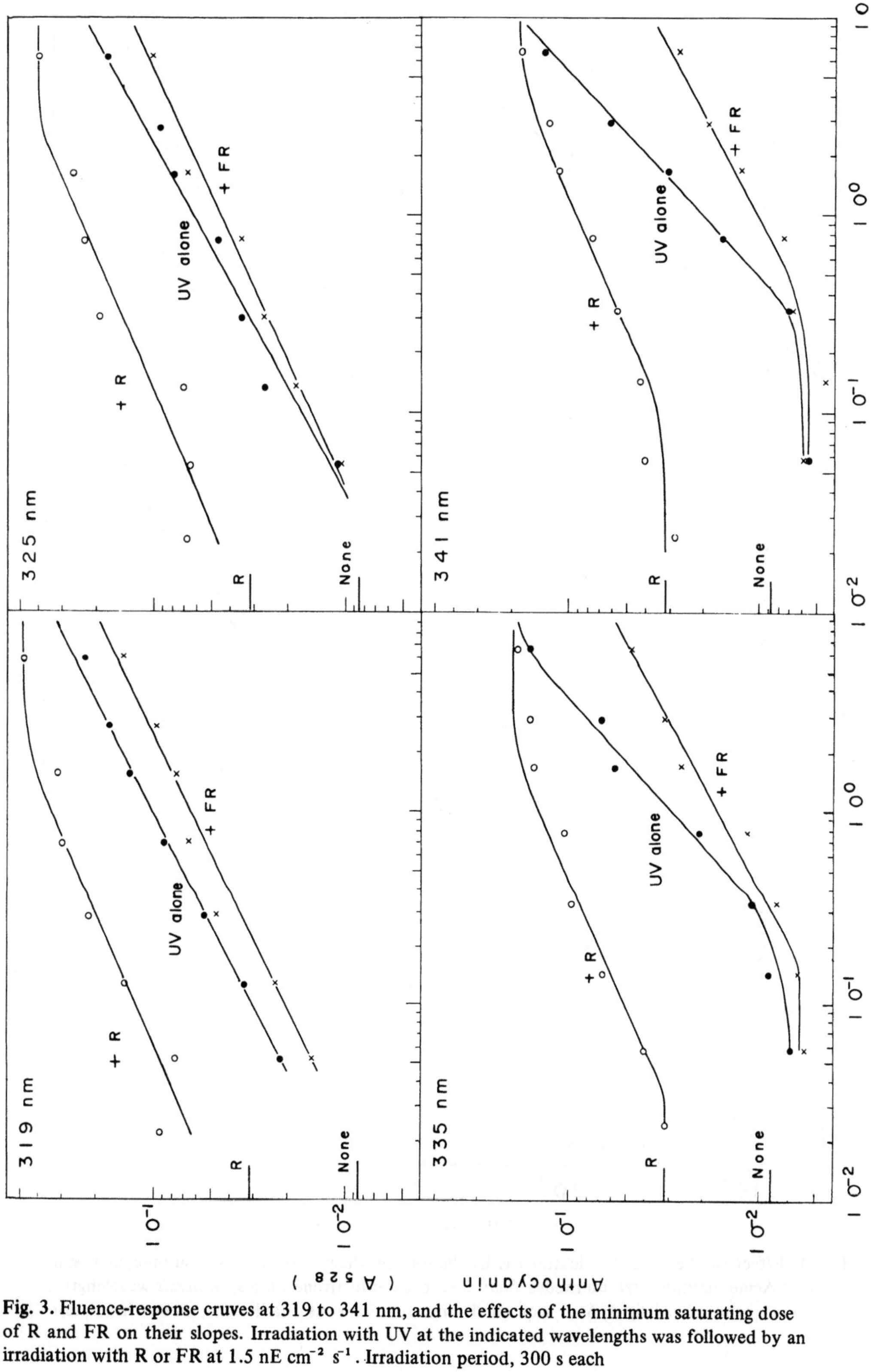

Fig. 3. Fluence-response cruves at 319 to 341 nm, and the effects of the minimum saturating dose of R and FR on their slopes. Irradiation with UV at the indicated wavelengths was followed by an irradiation with R or FR at 1.5 nE cm^{-2} s^{-1}. Irradiation period, 300 s each

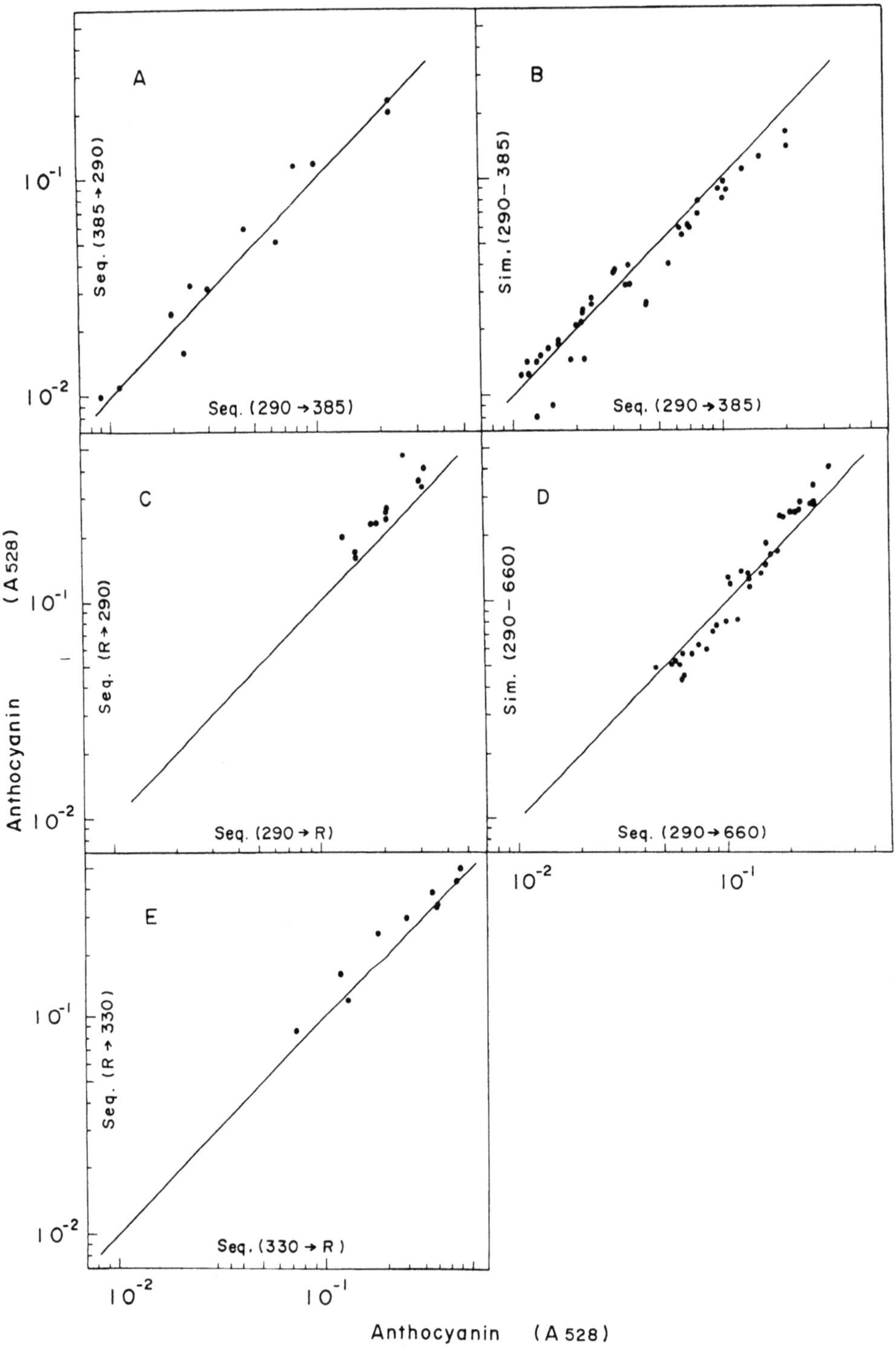

Fig. 4. Effect of the order of dichromatic irradiations on their effectiveness in anthocyanin synthesis of Acme sorghum first internodes. *Slant lines* equal effectiveness lines; *numerals* wavelengths; *R* a broad band of red light; *Seq.* sequential; *Sim.* simultaneous irradiations; *arrows* irradiation order

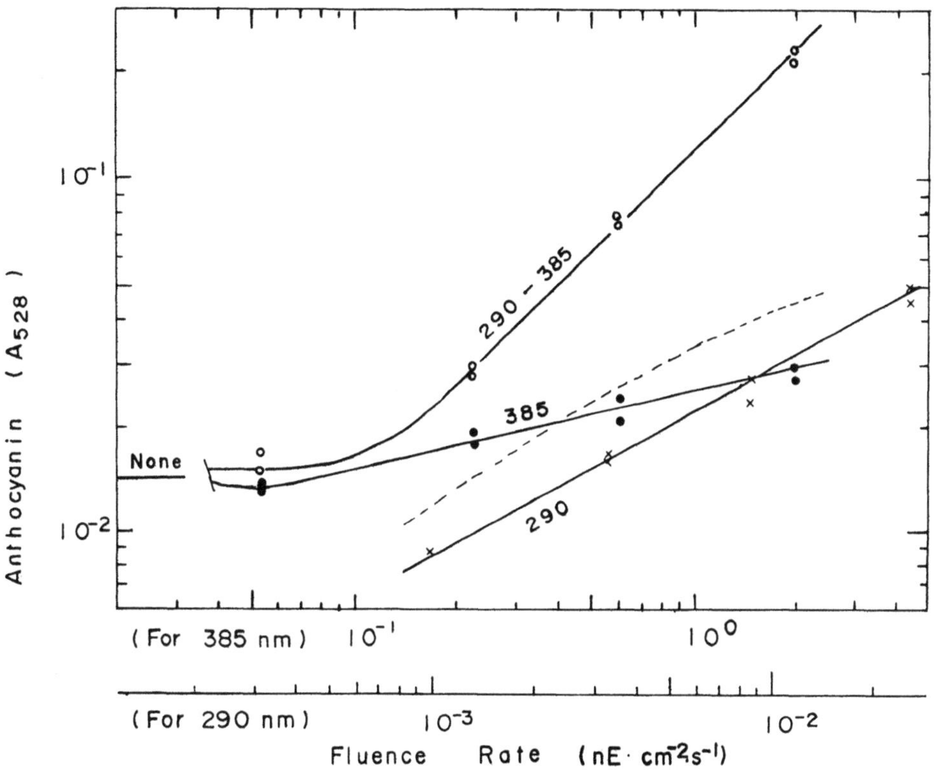

Fig. 5. Synergistic effect of dichromatic irradiation on anthocyanin synthesis in Acme sorghum first internodes. In 290–385, irradiation at 290 nm was followed by that at 385 nm. *Dotted line* sum of the actions of a single irradiation at each wavelength. Each irradiation, 100 s

Considering from the variation of the slope with the irradiated wavelengths (Fig. 3) this must be so. This problem is subject to further work.

Coiling and Curvature. A unilateral UV irradiation caused coiling at the part of the first internodes grown after the exposure to UV. The coiling response continued for 72 h or more, and the UV action seemed to be transmitted to newly produced cells. This is in contrast to anthocyanin synthesis, which is limited to the exposed part and does not last more than 24 h. An action spectrum for coiling is shown in Fig. 7, which shows that the action peak is at or below 257 nm with a shoulder at 287 nm. At wavelengths longer than 310 nm no coil was formed (the angle, less than 180°), however high the fluence of light given, but a sort of phototropic curvature was caused. The curvature increased during 24 h or a little longer after irradiation and did not vanish, unlike the phototropic curvature so far reported.

The coiling induced by monochromatic light below 300 nm was not affected by a subsequent exposure to FR, whereas the curvature due to UV irradiation at 310 to 400 nm was markedly suppressed (Ito et al., unpublished data). Accordingly it is considered that the curvature is not a weak coiling but is a photophysiologically different phenomenon.

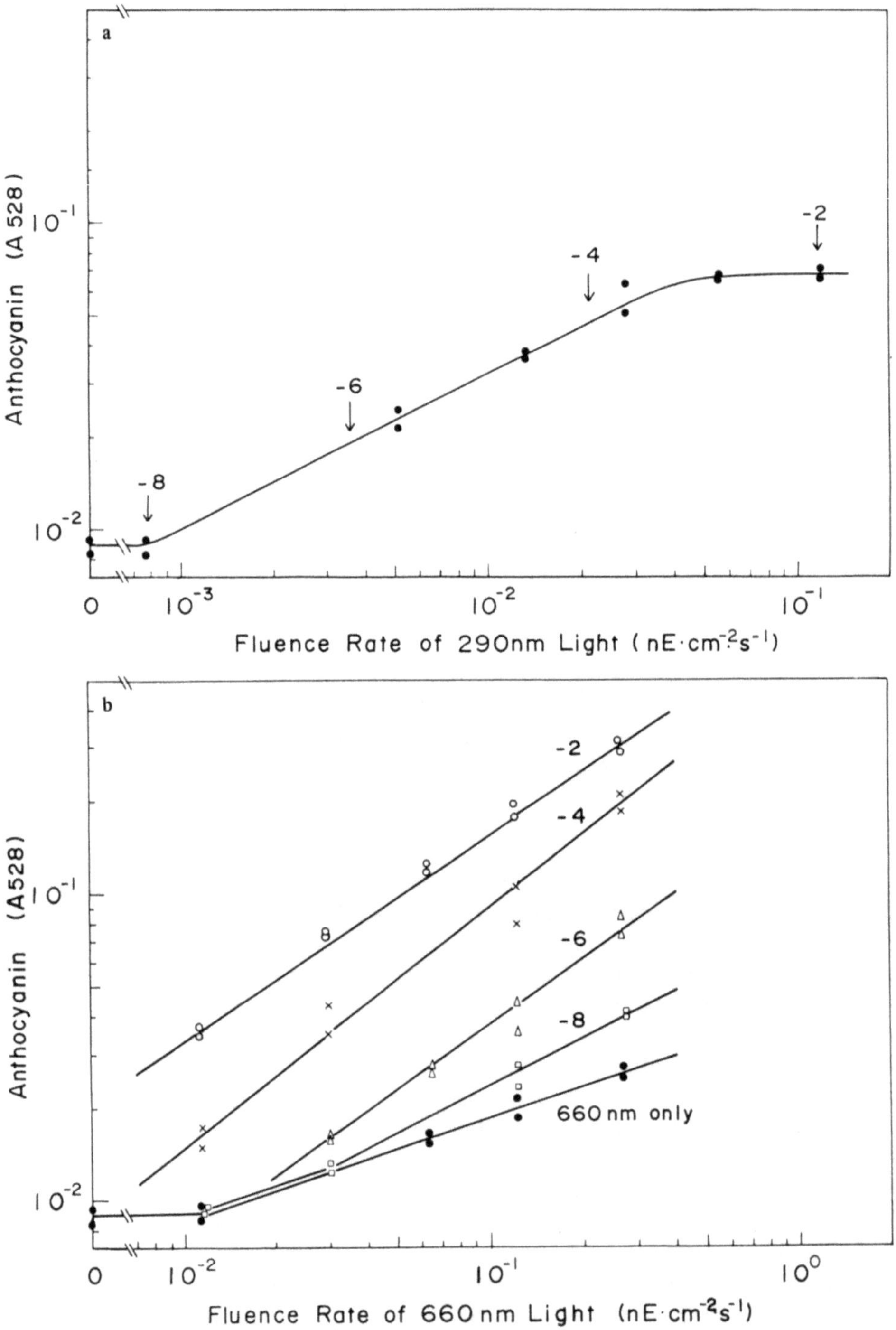

Fig. 6a,b. Effect of dichromatic irradiation at 290 and 660 nm on anthocyanin synthesis in Acme sorghum first internodes. **a** Fluence-response curve for a single irradiation at 290 nm; **b** Fluence-response curves of the combined lights. The *minus numeral* by each of the curves indicates that 660 nm light of 2.6×10^{-1} nE cm^{-2} s^{-1} was combined with the 290 nm light of the fluence rate indicated by the corresponding minus numerals in **a**. The fluence rate of the 290 nm light thus combined was reduced in proportion that that of the 660 nm light was reduced for irradiation. Irradiation, 100 s

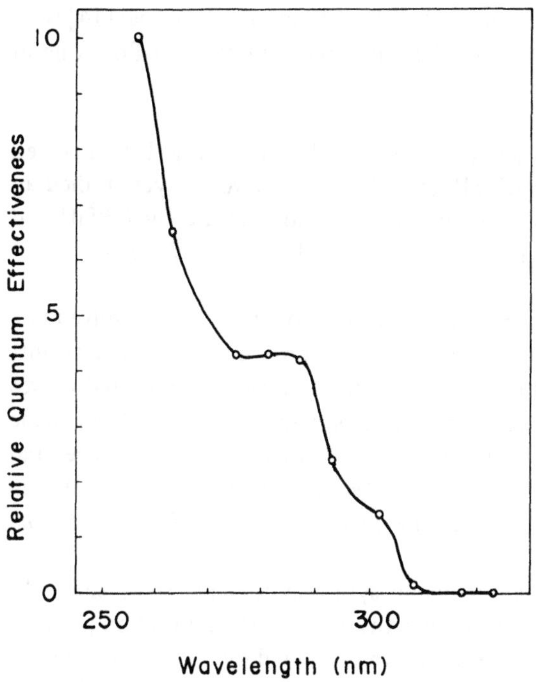

Fig. 7. Action spectrum for coiling of Sekishokuzairaifukuyama sorghum first internodes. The reciprocals of the photon fluence rates required for a curvature of 180 degrees were plotted as relative quantum effectiveness. Acme sorghum gave a similar result

Discussion

UV-B Photoreceptor. The 290 nm peak in the action spectrum (Fig. 1) for anthocyanin synthesis was assigned to a photoreceptor P-290 different from phytochrome based on its very high quantum effectiveness, lack of FR nullification response, and synergism with R (Figs. 1, 6b). Although purified phytochrome is converted to either P_{fr} or P_r by UV absorbed by its protein moiety [15], no symptom of the phytochrome involvement in the peak action was not found. Similar evidence (Figs. 1, 5) shows that the action of the 290 nm is different from that of the 385 nm peak. For coiling a completely different action spectrum was obtained (Fig. 7). This excludes that the 290 nm peak is a part of some UV-C-absorbing photoreceptor, the rest being screened. Accordingly P-290 may be concluded to be a UV-B photoreceptor.

UV-C Photoreceptor. A photoreceptor involved in the action peak for coiling at 257 nm or less was tentatively named Puv-c. Although the action of this peak has not been well characterized yet, coiling seems to be a UV-induced malformation, and it is possible that Puv-c be nucleic acid.

Near-UV Photoreceptor. Drumm-Herrel and Mohr [7] have claimed the presence of UV-A (including blue)-absorbing photoreceptor which co-operates with UV-B photoreceptor in anthocyanin synthesis of milo sorghum. The 385 nm peak shown in Fig. 1 is considered ascribable to phytochrome in every respect of its peak wavelength, complete nullification by FR, complete masking by phytochrome-saturating R (lack of interaction with R), and synergism between 290 and 385 nm lights (Fig. 5, Yatsuhashi et al. [23]). A similar conclusion has been reached by Ohtani and Kumagai [14] with a 380 nm action peak for reversal of a blue light-induced inhibition of flowering in *Lemna paucicostata* due to phytochrome.

Blue-Light Photoreceptor. The very small action peak at 480 nm in Fig. 1 disappeared on irradiation with a saturating dose of FR or R [23]. We have not yet studied an interaction of 480 and 290 nm lights. In parsley suspension culture Duell-Pfaff and Wellmann [9] have found an interesting synergism between blue and UV lights.

Slope of Fluence-Response Curves. If one and the same photoreceptor is involved in a photoresponse at different wavelengths, the fluence-response curves plotted against the log of the fluences must theoretically be parallel. However, the reported curves are not necessarily so (e.g., [2, 12, 16]), and no adequate analysis has been made except for those in the neighbourhood of 700 nm by Hartmann [10] and Wagner and Mohr [19]. Shropshire and Withrow [16] suggested "some sort of interaction between two pigment systems" for the steep slope of the curves at 370 to 375 nm in *Avena* coleoptile curvature.

In this article we showed with anthocyanin synthesis that (1) steep slopes of curves were observed in the intermediate wavelengths between the two action peaks at 290 and 385 nm (Figs. 2, 3); (2) the steep slopes were lowered by a saturating dose of FR following the UV irradiation (Fig. 3); (3) the steep slopes were established by dichromatic irradiations at 290 nm and either 385 or 660 nm (Figs. 5, 6). From these data it follows that in the region from 319 to 341 nm at least, the absorption bands of phytochrome and P-290 overlap, and even monochromatic irradiation excites both photoreceptors, which exert a synergistic action and give the steep slope to the fluence-response curves.

Independence of P-290 Excitation and Phytochrome Conversion. Difference in the order of phytochrome and P-290 absorbing light did not give difference in action (Fig. 4). The literature [5, 6, 7, 21] claims that an irradiation with UV-B, UV-A or blue light is a prerequisite for phytochrome action. In fact Drumm-Herrel and Mohr [8] have stated that "the effectiveness of phytochrome depends mainly on a preceeding (or simultaneous) light absorption in a blue/UV light photoreceptor". However, no data are available to show UV-B, UV-A or blue light irradiation must not be after R. It is assumed that P-290 and phytochrome act independently of each other, at least at the initial steps of the reactions leading to anthocyanin synthesis.

Synergistic Action Mechanism of Phytochrome and P-290 in Anthocyanin Synthesis. We propose the following scheme concerning the synergism of phytochrome and P-290 (Fig. 8). The excited P-290 promotes the reaction step $X_1 \dashrightarrow X_2$, and X_2 forms X_3 in proportion with the log of the fluence of light given. Similarly, P_{fr} accelerates step $Y_1 \dashrightarrow Y_2$, and Y_2 forms Y_3. Each of X_2 and Y_2 opens the step $A_1 \dashrightarrow A_2$ to form an ample amount of A_2. X_3 and Y_3 promote, respectively, $A_2 \dashrightarrow A_3$ and $A_3 \dashrightarrow A_4$, which eventually lead to anthocyanin synthesis. In this scheme the P-290-controlled system $X_1 \dashrightarrow X_2 \dashrightarrow X_3$ and the phytochrome-controlled system $Y_1 \dashrightarrow Y_2 \dashrightarrow Y_3$ can be exchanged. In broom sorghum used in this study as well as in apple [1, 18], soy-

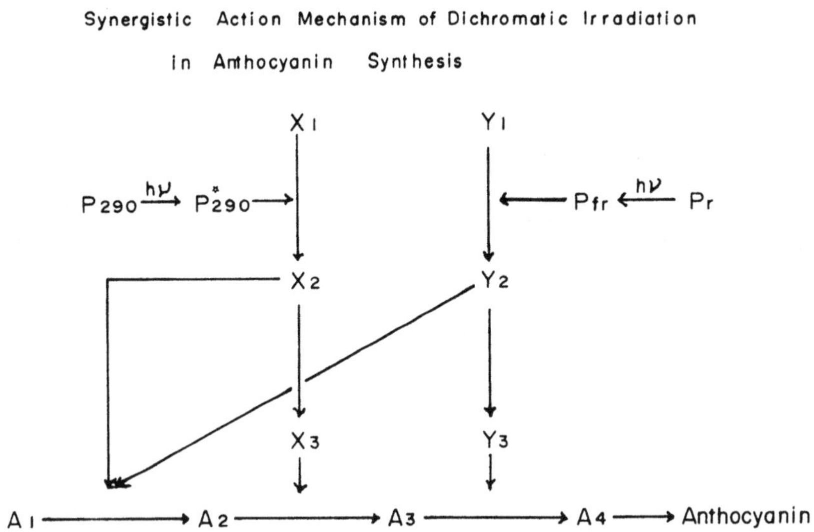

Fig. 8. Proposed synergistic action mechanism of dichromatic irradiation in synthesis of flavonoids including anthocyanins

bean [11] and tomato [8], nonirradiated seedlings or apple skin have a limited amount of X_3 and Y_3. The excitation of P-290 (or phytochrome) allows $A_2 \dashrightarrow A_3 \dashrightarrow A_4$ to proceed depending on the available amounts of Y_3 (or X_3), and the resulting amount of anthocyanin is proportional to the log of the fluence of light and plateaus depending on the amount of Y_3 (or X_3). If phytochrome (or P-290) also absorbs a fixed dose of light in addition to P-290 (or phytochrome), Y_3 (or X_3) increases the anthocyanin synthesis controlled by P-290 (or phytochrome) increases multiplicatively by the increase of Y_3 (or X_3), giving a curve parallel to the original fluence-response curve based on P-290 (phytochrome) and the plateau is raised. This situation can be seen as parallel curves of +R and +FR in Fig. 3. When the lights absorbed by P-290 and phytochrome vary together, anthocyanin synthesis is affected by the product of both variations of the lights, thus giving a steep slope to the fluence-response curve.

In synthesis of anthocyanin in milo sorghum [7] and of flavone glycoside in parsley suspension culture [9, 21, 22] where R is ineffective unless UV-B is absorbed, it is assumed that nonirradiated plants do not have X_3 at all. In red cabbage [17] and mustard [13] and ample amount of X_3 is assumed to be present in a nonirradiated state, although UV has not yet been tested.

In conclusion, action spectra for anthocyanin synthesis, as well as phototropic curvature and coiling of the first internode of broom sorghum (*Sorghum bicolor* Moench, cv. Acme Broomcorn and Sekishokuzairaifukuyama), showed the presence of phytochrome, P-290 and Puv-c as UV photoreceptors. So-called near-UV photoreceptor is phytochrome in this plant. In anthocyanin synthesis phytochrome and P-290 exert a synergistic action, and this gives steep slopes and high plateaux at the intermediate wavelengths where the absorption bands of phytochrome and P-290 overlap. An action mechanism of phytochrome and P-290 was proposed which can adequately explain the UV- and R-induced synthesis of anthocyanin and flavone glycoside in this plant and others as well.

Abbreviations

R, red light; FR, far-red light; UV, ultraviolet light; UV-A, UV-B and UV-C, respectively, 400–315, 315–280, and less than 280 nm.

Acknowledgments. We acknowledge that the most of our data presented here were obtained with the Okazaki Large Spectrograph, NIBB, Okazaki, supported by a grant (No. 82A09) for research with it. We also thank Dr. M. Watanabe and Mr. M. Kubota, NIBB, for operating the spectrograph, Dr. I. Tarumoto, MAFF, Tokyo for seeds, and Miss Sumiko Ito, ICPR, Saitama for her unpublished data cited here.

References

1. Arthur JM (1932) Red pigment production in apples by means of artificial light sources. Contrib Boyce Thompson Inst 4:1–18
2. Beggs CJ, Holmes MG, Jabben M, Schäfer E (1980) Action spectra for the inhibition of hypocotyl growth by continuous irradiation in light and dark-grown *Sinapis alba* L. seedlings. Plant Physiol 66:615–618
3. Butler WL, Hendricks SB, Siegelman HW (1964) Action spectra of phytochrome in vitro. Photochem Photobiol 3:521–528
4. Dobrowolski JA (1978) Coatings and filters. In: Driscoll WG, Vaughan W (eds) Handbook of optics. McGraw-Hill, New York
5. Downs RJ, Siegelman HW (1963) Photocontrol of anthocyanin synthesis in milo seedlings. Plant Physiol 38:25–30
6. Drumm H, Mohr H (1978) The mode of interaction between blue (UV) light photoreceptor and phytochrome in anthocyanin formation of the *Sorghum* seedling. Photochem Photobiol 27:241–248
7. Drumm-Herrel H, Mohr H (1981) A novel effect of UV-B in a higher plant (*Sorghum vulgare*). Photochem Photobiol 33:391–398
8. Drumm-Herrel H, Mohr H (1982) The effect of prolonged light exposure on the effectiveness of phytochrome in anthocyanin synthesis in tomato seedlings. Photochem Photobiol 35:233–236
9. Duell-Pfaff N, Wellmann E (1982) Involvement of phytochrome and a blue light photoreceptor in UV-B induced flavonoid synthesis in parsley (*Petroselinum hortense* Hoffm.) cell suspension cultures. Planta 156:213–217
10. Hartmann KM (1966) A general hypothesis to interpret 'High Energy Phenomena' of photomorphogenesis on the basis of phytochrome. Photochem Photobiol 5:349–366
11. Hashimoto T, Tajima M (1980) Effects of ultraviolet irradiation on growth and pigmentation in seedlings. Plant Cell Physiol 21:1559–1571
12. Jose AM, Vince-Prue D (1977) Action spectra for the inhibition of growth in radish hypocotyls. Planta 136:131–134
13. Mohr H (1957) Der Einfluß monochromatischer Strahlung auf das Längenwachstum des Hypocotyls und auf die Anthocyanbildung bei Keimlingen von *Sinapis alba* L. (= *Brassica alba* Boiss). Planta 49:389–405
14. Ohtani T, Kumagai T (1981) Phytochrome-mediated effects of near-ultraviolet radiation in the induction of flowering in etiolated *Lemna paucicostata* T-101, a short-day plant. Planta 153: 543–546
15. Pratt LH, Butler WL (1970) Phytochrome conversion by ultraviolet light. Photochem Photobiol 11:503–509
16. Shropshire Jr W, Withrow RB (1958) Action spectrum of phototropic tip curvature of Avena. Plant Physiol 33:360–365

17. Siegelman HW, Hendricks SB (1957) Photocontrol of anthocyanin formation in turnip and cabbage seedlings. Plant Physiol 32:393–398
18. Siegelman HW, Hendricks SB (1958) Photocontrol of anthocyanin synthesis in apple skin. Plant Physiol 33:185–190
19. Wagner E, Mohr H (1966) Kinetic studies to interpret 'High-Energy Phenomena' of photomorphogenesis on the basis of phytochrome. Photochem Photobiol 5:397–406
20. Watanabe M, Furuya M, Miyoshi Y, Inoue Y, Iwahashi I, Matsumoto K (1982) Design and performance of the Okazaki Large Spectrograph for photobiological research. Photochem Photobiol 36:491–498
21. Wellmann E (1971) Phytochrome-mediated flavone glycoside synthesis in cell suspension cultures of *Petroselinum hortense* after preirradiation with ultraviolet light. Planta 101:283–286
22. Wellmann E (1974) Regulation der Flavonoidbiosynthese durch ultraviolettes Licht und Phytochrom in Zellkulturen und Keimlingen von Petersilie (*Petroselinum hortense* Hoffm.). Ber Dtsch Bot Ges 87:267–273
23. Yatsuhashi H, Hashimoto T, Shimizu S (1982) Ultraviolet action spectrum for anthocyanin formation in broom sorghum first internodes. Plant Physiol 70:735–741

Multiple Effects of the Flavin Quencher Potassium Iodide on Light- and Dark-Processes in the Green Alga Mougeotia

E. SCHÖNBOHM and E. SCHÖNBOHM [1]

Introduction

Blue light, as well as near-UV, induces and controls many physiological processes [5, 13, 14, 15]. Mostly it is an open question whether the photoreceptor is a carotenoid, a flavin or another unknown photoreceptor. Various experiments have been attempted to test the role of flavins as blue/UV-photoreceptor pigments. Often the effect of potassium iodide (KI), known as a very effective quencher of the triplet excited state of flavins in vitro [25, 26] has been tested also in vivo [3, 6, 7, 8, 9, 10, 12, 16, 17, 22, 23, 24, 27, 28]. Most of the authors concluded that KI affected the blue light photoreceptor rather than secondary processes, and that an iodide inhibition was consistent with a flavin as the photoreceptor pigment. But in *Mougeotia* it was shown that KI not only inhibited the blue light-dependent strong light movement but also the weak light response, which is mediated by phytochrome; this inhibition could also be achieved if KI was given after the red light induction phase [17, 18]. Recently we have demonstrated that this inhibitory KI-effect on the chloroplast movement of *Mougeotia* under weak light as well as under strong light can be abolished completely by ATP or ADP given simultaneously with KI but not by ITP or IDP [21].

In some further experiments the effect of 0.1 M KI on different processes in *Mougeotia* has been investigated. Some of the results we obtained are given here.

KI and the Chloroplast Movements

KI loses its severe inhibitory effect on all types of oriented chloroplast movements, if KI is dissolved not as usual in the nutriation medium but in aqua bidest or in 10^{-4} M sorbitol. KI regains its strong inhibitory effect completely if it is dissolved in 0.01 M $CaCl_2$; whereas an incubation of *Mougeotia* in KI dissolved in $MgCl_2$ brings a strong inhibition of the weak light movement but not of the strong light response [21].

---

1 Botanisches Institut, Universität Marburg, Fachbereich Biologie, Auf den Lahnbergen, D–3550 Marburg, FRG

Blue Light Effects in Biological Systems
ed. by H. Senger

The Effect of KI on the Light-Induced Decrease of Centrifugability of Chloroplasts Within Living Cells

In *Mougeotia* centrifugability of chloroplasts decreases in weak light (phytochrome/cryptochrome responses) and increases in strong light (cryptochrome response), in darkness as well as in far-red light [19]. Centrifugability also increases by a treatment with SH-blockers [19] as well as by decrease of the available ATP (Schönbohm in preparation). It is known that KI depolymerizes muscle F-actin [4] and on the other hand that it is an effective uncoupler of oxidative phosphorylation [11]. Multiple effects of KI in *Mougeotia* cannot be excluded. Our centrifugation experiments show (Fig. 1) that KI does not increase centrifugability of chloroplasts significantly in dark-sensitized material (compare c with a) but it dramatically prevents the realization of a weak light-inducible decrease of centrifugability (compare a with b and b with d). There is no hint that KI influences already active contractile filaments, but we suppose that the light-inducible activation and/or aggregation of actomyosin-like filaments are inhibited by KI.

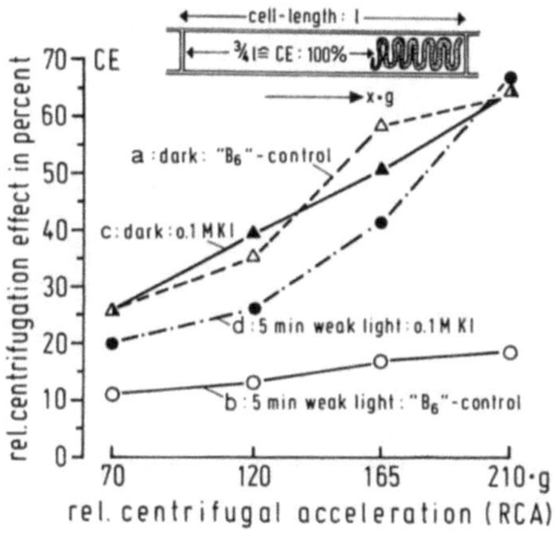

Fig. 1. The effect of 0.1 M KI dissolved in the culture medium (B6) on the weak light-induced decrease of centrifugability

The Effect of KI on the Oxygen Exchange During Dark and Light Phases

We have measured the oxygen exchange of *Mougeotia* (polarographic method!) continuously during red light and dark phases and have tested the effect of KI which was dissolved in the standard nutriation medium (B6) or in 10^{-4} M sorbitol. The following results were obtained: KI has a much higher inhibitory effect on the O_2-producing processes than on dark respiration (Fig. 2). Similarly as in the experiments on chloroplast movements KI loses also its strong inhibitory effect on oxygen evolution if it is given dissolved in 10^{-4} M sorbitol (Fig. 3b). Further experiments showed that with increasing incubation time also dark respiration becomes intensively inhibited by KI.

Figs. 2 and 3. Experiments on oxygen exchange. (After V. Opper unpublished) (KI uptake is increased by far-red)

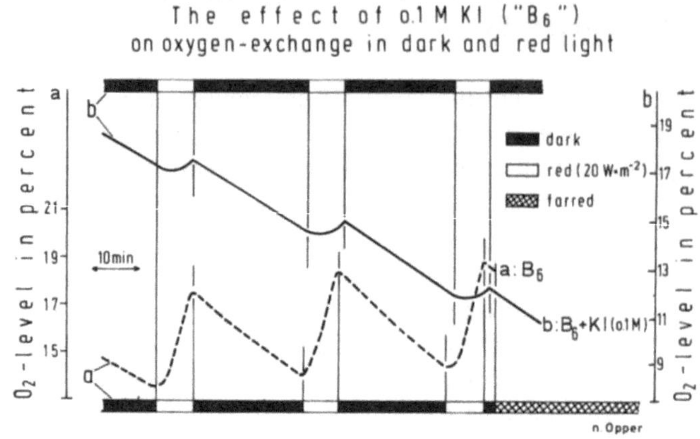

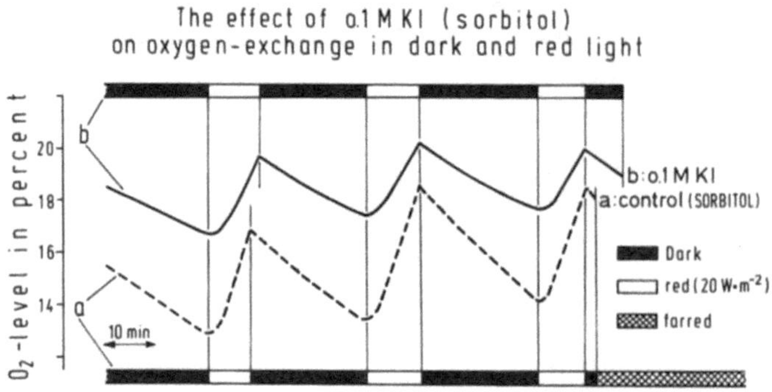

Fig. 3

The Effect of KI on Dark- and Light-Induced pH Changes

During light phases the pH of a *Mougeotia*-sample increases whereas it decreases during darkness (see pH curves B in Figs. 4, 5, 6). KI, dissolved in 0.01 M $CaCl_2$ (Fig. 4A) inhibits both the pH increase during the light phases as well as the pH decrease during the dark phases; however, the inhibition of pH decrease is much higher than of pH increase. Similarly as in other KI-sensitive processes the effect of KI on the pH changes also increases with increasing incubation time, and further, the inhibitory effect of KI dissolved in sorbitol is highly reduced (Fig. 5A).

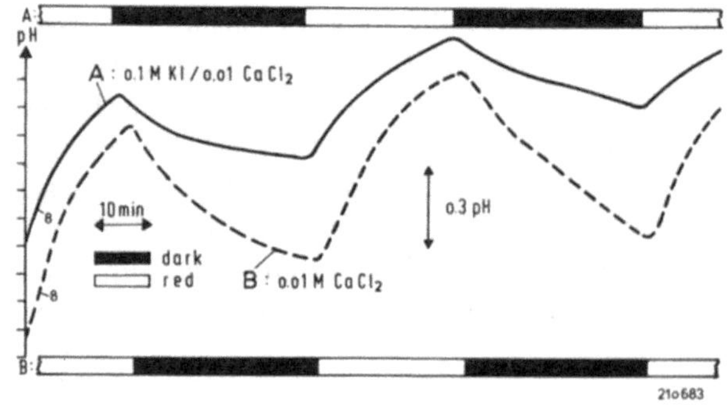

Figs. 4, 5 and 6. The effect of 0.1 M KI on the pH changes in dark and red light (15 W m^{-2})

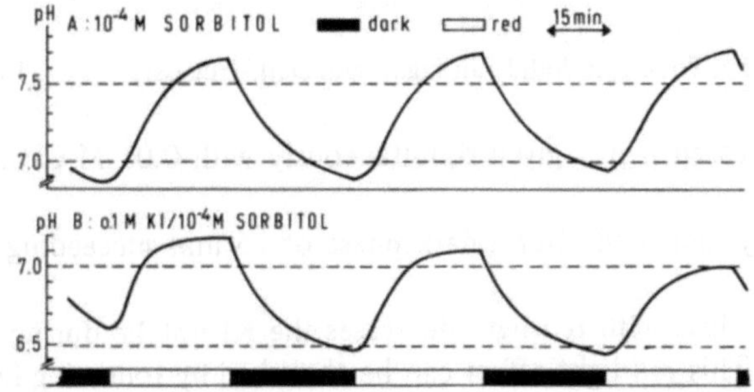

Fig. 5. Legend see p. 139

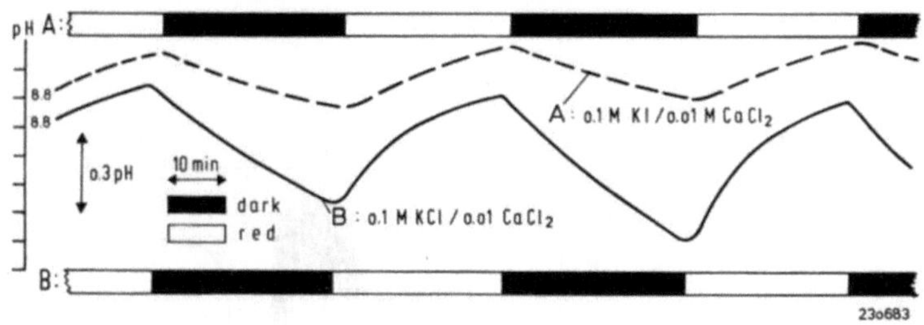

Fig. 6. Legend see p. 139

In contrast to KI (Fig. 6A), potassium chloride has no inhibitory effect on pH changes in *Mougeotia* Fig. 6B), this means that the effect of KI is a specific iodide effect.

Factors Influencing the Amount of KI Uptake

In the following we attempted to show that the amount of KI uptake in the cells becomes influenced by many external factors. In order to quantify the amount of KI uptake we run "difference spectra" of the incubation solutions after removing the algae (by centrifugation) against the fresh incubation solution. Some of the results we obtained are:

a) During an incubation time of 15 min KI uptake in the blue light is much higher than in the far-red light or in the red light (Fig. 7).

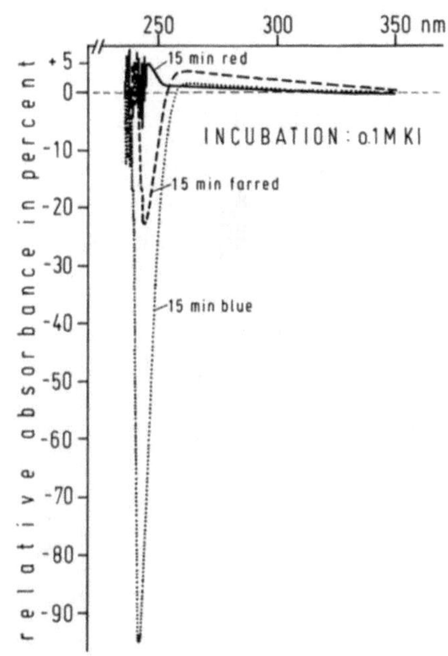

Fig. 7. The effect of light on the amount of KI uptake during 15 min (difference-spectra of 1:100 diluted incubation solutions; photometer sensitivity: 10 ×

b) KI uptake is highly increased if it is given simultaneously with 0.01 M $CaCl_2$ (Fig. 8).

c) KI uptake is dramatically increased after a dark phase of 15 min succeeding a red or blue light phase (Fig. 9).

d) A preirradiation of the algae with red light decreases the KI uptake during a following KI incubation phase. This red light effect can be abolished by following far red.

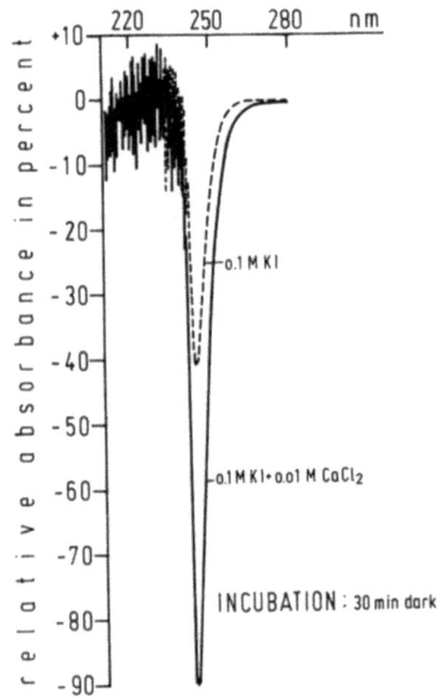

Fig. 8. The increasing effect of $CaCl_2$ on KI uptake (difference spectra of 1:100 diluted incubation solutions; photometer sensitivity: 10 ×

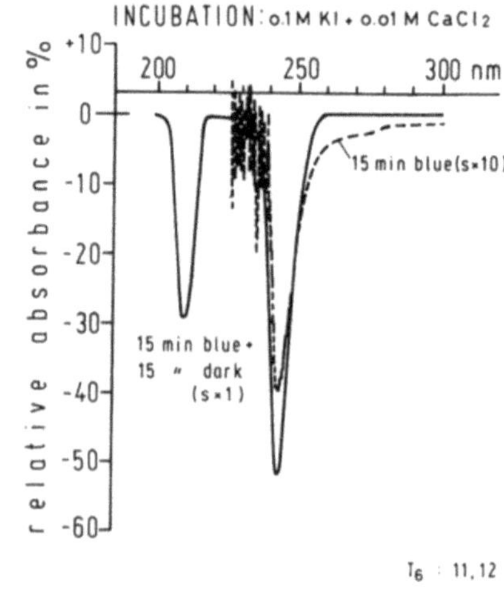

Fig. 9. KI uptake after 15 min blue (22 W m^{-2}) as well as after a succeeding dark phase of 15 min (difference spectra of 1:100 diluted incubation solutions; photometer sensitivity in brackets)

The Effect of KI on the Efflux of Phenolic Compounds During Dark and Light Phases

In the context of this paper all compounds of the efflux which could be bound to insoluble polyvinylpyrrolidone [1] are called phenolic compounds. In this analysis qualitative differences of the composition of the total effluxes are neglected; we can be sure that a high amount of the total phenolic compounds we measure in the efflux solution comes from mobile "tannin" vesicles [20]: the diameter of the vesicles changes inversely proportional to the amount of phenolic compounds in the efflux.

The curves of Fig. 10 demonstrate clearly a dramatic increase of the efflux of phenolic compounds by 0.1 M KI (c = bidest-control) and an inhibition of the KI-effect by simultaneously given $CaCl_2$. There is no further increased efflux if the incubation time (in white light) is strongly prolonged (Fig. 11). Comparing the curves, one can see that additionally given $CaCl_2$ but not $MgCl_2$ decreases the effect of KI. Short-term experiments performed with dark-adapted material (30 min dark in 10^{-4} M sorbitol) show (Fig. 12), that the efflux of phenolic compounds during a $KI/CaCl_2$-incubation-phase is much higher if red or blue light was given than in darkness or after a farred-phase. These results are typical for material with high sensitiveness in chloro-

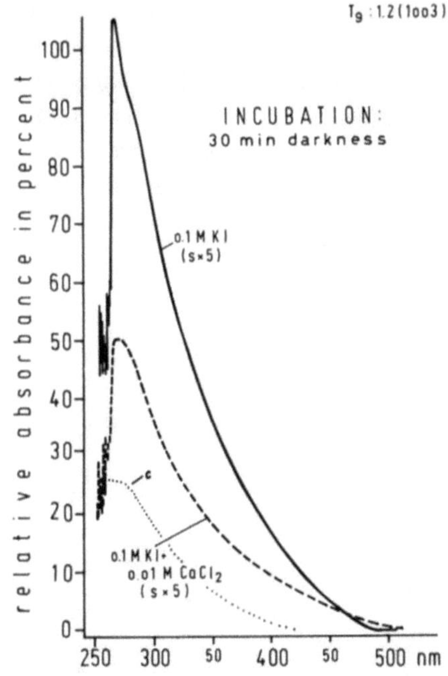

Figs. 10 and 11. Total efflux of phenolic compounds measured after different incubation conditions (difference spectra of nondiluted incubation solutions; photometer sensitivity: 5 ×

Fig. 11. Legend see p. 142

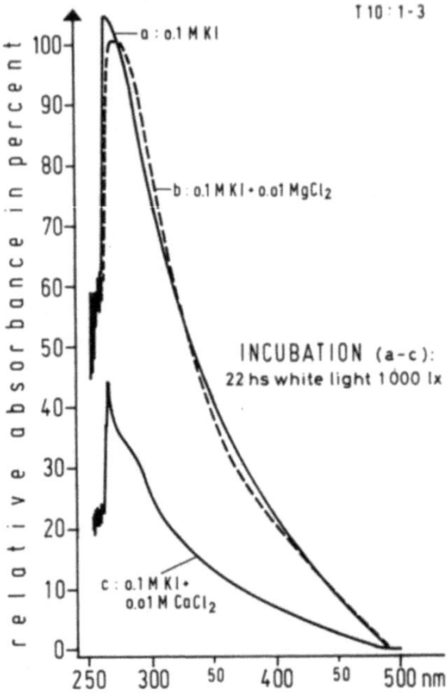

plast movement experiments. In so-called old material of *Mougeotia* the number and size of "tannin"-vesicles is highly increased; such cells have a diminished sensitiveness in experiments on light-oriented chloroplast movements. Old *Mougeotia*-material shows an increased efflux of phenolic compounds not only after red or blue, but also after far-red irradiations (see Fig. 13: photometer sensitivity: 5 ×). We suppose an involvement of phenolic compounds in the reaction-chain of light-

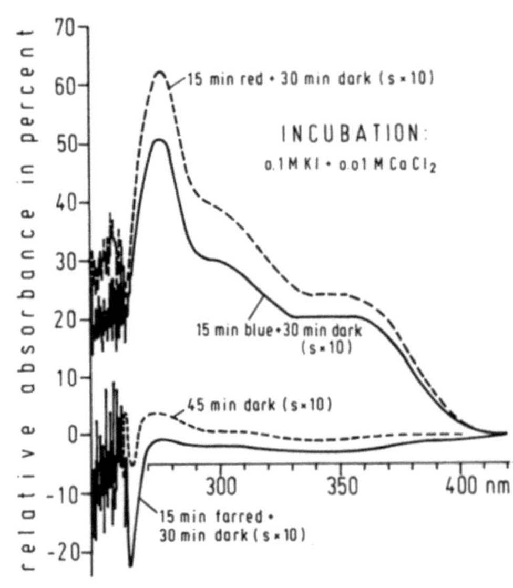

Figs. 12 and 13. Total efflux of phenolic compounds measured after an incubation time of 45 min (red: 15 W m^{-2}; blue: 22 W m^{-2}; far-red; 13.5 W m^{-2}; difference spectra of nondiluted incubation solutions). Material: "young" algae: **12** (photometer sensitivity: 10 ×), "old" algae: **13** (photometer sensitivity: 5 ×)

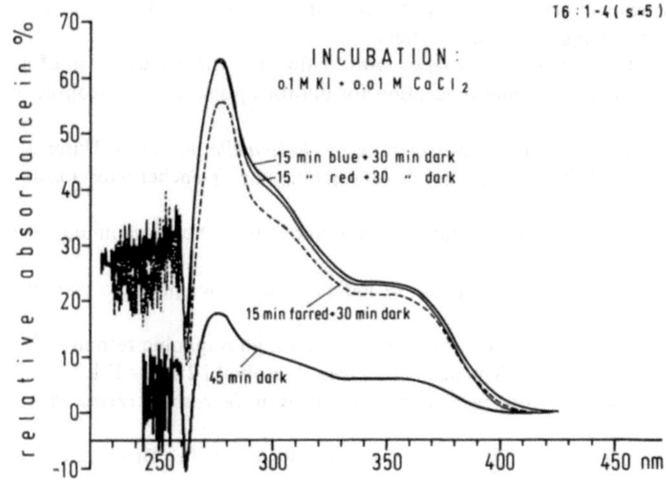

16 : 1-4 (s=5) **Fig. 13.** Legend see p. 143

oriented chloroplast movements in *Mougeotia.* This supposition is supported, for instance, by the results of an inhibition of chloroplast movements by one separated fraction of phenolic compounds of *Mougeotia* [21a].

We conclude from our experiments on the effect of KI on different processes in *Mougeotia* that chloroplast movement becomes inhibited by KI not by quenching flavins but rather via inhibiting secondary processes, as, for instance, ATP-production and/or the availability of ATP for chloroplast movements. It cannot be excluded that KI also disturbs the postulated role of phenolic compounds in chloroplast movements.

We think that most of the results from in vivo experiments with KI must be interpreted very cautiously, since iodide seems to have a broad field of activity in light as well as in dark processes and since the amount of KI penetrating the cell membrane depends on many factors.

References

1. Andersen RA, Sowers IA (1968) Optimum conditions for bonding of plant phenols to insoluble polyvinyl pyrrolidone. Phytochemistry 7:293–301
2. Colombetti G, Häder DP, Lenci F, Quaglia M (1982) Phototaxis in *Euglena gracilis:* Effect of sodium azid and triphenylmethyl phosphonium ion on the photosensory transduction chain. Curr Microbiol 7:281–284
3. Diehn B, Kint B (1970) The flavin nature of the photoreceptor molecule for phototaxis in *Euglena.* Physiol Chem Phys 2:483–488
4. Guba F (1950) Effect of halogen ions on F-actin. Nature (London) 165:439
5. Haupt W, Schönbohm E (1970) Light-oriented chloroplast movements. In: Halldal P (ed) Photobiology of microorganisms. Wiley-Interscience, London New York Sydney Toronto, pp 283–307
6. Kowallik W (1968) Über die Wirkung von Kaliumjodid auf die lichtstimulierte endogene Atmung von Algen. Planta 79:122–127

7. Lenci F, Colombetti G, Häder D-P (1984) Primary events in the sensory transduction of the negative phototaxis in flagellate, *Euglena gracilis*. In press
8. Leong TY, Briggs WR (1982) Evidence from studies with acifluorfen for participation of a flavin-cytochrome complex in blue light photoreception for phototropism of oat coleoptiles. Plant Physiol 70:875–881
9. Mayer F (1966) Lichtinduzierte Chloroplastenverlagerung bei *Selaginella martensii*. Untersuchungen zur Identifizierung des Photoreceptors durch Anwendung von Quencher-Substanzen. Z Pflanzenphysiol 55:65–70
10. Meyer AM (1969) Versuche zur positiven und zur negativen phototropischen Krümmung der *Avena*koleoptile. Z Pflanzenphysiol 60:418–433
11. Middlebrook M, Szent-Györgyi A (1955) The action of iodide on oxidative phosphorylation. Biochim Biophys Acta 18:407–410
12. Mikolajczyk E, Diehn B (1975) The effect of potassium iodide on photophobic responses in *Euglena*: Evidence for two photoreceptor pigments. Photochem Photobiol 22:269–271
13. Muñoz V, Butler WL (1975) Photoreceptor pigment for blue light in *Neurospora crassa*. Plant Physiol 55:421–426
14. Schmid GH (1970) Photoregulation von Flavinenzymen durch Blaulicht. Ber Dtsch Bot Ges 83:399–415
15. Schmidt W (1980) Physiological bluelight reception. In: Dunitz JD, Goodenough JB, Hemmerich P, Ibers JA, Jorgensen CK, Neilands JB, Reinen D, Williams RJP (eds) Structure and bonding, vol 41. Springer, Berlin Heidelberg New York, pp 1–44
16. Schmidt W, Hart J, Filner P, Poff KL (1977) Specific inhibition of phototropism in corn seedlings. Plant Physiol 60:736–738
17. Schönbohm E (1967) Die Hemmung der positiven und negativen Phototaxis des *Mougeotia*-Chloroplasten durch Iodid-Ionen. Z Pflanzenphysiol 56:366–374
18. Schönbohm E (1969) Untersuchungen über den Einfluß von Photosynthesehemmstoffen und Halogeniden auf die Starklicht- und Schwachlichtbewegung des Chloroplasten von *Mougeotia* spec. Z Pflanzenphysiol 60:255–269
19. Schönbohm E (1972) Die Wirkung von SH-Blockern sowie von Licht und Dunkel auf die Verankerung der *Mougeotia*-Chloroplasten im cytoplasmatischen Wandbelag. Z Pflanzenphysiol 66:113–132
20. Schönbohm E (1975) Der Einfluß von Colchicin sowie von Cytochalasin B auf fädige Plasmastrukturen, auf die Verankerung der Chloroplasten sowie auf die orientierte Chloroplastenbewegung. Ber Dtsch Bot Ges 88:211–224
21. Schönbohm E, Schönbohm E (1984) Zur Problematik des in-vivo-Einsatzes von Kalium-Iodid und Natrium-Azid als spezifische Triplett-Quencher von Flavinen. Biochem Physiol Pflanz 179: 95–108
21a. Schönbohm E, Schönbohm E (1984) Biophenole: Steuernde Faktoren bei der lichtorientierten Chloroplastenbewegung? Biochem Physiol Pflanz, in press
22. Scholz A (1976) Lichtorientierte Chloroplastenbewegung bei *Hormidium flaccidum*: Verschiedene Methoden der Lichtrichtungsperzeption und die wirksamen Pigmente. Z Pflanzenphysiol 77:422–436
23. Seitz K (1967) Eine Analyse der für die lichtabhängigen Bewegungen des Chloroplasten verantwortlichen Photoreceptorsysteme bei *Vallisneria spiralis ssp. torta*. Z Pflanzenphysiol 57:96–104
24. Seitz K (1970) Zur Frage der Jodid-Wirkung auf die Starklicht-Bewegung der Chloroplasten von *Vallisneria spiralis ssp. torta*. Z Pflanzenphysiol 62:63–69
25. Song PS, Metzler DE (1967) Photochemical degradation of flavins – IV. Studies of the anaerobic photolysis of riboflavin. Photochem Photobiol 6:691–709
26. Song PS, Moore TA (1968) Mechanism of the photodephosphorylation of menadiol diphosphate. A model for bioquantum conversion. J Am Chem Soc 90:6507
27. Widell S, Björn LO (1976) Light-induced absorption changes in etiolated coleoptiles. Physiol Plant 36:305–309
28. Zenk MH (1967) Untersuchungen zum Phototropismus der *Avena*-Koleoptile: I: Photooxidationen in vivo. Z Pflanzenphysiol 56:57–69

Molecular Interaction Between Riboflavin and Salicylic Acid Derivatives in Non Polar Solvents

B. S. YU, S. J. LEE, S. J. LEE and H. H. CHUNG [1]

Introduction

Salicylate is one of the oldest synthetic drugs and remains a widely used medical agent as analgetic and antipyretic. It is commonly known that the hydrogen bonding of salicylates in a biological system is related to their drug action [28]. It has been shown experimentally that higher concentration of salicylates results in marked stimulation of respiration, while low concentration depresses this function [2, 5, 10, 16, 25, 26]. It was reported that salicylic acid derivatives act as uncoupling agents on the isolated mitochondrial respiration, and lower the respiration rate of dinitrophenol uncoupled mitochondria, namely inhibition by them of state 3 mitochondrial respiration is accompanied by an increase in the oxidized state of all electron transports [6].

The electron transfer from nicotinamide adenine dinucleotide (NAD) to flavoprotein or the charge-transfer complex formed by them were studied by a number of authors to give an account of the function of respiratory chain [9, 20, 21, 22, 30]. Particularly, Honda has described that NADred-coupling enzyme (enzyme coupled-reduced nicotinamide adenine dinucleotide) complex convert spontaneously to the hypothetical intermediate as NADox-coupling enzyme (enzyme coupled oxidized nicotinamide adenine dinucleotide), which is considered indispensable to the formation of adenosine triphosphate (ATP) in the respiratory chain [12]. Simultaneously, the electrons in $(NADox)^{-2}$ can be transfered to flavoprotein.

On the basis of these facts, it is generally agreed that salicylates, uncoupling agents, act by causing the breakdown of some high energy intermediate involved in the phosphorylation process, but the most reliable mechanism has not been found yet.

However, the mechanism of action of salicylate may be described as follows. The first, salicylate, associates with the adenine moiety of flavin adenine dinucleotide (FAD) or NAD, the second, inhibits the interaction of the flavin moiety with the adenine moiety of FAD or NAD, the third affects the electronic environment due to the association with flavin, and the fourth inhibits the interactions of FAD and flavin mononucleotide (FMN) with apoprotein. So it seems worthwhile to examine the molecular interaction between salicylate and riboflavin, which concerns the respiratory chain metabolism.

1 Department of Physical Pharmacy, College of Pharmacy, Seoul National University, San 56-1, Shinrim-Dong, Kwanak-Ku, Seoul 151, Korea

Blue Light Effects in Biological Systems
ed. by H. Senger

In this paper a detailed analysis of the NMR and absorption spectra of the complex will be presented and the structure of the complex be proposed. The present results might provide a basis for interpreting their mode of action.

Experimental

Materials. Riboflavin-2′,3′,4′,5′-tetrabutyrate (Dae Woong Pharm. Co. Ltd., Seoul) (I) was obtained commercially. It was recrystallized from chloroform and its purity was confirmed by thin layer chromatography. Salicylic acid (Merck, Darmstadt) (II), aspirin (Il Sung Corp., Seoul) (III) and salicylamide (Il Sung Corp., Seoul) (IV) were used after recrystallization from chloroform. Deuterochloroform was purified by passing through alumina gel column, 5 cm in length. Carbon tetrachloride was treated with methanolic potassium hydroxide solution, washed with water several times, dried with calcium chloride overnight and distilled fractionally over phosphorous pentoxide through a 120 cm column packed with glass helices. The distillate was refluxed and redistilled fractionally. Benzene was treated with concentrated sulfuric acid several times. After repeated potassium hydroxide solution washing, it was dried with calcium chloride overnight and fractionally distilled over phosphorous pentoxide through a 120 cm column packed with glass helices and then crystallized two times. It was refluxed and again fractionally distilled.

Methods. ^{1}H-NMR spectra were recorded on a 90 MHz spectrometer equipped with a temperature control unit (Perkin-Elmer R32 NMR Spectrometer). For the measurement of NMR spectra, the samples of various mole ratio of I and salicylate were dissolved in deuterochloroform. Sample volumes of 5 ml in 5 mm diameter tubes with a Teflon cap were used. Tetramethylsilane included at 0.3% in deuterochloroform was used to provide the field lock signal and chemical shifts were measured from the signals. The probe temperature was 35°C. Absorption spectra were measured in a ultraviolet-visible spectrophotometer (Unicam SP 1750 UV Spectrophotometer) connected to a linear record (Unican AR 25 Linear Recorder). Fused quartz cells of 10 mm light path were used, each of which was fitted with a Teflon stopper. The slit width was chosen in such a way that the effective band width of exit beam might be sufficiently narrow as compared with the width of absorption bands. In order to determine the association constant of complex of the absorption spectra as uniquely as possible, we used ternary solution which contains a small amount of I and II (or III) in nonpolar solvents such as carbon tetrachloride and benzene. For each set of I and II (or III), several solutions containing equal quantity of I and different amount of II (or III) were prepared. The actual concentration of the former was of the order of 10^{-5} M (3×10^{-5}, 5×10^{-5} and 8×10^{-5} M), while that of the latter ranged from 0 to 10^{-2} M (or 2×10^{-2} M). The concentration of I is small enough to neglect safely the effect of self-association of I molecules. The absorbance changes of I were measured at 436 and 446 nm and the association constants were calculated. The experimental data were treated statistically.

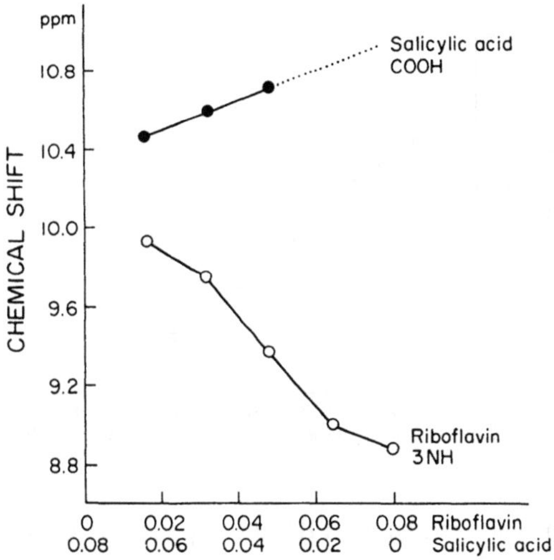

Fig. 1. Stoichiometric pairing of I and II. Dependence of the chemical shifts of the riboflavin (I) 3-NH and salicylic acid (II) carboxylic hydroxyl protons in a mixing experiment of I and II. Total concentration is constant at 0.08 M

Result

NMR Spectra. The proton magnetic resonance spectra of riboflavin derivative and salicylic acid derivatives in deuterochloroform are relatively simple in the downfield region. In the spectrum of II in deuterochloroform, the chemical shift of the carboxylic hydroxyl proton signal is observed at low field region (below 10 ppm), the phenolic hydroxyl proton signal does not appear at low concentration but at high concentration it is observed weakly, and the benzene ring protons are observed in 6.9–8.1 ppm region. The carboxylic hydroxyl signal of III does not appear. In the spectrum of IV, the chemical shift of the amino proton signal observed at about 6 ppm and that of hydroxy proton signal observed in the low field region (below 12 ppm). As the concentration of IV increases, the hydroxy proton signal moved slightly but the amino proton signal shifted downfield. The 3-NH proton of I is the most unshielded proton and occurs further downfield than all the other resonances. The 6,9-CH protons of I are both downfield from the residual nondeuterated chloroform peak. These two protons show concentration independence. When these protons take part in hydrogen bonding to form cyclic self-association dimers or I-salicylate pairs or triplets, they become less shielded and their resonances shift downfield. Thus the exact positions of the hydroxy, amino and imino resonances depend on the degree of association and hydrogen-bond formation, and therefore, both vary with temperature and concentration.

In order to obtain information on the characteristics of complex in deuterochloroform, experiments were performed at constant total concentration. Unlike the results obtained from optical methods, these curves do not show a maximum (or minimum) at the relative concentrations indicative of the stoichiometry of the reaction [14]. Figure 1 shows the dependence of the chemical shifts for the 3-NH proton of I and the carboxylic hydroxyl proton of II when I is mixed with II. The 3-NH proton signal of I shifts downfield as the relative concentration of II increases, and the carboxylic hydroxyl proton signal of II also moves downfield as that of I increases. From these observations we infer that the association between I and II is stronger than the self-

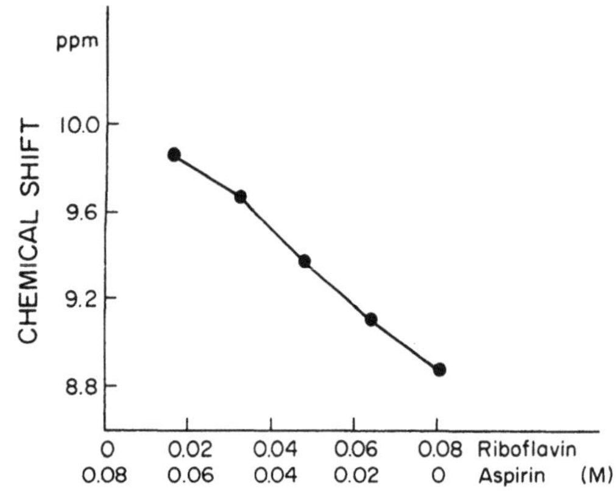

Fig. 2. Stoichiometric pairing of I and III. Dependence of the chemical shifts of the riboflavin (I) 3-NH and aspirin (III) carboxylic hydroxyl protons in a mixing experiment of I and III. Total concentration is constant at 0.08 M

association of these two compounds. When I is mixed with III, the dependence of the chemical shifts on the 3-NH proton of I is similar to the case of II (Fig. 2), while in the case of IV, the 3-NH proton signal of I changes slightly but the amino proton signal of IV moves downfield a little as the concentration of I increases and that of IV decreases (Fig. 3). We infer that the association between I and IV is stronger than the self-association of IV, and approximately the same as the self-association of I. It may be assumed that the shape of the slope of all the curves in three of these cases indicates that 1:1 complexes are formed according to Katz [14, 15].

The chemical shifts of the 3-NH proton of I are plotted against the concentration of the three salicylic acid derivatives at 35°C, keeping the concentration of I constant (0.04 M) (Fig. 4). The slopes of the 3-NH proton curves of I for II and III are much greater than that of the chemical shift due to the self-association of I, while that for IV is slightly greater. One can immediately interpret these data to indicate that I associates somewhat more strongly with II or III than with IV. These data are consistent with the former data. The corboxylic hydroxyl proton signal of II is more upfield as the concentration of II increases, keeping the concentration of I constant (not shown). From this fact alone, one can infer that the association between I and II is stronger

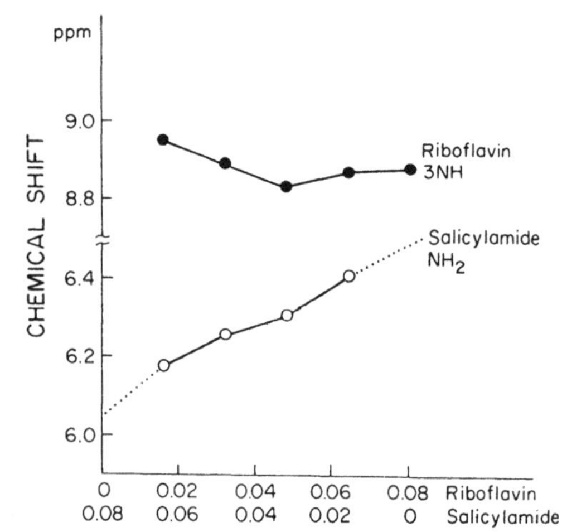

Fig. 3. Stoichiometric pairing of I and IV. Dependence of the chemical shifts of the riboflavin (I) 3-NH and salicylamide (IV) amino protons in a mixing experiment of I and IV. Total concentration is constant at 0.08 M

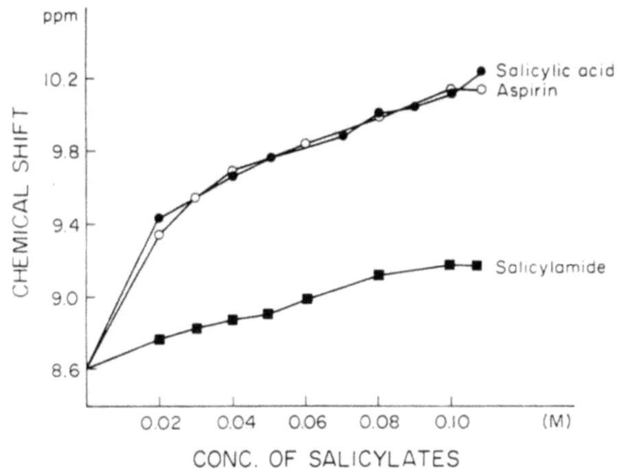

Fig. 4. Effects of the concentrations of salicylates on the chemical shifts of the riboflavin (I) 3-NH proton in deuterochloroform, keeping the concentration of I constant (0.04 M)

than the self-association of II and the mole ratio of complex is 1:1 type. Figure 5 shows that the chemical shift of the carboxylic hydroxyl proton and phenolic hydroxyl protons of II are plotted against the concentration of I, keeping the concentration of II constant (0.1 M). The curve of the carboxylic hydroxyl proton of II is convex upward, while the curve of the phenolic proton is concave upward. As the concentration of I increases, the carboxylic hydroxyl proton signal appears to broaden, whereas the phenolic hydroxy proton signal sharpens. The downfield shift and broadening effect of the carboxylic hydroxyl proton are assumedly due to the fact that this proton takes part in hydrogen bonding to form a I-II complex. The upfield shift and sharpening effect of the phenolic hydroxyl proton indicate the possibility of stacked configuation or two components of the complex being not in one plane. The 3-NH proton signal of I is more upfield as the concentration of I increases, keeping the concentration of II constant (not shown). From this fact, one can suggest that the associa-

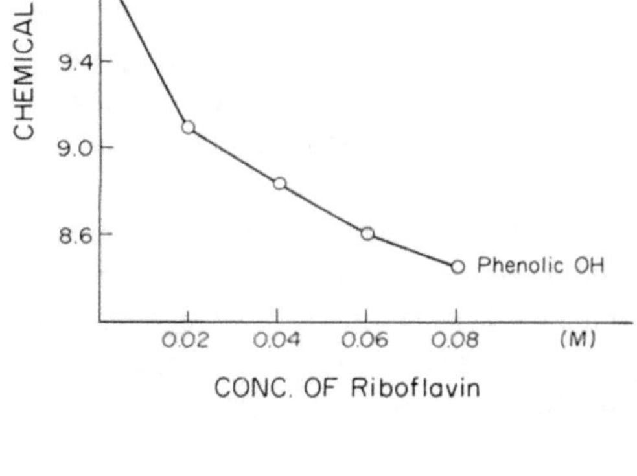

Fig. 5. Effects of the concentration of I on the chemical shifts of the carboxylic hydroxyl and the phenolic hydroxyl protons of II in deuterochloroform, keeping the concentration of II constant (0.1 M)

Fig. 6. Effects of the concentration of I on the chemical shifts of the salicylamide (IV) amino protons in deuterochloroform, keeping the concentration of IV constant (0.05 M)

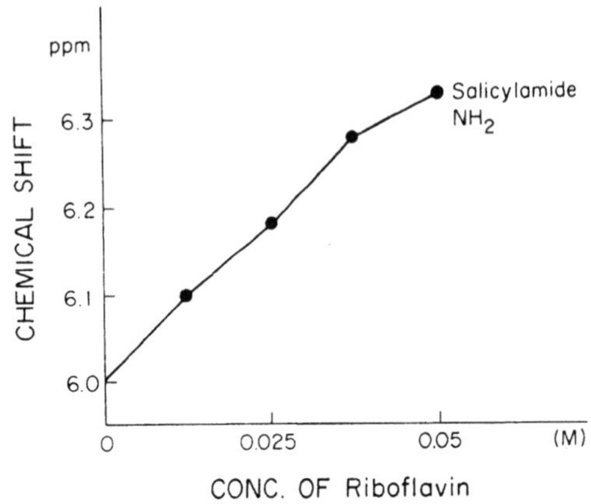

tion between I and II is stronger than the self-association of I and the mole ratio of the complex is 1:1 type. A similar phenomenon was observed with III upon addition of I, keeping the concentration of III constant (0.06 M) (not shown). In the case of IV, the curve of the amino proton is straight upward (Fig. 6), whereas the phenolic hydroxyl proton signal decreases in intensity and then disappears without shielding effect as the concentration of I increases (not shown). We infer that only the amino protons take part in hydrogen bonding to form a I-IV complex. Besides the direct observation of complex formation in the downfield shift of the protons directly involved, a secondary effect may be also possible due to alteration of the electron density at sites where several bonds are removed from donor or acceptor atoms for the hydrogen bond. Small changes may be produced in the ring currents and the electron densities in the molecules accepting the hydrogen bonds, resulting in shifts in the resonances of the other protons attached to the molecules. It is also possible to produce small changes in the ring resonance positions due to neighboring unsaturated ring systems. The resonances of the 6,9-CH protons of I are not shielded. But small upfield shifts are observed in all of the ring protons of II and III. In the case of IV, some of the ring protons move downfield slightly and others shift upfield.

Absorption Spectra. As shown in Fig. 7, a marked spectral change was produced upon adding II to I in nonpolar solvent, carbon tetrachloride. It may safely be assumed that this spectral change is due solely to the formation of a hydrogen bond between I and II. This view is further substantiated by the existence of several isosbestic points in each set of spectra. The shorter wavelength band shifts to the red, while the longer one shifts to the blue, and hyperchromism of the longer one is observed. Frequency shifts due to hydrogen bonding were determined from the spectra of the free and hydrogen-bonded species (resulting from the difference between the energy of the hydrogen bond in the excited state and that in the ground state). Similar phenomena were observed with III or IV in carbon tetrachloride. In the other nonpolar solvent, benzene, similar phenomena were also observed. Determination of the association constant of the complex is possible from sprectral data if the complex shows significantly different spectrum from that of its component. The association constant of the hydrogen bonding complex was obtained by using the following equation [1]:

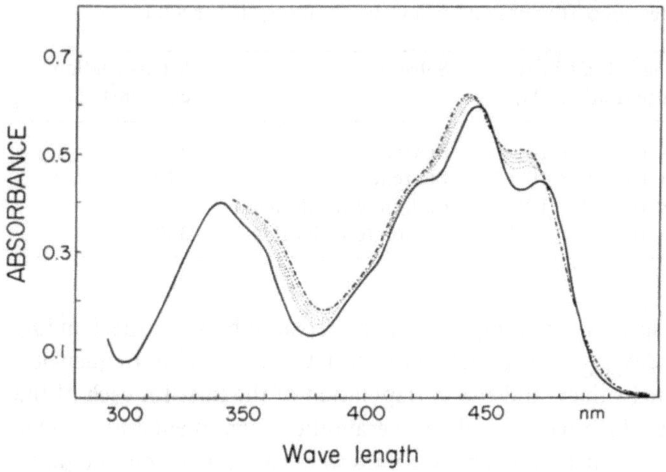

Fig. 7. Effects of salicylic acid on the absorbance spectrum of I (5×10^{-5} M) in carbon tetrachloride. II is added from 0 to 1×10^{-2} M. *Solid curve* free molecule; *dotted curve* spectra in the presence of II in order of increasing concentration of II; *broken curve* in the presence of 1×10^{-2} M salicylic acid

$$\frac{1}{\Delta\epsilon_{obsd}} = \frac{1}{K(\epsilon_c - \epsilon_a)} \frac{1}{C_b} + \frac{1}{\epsilon_c - \epsilon_a} \qquad (1)$$

Where $\Delta\epsilon_{obsd} = \epsilon_{obsd} - \epsilon_a$, ϵ_a is the molar absorptivity of I in uncomplexed form, ϵ_c is that of I in pure complex, ϵ_{obsd} is the observed absorptivity of I in complexed media, C_b is the concentration of salicylates on some scale defined in experimental section and K is the association constant of the complex. All these absorptivity refer to a given frequency. Equation (1) can be easily derived on the assumption that the concentration of I is negligibly small compared to that of salicylate; such is actually the case in the present study. From Eq. (1), a plot of $1/\Delta\epsilon_{obsd}$ vs $1/C_b$ is expected to give a straight line. When both the concentration of I and the cell length are kept con-

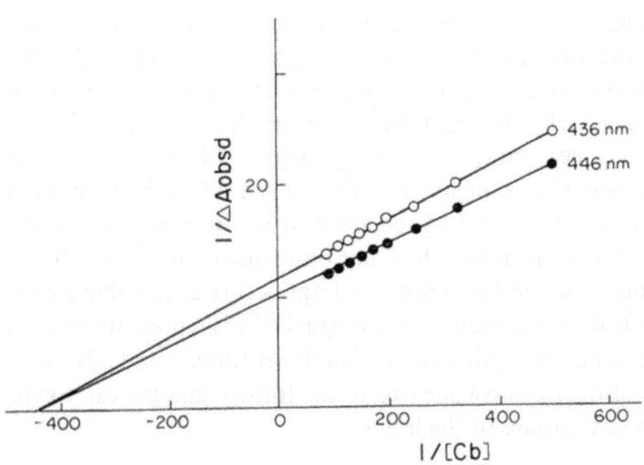

Fig. 8. Relation between $1/A_{obsd}$ and $1/C_b$ values of riboflavin (I)-salicylic acid (II) system in benzene obtained according to Eq. (1)

Table 1. Measured and calculated properties of I and salicylates on absorption spectra

Salicylates	Range of salicylate concentration (M)	Solvent	K (Association constant) M^{-1}
Salicylic acid (II)	$1 \times 10^{-3} - 1 \times 10^{-2}$	Benzene	450
Aspirin (III)	$1 \times 10^{-3} - 1 \times 10^{-2}$	Benzene	370
Salicylic acid (II)	$1 \times 10^{-3} - 1 \times 10^{-2}$	Carbon tetrachloride	435
Aspirin (III)	$2 \times 10^{-4} - 1 \times 10^{-3}$	Carbon tetrachloride	350

stant throughout a set of spectra, ϵ's in Eq. (1) may be replaced by the corresponding absorbances A's. Thus $1/\Delta A_{obsd}$ was plotted against $1/C_b$ for various frequencies. These plots gave good straight lines, and the extrapolation of the lines for each of the salicylates led to K values which were equal to one another experimental error. One example of the plots is shown in Fig. 8. The K values were calculated from the plots, as summarized in Table 1. These association constants were obtained by averaging the values from three different concentrations of I (3×10^{-5}, 5×10^{-5} and 8×10^{-5} M).

Discussion

High-resolution proton magnetic resonance techniques provide a direct observation of the hydrogen bonded association in solution. As shown in the NMR spectra, it is easily noticed apparently that the 3-N imino proton of I, the carboxylic hydroxyl protons of II and III, and amino protons of IV seem to participate in the association. The NMR method utilized here measures the chemical shifts of the donor proton in hydrogen-bond formation. The acceptor atoms are, of course, protonless, and do not give rise to resonances. Hence no direct information can be gained on the identity of these acceptors. By the previous report [13] using IR spectra, it could be concluded which carbonyl group at the 2-C or 4-C position of riboflavin is used. In the 6 μ region of infrared spectra, it was easily noticed that only the association through the 2-C carbonyl group of the isoalloxazine ring is predominent. This is consistent with the fact that in the sequence of the hydrogen bonding on the isoalloxazine ring, the 2-C carbonyl group is preferred to the 4-C carbonyl group, as suggested by Yagi et al. [29]. The suggestions can be illustrated by the fact that the proton affinities of heteroatoms in the isoalloxazine ring are actually different from one another [12, 23. 25, 27]. According to what Honda [12] determined, the charge density of the 2-C carbonyl oxygen atom in the isoalloxazine ring is higher than that of the 4-C carbonyl oxygen atom and the 3-NH proton is very labile and a good electron acceptor since the charge density of 3-N imino nitrogen atom is higher than any other heteroatom. The change of the absorption spectra supports that hydrogen bondings occurs at the above-mentioned binding sites [19]. Thus the most probable hydrogen bond is represented by the following. I and II form the 1:1 cyclic hydrogen bonded dimer through the 3-N imino and the 2-C carbonyl groups of the isoalloxazine ring of the former, and the carboxylic hydroxyl and carboxylic carbonyl groups of the latter.

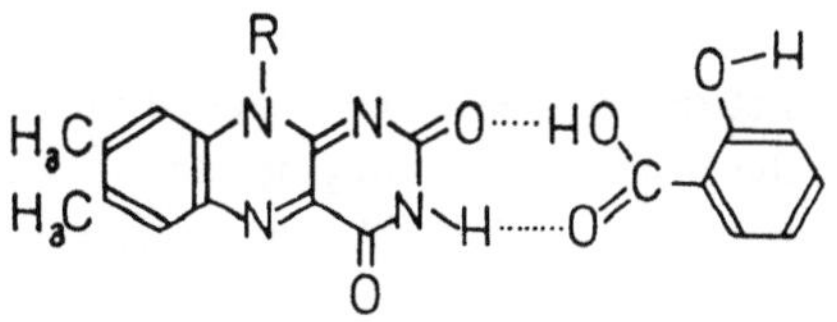

I and III form the 1:1 cyclic hydrogen bonded dimer by the same mode.

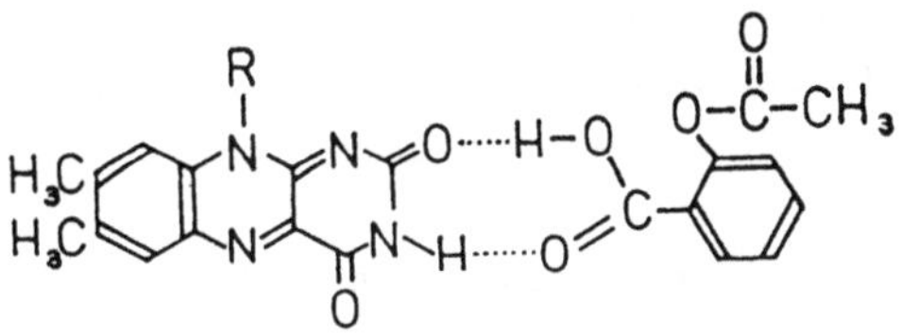

IV forms the 1:1 cyclic hydrogen bonded dimer with I through the 3-N imino and the 2-C carbonyl groups of the isoalloxazine ring, and the amino and the carbonyl groups of IV. These binding ratios are reasonable.

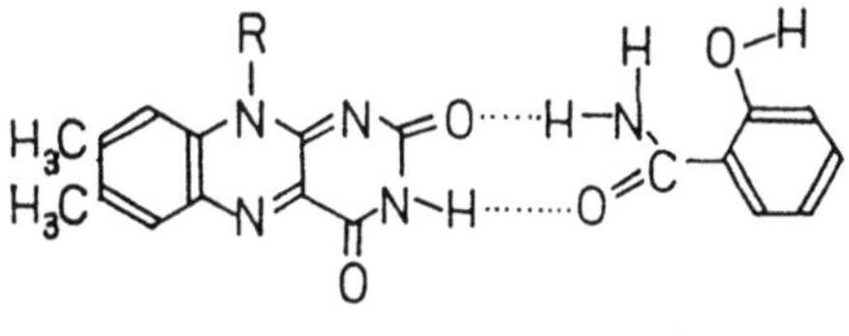

Song et al. demonstrated that the frontier orbital density of N-5 is the highest of all the atoms in the isoalloxazine ring [23]. Therefore, N-5 seems to be the entrance of electrons. The formation of various types of hydrogen-bonding affects the frontier orbital density of N-5 of the isoalloxazine ring [19]. When we consider that salicylates interact with riboflavin at the above-mentioned sites, the electron affinity of the isoalloxazine ring increases and the frontier electron density of N-5 increases, which accelerate the electron flow from the substrate to coenzyme [3, 4, 11, 19, 23, 24, 27, 29]. The hydrogen bonding involved in flavoproteins may be considered to be significant for their catalytic activity.

In nonpolar solvents, carbon tetrachloride and benzene, marked absorption spectral changes due to hydrogen bonding were produced upon adding salicylates to I. Especially, frequency shifts were observed. It is known that the energy of the hydrogen bond in the excited state is differentiated from that in the ground state. The association constants of I with II or with III in nonpolar solvents are very great, but with IV is too small to be calculated from the absorption spectral data. Similar results could be obtained from the NMR data. On the other hand, in chloroform solution, the absorption spectral change due to hydrogen bonding was produced weakly (not shown). We infer that this fact is attributed to the solute-solvent interaction.

Although II, III and IV are structurally very similar, some of their effects on the body may be quite diverse. Sodium salicylate and III effectively depress oxidative phosphorylation while IV is a weak or ineffective uncoupling agent. III is more potent than sodium salicylate as an analgetic and antipyretic, and IV is much less effective than either [7, 8]. These facts are consistent with our experimental results. The use of deuterochloroform, carbon tetrachloride and benzene as solvents in these experiments enables us to observe these interactions free of strong solute-solvent association effect. This environment may mimic in some ways the inside of the enzyme-substrate complex from which water may be excluded. It is suggested that FAD in enzyme is surrounded by a hydrophobic environment in even aqueous phase [17, 18]. If one considers that oxidation-reduction process in these systems occurs in a lipophilic environment rather than in an aqueous environment, it may be worthwhile to research concretely the direct interaction between salicylates and riboflavin moiety in a nonpolar system. Further quantitative analysis of these data should lead to a more precise fundamental understanding of riboflavin-salicylate complex formation.

Acknowledgment. This research was supported by the 1981 grant from the Korea Science and Engineering Foundation.

References

1. Baba H, Suzuki S (1961) J Chem Phys 35:1118
2. Brody TM (1956) J Pharmacol Exp Ther 117:39
3. Bruice TC (1976) Prog Bioorg Chem 4:2
4. Edmonson DE, Barman B, Tollin G (1972) Biochemistry 11:1133
5. Fishgold JT, Field T, Hall VE (1951) Am J Physiol 164:727
6. Galen FX, Truchot R, Michel R (1974) Biochem Pharmacol 23:1379
7. Gilman AG, Goodman LS, Gilman A (1980) In: The pharmacological basis of therapeutics. Macmillan, New York, p 668
8. Goth A (1976) In: Medical pharmacology. CV Moshy, Saint Louis, p 328
9. Grabe B (1958) Biochim Biophys Acta 30:560
10. Graham JDP, Parker WA (1948) Q J Med 17:153
11. Hemmerich P (1976) Fortschr Chem Org Naturst 33:451
12. Honda M (1971) J Phys Soc Jpn 31:1196
13. Huh JW, Yu BS (1976) J Pharm Soc Korea 20:130
14. Katz L (1969) J Mol Biol 44:279
15. Katz L, Penman S (1966) J Mol Biol 15:220
16. Levy J (1946) Bull Soc Chim Biol 28:338
17. Massey V, Gather H (1965) Biochemistry 4:1161
18. Muller F, Mayhew SG, Massey V (1973) Biochemistry 12:4654
19. Nishimoto K, Watanabe Y, Yagi K (1978) Biochim Biophys Acta 526:34
20. Pullman B, Pullman A (1959) Proc Natl Acad Sci USA 45:136
21. Sakurai T, Hosoya H (1966) Biochim Biophys Acta 112:459
22. Shifrin S (1964) Biochim Biophys Acta 81:205
23. Song P-S, Choi JD, Fugate RD, Yagi K (1976) In: Singer TP (ed) Flavins and flavoproteins. Elsevier, Amsterdam, p 381
24. Song P-S, Sun M (1974) In: Pullman B, Bergmann E (eds) Chemical and biochemical reactivity. Isr Acad Sci, Jerusalem, p 407
25. Spenny JG, Brown M (1977) Gastroenterology 73:995
26. Sproull DH (1954) Br J Pharmacol 9:262
27. Sun M, Song P-S (1973) Biochemistry 12:4663
28. Wilson CO, Gisvold O, Doerge RF (1977) In: Textbook of organic medicinal and pharmaceutical chemistry. Lippincott, Philadelphia, p 47
29. Yagi K, Ohishi N, Nishimoto K, Choi JD, Song P-S (1980) Biochemistry 19:1553
30. Yomasa S (1967) Prog Theor Phys Suppl 40:249

ESR of Transition Metal "Chromophores" in RuBPCase. Supporting Neutron Activation and Optical Data. Concepts of UV Light Activation Through Transition Metal Chromophores

L. S. DALEY[1], J. E. DRUMHELLER[2], M. OTTE[3], S. AMANI[1], and L. J. THERIOT[1]

Introduction

Objectives of Research

Our objective is an understanding of the chromophore through which the long wavelength ultraviolet (UV) activation of ribulose-1,5-bisphosphate carboxylase [19, 20, 21, 71, 74] (RuBPCase) is mediated. The part of the investigation reported here deals with the possibility that this chromophore could be a result of interaction of transition metal ions with the enzyme, RuBPCase; or its phosphorylated substrate, ribulose-1,5-bisphosphate (RuBP). Although there have been reports to the contrary, [12, 40, 43] transition metal ions are found bound to RuBPCase, as reported originally, [75, 77, 78] confirmed by us [67] and found associated by others [5, 35, 62].

The Existence of Transition Metals Bound to RuBPCase

Transition metals (or elements) are those elements with partially filled d (or f) shells in their different oxidation states. Transition metal cations can form compounds that have redox and absorbance properties that make them potentially suitable chromophores for UV and visible light effects. For these reasons transition metal cations have been suggested as chromophores in biological processes.

The first mention that we could find of this hypothesis applied to RuBPCase states: . . ."bound Cu^{2+} may be involved only in the maintenance of a particular conformation of ribulose 1,5-diphosphate carboxylase, (but) it is tempting to envisage additional roles for this bound metal ion." In this context, the light activation factor for this enzyme described by Wildner and Criddle [71] may be of interest. The bound Cu^{2+} might provide a part of the binding site for this factor and could also be involved in the mechanism for transmission of the light signal since Cu^{2+} is a good electron acceptor which is readily reduced to Cu^{1+}" [77]. However, after the first reports [75, 77, 78] of the presence of copper in RuBPCase other reports followed [12, 40, 43] that denied the presence of copper in RuBPCase. These other reports [12, 40, 43] lead to difficulties in theoretical considerations of RuBPCase activation [67], and; these

1 Chemistry, NTSU, Denton, TX 76203-5068, USA. Present address: Horticulture, OSU, Corvallis, OR, 97331
2 Physics, MSU, Bozeman, MT 59715, USA
3 Texas Engineering Experiment Station, Texas A & M, College Station, TX 77843, USA

Blue Light Effects in Biological Systems
ed. by H. Senger

Table 1. Summary of copper analysis of fraction I protein. Comparison with values reported in the literature. (After [67])

Source of fraction I (RuBPCase)		Copper content reported	Reference
Atomic absorption data			
	Matrix effect uncorrected	Cu/E[c]	
Spinach		0.14	[40]
Tobacco[a]		0.15	[12]
Parsley		0.18	[43]
Spinach[a]		0.05	[34]
Spinach		0.10	[67]
Tobacco[a]		0.07	[35]
Alfalfa[a]		0.13	[35]
Potato[a]		0.09	[35]
Tomato[a]		0.40	[35]
		0.146 ± 0.1038	
Minus tomato		0.114 ± 0.044	
	Method of matrix effect compensation attempted	Cu/E	
Spinach	By dilution	0.08	[67]
Spinach	With 2% NaOH	0.26	[67]
Spinach	By standard addition and double reciprocal plot	1.0	[67]
Spinach	With 2% NaOH and gel chromatographic separation	1.10	[67]
Spinach	Matrix compensated in original reference (uncorrected value not given)	0.95	[77, 78]
ESR data			
	% Reported by atomic absorption	Cu/E	
Spinach	Copper calculated as Cu^{2+} . . . 80%	0.75	[77, 78]
Parsley	Total copper not calculated because of different signals	+++[b]	[5]
Parsley	No signal observed . . . 0%	0.00	[43]
Spinach	Total copper not calculated because of complex signal	+++[b]	See Fig. 3
Other Methods			
	% Reported by atomic absorption	Cu/E	
Tobacco[a]	By emission no matrix compensation . . . 213%	0.32	[12]
Parsley	Biquinoline reaction, (Pooled oxidase and carboxylase)	0.20	[5]
Spinach	Neutron activation . . . 121%	1.15	[77, 78]
Tobacco[a]	Neutron activation . . . 133%	0.20	[12]

[a] Crystalline preparation
[b] Present in significant amounts
[c] Atom copper/molecule fraction I

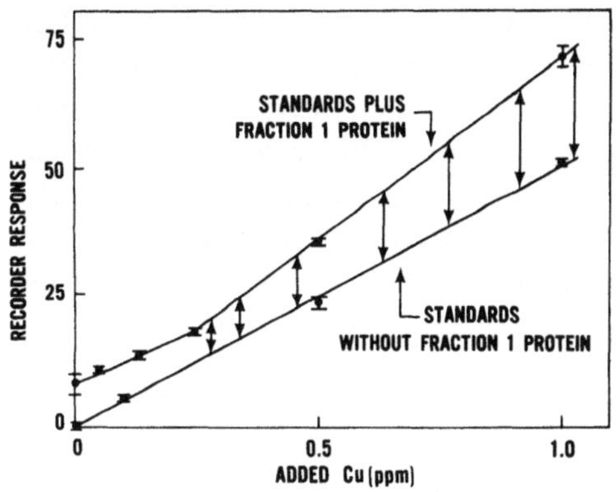

Fig. 1. Effect of standard additions of copper chloride on atomic absorption detector response to copper content of RuBPCase. The data points reflect maximum and minimum instrument response for triplicate sample analysis. The vertical double-headed arrows indicate the differing and increasing response of the detector to the same amount of RuBPCase (10 mg/ml), at different and increasing amounts of added $CuCl_2$. This constitutes evidence for a matrix effect which causes underestimation of the amounts of copper present. This is attributed to the matrix (the other material, such as protein, in which the copper is found). This figure is modified from previous work (Theriot et al. 1980)

theoretical difficulties, plus some unexplained data [67] including copper interactions with UV activation of RuBPCase, [19, 20, 21] suggested a reinvestigation [67] of the matter of copper content of this enzyme. All known data on copper in RuBPCase is tabulated (Table 1), once this is done it becomes apparent that the experimental question is not, "is copper present in RuBPCase?", but "how much copper is present in RuBPCase?". Atomic absorption reports on copper content of RuBPCase can be divided into two groups; (a) values of 1 atom per molecule of enzyme, and (b) lower values. The lower values were evaluated statistically (Table 1) and had a mean of 0.11 atoms per molecule, but with a very small standard deviation 0.04 atoms per molecule. This small standard deviation strongly suggested a systemic error. To investigate this possibility we tested for matrix effects of RuBPCase on bound copper [67]. Matrix effects have been reported, previously, for copper in protein-containing mixtures [37]. Figure 1 demonstrates the presence of a systemic error in the atomic absorption analysis of copper bound to RuBPCase. When RuBPCase is analyzed for copper, without standard additions, the amount of copper is seriously underestimated. Standard additions of $CuCl_2$ increase the amount of copper found bound to RuBPCase (Fig. 1). Double reciprocal plots of this change lead to a value of about one atom per molecule (Table 1); a value consistent with analysis of this preparation obtained after sodium hydroxide degradation and gel chromatography (Table 1). However, these corrections do not account for the Electron Spin Resonance (ESR) and Neutron Activation (NA) data reported. This data is also very variable (Table 1). ESR signal intensity of transition metals bound to RuBPCase is much reduced. Free Mn^{2+} in solution has intense isotropic ESR spectra; however, when Mn^{2+} binds to small molecules or proteins the ESR signal is often broadened and difficult to detect [16]. Thus, when RuBPCase is added to a solution of Mn^{2+} the intensity of the aquo complex is much reduced, but no changes in ESR spectra were reported. Thus, ESR is not a simple means of analysis of these metals when they are bound to RuBPCase. A preliminary report of ESR signals of RuBPCase included in the first Blue Light Symposia, showed that the signals were complex [21]; more detailed spectra are presented here. Given these considerations the value of ESR to *quantify* transition metals on RuBPCase, without the proper controls is somewhat inadequate. Neutron activation data presented in Table 1 is also difficult to interpret, since, not only does this data vary between reports, but it also varies from different methods presented in the same report. An investigation of this matter and a study of the effects of the geometry of the sample used for neutron activation is also included in this report.

Chromophores of RuBP/RuBPCase

Information on chromophore characteristics of RuBP, RuBPCase and transition metals-RuBPCase/RuBP complexes is scattered throughout the literature: (a) A 290 nm shoulder on RuBPCase has been observed [12]. (b) RuBP forms a UV (280 nm) chromophore on association with RuBPCase [56]. (c) This chromophore is proportional to added RuBP, or 2-carboxy-D-ribitol-1,5 diphosphate (CRDP) [65]. CRDP is an analogue of the six carbon intermediate of the RuBPCase reaction. This chromophore is saturated at a ratio of about ten atoms of CRDP per molecule of RuBPCase. (d) Mn^{2+} forms colored complexes with RuBP [15], and iron a colored complex with RuBPCase [27].

Influence and Binding of Transition Metal Cations on RuBPCase

In addition, the binding of transition metal cations to RuBPCase is frequently reported; [8, 13, 14, 16, 27, 41, 44, 46, 47, 48, 58, 61, 65, 75, 77, 78] and the binding of these metal cations to RuBPCase is a result of complex formations which seem to include substrate contributions. These metals influence the oxygenase and carboxylase activities of RuBPCase (see Table 2). Changes in the redox conditions of these metals influence RuBPCase activity. Redox reactions, involving thioredoxin [79], have also been shown to influence RuBPCase. In addition, copper^{3+} complexes of simple peptides have UV spectra (λ max 278 nm, 395 nm; ϵ, 12 and 5.3 mM^{-1} cm^{-1}), and are reduced by Fe^{2+} or light [38]. Thus, such complexes of transition metals, RuBP and RuBPCase exist, and may have appropriate properties for our purposes, while other types of chromophores are less applicable [63, 71].

On UV Light Activation In Vivo

The in vivo light activation of RuBPCase is well documented [4, 22, 45, 50, 53, 64] and may be related to blue light inhibition of RuBPCase [17, 18, 69, 70]. The rational for such a mechanism is readily apparent [23]. However, although the photosynthetic contribution of near UV is not negligible (Cadwell this Vol.), UV light activation in vivo is rarely reported and apparently only occurs under conditions of strong photosynthetically active radiation (PAR) [1, 2, 10, 66]. It is also complicated by blue light effects [17, 18, 69, 70]. A theoretical concept of seasonal UV signals in the lower and middle latitudes of our planet, can be inferred (Caldwell 1983, personal comm.) from data showing seasonal fluctuations of UV intensity in these regions (Caldwell this Vol.). In the higher latitudes these seasonal UV fluctuations are less apparent (Caldwell this Vol.), but could be of more importance given the greater weather variation in the yearly cycle at these higher latitudes.

Material and Methods

Spinach RuBPCase, of documented activity [36] was provided by Sigma Chemical Co. (St. Louis, MO). The enzyme was further purified, amino acid composition verified [59], etc. as described previously [67]. *ESR analysis* were performed on a Varian (Palo Alto, California) E-1009 X-Band EPR provided with an EPR data acquisirion system. The ESR analyses were performed by one of us (J.E.D.) at the Montana site. The *UV spectra* were performed by another of us (S.A.) at the North Texas site. *The neutron activation* analysis was performed by a third coauthor (M.O.) at the College Station, Texas site. Irradiation (output 1 Megawatt) was performed with a TRIGGER reactor (General Atomic, San Diego, California). Manganese and copper were irradiated at a flux of 1.2×10^{13} neutrons cm^{-2} s^{-1} for 15 min. Iron and zinc were irradiated at a flux of 1×10^{13} neutrons cm^{-2} s^{-1} for 14 h. Radiation was detected with a Scorpio multichannel analyzing computer system with a Digital Equipment 11 Central processing unit (Canberra Comp., Meridian, Connecticut). The materials were placed unirradiated vials, after irradiation, and before analysis and weighed at the start and end of analysis.

Results and Discussion

Transition Metals Exist and Can Control Function in RuBPCase

Figure 2 shows neutron activation analysis of RuBPCase metal content. The analyses were performed with various amounts of RuBPCase as illustrated (Fig. 2). The results show that the weight of the sample influences the values obtained. The apparent explanation of this is: since protein powders have relatively low densities, and the vials used for analysis are relatively tall, adding more protein to the sample vial increases the height of the packed protein and, thus, the mean distance from the detector. The increased distance from the detector decreases the radiation received at the detector by a function proportional to the inverse square of the distance. When the values obtained are plotted (Fig. 2), the absolute proportions of the elements are one atom of copper, one of manganese and ten of iron per molecule (540,000 M.W./unit) of RuBPCase.

Many enzymes are activated by several different cations (e.g. 49). There are five divalent cations that activate RuBPCase: Mg^{2+}, Mn^{2+}, Fe^{2+}, Co^{2+}, and Ni^{2+} [68, 76] (Table 2). Cu^{2+} is reported either not to be active, or to inhibit RuBPCase (especially in the dark). Table 2 calculations based on the data of Johal et al. [35] show that the % RuBP-oxidase activity of crystalline RuBP is a function of copper but not iron, content. This activation is dependent on the oxidation state of the metals, for example, Fe^{3+} and Co^{3+} are inactive or inhibitory (Table 2). Moreover, the oxidation states of these elements are not independent of other cations, e.g., Ni^{3+}, Cu^{3+} peptides, which are reduced by long wavelength UV and blue light are also reduced by Fe^{2+} [38]. Moreover,

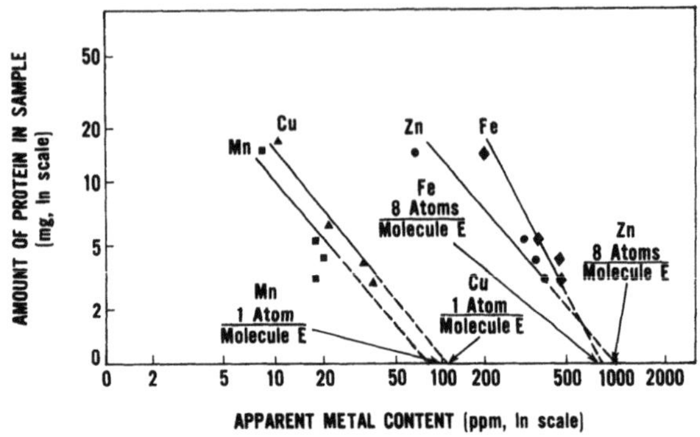

Fig. 2. Plot of neutron activation data. This plot shows apparent manganese, copper, zinc, and iron content (ppm) with different RuBPCase sample size (mg). This plot shows an apparent increase in Mn, Cu, Zn, and Fe content as sample size decreases. This kind of data analysis yields corrected values of about 1:1:10:10 atoms of Mn:Cu:Zn:Fe per molecule of RuBPCase. Details of analysis are found in Materials and Methods section

the oxidation states of these elements are not independent of other cations, e.g., Ni^{3+}, Cu^{3+} peptides, which are reduced by long wavelength UV and blue light are also reduced by Fe^{2+} [38]. The transition metal cations have differing ability to participate in the catalysis of the carboxylation and oxidation reactions of RuBPCase (Table 2). Thus, the reaction rate of in vivo carboxylation and oxidation of RuBP via RuPBCase is, in part, controlled by the nature and ratio of the cations that are bound to, or interact with, RuBPCase in situ. In addition, some of these metals have been localized on, or very close to RuBPCase in vivo [62]. Light can reduce Cu^{2+}, as well as, Cu^{3+} and, thus, can modulate RuBPCase activity in vivo if Cu^{2+} and Cu^{3+} are in appropriate "contact" with these other cations on, or "near", RuBPCase.

An interesting aspect of this proposed mechanism is that it does not require tightly bound cations in constant stochiometric ratios to be found on RuBPCase; but rather requires that the composition and ratio of these cations, and their oxidation states in vivo be appropriate to the RuBP carboxylation and oxidation requirements of the variable physiological functionality of each particular plant at every different environmental situation. In addition, since the cations have different binding affinities for RuBPCase (Table 2), Mg^{2+} being less actively bound than others, competition between different cations for sites on RuBPCase is dependent, not only on the amount, but the nature of the cations that are in contact with RuBPCase.

Transition Metals Interact on RuBPCase

A supporting, but not necessarily required, item for the hypothesis put forth in Section Introduction, is that the cations interact on RuBPCase. The criterion chosen for this is "coupling" or interaction of ESR signals of paramagnetic ions on RuBPCase. The experimental question asked is: does the ESR signal support interaction of the paramagnetic cations found by neutron activation on RuBPCase?

The ESR signal is complex (Fig. 3a); not only at the $G \cong 2$ region, (high field), but also at the $G \cong 4$ region (low field). The absorbance at $G \cong 4$ (low field) is considered to indicate interaction "coupling" between ESR active moieties [51]. Examination of

Table 2. Summary of known metal cation interactions with ribulose-1,5-bisphosphate carboxylase

Metal ions	Mg^{2+}	Mn^{2+}	Cu^{2+}	Fe^{2+}	Co^{2+}	Ni^{2+}
Other pertinent oxidation states	–	Not known if active	Cu^{1+}, Cu^{3+} not known if active	Fe^{3+} is inactive[a]	Co^{3+} formation may be inhibitory[k]	Ni^{3+} not known if active
Activities: Carboxylase						
K_a, $K_{1/2}$, K_m (μM metal ion)	1100[b] 1150[i] 843[n]	39[b] 110[i] 22[n]	Not active[c,n] inhibitory in dark[l]	80[n]	39[n]	350[n]
V_{max} $\frac{\mu \text{mol } CO_2}{\text{min} \times \mu \text{mol enz}}$	1060[b] 950[h] 491[i] 918[j] 891[n]	145[a] 45[h] 33[i] 156[j] 162[n]	Either not active[c,n] or inhibitory in dark[l] See also [9]	122[a] 103[n]	394[a] 97[n]	874[a] 378[n]
Oxygenase						
K_a, $K_{1/2}$, K_m (μM metal ion)	5500[a] 5900[h] 6000[i]	300[i]	Not active[n]	Oxidizes to Fe^{3+}[n]	–	–
V_{max} $\frac{\mu \text{mol } O_2}{\text{min} \times \mu \text{mol enz}}$	49[h] 45[i] 329[j] 351[l]	64[g] 30[i] 156[j] 157[l]	% Oxygenase increases with bound copper[m]	Oxidizes to Fe^{3+}[n]	20[g] 373[n]	81[n]
Binding: Number of sites per molecule	6.4[c] 10.0[d]	6.7[c] 3[e] 8[f] 9.6[o]	1.3[c] 1.1[l]	10[l]	11.2[k] 8.8[o]	4.1[c]
K_d (μM metal ion)	540[e] 1065[n]	60[e] 10[f] 57[n]	–	158[n]	75[n]	560[n]

[a] [68] The V_{max} normalized to mean carboxylase V_{max} for Mg^{2+} of all reports. [b] [52]. [c] [77]. [d] [65]. [e] [16]. [f] [48]. [g] [60]. V_{max} normalized to mean oxidase V_{max} for Mg^{2+} of all reports. [h] [61]. [i] [72]. K_a used, since activation and catalysis are very closely linked: and the cation may not need to rebind with each catalytic cycle. [j] [15]. [k] [58]. [l] [67]. [m] [35]. Recalculated from activities and copper content presented. [n] [13]. $K_{1/2}$ used as estimate of K_m. [o] [47]

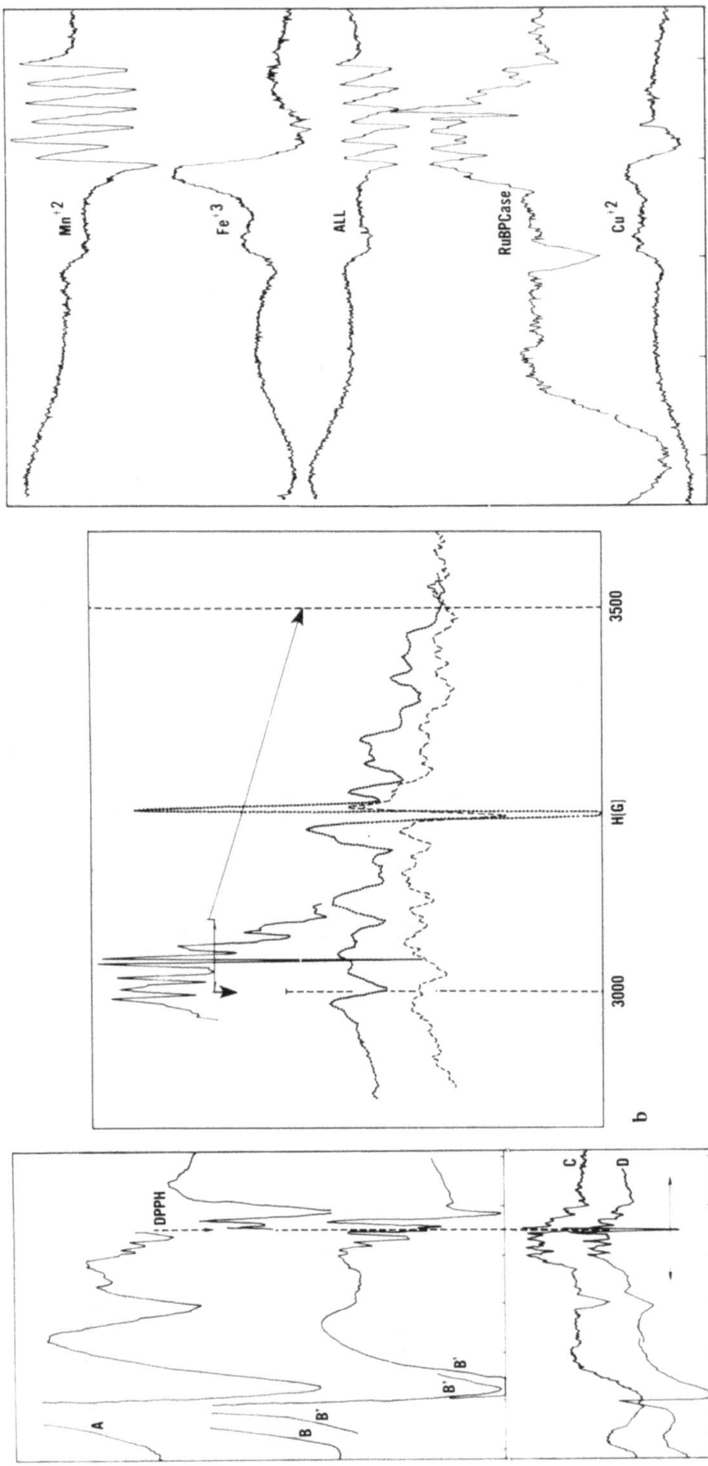

Fig. 3a–c. Electron spin resonance spectroscopy (ESR) of RuBPCase sample. **a** Spectra of: *A* room temperature powder; *B* −162° powder; *C* liquid sample room temperature; *D* frozen liquid sample at −162° C. Powder samples were lyophylized RuBPCase; liquid samples were in NH_4HCO_3 buffer and contained 10 mg/ml RuBPCase. The NH_4HCO_3 was found free of metal contaminants when analyzed by neutron activation. The microwave frequencies were: *A* 9.060; *B* 9.064; *C* 9.055; *D* 9.059 GH. The microwave power was 12.5, 12.5, 11.5, 12.5 mW. The receiver gain was 1.25 ×, 25 ×, 3.2 × and 1.25 × 10^4. The modulation frequency was 100 KHz. The modulation amplitude was 20 G. The time constant was 1/2 s. The field set was 3400 G. The scan range was 5000 G. The scan times were 8, 4, 16 and 16 mm. Note absorption (the ordinate indicates derivative of absorption) at G ≅ 4 or near 1500 G. This is taken to indicate coupling as described in Results and Discussion section. **b** Details of liquid and frozen samples of RuBPCase, as in **a**, at G ≅ 2 or near 3200 G. These details illustrate complexity of spectra and thus interactive phenomena as described in Results and Discussion section. This is taken to support the concept of coupling between Mn, Cu and Fe cations on RuBPCase. **c** Spectra of comparable doped bovine serum albumen (BSA). BSA doped with comparable amounts of Mn^{2+}, Cu^{2+}, Fe^{3+} separately and combined (All) is compared to an equivalent liquid room remperature sample of RuBPCase. This supports the concept that the binding of these cations to RuBPCase as opposed to BSA, is not one of random adhesions, but constitutes a coupled interactive system

the number, intensity, and spacing of peaks at G ≅ 2 (Fig. 3b) indicates that interaction does occur between different paramagnetic species. Figure 3c demonstrates that merely doping a protein, BSA, with appropriate cations, following the neutron activation analysis of RuBPCase, does not result in a similar spectra. Interpretation of this ESR spectra with its, apparently, interacting components is difficult, and will be reported in detail in following publications [7]. However, the Cu^{2+}, Mn^{2+}, and Fe^{3+} spectra illustrated (Fig. 3c) cannot explain the RuBPCase spectral characteristics found. The ESR spectra of the combination of cations (ALL in Fig. 3c), also, does not replicate the RuBPCase spectra. The *ALL* spectra in Fig. 3c is different from the sum of the spectras for the other cations, suggesting an interaction that results in a distinct and less complex spectra than that found with RuBPCase. It is important to note that, using the hypothesis of cation control of RuBPCase, where the amount and nature of the cations surrounding RuBPCase in vivo controls RuBPCase activity, would result in different ESR spectra of RuBPCase, depending on the physiology of the plant and the extraction procedure followed. Thus, for example, it is consistent with this hypothesis that the relatively simple ESR spectra of Wishnick et al. [78] only reflects a condition or modality of RuBPCase that is different from what we report. In this view, the ESR spectra of RuBPCase will differ depending on its particular metal cation "content", which in turn is a reflection in its catalytic ability.

In certain circumstances, when the ESR spectra of the element are lost through oxidation or reduction, e.g., Cu^{2+} (ESR detectable) becomes Cu^{1+} (not detectable by ESR); thus, no ESR spectra might be obtained. This lack of ESR spectra has been noted in preparations, where low but detectable amounts of potentially ESR active elements have been reported (Table 1).

Transition Metals Form Chromophores with RuBPCase

The chromophores formed between RuBP and/or RuBPCase and transition metals are discussed in the Introduction. There is reason to believe (Amani, Daley and Theriot unpublished) that other reported variable RuBPCase chromophores [24, 25] are a result of transition metal interaction. Figure 4a illustrates the pH dependency of the Cu^{2+} RuBP complex. The chromophore is distinct from $CuCl_2$ [and the $Cu(OH)_2$] spectra. $Cu(OH)_2$ forms as a result of OH^- ions found at higher pH and is expected to be the form that absorbs 280 nm radiation. The ratio of RuBP: Cu is 1:1 and 3 PGA (3 phosphoglyceric acid): Cu is 1:1 as demonstrated by varying the proportions of phosphorylated ligand without significantly changing the extinction coefficient (Fig. 4a). When variable proportions of RuBP and 3 PGA are mixed with $CuCl_2$, the spectra formed (Fig. 4b) varies somewhat at shorter wavelengths, but not in the fashion that would explain the reported chromophores. Please note that the conditions and proportions of ligands vary somewhat between Fig. 4a and 4b. Figure 4b was done with a different sample of RuBPCase. The chromophores formed with Cu^{2+}, as expected, can only explain part of the light activation spectra of RuBPCase. Further studies with combinations of Cu^{2+}, Mn^{2+}, and Fe^{2+}/Fe^{3+} are expected to better approximate the action spectra of Wildner and Criddle [71, 74].

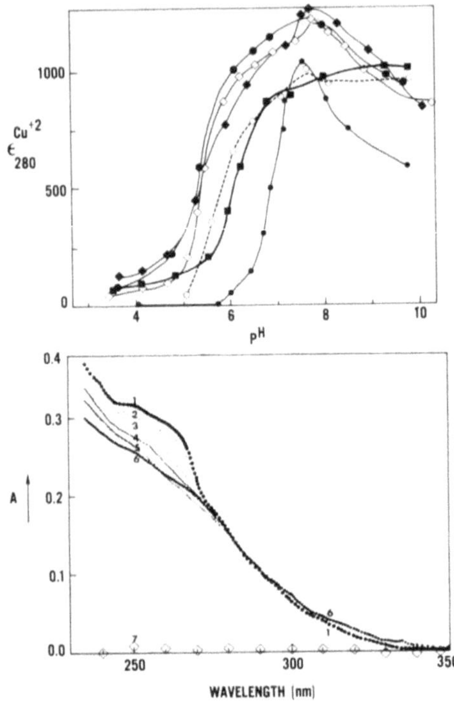

Fig. 4. Optical properties of Cu^{2+}-RuBP and Cu^{2+}-3 PGA complexes. **a** Effect of pH and ligand concentration on 280 nm, Cu^{2+} extinction coefficient –◇–, –●–, –◆– represent 3 PGA ϵ at a molar ratio of 1, 2, and 4 to Cu^{2+}; –□–, –○– represent RuBP ϵ at a molar ratio of 1, and 4 to Cu^{2+}; and –•– represents $CuCl_2$ ϵ characterisitcs. **b** Spectra of 3 PGA and RuBP mixes in presence of $CuCl_2$. Molar ratios (3 PGA; RuBP; $CuCl_2$) are *1* 0:4:1; *2* 2.4; 2.8:1; *3* 4:2:1; *4* 5.6:1.2:1; *5* 6.4:0:8:1, *6* 8:0:1; *7* is reference blank. Concentration of copper is constant (25 μM). The RuBP is from a different lot than that used in **a**. These figures illustrate optical properties of RuBPCase substrate and product, RuBP and 3 PGA, when they act ligands to Cu^{2+}, a transition metal cation

Summary and Conclusions

Since: (a) transition metals exist on RuBPCase; (b) the transition metal cations on RuBPCase have ESR spectra consistent with interaction, (c) RuBP/RuBPCase complexes with transition metal cations have the appropriate spectra; (d) transition metals are found near RuBPCase in vivo and can control RuBPCase activity. And, (e) the above discussion, it is appropriate to propose an in vivo transition metal cation complex that can respond to UV, in a way consistent with the Wildner and Criddle experiment [71, 74].

This hypothesis can be considered merely as a *generalization* and *extension* of the accepted proton and Mg^{2+} flux control mechanism for in vivo light activation of RuBPCase [53]. It is *more general* in that it considers the totality of divalent cation flux on illumination. And it *extends* the accepted mechanism in that, redox reaction, replacement kinetics, and the effects of interaction of RuBPCase with inhibitory, non-activating or limited activating divalent cations, and also considered to contribute to RuBPCase activity. We note that this hypothesis includes concepts previously proposed by others [13, 14, 28, 42, 53, 54, 57, 68, 72, 73, 80]. The putative metal cation requirement for oxidase activity is not considered here. The controversy on amounts of metal ions found, but not number of sites on RuBPCase, are moot in this hypothesis, since variable amounts of cations would be found bound to RuBPCase as a function of the in vivo physiological situation at the time of testing.

The variable in vivo response of RuBPCase to light activation would be a function of the metal cations near RuBPCase at the time of irradiation. And the diverse classes of inactive RuBPCase [3, 26, 30, 31, 32, 33, 39, 55] (R. Schmid pers. comm.). The possibility that such a mechanism can give rise to the diverse products of the RuBPCase reaction, previously give rise to the diverse products of the RuBPCase reaction, previously found [19, 21] and reported by others [6, 29] is also considered.

Acknowledgments. This work was supported by Daley Research Institute, East Texas State University Competitive Grant Program, Sigma Chemical Company, Reactor Sharing Program (Department of Energy), and Welch Foundation. We thank N. Daley for editorial work and M. Nash for figure design and preparation.

References

1. Aspeden WC, Wells GN (1980) Photosynthetic carbon reduction by sea grasses exposed to ultraviolet-A radiation. Annu Meet South Sect Am Soc Plant Physiol, Greenville, MS
2. Aspeden WC, Wells GN (1980) Photosynthetic carbon metabolism in sea grasses. Plant Physiol (Suppl) 65:88
3. Bowien B, Gottschalk E-M (1982) Influence of the activation state on the sedimentation properties of ribulose bisphosphate carboxylase from *Alcaligenes eutrophus.* J Biol Chem 257: 11845–11847
4. Boyle FA, Keys AJ (1982) Regulation of ribulose bis phosphate carboxylase activity-associated with photo inhibition of wheat *Triticum aestivum* cultivar Kolibri/b. Photosynth Res 3:105–112
5. Branden R (1978) Ribulose-1,5-diphosphate carboxylase and oxygenase from green plants are two different enzymes. Biochem Biophys Res Commun 81:539–546
6. Branden R, Nilsson T, Styring S (1981) Carbon-dioxide and hydrogen ions regulate the portion of L-3 phospho-glycerate that is formed in the ribulose bis phosphate carboxylase reaction. A proposed relation to photorespiration. In: Akoyunoglou G (ed) Photosynthesis, vol IV. Regulation of carbon metabolism. Balaban, Philadelphia, pp 57–62
7. Brooks K, Drumheller JE, Daley LS, Theriot LJ (1982) ESR signals of ribulose-bis-phosphate carboxylase. Character of ultrafine signals as detected by ESR and some comparisons with ESR signals of ceruloplasmin. Plant Physiol (Suppl) 69:91
8. Brown HM, Chollet R (1982) Isolation of a stable enzyme carbon-14 dioxide magnesium II carboxy arabinitol bis phosphate complex with ribulose bis phosphate carboxylase EC-4.1.1.39 from *Chromatium vinosum.* J Bacteriol 149:1159–1161
9. Bush KJ, Sweeney BM (1971) Ribulose diphosphate carboxylase from Gonyaulax. Plant Physiol (Suppl) 47:10
10. Chassagne NM, Gaudillere JP, Monties B (1981) Effect of ultraviolet radiations on plants grown under natural and artifical-light 2. Effects of photosynthetic ability of tomato and sweet pepper. Acta Oecologica-Oecologia Plant 2:327–338
11. Chiai EIO (1977) Bioinorganic chemistry – An introduction. Allyn and Bacon, Boston
12. Chollet RL, Anderson L, Hovsepian LC (1975) The absence of tightly bound copper, iron, and flavin nucleotide in crystalline ribulose 1,5-bisphosphate carboxylase oxygenase from tobacco. Biochem Biophys Res Commun 64:97–107
13. Christeller JT (1981) The effects of bivalent cations on ribulose bisphosphate carboxylase/ oxygenase. Biochem J 193:839–844
14. Christeller JT (1982) Effects of divalent cations on the activity of ribulose-bisphosphate carboxylase: interactions with pH and with D_2O as solvent. Arch Biochem Biophys 217:485–490
15. Christeller JT, Laing WA (1979) Effects of manganese ions and magnesium ions on the activity of soya-bean ribulose bisphosphate carboxylase/oxygenase. Biochem J 183:747–750

16. Chu DK, Chang R, Vickery LE (1974) An electron paramagnetic resonance study of the binding of manganese to ribulose-1,5-diphosphate carboxylase. Inhibition by magnesium. Biochem Biophys 334:438–441
17. Codd GA, Stewart R (1980) The photoinactivation of microalgal ribulose bis-phosphate carboxylase: Its physiological and ecological significance. In: Senger H (ed) The blue light syndrome. Proc Life Sci. Springer, Berlin Heidelberg New York, pp 391–400
18. Codd GA, Stewart R (1980) Photoinactivation of ribulose-bisphosphate carboxylase from green algae and cyanobacteria. FEMS Microbiol Lett 8:237–240
19. Daley LS (1975) Light activation of RuBPCase: conditions and products of the reaction. Dissertation, Univ Calif, Davis, CA
20. Daley LS, Dailey F, Criddle RS (1978) Light activation of ribulose bisphosphate carboxylase. Purification and properties of the enzyme in tobacco. Plant Physiol 62:718–722
21. Daley LS, Tibbals HF, Theriot LJ (1980) Effect of 360 nm light on RuBPCase products. In vitro role of copper in the reaction. In: Senger H (ed) The blue light syndrome. Proc Life Sci. Springer, Berlin Heidelberg New York, pp 381–391
22. DiMarco Giovannozzi-Sermani GG, Grego S, Veri G, Petrosanti T, Tricoli D, Catarcioni M, DiPietro A (1976) Daily fluctuations of some metabolic characters of durum wheat under field conditions. Crop Sci 16:343–348
23. Gauhl E (1976) Photosynthetic response to arying light intensity of ecotypes of *Solanum ducamara* L from shaded and exposed habitats. Oecologia 22:275-285
24. Gray GR (1971) An examination of D-fructose 1,6-diphosphate and related sugar phosphates by Fourier transform ^{31}P nuclear magnetic resonance spectroscopy. Biochemistry 10:4705–4711
25. Gray G, Barker R (1970) Studies on the substrates of D-fructose 1,6 diphosphate aldolase in solution. Biochemistry 9:2454–2462
26. Gutteridge S, Parry MAJ, Schmidt CNG (1982) The reactions between active and inactive forms of wheat ribulosebisphosphate carboxylase and effectors. Eur J Biochem 126:597–602
27. Hall NP, McCurry SD, Tolbert NE (1981) Storage and maintaining activity of ribulose-bisphosphate carboxylase (EC-4.1.1.39) oxygenase. Plant Physiol 67:1220–1223
28. Henkel J, Sane PV, Wildner GF (1980) The oxygen-dependent deactivation and reactivation of spinach ribulose-1,5-bisphosphate carboxylase. Biochim Biophys Acta 614:564–576
29. Hsu TC, Kuehn GD (1978) Active site directed, paracatalytic modification of ribulose 1,5 bisphosphate carboxylase-oxygenase by activation of the carbonionic enzyme substrate intermediate with the oxidant hexacyanoferrate. Proc Fed Am Soc Exp Biol 37:1511
30. Huner NPA, Carter JV, Wold F (1982) Effects of reducing agent on the conformation of the isolated subunits of ribulose bis phosphate carboxylase oxygenase EC-4.1.1.39 from cold hardened and unhardened rye *Secale cereale* cultivar pumala. Z Pflanzenphysiol 106:69–80
31. Huner NPA, Hayden DB (1982) Alterations in heterogeneity of ribulose bis phosphate carboxylase associated with growth at low temperatures. Plant Physiol 69:107–112
32. Huner NPA, Hayden DB (1982) Changes in the heterogeneity of ribulose bis phosphate carboxylase-oxygenase in winter rye induced by cold hardening. Can J Biochem 60:897–903
33. Huner NPA, Palta JP, Li PH, Carter JV (1981) Comparison of the structure and function of ribulosebisphosphate carboxylase-oxygenase from a cold hardy and nonhardy potato species. Can J Biochem 59:280–289
34. Johal S, Bourque DP (1979) Crystalline ribulose 1,5-bisphosphate carboxylase-oxygenase from spinach. Science 204:75–77
35. Johal S, Bourquet DP, Smith WW, Suh SW, Eisenberg D (1980) Crystallization and characterization of ribulose 1,5-bisphosphate carboxylase/oxygenase from eight plant species. J Biol Chem 255:8873–8880
36. Kent SS, Young JD (1980) Simultaneous kinetic analysis of ribulose 1,5-bisphosphate carboxylase/oxygenase activities. Plant Physiol 65:465–470
37. Kirkbright GF, Sargent M (1974) Atomic absorption and fluorescent spectroscopy. Academic Press, London New York
38. Kirksey ST, Jr, Neubecker TA, Margerum DW (1979) Thermally stable copper III and nickel (III)-tripeptide complex and their photochemical decomposition in acid solutions. J Am Chem Soc 101:1631–1633

39. Lanaras T, Codd GA (1981) Structural and immunoelectrophoretic comparison of soluble and particulate ribulose bis phosphate carboxylases from the cyanobacterium *Chlorogloeopsis fritschii.* Arch Microbiol 130:213–217
40. Lorimer GH, Andrews TJ, Tolbert NE (1973) Ribulose diphosphate oxygenase. II. Further proof of reaction products and mechanism of action. Biochemistry 12:18–23
41. Lorimer GH, Miziorko HM (1980) Carbamate formation on the ϵ-amino group of a lysyl residue as the basis for the activation of ribulose bisphosphate carboxylase by CO_2 and Mg^{2+}. Biochemistry 19:5321–5328
42. Martin MN, Tabita FR (1982) Selective inhibition of the oxygenase activity of ribulose-1,5-bis phosphate carboxylase oxygenase from *Rhodospirillum rubrum*. Fed Proc 41: Abstr 5149
43. McCurry SD, Hall NP, Pierce J, Paech C, Tolbert NE (1979) Ribulose-1,5-bisphosphate carboxylase oxygenase from parsley. Biochem Biophys Res Commun 84:895–900
44. McCurry SD, Pierce J, Tolbert NE, Orme-Johnson WH (1981) Mechanism of effector mediated activation of ribulose bis phosphate carboxylase oxygenase. J Biol Chem 256:6623–6628
45. Maechler F, Noesberger J (1980) Regulation of ribulose-bis-phosphate carboxylase activity in intact wheat (*Triticum aestrivum* co. kolibri) leaves by light, carbon-dioxide and temperature. J Exp Bot 31:1485–1492
46. Miziorko HM (1974) Studies of the role of the metal activator of ribulose diphosphate carboxylase by EPR, ^{1}H and ^{13}C NMR. Fed Proc 30:1312
47. Miziorko HM, Behnke CE, Houkom EC (1982) Protein liganding to the activator cation of ribulosebisphosphate carboxylase. Biochemistry 21:6669–6674
48. Miziorko HM, Mildvan AS (1974) Electron paramagnetic resonance, ^{1}H, and ^{13}C nuclear magnetic resonance studies of the interaction of manganese and bicarbonate with ribulose 1,5-diphosphate carboxylase. J Biol Chem 249:2743–2750
49. Miziorko HM, Nowak T, Mildvan AS (1974) Spinach leaf phosphoenolpyruvate carboxylase: purification, properties, and kinetic studies. Arch Biochem Biophys 163:378–389
50. Oliver DJ (1980) Light activation of RuBP carboxylase in situ. Plant Physiol (Suppl) 65:72
51. Pauley CR, Theriot LJ (1974) Synthesis and magnetic properties of copper (II) complexes on N-(Hydroxyalkyl)pyrole-2-carboxaldimines. Inorg Chem 13:2033–2034
52. Paulsen JM, Lane MD (1966) Spinach ribulose diphosphate carboxylase. I. Purification and properties of the enzyme. Biochemistry 5:2350–2357
53. Perchorowicz JT, Raynes DA, Jensen RG (1981) Light limitation of photosynthesis and activation of ribulose bisphosphte carboxylase in wheat seedlings (CO_2 exchange/carbon reduction cycle). Proc Natl Acad Sci USA 78:2985–2989
54. Raghavendra AS, Carrillo NJ, Vallejos RH (1981) Differential modulation of carboxylase and oxygenase activities of ribulose-1, 5-bis-phosphate carboxylase oxygenase · (EC-4.1.1.39) released from freshly ruptured spinach (*Spinacia oleracea*) chloroplasts. Plant Cell Physiol 22: 1113–1118
55. Raghavendra AS, Carillo N, Vallejos RH (1981) The pattern and characteristics of membrane bound ribulose-bis-phosphate EG-4.1.1.39 in spinach (*Spinacia oleracea*) chloroplasts. In: Akogunoglou G (ed) Photosynthesis, vol IV. Regulation of carbon metabolism. Balaban, Philadelphia, pp 31–38
56. Rice SC, Pon NG (1978) Direct spectrophotometric observation of ribulose-1,5-bisphosphate carboxylase activity. Anal Biochem 87:39–48
57. Rivkin RB, Seliger HH, Swift E, Biggley WH (1982) Light shade adaptation by the oceanic dinoflagellates *Pyrocystis noctiuca* and *Pyrocystis fusiformis.* Mar Biol 68:181–192
58. Robison PD, Tabita FR (1980) Formation of stoichiometric cobalt, (III) CO_2 complexes with spinach ribulose bisphosphate carboxylase/oxygenase. FEBS Lett 111:369–372
59. Rutner AC, Lane MD (1967) Nonidentical subunits of ribulose diphosphate carboxylase. Biochem Biophys Res Commun 28:531–537
60. Ryan FJ, Tolbert NE (1975) Ribulose diphosphate carboxylase/oxygenase (III) Isolation and properties. J Biol Chem 250:4229–4233
61. Ryan FJ, Tolbert NE (1975) Ribulose diphosphate carboxylase/oxygenase (IV) Regulation by phosphate esters. J Biol Chem 250:4234–4238
62. Salsbury JL, Floyd GL (1978) Molecular, enzymatic and ultrastructure characterization of the pyrenoid of the scaly green monad *Micromonas squamata.* J Phycol 14:362–368

63. Schmid GH (1980) Conformational changes caused by blue light. In: Senger H (ed) The blue light syndrome. Proc Life Sci. Springer, Berlin Heidelberg New York, pp 198–204
64. Sicher RC (1982) Reversible light activation of ribulose bis phosphate carboxylase oxygenase in isolated barley *Hordeum vulgare* cultivar pennrad, protoplasts and chloroplasts. Plant Physiol 70:366–369
65. Siegel MI, Lane MD (1972) Interaction of ribulose diphosphate carboxylase with 2-carboxyribitol diphosphate, an analogue of the proposed carboxylated intermediate in the CO_2 fixation reaction. Biochem Biophys Res Commun 48:508–516
66. Teramura AH, Kossuth SV, Briggs RH (1978) Effects of UV-B enhancement under contrasting PAR growth regimes on NCE, dark respiration, and growth in soybeans (*Glycine max*). Plant Physiol (Suppl) 61:74
67. Theriot LR, Askins R, Chang C, Davis T, Gupta S, Talent J, Tibbals HF, Daley L, Wendel C (1981) Matrix effect in atomic absorption determination of copper in fraction I protein possible role of copper in light sensitive control of ribulose bis phosphate carboxylase carboxylase EC-4.1.1.39 activity. Tex J Sci 33:235–252
68. Weissbach A, Horecker BI, Hurwitz (1956) The enzymic formation of phosphoglyceric acid from ribulose diphosphate and carbon dioxide. J Biol Chem 218:795–810
69. Whitelam GG, Codd GA (1983) Photoinactivation of *Microcystis aeruginosa* ribulose 1,5-bisphosphate carboxylase: effects of endogenous and added sensitizers and the role of oxygen. FEMS Microbiol Lett 16:269–272
70. Whitelam GC, Codd GA (1983) Photoinhibition of photosynthesis in the cyanobacterium *Microcystis aeruginosa*. Planta 157:561–566
71. Wildner GF, Criddle RS (1969) Ribulose diphosphate carboxylase I. A factor involved in the light activation of the enzyme. Biochem Biophys Res Commun 37:952–960
72. Wildner GF, Henkel J (1978) Differential reactivation of ribulose 1,5-bisphosphate oxygenase with low carboxylase activity by Mn^{2+}. Febs Lett 91:99–103
73. Wildner GF, Henkel J (1979) The effect of divalent metal ions on the activity of Mg^{++} depleted ribulose-1,5-bisphosphate oxygenase. Planta 146:223–228
74. Wildner GF, Zilg H, Criddle RS (1969) Light effect on isolated ribulose diphosphate carboxylase activity. 2nd Int Congr Photosynth, Strega, pp 1825–1830
75. Wishnick M (1970) The interaction of metals with ribulose diphosphate (RuDP) carboxylase. Fed Proc 29:531
76. Wishnick M, Lane MD (1971) Ribulose-di-phosphate carboxylase from spinach leaves. Methods Enzymol 23:570–577
77. Wishnick M, Lane MD, Scrutton MC (1970) The interaction of metal ions with ribulose 1,5-diphosphate carboxylase from spinach. J Biol Chem 245:4939–4947
78. Wishnick M, Lane MD, Scrutton MC, Mildvan AS (1969) The presence of tightly bound copper in ribulose diphosphate carboxylase from spinach. J Biol Chem 244:5761–5763
79. Wu G-Y, Den Y-F, Wu X-Y (1980) Activation of ribulose-bisphosphate carboxylase by thioredoxin. Acta Bot Sin 22:241–246
80. Yamada T, Idawa T, Nisizawak (1978) Partial purification and some properties of ribulose 5 diphosphate carboxylase from a green alga *Bryopsis maxima*. Jpn J Phycol 26:49–52

Primary Events and Signal Transduction

A Light-Inducible Cytochrome b Reduction in the Green Alga *Acetabularia*

R. CAUBERGS,[1] T. VANDEN DRIESSCHE,[2] and J. A. DE GREEF[1]

Blue light induces an absorbance change (LIAC) in a membrane fraction derived from higher plant systems such as corn coleoptiles and cauliflower inflorescences. It is believed that this LIAC is directly related with the photoreception performed by a flavoprotein. Blue light is obligatory for normal growth and cap morphogenesis [5] in the green alga *Acetabularia mediterranea* which makes it a valuable object for investigations in relation to the blue light-inducible absorbance change.

In a membrane fraction from *Acetabularia* obtained by an aqueous two-phase system separation [3], a light-inducible absorption change has been observed. This

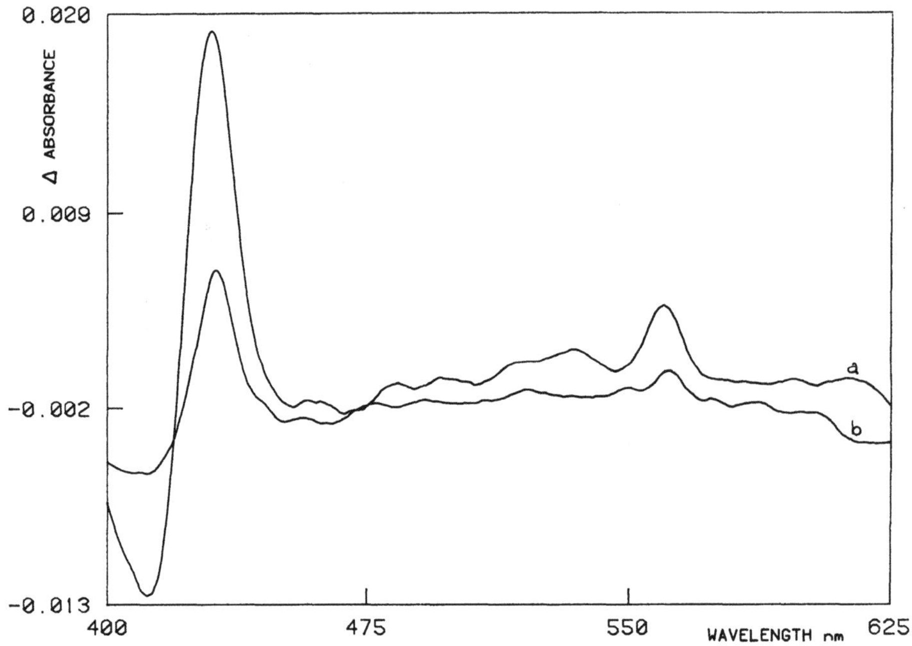

Fig. 1. Light minus dark difference spectrum (LIAC) obtained after actinic blue irradiation (1 min) of presumptive plasmamembrane fractions of cauliflower inflorescences (*a*) (protein: 0.86 mg ml^{-1}) and *Acetabularia* (*b*) (protein 0.47 mg ml^{-1})

1 Department of Biology, University of Antwerpen, RUCA-UIA, Universiteitsplein 1, B–2610 Wilrijk, Belgium
2 Laboratory of Cytology and Molecular Embryology, Free University, Brussels, Belgium

Blue Light Effect in Biological Systems
ed. by H. Senger

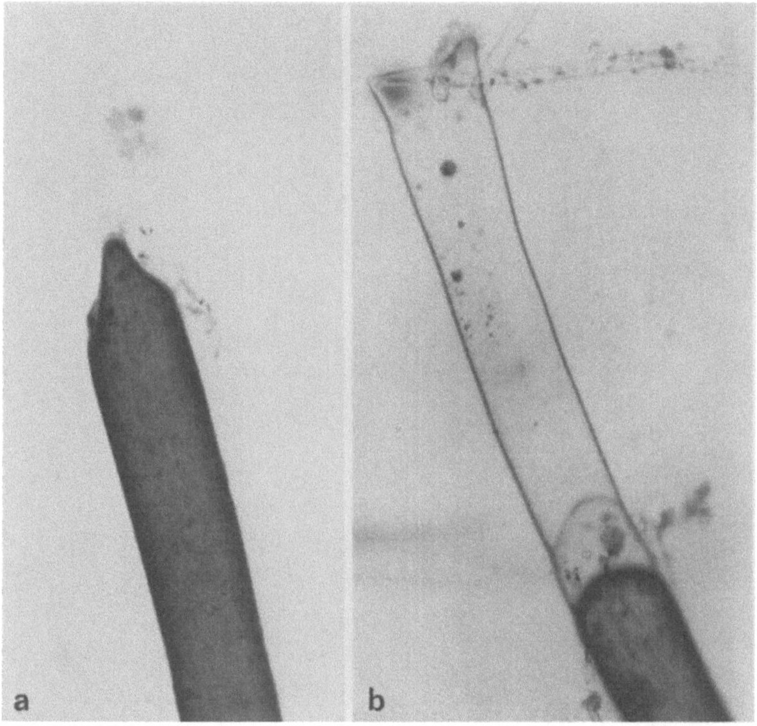

Fig. 2a,b. The influence of SHAM (112 μM) on the regeneration under continuous blue light of the nucleated fragments, 2 days after fragmentation. **a** control; **b** after SHAM treatment

LIAC has the same spectrophotometrical characteristics as those of the cytochrome b reduction measured in the plasmamembrane fraction obtained from corn or cauliflower (Fig. 1) [2, 4]. Salicylhydroxamic acid (SHAM) at low concentrations shuts off the spectral response in the membrane fraction of higher plants [1]. Therefore this substance could be, in theory, an inhibitor of other blue light-induced phenomena as well. No effect, however, has been obtained on phototropism of corn coleoptiles. The

Table 1. SHAM (150 μg ml^{-1}) influence on the regeneration and cap-formation of nucleated fragments

	After 5 days	After 6 days	After 23 days	Cap formation after 23 days % of total fragments
Controls: increment (mm) versus day 2 after fragmentation	1.30	1.51	8.10	54%
SHAM-treated: increment (mm) versus day 2 after fragmentation	0.37	0.58	6.23	18%

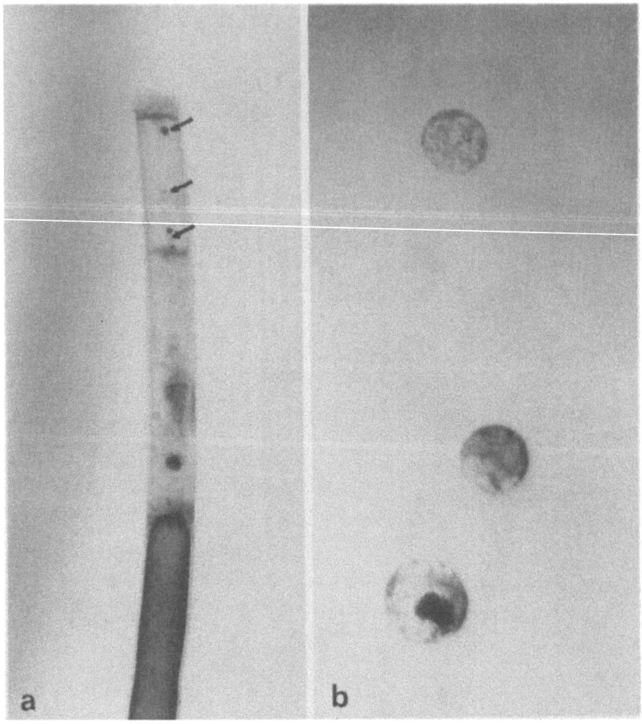

Fig. 3a,b. The release of cytoplasts by fragmented algae after SHAM treatment (*arrows*) (**a**), and cytoplasts present in the medium (**b**)

reason for this failure is probably related with problems of uptake. Regeneration under blue light of nucleated *Acetabularia* fragments on the other hand is, in the presence of SHAM, temporarely inhibited (Fig. 2, Table 1).

These data give circumstantial evidence that a flavin-cytochrome b protein complex might be the photoreceptor for this process. Rather surprising is the fact that SHAM does not inhibit the LIAC in membrane fractions of *Acetabularia.*

To reconcile this result, apparently conflicting with those of higher plant systems, we postulate the existence of a membrane-bound factor, sensitive to SHAM and affecting directly the LIAC in one case and not in the other, but nevertheless functionally related to the photoreceptor. In support is the observation that SHAM application results in the release of cytoplasts indicative for a probable action of SHAM on the plasmamembrane (Fig. 3).

Acknowledgment. We are grateful to Dr. H. Alexandre for making the photographs.

References

1. Caubergs RJ, Goldsmith MHM, Briggs WR (1979) Light-inducible cytochrome reduction in membranes from corn coleoptiles: fractionation and inhibitor studies. Carnegie Inst Yearb 78: 121–125
2. Caubergs RJ, De Greef JA (1982) Light inducible cytochrome reduction from cauliflower (*Brassica oleraceae*) inflorescences. Photochem Photobiol 35:405–410
3. Clauss H (1968) Beeinflussung der Morphogenese, Substanzproduktion und Proteinzunahme for *Acetabularia mediterranea* durch sichtbare Strahlung. Protoplasma 65:49–80
4. Goldsmith MH, Caubergs RJ, Briggs WR (1980) Light-inducible cytochrome reduction in membrane preparations from corn coleoptiles. Plant Physiol 66:1067–1073
5. Widell S, Larsson C (1981) Separation of presumptive plasmamembranes from mitochondrial by partition in an aqueous polymer two-phase system. Plant Physiol 51:368–378

Blue Light Effects and the Role of Membranes

S. WIDELL[1] and CH. LARSSON[2]

The information from perceived light, especially from the blue light region, is often transmitted into a change of orientation, as occurs in phototropism. Such a capacity implies that the photoreceptor is membrane-bound, although it does not necessarily mean that it is bound in darkness. The debate concerning possible receptors for phototropism and other blue light-induced phenomena has taken this aspect into account. A search has been going on for a long time for potential photoreceptors in different membranes of the cell.

Poff and Butler [17] described the existence of a blue light-induced absorbance change in certain fungi, reflecting the reversible reduction of a b-type cytochrome. Brain et al. [3] studied the occurrence of such absorbance changes (later termed LIAC), in different centrifuge fractions from homogenates of both Neurospora and corn coleoptiles. They found that LIAC was predominantly associated with a particulate fraction that was neither enriched in mitochondria nor in endoplasmic reticulum. Further studies showed that LIAC also could be elicited by red light in the presence of methylene blue [4], and that methylene blue could substitute for an endogenous photoreceptor to induce phase shift in the circadian rhythm of conidiation found in the *bd* strain of *Neurospora* (Brain, unpublished results, see [21]) and to stimulate carotenoid synthesis in fungi [8]. Endogenously sensitized (i.e., blue light-induced) LIAC could not be obtained with higher plant material, until it was discovered that a lowered oxygen tension was needed in the assay in order to obtain photoreduction. With the addition of glucose and glucose oxidase to the assay medium this requirement was met [5].

In order to find out whether LIAC was related to more than one type of membrane, and to better characterize the b-cytochrome itself, it became necessary to further purify the membrane(s) containing LIAC activity. Leong and Briggs [10] used sucrose gradient centrifugation followed by centrifugation in a Renografin gradient. They obtained a membrane fraction that was enriched in LIAC, containing hardly any other cytochromes [12]. This fraction was identified as plasma membrane since it coincided with Mg^{2+}-ATPase activity in the presence of oligomycin, and with glucan synthetase II activity.

One major problem with the characterization of LIAC and its location has been to identify the different membranes of the cell. Markers such as K^+-stimulated Mg^{2+}-

1 Department of Plant Physiology, P.O.B. 7007, S–220 07 Lund, Sweden
2 Department of Biochemistry, P.O.B. 740, S–200 07 Lund, Sweden

Blue Light Effects in Biological Systems
ed. by H. Senger

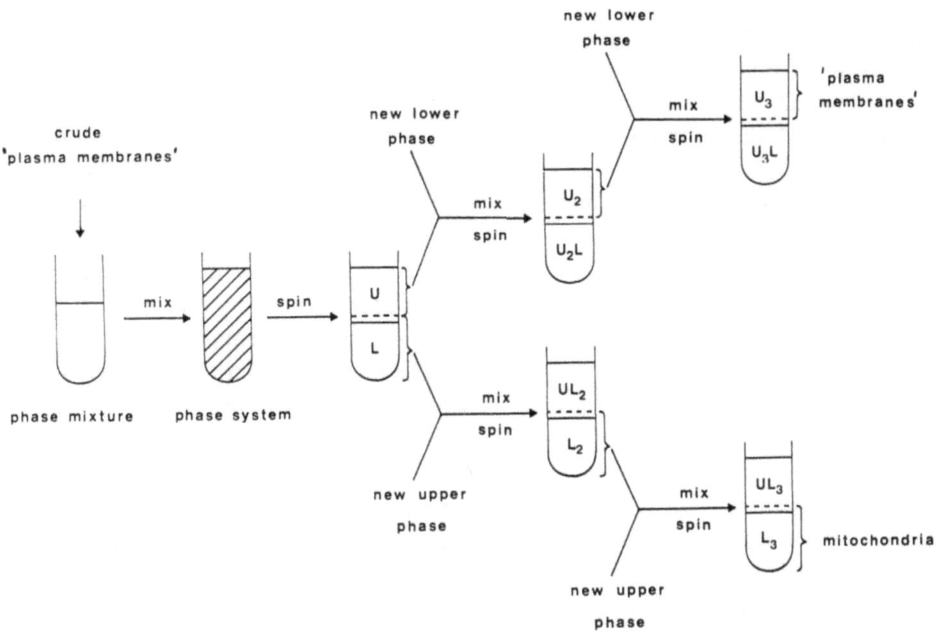

Fig. 1. Batch preparation of plasma membranes by partition in a dextran-PEG two-phase system. One ml of a suspension of a microsomal fraction in 0.25 M sucrose, 5 mM K-phosphate, pH 7.8, was added to a 7.0 g phase mixture to produce an 8.0 g phase system with a final concentration of 6.3% (w/w) dextran T 500, 6.3% (w/w) polyethylene glycol 4000, 0.25 M sucrose, 5 mM K-phosphate, pH 7.8. The phase system was mixed by 20 inversions of the tube, and was centrifuged at 1000 *g* for 4 min to facilitate phase settling. Ninety percent (indicated by *dashed line*) of the upper phase, U_1, was removed with a Pasteur pipette and added to a tube containing 4 ml new lower phase. The remaining lower phase, interface and 10% upper phase, L_1, received 4 ml new upper phase. The two new phase systems so obtained were treated as above to finally produce a twice-washed upper phase, U_3, enriched in plasma membranes, and a twice-washed lower phase, L_3, containing other membranes

dependent ATPase, glucan synthetase II, naphtyl phtalamic acid binding, and staining with silicotungstic acid have been proposed for the plasma membrane, but none of them has been applicable to all plants tested [18].

We chose to base the separation on surface properties of the membranes rather than on size and density. This method, called aqueous polymer two-phase partition [1] had earlier been successfully used for the purification of animal plasma membranes, chloroplasts, mitochondria, etc. (for reviews see [2] and [9]). A suitable two-phase system was obtained by mixing water solutions of the polymers polyethylene glycol 4000 and dextran T 500 to final concentrations of 6.3% (w/w) of each polymer. The system was buffered with 5 mM potassium phosphate, pH 7.8 and made isotonic with sucrose. The upper phase formed will be rich in polyethylene glycol and the lower phase rich in dextran.

Intact protoplasts have been shown to have a high partition ratio in these phase systems, i.e., a high affinity for the polyethylene glycol-rich upper phase compared to other cell organelles and membranes [6]. Therefore we thought that the problem with finding a suitable membrane marker could at least partially be overcome by looking for membranes with a high partition ratio. It was supposed that at least right-side-out vesicles of plasma membranes would behave similarily to protoplasts, since they expose the same surface to the phase system.

We found that microsomal fractions of corn coleoptiles, cauliflower inflorescences, barley shoots, wheat roots, oat shoots, and oat roots [13, 23, 25] indeed all gave a membrane fraction (termed U_3) which had a high partition ratio and was rich in LIAC. The separation procedure is outlined in Fig. 1. However, another fraction (termed L_3) also showed LIAC activity. This fraction contained mitochondria, endoplasmic reticulum and other organelles, and was characterized by having a low partition ratio. It was later found with oats that the presumptive plasma membrane fraction (U_3), besides containing LIAC, also was enriched in K^+-stimulated Mg^{2+}-ATPase and glucan synthetase II, and that most of the vesicles stained well with silicotungstic acid [25], all supporting its identification as a plasma membrane fraction. Thus, there seemed to be one LIAC related to the plasma membrane and another LIAC which might be related to some other membrane or possibly to plasma membranes with different surface properties (e.g., plasma membrane vesicles turned inside-out). Since the presence of LIAC in more than one type of membrane could be an argument against its having a photobiological role, both LIAC's needed further characterization.

In Table 1, data from U_3 and L_3 fractions prepared from three different centrifuge fractions of cauliflower are shown. The ratios between LIAC (measured with endogenous photoreceptor) and cytochrome c oxidase and NADPH-dependent cytochrome c reductase, respectively, have been calculated to see whether the distribution of LIAC in the L_3 fractions coincided with the distribution of either mitochondria or endoplasmic reticulum. It was found that the ratio of LIAC to cytochrome c oxidase increased with lighter centrifuge fractions, whereas the ratio of LIAC to cytochrome c reductase was similar in the different L_3 fractions (with the exception of 9 KP L_3 – "intact mitochondria") and the 50 KP fraction ("endoplasmic reticulum"). This indicates that the LIAC recovered in the L_3 fraction may be related to endoplasmic reticulum rather than to mitochondria or to inside-out vesicles of the plasma membrane.

Low temperature spectra [22] of U_3 and L_3 fractions from cauliflower also showed characteristics which supports an association of L_3 LIAC with the endoplasmic reticulum. In the U_3 fraction (Fig. 2) only one cytochrome was seen, independent of whether the reduction was obtained chemically or by light. The α-band of this cytochrome was found at 556 nm, which may be compared to a wavelength of 555 nm for the α-band of LIAC in corn plasma membranes [12]. In the L_3 fraction (Fig. 3), chemical reduction revealed several cytochromes, and at least three bands were also obtained upon light reduction. These were found at 599 nm, 551 nm and 557 nm, respectively.

The band at 599 nm can be ascribed to cytochrome c oxidase, whereas the other two bands are not readily identified. Rich and Bendall [20] found at least five different b-type cytochromes in a microsomal fraction from cauliflower. One of these was spectrally very similar to the light-reducible cytochrome in our U_3 fraction (both have peaks at 428 nm and 556 nm and a shoulder at 445 nm). They identified this species as cytochrome P 420/450, a b-type cytochrome supposed to be bound to the endoplasmic reticulum. Thus, it is possible that the light-reducible cytochrome in our U_3

Table 1. Distribution of protein, LIAC and cyt c oxidase (total values) among different membrane fractions obtained from 90 g of cauliflower inflorescences. Fractionation was achieved by differential centrifugation – giving 9 KP, 21 KP and 50 KP fractions – followed by partition in an aqueous polymer two-phase system – giving U_3 and L_3. 9 KP L_3 was further fractionated into a light and heavy band on a Percoll gradient (corresponding to densities of 1.04 and 1.08 g ml^{-1}, respectively). The total amounts in the 50 KP fraction are also shown.
End-LIAC: blue light-induced absorbance change $\Delta(A_{428}-A_{410})$ mediated by the endogenous photoreceptor; MB-LIAC: red light-induced absorbance change $\Delta(A_{428}-A_{410})$ mediated by methylene blue; U_3 and L_3: see Fig. 1; 9 KP: 1,000–9,000 g, 20 min pellet; 21 KP: 9,000–21,000 g, 20 min pellet; 50 KP: 21,000–50,000 g, 45 min pellet [24]

Fraction	End-LIAC $\Delta(\Delta A)\ 10^3$	MB-LIAC $\Delta(\Delta A)\ 10^3$	cyt c oxidase nmol min^{-1}	NADPH-cyt c reductase nmol min^{-1}	Protein mg	End-LIAC / Cyt c ox	End-LIAC / Cyt c red
9 KP U_3	11.6	17.7	20	89	1.70	0.58	0.13
21 KP U_3	19.0	27.0	8	64	1.44	2.5	0.30
50 KP U_3	27.5	46.8	15	46	1.76	1.8	0.59
9 KP L_3							
light	8.0	8.1	2250	120	2.70	0.004	0.067
heavy	9.5	neg	8150	500	4.44	0.001	0.019
21 KP L_3	14.8	15.2	1920	178	2.96	0.008	0.083
50 KP L_3	4.2	6.4	382	76	1.16	0.011	0.055
50 KP	84.5	125	497	1030	18.2	0.17	0.082

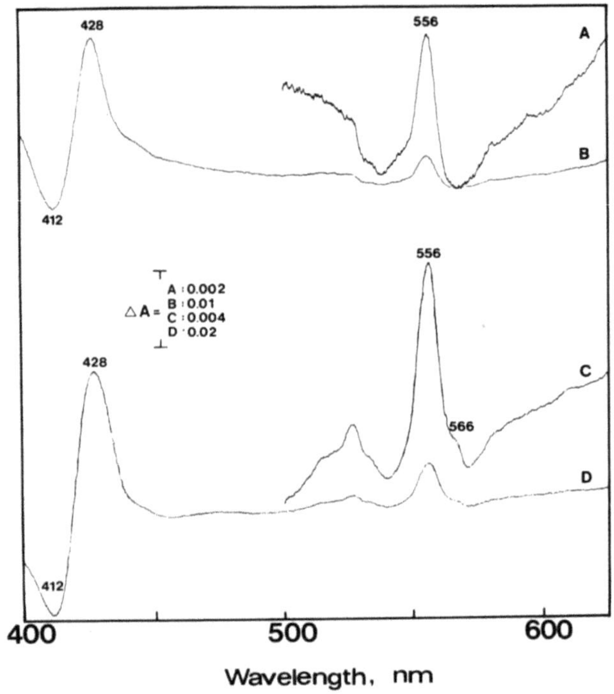

Fig. 2. Blue light minus dark (*A, B*) and dithionite reduced minus oxidized (*C, D*) difference spectra of a plasma membrane-enriched fraction (21 + 50 KP, U_3) from cauliflower. Temperature 77 K. Protein concentration: 1 mg ml^{-1} [22]

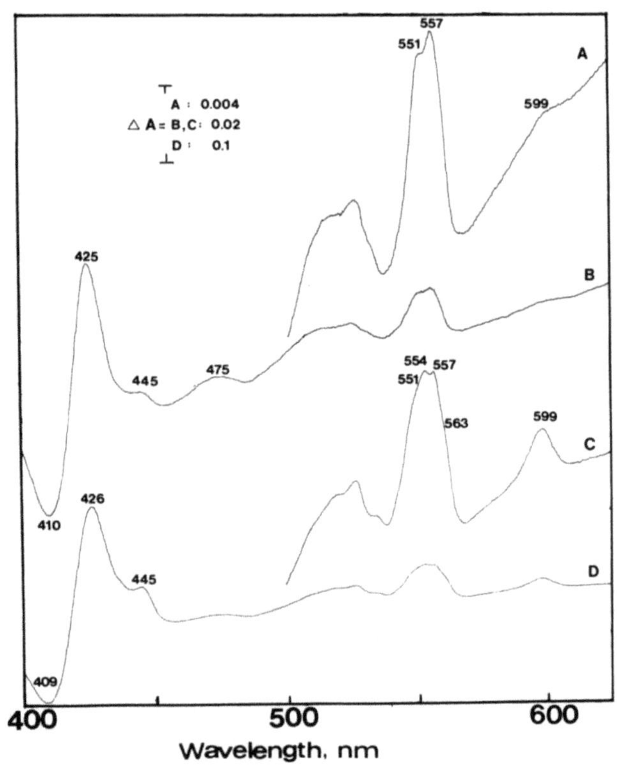

Fig. 3. Blue light minus dark (*A, B*) and dithionite reduced minus oxidized (*C, D*) difference spectra of a fraction containing mitochondria, endoplasmic reticulum, etc. (21 + 50 KP, L_3) from cauliflower. Temperature 77 K. Protein concentration: 10 mg ml^{-1} [22]

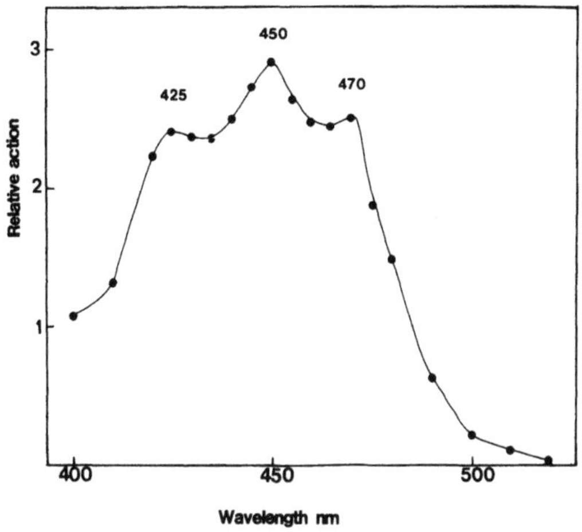

Fig. 4. Action spectrum of LIAC (blue light-induced absorbance change $\Delta(A_{430} - A_{410})$ in a plasma membrane fraction (50 KP, U_3) from cauliflower. Protein concentration: 0.4 mg ml^{-1} [22]

fraction is identical with cytochrome P 420/450 of Rich and Bendall. Indeed, in recent experiments in our laboratory it was found that the fraction showed a CO-binding spectrum characteristic for cytochrome P 420/450 (C. Schelin, P. Kjellbom, S. Widell and C. Larsson, unpublished). However, our U_3 LIAC cannot be located in the endoplasmic reticulum since cytochrome b_5 was not detected in the U_3 fraction.

Cytochrome b_5 of cauliflower has a split α-band at liquid nitrogen temperature (peaks at 552 and 558 nm [20]) which in fact corresponds fairly well with the main peaks of the light reducible component(s) found in L_3. This suggests that LIAC in the L_3 fraction may be identical to cytochrome b_5 of the endoplasmic reticulum. Alternatively, the shorter wavelength peak could be related to cytochrome c_1 of mitochondria [15], and the 557 nm peak could be identical to our U_3 component, but be located in inside-out vesicles of the plasma membrane. The latter suggestion is, however, less probable, since it does not fit with the enzyme ratios discussed above (Table 1). Furthermore, in the work by Leong and Briggs [10] on corn membranes, a certain fraction of LIAC may be associated with the endoplasmic reticulum (cf. their Fig. 3). Their major part of LIAC was, as mentioned above, not related to the endoplasmic reticulum, but rather to the plasma membrane.

The action spectrum of the plasma membrane-bound LIAC (Fig. 4) corresponds well with flavin spectra and with action spectra of phototropism and other blue light induced phenomena. Unfortunately, no action spectra could be obtained with the L_3 fractions since the signals were too unstable, possibly due to mitochondrial respiration. Thus, it seems that at least the plasma membrane-bound LIAC may be associated with light effects. However, also other LIAC may be involved here. Flavin-binding sites are found in both the endoplasmic reticulum and the plasma membrane [7]. Furthermore, carotenoid synthesis, which is a photostimulated process especially in fungi, has been related both to LIAC [19] and to the endoplasmic reticulum [14].

More direct evidence for the association of LIAC with light effects comes from the work by Leong and Briggs [11]. They studied the effect of acifluorfen on oat phototropism and on LIAC. In the assay for LIAC, they used a detergent-solubilized microsomal fraction which could contain a mixture of all LIACs. They found that acifluorfen enhanced phototropism and LIAC to about the same extent (Fig. 5), and

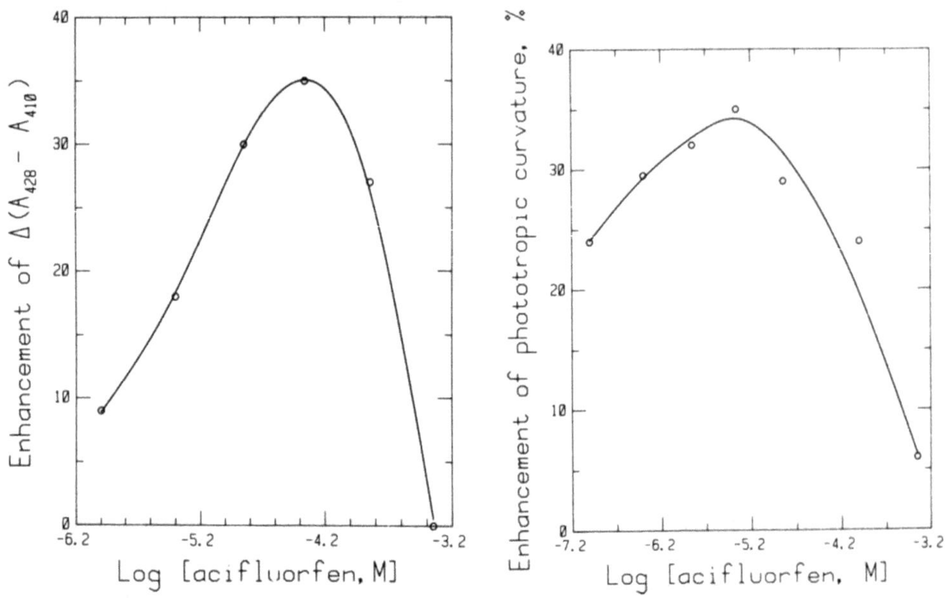

Fig. 5. **A** Effects of acifluorfen on LIAC of Triton-solubilized 21 KP from oat coleoptiles. Results expressed as percent buffer control. Each *point* is the average of three experiments [11]. **B** Effects of different concentrations of acifluorfen on phototropism of dark-grown oat coleoptiles [11]

they could show that the increase in LIAC was due to a slower dark reoxidation. No other cytochromes of the microsomal fraction seemed to interact with acifluorfen.

Finally, the great diversity of blue light-sensitive processes implicates more than one receptor. Our finding that the plasma membrane-bound LIAC may have cytochrome P-450 properties could be a first step towards an understanding of how LIAC exerts its effects, and perhaps also of how phytochrome and the blue light receptor may be coordinated in processes like pigment syntheses that in many cases include both phytochrome and cytochrome P-450.

Acknowledgments. The authors wish to thank Miss Helga Krook for typing the manuscript and B.Sc. Janet F. Bornman for linguistic help.

References

1. Albertsson P-Å (1971) Partition of cell particles and macromolecules, 2nd edn. Wiley, New York
2. Andersson B, Åkerlund H-E, Larsson C (1981) Phase partition – a method for analysis and purification of cell organelles and membrane vesicles. In: Glick D (ed) Methods of biochemical analysis, vol 28. Wiley Interscience, New York, pp 115–150
3. Brain RD, Freeberg JA, Weiss C, Briggs WR (1977) Blue light induced absorbance changes in membrane fractions from corn and Neurospora. Plant Physiol 55:948–952
4. Britz SJ, Schrott E, Widell S, Briggs WR (1979) Red light induced reduction of a particle associated *b*-type cytochrome from corn in the presence of methylene blue. Photochem Photobiol 29:359–365

5. Goldsmith MH, Caubergs RJ, Briggs WR (1980) Light inducible cytochrome reduction in membrane preparations from corn coleoptiles. Plant Physiol 66:1067–1073
6. Hallberg M, Larsson C (1981) Compartmentation and export of $^{14}CO_2$ fixation products in mesophyll protoplasts from the C_4 plant Digitaria sanguinalis. Arch Biochem Biophys 208: 121–130
7. Hertel R, Jesaitis AJ, Dohrmann U, Briggs WR (1980) In vitro binding of riboflavin to subcellular particles from maize coleoptiles and Cucurbita hypocotyls. Planta 147:312–319
8. Lang-Feulner J, Rau W (1975) Redox dyes as artificial photoreceptors in light dependent carotenoid synthesis. Photochem Photobiol 21:179–183
9. Larsson C (1983) Partition in aqueous polymer two-phase systems – a rapid method for separation of membrane particles according to their surface properties. In: Hall JL, Moore AL (eds) Isolation of membranes and organelles from plant cells. Academic Press, London New York, pp 277–307
10. Leong T-Y, Briggs WR (1981) Partial purification and characterization of a blue light-sensitive cytochrome-flavin complex from corn membranes. Plant Physiol 67:1042–1046
11. Leong T-Y, Briggs WR (1982) Evidence from studies with acifluorfen for participation of a flavin-cytochrome complex in blue light photoreception for phototropism of oat coleoptiles. Plant Physiol 70:875–881
12. Leong T-Y, Vierstra RD, Briggs WR (1981) Blue light-sensitive cytochrome-flavin complex from corn coleoptiles. Further characterization. Photochem Photobiol 34:696–703
13. Lundborg T, Widell S, Larsson C (1981) Distribution of ATPases in wheat root membranes separated by phase partition. Physiol Plant 52:89–95
14. Mitzka-Schnabel U, Rau W (1981) Subcellular site of carotenoid biosynthesis in Neurospora crassa. Phytochemistry 20:63–69
15. Ninnemann H, Strasser RJ, Butler WL (1977) The superoxide anion as electron donor to the mitochondrial electron transport chain. Photochem Photobiol 26:41–47
16. Poff K, Butler WL (1974) Absorbance changes induced by blue light in Phycomyces blakesleeanus and Dictyostelium discoideum. Nature (London) 248:799–801
17. Poff K, Butler WL (1975) Spectral characterization of the photoreducible *b*-type cytochrome of Dictyostelium discoideum. Plant Physiol 55:427–429
18. Quail PH (1979) Plant cell fractionation. Annu Rev Plant Physiol 30:425–484
19. Rau W (1980) Blue light-induced carotenoid biosynthesis in microorganisms. In: Senger H (ed) The blue light syndrome. Springer, Berlin Heidelberg New York, pp 283–298
20. Rich PR, Bendall DS (1975) Cytochrome components of plant microsomes. Eur J Biochem 55: 333–341
21. Senger H, Briggs WR (1981) The blue light receptor(s): Primary reactions and subsequent metabolic changes. In: Smith K (ed) Photochemical and photobiological reviews, vol VI. Plenum Press, New York, pp 1–38
22. Widell S, Caubergs RJ, Larsson C (1983) Spectral characterization of light reducible cytochrome in a plasma membrane-enriched fraction and in other membrane from cauliflower influorescences. Photochem Photobiol 38:95–98
23. Widell S, Larsson C (1981) Separation of presumptive plasma membranes from mitochondria by partition in an aqueous polymer two-phase system. Physiol Plant 51:368–374
24. Widell S, Larsson C (1983) Distribution of cytochrome *b* photoreductions mediated by endogenous photosensitizer or methylene blue in fractions from corn and cauliflower. Physiol Plant 57:196–202
25. Widell S, Lundborg T, Larsson C (1982) Plasma membrane from oats prepared by partition in an aqueous polymer two-phase system. On the use of light induced cytochrome *b* reduction as a marker for the plasma membrane. Plant Physiol 70:1429–1435

Localization of Light-Induced Phenylacetic Acid-Binding Sites in Corn Coleoptile Membranes: Possible Association of the Blue Light Photoreceptor Pigment with the Plasma Membrane

R. D. VIERSTRA[1,2] and K. L. POFF[1]

Introduction

A wide range of physiological responses in plants is controlled by a pigment which absorbs blue light (reviews: [14, 17]. Experimental evidence indicates that a flavoprotein is the photoreceptor pigment for a majority of these blue light responses. Particularly notable are results involving the phototropic responses in *Phycomyces* [13] and corn [16, 19] and phototaxis in *Euglena* [4]. The isolation of the particular flavoproteins involved in these photoresponses has proven difficult because there are so many cellular flavoproteins and because there is no in vitro assay specific for the photoreceptor pigment at present.

Recently, an indirect attempt has been made toward isolation of the photoreceptor responsible for phototropism in corn seedlings through the partial purification of a novel blue light photoreducible b-type cytochrome [10, 11, 20]. It has been postulated that the flavoprotein involved in this photoreduction is the same as that involved in phototropism [2, 12]. However, such conclusions will remain ambiguous until a direct causal link between the photoreduction of this unique b-type cytochrome and the photoresponse has been accomplished.

We have attempted to develop a direct assay for the flavoprotein involved in the phototropic response of corn seedlings using phenylacetic acid (PAA) as a photoaffinity label. PAA which photoreacts with flavins producing a stable covalent adduct [8], specifically inhibits phototropism using the effects of PAA on geotropism as a control [16, 18]. This specific inhibition demonstrates that the flavoprotein involved in phototropism is highly photoreactive with PAA, probably more so than most other cellular flavins. Although it is likely that other flavoproteins in addition to the photoreceptor pigment will be derivatized, this photoaffinity labeling approach should allow us to identify possible candidates and eliminate a majority of the flavoproteins that are weakly or not photoreactive.

In this paper, we report our initial attempts to localize the flavoprotein involved in phototropism using tritiated PAA to identify corn coleoptile membranes that contain photoreactive flavins. We have restricted our study to membrane-bound flavoproteins because several lines of evidence indicate that the photoreceptor pigment is membrane-bound (review: [7]).

1 MSU-DOE Plant Research Laboratory, Michigan State University, East Lansing, MI 48824, USA
2 Present Address: Department of Horticulture, 1575 Linden Drive, University of Wisconsin, Madison, WI 53706, USA

Blue Light Effects in Biological Systems
ed. by H. Senger

Materials and Methods

Plant Material. Etiolated corn seedlings (*Zea mays*) hybrid MS WFg × Bear 38 (National Starch and Chemical Co., Decatur, IL) were grown as previously described [18]. The apical centimeter of 5-day-old shoots were harvested, the coleoptiles separated from the primary leaves and put on ice. All tissue and subsequent extracts were kept under dim green safelights at ice temperature unless otherwise noted.

Homogenization and Fractionation. Coleoptiles were chopped with a razor blade and homogenized in a mortar with extraction buffer (0.5 ml g^{-1} fresh weight) consisting of 0.25 M sucrose, 50 mM N-2-hydroxyethylpiperazine-N-2-ethanesulfonic acid (HEPES), 50 mM 2-(N-morpholino)ethanesulfonic acid (MES), 1 mM disodium ethylenediaminetetraacetic acid (Na_2EDTA) and 0.1 mM $MgCl_2$ [pH 8.0 (23°C)]. The homogenate was squeezed through a 100-μm nylon mesh and the remaining cake reground in a second volume (0.5 ml g^{-1} fr.wt.) of extraction buffer and again squeezed through the nylon mesh. The combined filtrates were centrifuged at 500 *g* for 10 min. The resulting 500 g supernatant (0.5 KS) was layered onto a linear 15 to 45% (w/w) sucrose gradient. Linear 15 to 45% sucrose gradients (25 ml) were constructed over 2 ml of 45% sucrose buffer in a 40 ml cellulose nitrate tube and a layer of 15% sucrose buffer applied to the top. The sucrose buffers contained 10 mM HEPES, 10 mM MES, 1 mM Na_2EDTA, 1 mM KCl and 0.1 mM $MgCl_2$ [pH 7.0 (23°C)]. The 0.5 KS was layered over the 15% sucrose and the tube centrifuged at 95,000 *g* for various time periods. For isopycnic gradients, 3 h was found to be sufficient [15].

Assays. Protein content was determined using the procedure described by Bradford [1] with bovine serum albumin as a standard. Cytochrome-c oxidase, NADH-dependent cytochrome-c reductase and 1-N-naphthylphthalamic acid (NPA)-binding activities were measured according to Jesaitis et al. [9]. Glucan synthetase I and II activities were determined according to [15]. Carotenoid content was estimated from the sample's absorbance at 480 nm [5]. Cytochrome content was analyzed using low temperature (77 K) reduced-minus-oxidized absorbance spectroscopy. Absorbance maxima were determined by fourth derivative analysis [6].

Flavin concentrations were determined by the lumiflavin fluorescence assay using FMN as a standard as described by Jesaitis et al. [9]. The presence of PAA in the samples did not interfere with the assay as long as the samples were not irradiated with light prior to the addition of NaOH. Additionally, flavins permanently photoreduced with PAA [18] were stable under the assay conditions and as a result not converted to lumiflavin and not fluorescent at 530 nm.

In Vivo ^{3}H-PAA Binding. Coleoptile tips (6 g) were harvested and chopped into approximately 3-mm sections. The sections were added to 4 ml of water containing 7.4×10^6 Bq (27 nM) (ring-4-^{3}H) PAA (sp. act. 6.85×10^{11} Bq/mmol, IRE, Belgium) and kept on ice in darkness or in an ice bath while being irradiated from the top and bottom with blue light (5 or 0.1 mW cm^{-2}) from two slide projectors in combination with 3 cm aqueous 10% $CuSO_4$ and a DT Blau filter [370–470 nm transmission (Blazer)]. The sections were irradiated with blue light or kept in the dark for 1 h. Afterwards, the sections were washed three times for a total of 30 min with 25 ml aliquots of unlabeled 50 mM PAA (pH 7.0), and homogenized and fractionated according to the aforementioned procedures with the addition of 50 mM PAA and 0.1 mM IAA to the extraction buffer and 10 mM PAA and 0.1 mM IAA to the gradient media. The addition of PAA and IAA to these buffers had no detectable effect on the sedimentation behaviour of corn coleoptile membranes nor on the marker activities used in this study. Radioactively labeled PAA bound to each membrane fraction was determined by scintillation counting. Light-induced PAA binding was calculated from the difference in counts per min (cpm) between comparable membrane fractions obtained from coleoptile sections irradiated with blue light or kept in darkness.

In Vitro Photoreaction of Membrane-Bound Flavins with PAA. After fractionation by isopycnic sucrose gradient centrifugation, corn coleoptile membranes were made 50 mM in PAA and the membrane fractions either kept in darkness or irradiated with 1 mW cm^{-2} blue light. Afterwards, the flavins that had not photoreacted with PAA were assayed by the lumiflavin fluorescence method (see above).

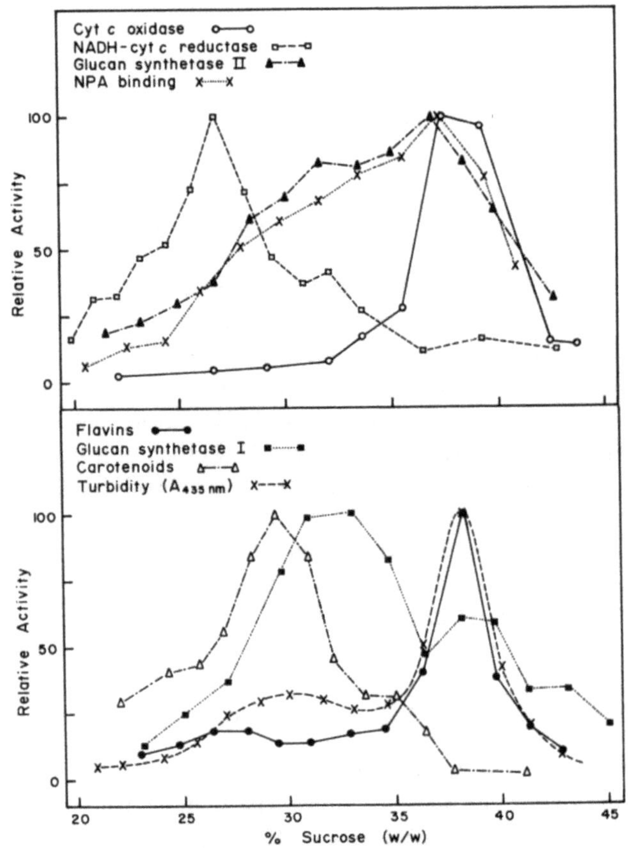

Fig. 1. Distribution of marker enzymes, carotenoids, flavins and turbidity after isopycnic sucrose gradient centrifugation of a crude corn coleoptile homogenate. The 0.5 KS from a homogenate of 6 g of coleoptiles was layered directly on a 15 to 45% linear sucrose gradient and centrifuged for 3 h at 95,000 *g*. Afterwards, the gradient was fractionated and relative activity of each marker determined and expressed as a function of the sucrose concentration

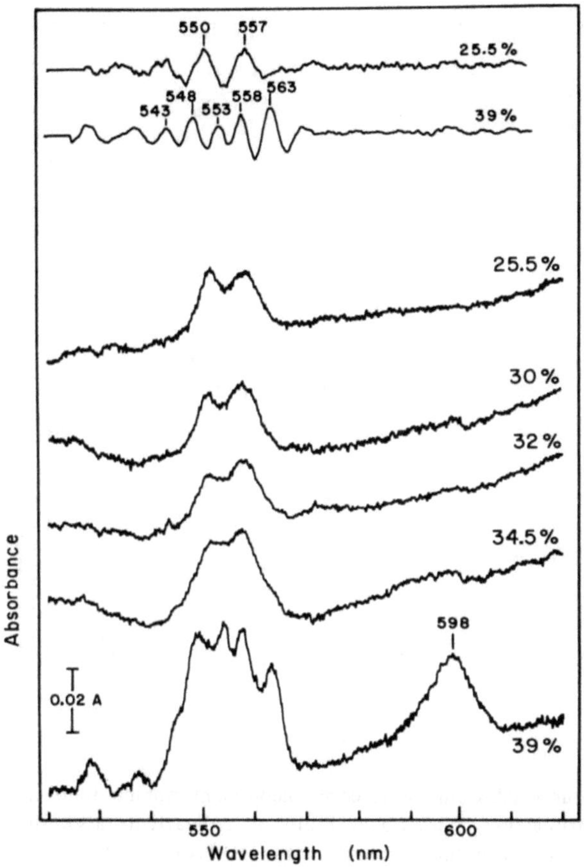

Fig. 2. Low temperature reduced-minus-oxidized absorbance spectra of various linear sucrose gradient fractions (designated by their percentage sucrose). The fractions were prepared after isopycnic centrifugation of a crude corn coleoptile homogenate as in Fig. 1. Oxidized absorbance spectra were obtained from 0.5 ml of sample frozen to 77 K. Reduced spectra were obtained from the same sample after thawing, adding several grains of dithionite, and refreezing. Absorbance maxima were determined by fourth derivative analysis (*top*)

Results

After isopycnic gradient centrifugation, several distinct corn coleoptile membrane types were resolved according to their density (Fig. 1). Mitochondria (∿ 38% sucrose) and endoplasmic reticulum (∿ 27% sucrose) were located by the profiles of cytochrome-c oxidase and NADH-dependent cytochrome-c reductase activities, respectively. The position of mitochondria was further substantiated by the presence of cytochrome-c (λ_{max} 548 nm) and cytochrome oxidase (λ_{max} 598 nm) in low temperature-reduced-minus-oxidized absorbance spectra [6] (Fig. 2). The two b-type cytochromes at 550 and 557 nm in the region enriched for endoplasmic reticulum (25.5% sucrose) are similar to those observed by Jesaitis et al. [9] in comparable fractions. The plasma membrane, as assayed by the markers, NPA binding and glucan synthetase II activity, sedimented over a broad range of sucrose concentrations with a maximum at 35–36% sucrose (Fig. 1). Carotenoid content, indicating the presence of etioplast membranes (mainly of prolamella body origin), was maximal at 29% sucrose and glucan synthetase I, characteristic of Golgi, peaked at 32% sucrose. Flavins, as determined by the lumiflavin fluorescence assay, were associated primarily with mitochondria

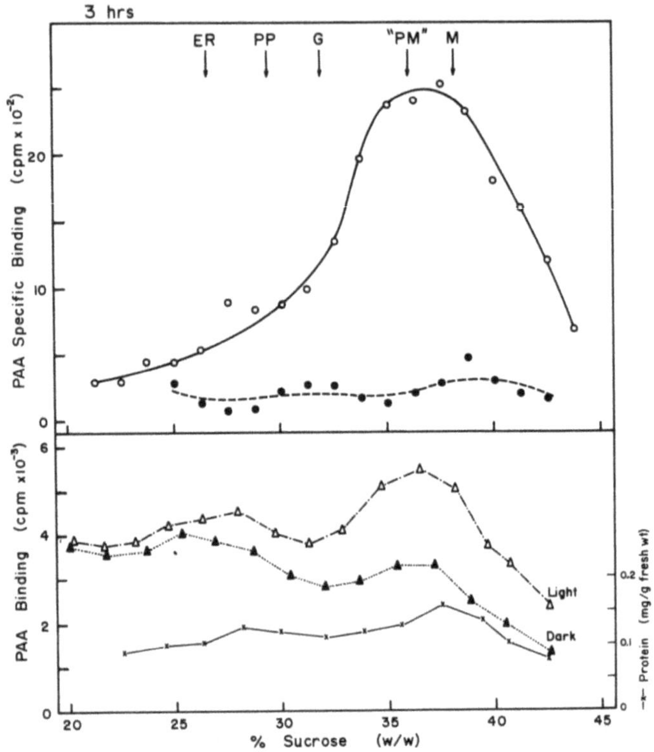

Fig. 3. Distribution of in vivo light-induced PAA binding to corn coleoptile membranes following isopycnic linear sucrose gradient centrifugation (3 h). *Lower* Profile of radioactivity associated with membrane fractions from 6 g of corn coleoptiles incubated with (^{3}H)-PAA and either irradiated with 5 mW cm^{-2} blue light (△) or kept in darkness (▲); and protein content of similar isopycnic gradients (X) plotted as a function of percent sucrose. *Upper* Distribution of in vivo light-induced PAA binding measured as the difference in radioactivity between membrane fractions from corn coleoptiles irradiated with either 5 mW cm^{-2} (○) or 0.1 mW cm^{-2} (●) and those kept in darkness. Arrows indicate the peaks of activity for the membrane markers from Fig. 1. (*ER* endoplasmic reticulum; *PP* proplastids; *G* Golgi; *PM* plasma membrane; *M* mitochondria)

(∼ 60% of the total) with a small peak in the region enriched for the endoplasmic reticulum. The locations of identifiable membrane types were further verified by electron microscopy (data not shown).

Using identical centrifugation procedures to those above, we attempted to identify specific coleoptile membranes that would display in vivo light-induced binding of PAA. For unirradiated tissue, ^{3}H-PAA binding was observed in all membrane fractions with small peaks at 37 and 26% sucrose (Fig. 3, lower). Irradiation of the coleoptiles with 5 mW cm^{-2} blue light significantly increased the radioactivity associated with membraneous material as compared with the dark controls (∼ 65% increase at 36% sucrose). When expressed as the difference in radioactivity between light irradiated and dark treatments (Fig. 3, upper), a broad peak of light-induced PAA binding was observed centered at 36% sucrose with significant activity from 28 to 43% sucrose. A 50-fold reduction in light intensity during the incubation abolished the light-induced

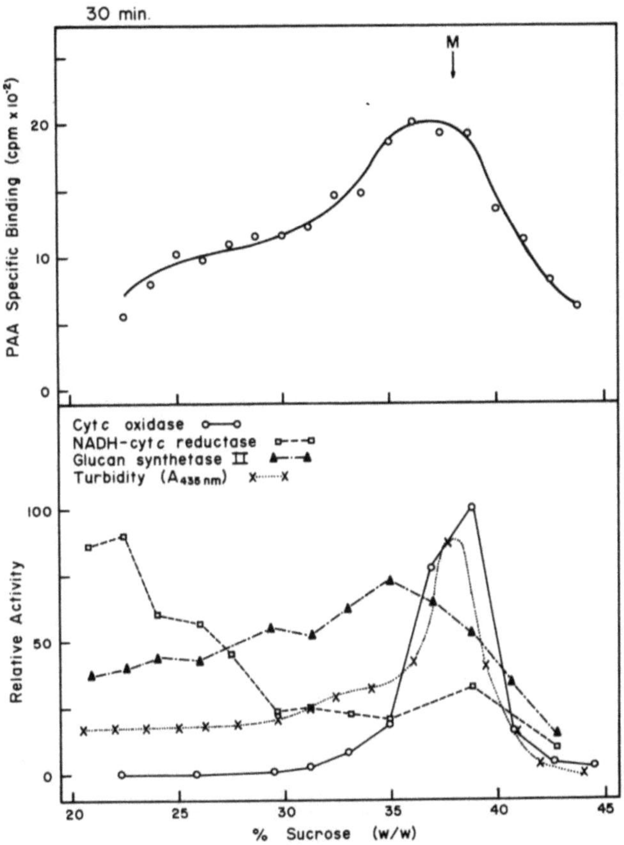

Fig. 4. Distribution of in vivo light-induced PAA binding and marker enzyme activities following a 30-min centrifugation. The 0.5 KS from 6 g of corn coleoptiles was layered onto a 15 to 45% linear sucrose gradient identical to that used in Fig. 1, but centrifuged at 95,000 *g* for 30 min. *Lower* Relative activities of enzymatic membrane markers after centrifugation for 30 min. Marker activities of each fraction are expressed as percentages of the maximal activity obtained from membrane fractions from crude homogenates of 6 g of coleoptiles centrifuged for 3 h (Fig. 1). *Upper* Profile after a 30 min centrifugation of in vivo light-induced PAA binding measured as the difference in radioactivity associated with membrane fractions from 6 g of coleoptiles irradiated with 5 mW cm^{-2} blue light and those kept in darkness. (*M* peak in activity of the mitochondrial marker; cytochrome-c oxidase)

difference, demonstrating that a high light intensity was required for the increased binding of PAA to the membranes. Based on the position of marker enzymes, this peak was not primarily associated with either endoplasmic reticulum, etioplasts or Golgi, but did sediment at sucrose concentrations characteristic for mitochondria and the plasma membrane. In addition, this in vivo light-induced binding did not correlate with the distribution of total flavins (Fig. 1).

In an effort to determine which membrane type, mitochondria or plasma membrane contained the observed light-induced PAA binding, the two membranes were partially separated using rate zonal centrifugation. After centrifugation at 95,000 *g*

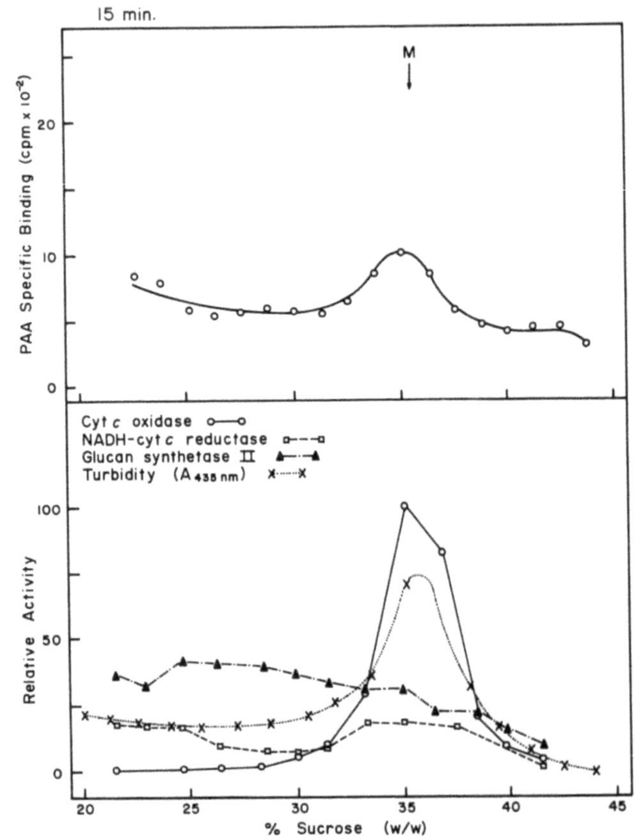

Fig. 5. Distribution of in vivo light-induced PAA binding and marker enzyme activities following a 15 min centrifugation. The 0.5 KS from 6 g of corn coleoptiles was layered onto a 15 to 45% linear sucrose gradient identical to that used in Fig. 1, but centrifuged at 95,000 *g* for 15 min. *Lower* Relative activities of enzymatic membrane markers after centrifugation for 15 min. Marker activities of each fraction are expressed as percentages of the maximal activity obtained from membrane fractions from crude homogenates of 6 g of coleoptiles centrifuged for 3 h (Fig. 1). *Upper* Profile after a 15 min centrifugation of in vivo light-induced PAA binding measured as the difference in radioactivity associated with membrane fractions from 6 g of coleoptiles irradiated with 5 mW cm^{-2} blue light and those kept in darkness. (*M* peak in activity of the mitochondrial marker; cytochrome-c oxidase)

for only 30 min, mitochondria were at isopycnic equilibrium (Fig. 4, lower) and were close to equilibrium after 15 min at 95,000 *g* (Fig. 5, lower). In contrast, only ∿ 70% of the activity associated with the plasma membrane had reached equilibrium after 30 min (when compared with gradient centrifuged for 3 h) with the remaining activity smeared toward lower sucrose concentrations (Fig. 1). Little plasma membrane has entered the gradient after 15 min.

Using identical centrifugation times, the distributions of light-induced PAA binding were assayed. As observed in Figs. 4 and 5, upper, decreasing the period of centrifugation from 3 h to 30 and 15 min, reduced the amount of light-induced PAA binding

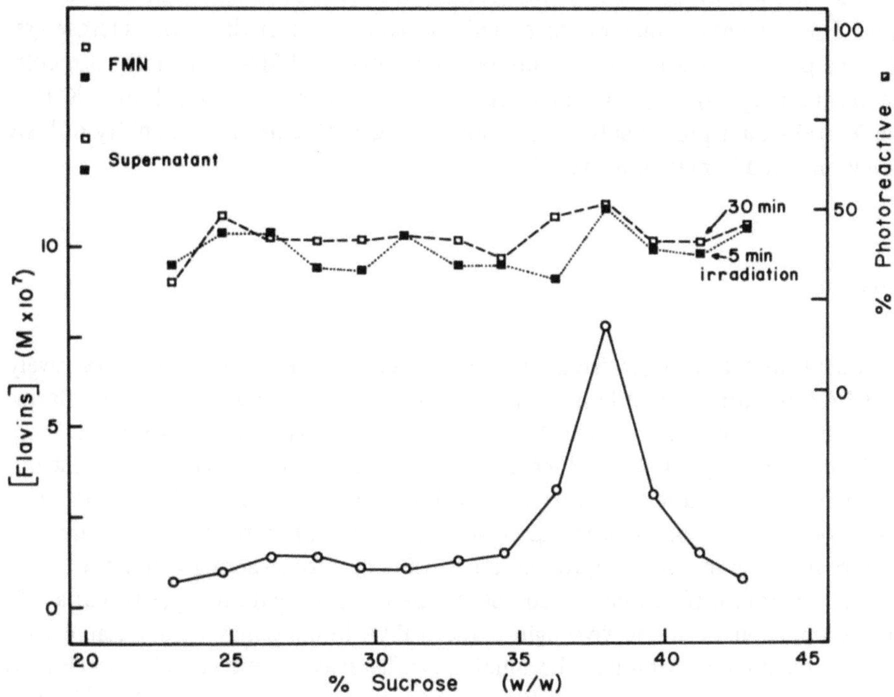

Fig. 6. In vitro photoreactivity of PAA with membrane bound flavins obtained after isopycnic sucrose gradient centrifugation. The 0.5 KS from 10 g of corn coleoptiles was layered onto a 15 to 45% linear sucrose gradient and centrifuged at 95,000 *g* for 3 h. Aliquots from each fraction were assayed for the ability of its flavoproteins to photoreact with PAA. (○) Distribution of flavin content before irradiation. Percentage photoreactive (□, ■) represents the amount of flavins undetected by the lumiflavin assay after irradiation for 5 (■) or 30 min (□) with 1 mW cm^{-2} blue light in the presence of 50 mM PAA as compared with an unirradiated aliquot. The photoreactivity of a 0.3 uM solution of FMN and an aliquot from the supernatant at the top of the sucrose gradient (5.4 uM flavin) were used for comparison

at 37% sucrose and concomitantly increased the binding activity at lower sucrose concentrations. After a 15-min centrifugation only a small peak of activity at 35% sucrose was observed. Comparison of the three centrifugation times suggests that a substantial proportion of the light-induced PAA binding was coincident with plasma membrane with a minor component associated with the mitochondria.

It is possible that the observed profiles of light-induced PAA binding represent the actual reactivity profile of membrane bound flavins and/or the profiles of accessibility of each membrane type to PAA. To test the former possibility, we assayed the ability of the flavins associated with membrane fractions to photoreact in vitro with PAA. After isopycnic centrifugation, PAA was added to the membrane fractions and the fractions irradiated with blue light. Afterwards, the flavins that did not photoreact with PAA were assayed by the lumiflavin method and compared to that of unirradiated samples. The flavin distribution of such a gradient prior to irradiation is shown in Fig. 6. When these fractions were irradiated with 1 mW cm^{-2} blue light, a substantial percentage of the flavoproteins photoreacted with PAA, 38% by 5 min and 43% by 30 min. The percentage that photoreacted was constant over the entire sucrose gradient. This percentage was considerably less than that for FMN or soluble corn coleoptile flavins (Fig. 6) and was unaffected by the addition of 0.1% Triton X 100. Triton X 100 would presumably have eliminated the problems of accessibility of PAA to flavins enclosed in membrane vesicles.

Discussion

It is evident that blue light irradiation does increase the binding of radioactively labeled PAA to corn coleoptile membranes. However, because affinity-labeled flavins have not been isolated from such membranes after light treatment, we cannot conclusively state that light-induced PAA binding is via a covalent attachment to flavins. The response does require high intensities of blue light, indicating that a photochemical reaction of a blue light-absorbing pigment is involved. In vitro studies also demonstrate that such membrane flavoproteins are capable of photoreacting with PAA.

Evidence from differential and sucrose gradient centrifugation suggests that a substantial proportion of the in vivo light-induced PAA binding may be associated with the plasma membrane. Although the binding profile at isopycnic equilibrium (Fig. 3) does not agree exactly with the profile for the plasma membrane (Fig. 1), it is clear that this binding does not correlate with the markers for endoplasmic reticulum, Golgi or etioplasts. In addition, a majority of the binding is not associated with mitochondria (even though the binding overlaps with the distribution of mitochondria marker activities after isopycnic gradient centrifugation) because (1) isopycnic sucrose gradient fractions at 35% sucrose, which contain substantially less mitochondria than the peak fraction at 38% sucrose (as determined by enzymatic activity, absorbance spectroscopy and electron microscopy), exhibit close to maximal amounts of light-induced binding (Fig. 3) and (2) fractions from rate zonal centrifugation containing maximal mitochondrial activity with little plasma membrane contamination exhibit a much reduced light-induced binding activity (Figs. 4 and 5) as compared to fractions from isopycnic gradients where both membrane activities are present (Fig. 3).

The profile for in vitro photoreactivity of membrane bound flavins with PAA was different from the profile obtained for in vivo labeling. When PAA was added to previously fractionated corn coleoptile membranes, all fractions had roughly the same percentage of flavins that photoreacted with PAA. As a consequence, the distribution of in vitro light-induced PAA binding should correlate with the distribution of flavins. It is possible that this difference between in vitro photoreactivity and in vivo labeling results from differences in accessibility of PAA to the various membranes of intact corn coleoptiles. Using intact tissue, the concentration of PAA may be highest at the plasma membrane and therefore PAA would photoreact with a greater percentage of plasma membrane flavins.

Since the procedure to detect in vivo light-induced PAA binding involved the use of intact corn coleoptiles, these results probably simulated the labeling patterns that occur during experiments where PAA specifically inhibits coleoptile phototropism

[16, 18]. If one assumes that the blue light photoreceptor pigment is a membrane-bound flavoprotein [7] and that PAA will photoreact with a substantial percentage of this pigment [18], one would expect the photoreceptor pigment to be associated with membranes that contain significant light-induced PAA binding activity. Therefore, it is possible that the blue-light photoreceptor pigment is localized on membrane fractions enriched for the plasma membrane. This possibility is in agreement with earlier reports suggesting that the blue-light photoreceptor pigment is plasma membrane-bound [7]. Recently, Leong and Briggs [10] and Widell and Larson [20] have reported that the blue-light-photoreducible b-type cytochrome, which is possibly involved in phototropism [12], is also located on the plasma membrane in corn coleoptiles. This flavin-mediated photoreduction is effectively inhibited by PAA [3].

It is of interest that approximately 60% of membrane-bound flavins do not photoreact strongly with PAA in vitro. Because of this and the fact that several membrane types do not display significant in vivo light-induced labeling, a majority of membrane-bound flavins may be eliminated as potential candidates for the photoreceptor pigment. Further identification of membrane-bound flavoproteins that are photoaffinity labeled by PAA should provide a promising approach for isolating possible candidates for the blue light photoreceptor pigment in corn seedlings.

Acknowledgments. This work was supported by the United States Department of Energy under Contract DE-AC02-76ER01338. Thanks are due to D. DeGaetano for his technical assistance, Dr. G. Gardner for his generous gift of radiolabeled NPA, Dr. T. Giddings for help with the electron microscopy and Dr. P.H. Quail for his review of the manuscript.

References

1. Bradford MM (1972) A rapid and sensitive method for the quantitation of microgram quantities of protein utilizing the principle of protein-dye binding. Anal Biochem 72:248–254
2. Britz S, Schrott E, Widell S, Briggs WR (1979) Red light-induced reduction of a particle-associated b-type cytochrome from corn in the presence of methylene blue. Photochem Photobiol 29:359–365
3. Caubergs R, Goldsmith MH, Briggs WR (1979) Light-inducible cytochrome reduction in membranes from corn coleoptiles: Fractionation and inhibitor studies. Carnegie Inst Wash Yearb 79:121–125
4. Checcucci A, Colombetti G, Ferrara R, Lenci F (1976) Action spectra for photoaccumulation of green and colorless *Euglena:* Evidence for the identification of receptor pigments. Photochem Photobiol 23:51–54
5. Davies BH (1976) Analysis of carotenoids. In: Goodwin TW (ed) Chemistry and biochemistry of plant pigments, vol I. Academic Press, London New York, pp 655–679
6. Davis K, Hatefi Y, Poff KL, Butler WL (1973) The *b*-type cytochromes of bovine heart mitochondria: Absorption spectra, enzymatic properties, and distribution of electron transfer complexes. BBA 325:341–356
7. Haupt W (1980) Localization and orientation of photoreceptor pigments. In: Lenci F, Colombetti G (eds) Photoreception and sensory transduction in aneural organisms. Plenum Press, New York, pp 155–172
8. Hemmerich D, Massey V, Weber G (1967) Photo-induced benzyl substitution of flavins by phenylacetate: A possible model for flavoprotein catalysis. Nature (London) 213:728–730
9. Jesaitis A, Heners P, Hertel R, Briggs WR (1977) Characterization of a membrane fraction containing a *b*-type cytochrome. Plant Physiol 59:941–947

10. Leong TY, Briggs WR (1981) Partial purification and characterization of a blue light-sensitive cytochrome-flavin complex from corn membranes. Plant Physiol 67:1042–1046
11. Leong TY, Vierstra RD, Briggs WR (1981) Blue light-sensitive cytochrome-flavin complex from corn coleoptiles: Further characterization. Photochem Photobiol 34:697–703
12. Leong T-Y, Briggs WR (1982) Evidence from studies with Acifluorfen for the participation of a flavin-cytochrome complex in blue light photoperception for phototropism of oat coleoptiles. Plant Physiol 70:875–881
13. Presti D, Hsu WJ, Delbrück M (1977) Phototropism in *Phycomyces* mutants lacking β-carotene. Photochem Photobiol 26:403–405
14. Presti D, Delbrück M (1978) Photoreceptors for biosynthesis, energy storage and vision. Plant Cell Environ 1:81–100
15. Ray PM (1977) Auxin-binding sites on maize coleoptiles are localized on membranes of the endoplasmic reticulum. Plant Physiol 59:594–599
16. Schmidt W, Hart J, Filner P, Poff KL (1977) Specific inhibition of phototropism in corn seedlings. Plant Physiol 60:736–738
17. Senger H, Briggs WR (1981) The blue light receptor(s): primary metabolic reactions and subsequent metabolic changes. Photochem Photobiol Rev 6:1–38
18. Vierstra RD, Poff K (1981) Mechanism of specific inhibition of phototropism in corn seedlings by phenylacetic acid. Plant Physiol 67:1011–1015
19. Vierstra RD, Poff K (1981) The role of carotenoids in the phototropic response in corn seedlings. Plant Physiol 68:798–801
20. Widell S, Larson C (1981) Separation of presumptive plasma membranes from mitochondria by partition in an aqueous polymer two-phase system. Physiol Plant 51:368–374

Wavelength Dependence of Nitrite Release and the Effects of Different Nitrogen Sources and CO_2 Tensions on *Chlamydomonas reinhardii* Inorganic Nitrogen Metabolism

P. J. APARICIO and M. P. AZUARA[1]

Introduction

In eukaryotic photosynthetic organisms, inorganic N metabolism, especially the conversion of NO_3^--N into organic-N is being gradually recognized as one of the most fundamental assimilatory processes comparable in green algae to CO_2 assimilatory reduction with respect to the utilization of photosynthetic reducing power [12, 24]. The fact that green algae growing on NO_3^- under the conditions prevailing in most natural environments, namely common air and sunlight, release NO_2^- and NH_4^+ to the medium adds new scopes to the interpretation of their photosynthetic metabolism [4].

The entrance of NO_3^- in the cytosol and its accumulation in cell organels, its reduction to NO_2^- catalyzed by the NAD(P)H-nitrate reductase, the reduction of NO_2^- to NH_4^+ catalyzed by the reduced ferredoxin-nitrite reductase and the incorporation of NH_4^+ to carbon skeletons to form mainly glutamate catalyzed by the glutamine synthetase-glutamate synthase system are the main steps currently consider in the assimilatory reduction of NO_3^-. The best-documented regulatory features of this pathway concerns nitrate reductase biosynthesis, assembly, degradation and interconvertible inactivation [12, 26].

NAD(P)H-nitrate reductase in these organisms is a multimeric enzyme (190,000–480,000 M.W.) with several subunits (3–8 according to different organisms) that includes FAD, cyt b_{557} and Mo as the main redox components [9, 12]. Molybdenum associated with pterins seems to be held in a special cofactor that appears to be shared by most molybdoproteins other than nitrogenase [14].

In the interconvertible inactivation of nitrate reductase, that only affects the enzyme ability to reduce NO_3^- to NO_2^- but no other enzyme activities, molybdenum cofactor domain seems to play a central role [13]. In vitro the enzyme inactivates when it is reduced in the absence of NO_3^-, even by its own physiological reductant NADH or NADPH [3, 20]. Alkaline pH, ADP, O_2^- and particularly CN^- have been proposed as physiological effectors that actively contribute to accelerate the inactivation of the enzyme [8, 16, 21, 23, 25]. Acetylene proved to be also a strong inactivating agent [18]. Incubation with the strong oxidant ferricyanide brings back full nitrate reductase activity [1, 3, 20, 25, 26]. Recently, it has been described that the strong oxidant tri-

1 Instituto de Biología Celular, C.S.I.C. Velázquez, 144, Madrid-6, Spain

Blue Light Effects in Biological Systems
ed. by H. Senger

valent manganese, which may be generated in the light by the photosynthetic apparatus, also reactivates nitrate reductase [17]. Nevertheless, the most outstanding agent that activates nitrate reductase in plants and fungi appears to be blue light [1, 3, 19, 22]. It has been proposed that photoexcited flavins, that become strong oxidants, are the actual activating agents removing electrons from the molybdenum domain which leads to the liberation of the bound inactivating agents [19]. In the case of the CN^--inactivated enzyme, excited flavins might alternatively remove CN^-, since in their excited triplet state are able to bind CN^- with very high affinity [19].

Recently it was reported that blue light reactivates *Chlamydomonas* nitrate reductase not only in vitro but also in vivo [4]. Moreover, in situ nitrate reductase photoactivation was found to be closely associated with NO_3^--dependent NO_2^- and NH_4^+ release to the culture medium in cells sparged with common air (0.03% CO_2). Consumption of NO_2^- under these conditions was also associated with NH_4^+ release, but it occurred at similar rates under blue than under red light. The physiological excretion of NO_2^- and NH_4^+ might have a relevant significance in the cellular balance between C and N metabolisms, since it appears to be one way of exempting the cells from superfluous, if not harmful, excess of the internal reductants eventually generated during the photosynthetically active intervals.

As will be reported elsewhere [5] CO_2 tensions and NO_3^- concentrations in conjunction with light quality and intensity proved to have a marked influence not only on the rates of NO_3^- consumption, but indeed on the nature and on the final allocation of the reduced combined N end products. Under air and with low NO_3^- concentrations (1 mM) in the medium, the cells did not release either NH_4^+ or NO_2^-, while under CO_2-free air the cells release only NH_4^+. However, at high concentration of NO_3^- (5 mM), the cells release both NO_2^- and NH_4^+, although at higher rates under CO_2-free air than under air. If the sparging air was eventually supplemented with 2% CO_2, the previously excreted reduced combined N species were rapidly assimilated. The study of the effect of monochromatic light on the release of NO_2^- by cells sparged with CO_2-free air shows that blue light of 460 nm was really effective, whereas light of 404 and 630 nm did not sustain substantial rates of NO_2 release. However, when the cells were illuminated simultaneously with both 404 and 630 nm light, they excreted NO_2^- at rates that were comparable to those obtained with 460 nm light. From these results and based on the well-established link between NO_2^- excretion and blue light activation of nitrate reductase [4], it was concluded that 404 nm more probably activates nitrate reductase, but conversely was not able to raise enough photosynthetic reducing equivalents to sustain NO_2^- excretion, whereas 630 nm red light, that was effective in activating the photosynthetic apparatus, was, however, mostly unable to activate nitrate reductase.

The results of this article show the effect of different N sources, namely NO_3^-, NO_2^- and NH_4^+, on the modulation of the inorganic N metabolism of *Chlamydomonas reinhardii* sparged with air or with CO_2-free air. Under air, when NO_3^- was present in the culture medium as the sole N source at the beginning of the light period, its consumption was associated with the release of both NO_2^- and NH_4^+. However, if prior to light exposure, NH_4^+ in addition to NO_3^- or to both NO_3^- and NO_2^- was present in the medium, NH_4^+ was preferentially assimilated. If besides NO_3^-, NO_2^- was present in the medium, NO_2^- concentration remained mostly unaltered, while NO_3^- consumption and NH_4^+ release proceeded normally. Along the successive light periods of their synchronous growth, in the first light period, starting with NO_3^- as the only N source, the cells consumed NO_3^- and excreted NO_2^- and NH_4^+; in the second light period, they assimilated NH_4^+ almost until its completion and in the third light period, they consumed both NO_3^- and NO_2^-, albeit NO_3^- at higher rates, reinitiating the excretion of NH_4^+.

The cells sparged with CO_2-free air consumed NO_3^- at lower rates than those kept under air, while NO_2^- excretion was similar in both cases, and NH_4^+ was excreted at higher rates and to greater extent by cells sparged with CO_2-free air. Utilization of NO_3^- saturated at lower white light intensity under CO_2-free air than under air. Under CO_2-free conditions, the pH optima for NO_3^- consumption and NO_2^- and NH_4^+ release ranged between 8 to 9. Under these conditions, NO_3^--dependent NO_2^- release saturated at much lower light intensity under blue (460 nm) than under red (630 nm) light. The present available data indicated that the more efficient radiations for NO_2^- release were between 400 and 500 nm with a pick at 480 nm. In the red region of the spectrum, it was observed a much smaller pick at 630 nm.

These results will be discussed in relation with the light-independent inactivation of nitrate reductase and the photoactivation of this enzyme, which might suggest that in green algae inorganic N metabolism is photochromically regulated by mainly blue light.

Materials and Methods

Chlamydomonas reinhardii 11-32b from the Alga Collection of Göttingen University was grown synchronically as previously described [4] but illuminated with 6 x 20 W Fluora Osram fluorescent lamps. Samples were removed from the growing culture in the last hour of the dark period for each experiment. For the determination of the pH optima, nutrient solutions were adjusted with K-phosphate for pH 6 to 8 and with tris-ClH for pH 9 and 10.

In the white light experiments, 35 ml of 20 μg chl ml^{-1} cell suspensions were held in 50 ml cylinders and illuminated (100 W m^{-2}) with 160-W model SBR-F, PAR 38 Eye, self-ballasted mercury lamps. In the monochromatic light experiments, actinic light was obtained fitting ordinary slide projectors with interference Oriel filters (10 nm bandwidth), and NO_2^- excretion values represent the difference between those observed in the light and the corresponding ones from the dark controls. Other experimental conditions were as described elsewhere [4].

Quantum fluence rates were measured with a LICOR LI-188B Integrating Quantum Photometer. Air was bubbled through saturated KOH and K-phosphate solutions to eliminate its CO_2. NO_3^- was monitored by measuring the absorbance differences between 210 and 250 nm [6] and NH_4^+ using glutamate dehydrogenase [7].

Results

In green algae growing with saturating CO_2 tensions it was well documented that NH_4^+ at low concentrations (μM) impaired almost immediately (1–5 min) NO_3^- assimilation, more likely by blocking the NO_3^- uptake system [11]. Moreover, in the long term (45–60 min) and a higher concentrations (mM), NH_4^+ led additionally to the repression of nitrate reductase biosynthesis [15]. However, when NO_3^- initially was the only N source in the culture medium, *Chlamydomonas reinhardii* kept in white light under common air excreted ordinarily NO_2^- and NH_4^+ steadily up to more than 0.5 mM [4]. Notwithstanding, as shown in Fig. 1, if prior to white light exposure, 0.8 mM NH_4^+ and 0.5 mM NO_2^- were present in the culture medium together with 5 mM NO_3^-, the cells utilized preferentially NH_4^+ almost until its completion. Subsequently, the cells started the reduction of NO_3^- and the release of NH_4^+, while NO_2^- concentration remained unchanged. Figure 2 shows that, if in addition to NO_3^-, only 0.5 mM NO_2^- was present in the medium before light exposure the cells release NH_4^+ associated exclusively with NO_3^- consumption. However, when NH_4^+ concentration in the medium reached values above 0.5 mM, reduction of NO_3^- stopped. Figure 3 shows that when 0.8 mM NH_4^+ together with 5 mM NO_3^- were present in the medium prior to light exposure, only NH_4^+ was assimilated until its completion. Subsequently reduction of NO_3^- coupled with NO_2^- release was iniciated. Indeed NH_4^+, or some of its amino-N assimilatory products have a relevant role, albeit weaker than in cells growing on saturating CO_2 tensions, impeding NO_3^- utilization.

The study of the inorganic nitrogen metabolism along the successive light-dark periods of the synchronized *Chlamydomonas* growth are in good agreement with above observations. For this experiment (Fig. 4), cells growing synchronically were collected by centrifugation at the end of an ordinary dark period and placed in a culture medium containing 10 mM NO_3^- as the only N source. During the first 7 h white light period, NO_3^- was steadily consumed and NO_2^- and NH_4^+ were released at compar-

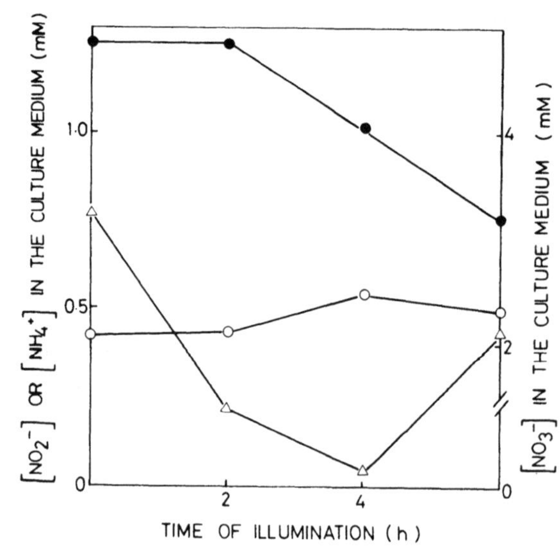

Fig. 1. Effect of both NH_4^+ and NO_2^- on NO_3^- utilization. NO_3^- (•), NO_2^- (○) and NH_4^+ (△)

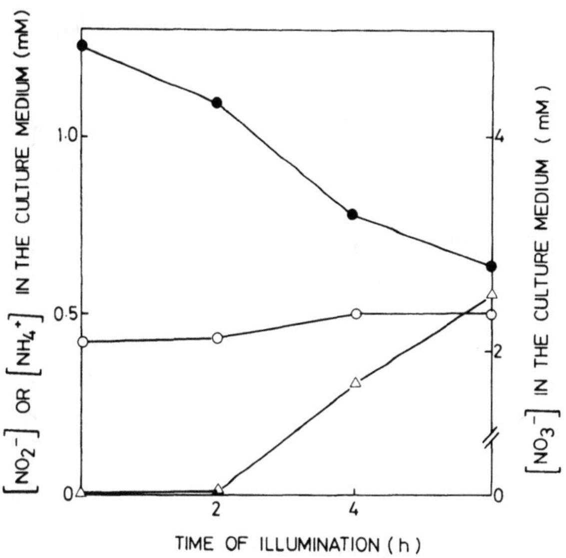

Fig. 2. Effect of NO_2^- on NO_3^- utilization. NO_3^- (•), NO_2^- (○) and NH_4^+ (△)

able rates up to 0.8 mM. In the subsequent second light period, when besides NO_3^-, NO_2^- and NH_4^+ were also present in the medium, the cells preferentially assimilated NH_4^+ almost until completion, while NO_3^- and NO_2^- concentrations remained mostly unchanged. In the third light period, when the main nitrogenous compounds in the medium were NO_3^- and NO_2^- both were simultaneously reduced, although NO_2^- at lower rate than NO_3^-, while the release of NH_4^+ reassumed. During the dark periods minor changes in the concentration of the different nitrogenous compounds were observed, thus indicating that inorganic nitrogen metabolism is mainly associated with the photosynthetically active intervals.

As it has been studied in more detail elsewhere [5], CO_2 tensions in the gas phase were one of the most critical factors affecting inorganic N metabolism in green algae.

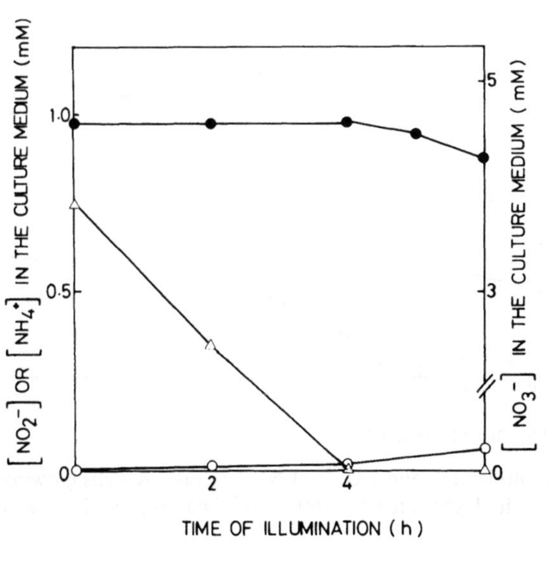

Fig. 3. Effect of NH_4^+ on NO_3^- utilization. NO_3^- (•), NO_2^- (○) and NH_4^+ (△)

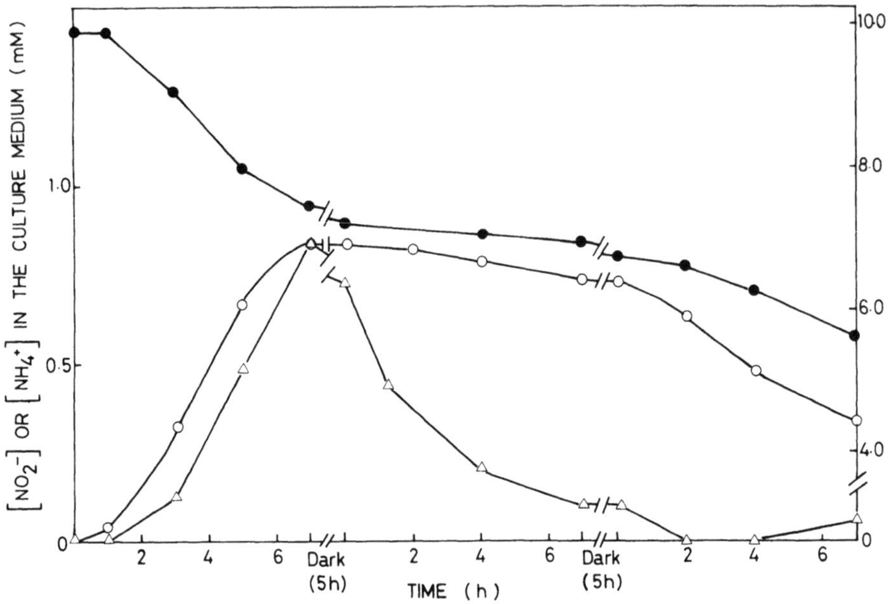

Fig. 4. Utilization and/or excretion of NO_3^-, NO_2^- and NH_4^+ along the synchronized *Chlamydomonas* growth. NO_3^- (●), NO_2^- (○) and NH_4^+ (△)

Hence, in order to standarized experimental conditions, that may also contribute to the reproducibility of the data, in all the experiments described below the cell suspensions were sparged with CO_2-free air.

Under this conditions, NO_3^- utilization rates, are ordinarily lower than under air (Fig. 5), and saturated at low white light intensity (in the order of 40 and 200 W m^{-2}

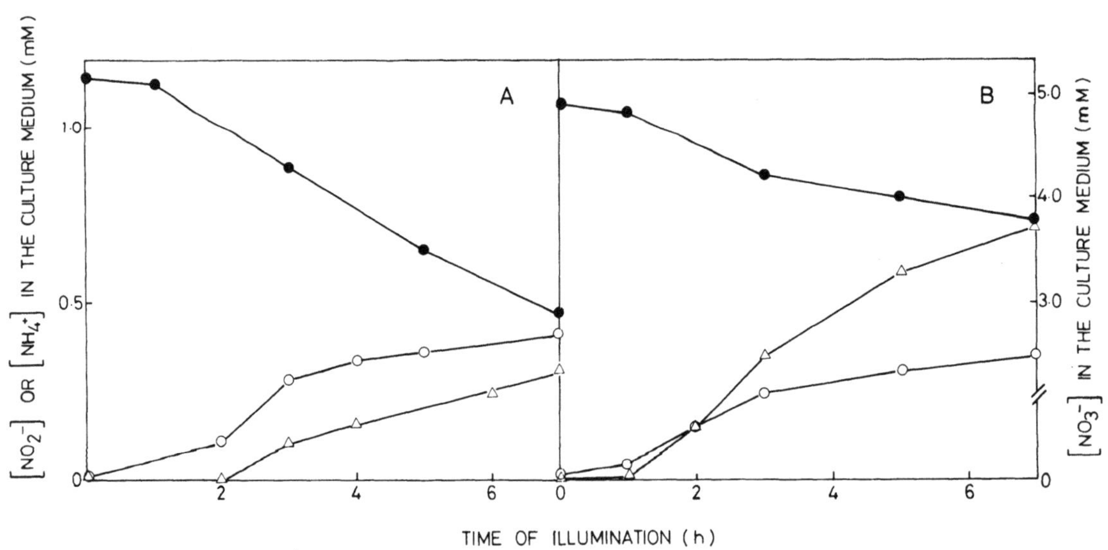

Fig. 5A,B. Effect of CO_2-free air on NO_3^- utilization and NO_2^- and NH_4^+ release. **A** Cells growing under air; **B** Cells growing under CO_2-free air. Light intensity 100 W m^{-2} NO_3^- (●), NO_2^- (○) and NH_4^+ (△)

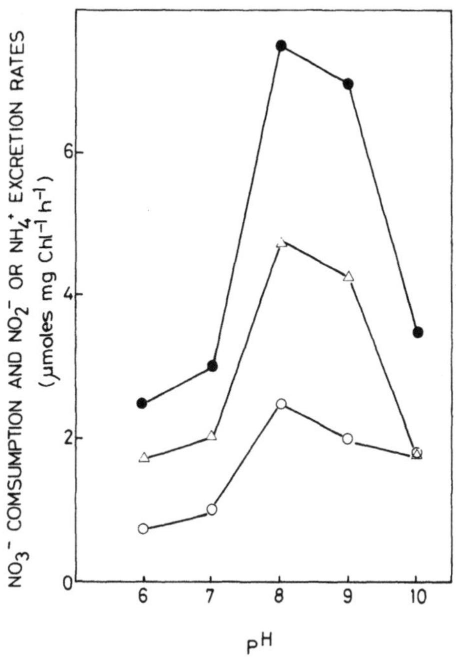

Fig. 6. Effect of pH on NO_3^- utilization and NO_2^- and NH_4^+ release rates. Cell suspensions were suspended in a 10 mM NO_3^- medium and sparged with CO_2-free air. NO_3^- (•), NO_2^- (○) and NH_4^+ (△)

for CO_2-free air and common air, respectively, data not shown). As expected, the cells, that under CO_2-free air should lack carbon skeletons, excreted all the generated NO_2^- and NH_4^+ (Fig. 5B). The pH of the culture medium proved to be also an important factor in the excretion of reduced combined nitrogen species. Figure 6 shows that the pH optimum for the NO_3^- consumption and its concomitant NO_2^- and NH_4^+ release under these conditions ranged between 8 to 9.

Under CO_2-free air, *Chlamydomonas* inorganic N metabolism was also photochromically modulated. The results of Fig. 7 show that NO_2^- excretion saturated at much lower photon fluence rates under 460 nm blue monochromatic light than under 630 nm red monochromatic light (10 nm half bandwidth). The present studies of the rates of NO_2^- release under different monochromatic lights of similar photon fluence rates revealed that the more efficient radiations were those of 460 and 480 nm in the blue region of the spectrum (Fig. 8). The small peak (630 nm) observed in the red region might reflect the role of the photosynthetic apparatus in generating the eventual reductants required for the reduction of NO_3^-, although other pigments might also be involved. Nevertheless, the pick in the blue light region is similar to that reported for the photoactivation of nitrate reductase of *Neurospora* [22].

Discussion

In vitro blue light activation of a key enzyme like nitrate reductase with such relevant regulatory features on the inorganic N metabolism in photosynthetic eukaryotes [26], deserved further experimental efforts to stablish if the photoactivation was really

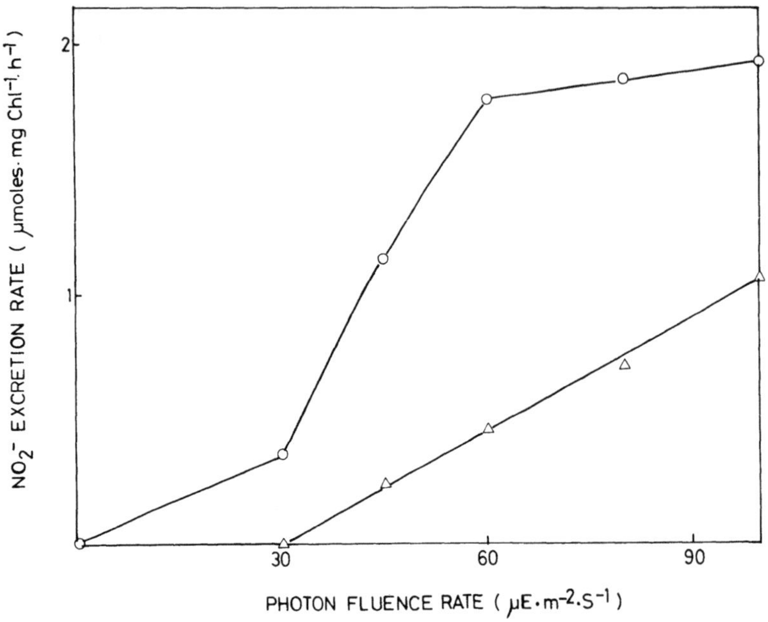

Fig. 7. Light saturation of NO_2^- release rates with 460 nm blue and 630 nm red light. Cells were suspended in a 10 mM NO_3^- medium and sparged with CO_2-free air. 460 nm (○) and 630 nm (△) light

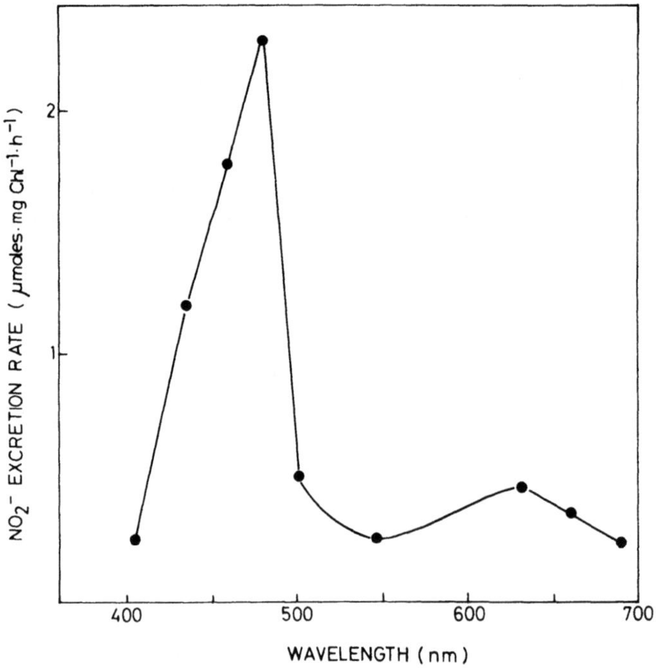

Fig. 8. Wavelength-dependent NO_2^- release in *Chlamydomonas reinhardii.* Photon fluence rate 60 $\mu E\ m^{-2}\ s^{-1}$. Other experimental conditions as in Fig. 7

opperative in vivo. Recent data from *Chlamydomonas reinhardii* cells growing in the light under air suggest that this was actually the case [4]. Blue light promoted a NO_3^--dependent NO_2^- and NH_4^+ release associated with the in situ activation of nitrate reductase. Additionally, in this organism blue light seemed to accelerate the biosynthesis and/or assembly of nitrate reductase provided that NO_3^- was present in the culture medium.

Since the reductive consumption of NO_3^-, besides its assimilatory purposes, may have other relevant regulatory roles [4], the study of the recently reported new physiological reactions like the excretion of NO_2^- and NH_4^+, might add valuable insights to the interpretation of the inorganic N metabolism. Inorganic-N assimilation greatly depends on the internal availability of carbon skeletons to form amino acids, thus CO_2 tensions and light intensity are some of the major factors involved in the process [5]. Additionally, the actual and relative concentrations of the different N sources, namely NO_3^-, NO_2^- and NH_4^+ modulates their distinct and preferential assimilation or excretion. *Chlamydomonas* synchronous growth under common air shows alternating pronounced patterns in the consumption of NO_3^- and in the excretion or assimilation of NO_2^- and NH_4^+.

What is contributing to increase the physiological interest of the inorganic nitrogen metabolism in photosynthetic eukaryotes certainly is the photochromic nature of its regulation [2]. Under certain environmental conditions nitrate reductase may become reversibly inactivated in a light-independent process [12] and preferentially under blue or white light it can be readily reactivated [1, 4, 19]. Under CO_2-free air, NO_3^--dependent NO_2^- release takes place at relatively low light intensities and is sustained much more efficiently by a number of different blue monochromatic radiations than by radiations in the red spectral region. However, a more detailed and expanded action spectrum, would be required to ascribe to a flavin-sensitized photoreaction the observed results. Nevertheless the available information on the subject [4, 19, 22] points out that blue light more likely activates in situ nitrate reductase in a flavin photoreaction rising simultaneously photosynthetic reducing power for the reduction of NO_3^-. In this respect, results that will be presented elsewhere [5], strongly suggest that NO_2^- release rely on the simultaneous occurrence of these two different physiological phenomena, since radiations of 404 and 630, that respectively activate mainly nitrate reductase and the photosynthetic apparatus, are able to sustain high NO_2^- excretion rates, only if acting simultaneously. Nevertheless, to arrive at a convincing interpretation for the small pick at 630 nm, further experimental work would be needed since chlorophyll, phytochrome or even flavins absorb light in this spectral region [10, 27].

Acknowledgments. The authors thank Drs. F.F. del Campo, J.M. Ramírez, and G. Giménez-Gallego for their useful discussions and suggestions. We are indebted to Mrs. M.C. Partearroyo and C.F. Cabrera for their secretarial assistance. This work was aided by grants from C.S.I.C. and CAICYT 611/124.

References

1. Aparicio PJ, Maldonado JM (1979) Regulation of nitrate assimilation in photosynthetic organisms. In: Hewitt EJ, Cutting CV (eds) Nitrogen assimilation of plants. Academic Press, London New York, pp 207–215
2. Aparicio PJ, Maldonado JM (1983) Photochromic regulation of nitrate reductase in green algae and higher plants. In: Montagnoli G, Erlanger BG (eds) Molecular models of photoresponsiveness. NATO Adv Study Inst Ser. Plenum Press, New York, pp 364–365
3. Aryan AP, Batt RG, Wallace W (1983) Reversible inactivation of nitrate reductase by NADH and the occurrence of partially inactive enzyme in wheat leaf. Plant Physiol 71:582–587
4. Azuara MP, Aparicio PJ (1983) In vivo blue-light activation of *Chlamydomonas reinhardii* nitrate reductase. Plant Physiol 71:286–290
5. Azuara MP, Aparicio PJ (1984) Effects of light quality on the regulation of the inorganic nitrogen metabolism of *Chlamydomonas reinhardii.* Photosynth Res, in press
6. Cawse PA (1967) The determination of nitrate in soil solutions by ultraviolet spectrophotometry. Analyst 92:311–315
7. Da Fonseca-Wollheim F, Bergmeyer HU, Gutmann I (1974) Methoden der enzymatischen Analyse, 2nd edn. Verlag Chemie, Weinheim, pp 1850–1853
8. De la Rosa MA, Gómez-Moreno C, Vega JM (1981) Interconversion of nitrate reductase from *Ankistrodesmus braunii* related to redox changes. Biochim Biophys Acta 662:77–85
9. De la Rosa MA, Vega JM, Zumft WG (1981) Composition and structure of assimilatory nitrate reductase from *Ankistrodesmus braunii.* J Biol Chem 256:5814–5819
10. Delbruck M, Katzir A, Presti D (1976) Responses of *Phycomyces* indicating optical excitation of the lowest triplet state of riboflavin. Proc Natl Acad Sci USA 73:1969–1973
11. Florencio FJ, Vega JM (1982) Regulation of the assimilation of nitrate in *Chlamydomonas reinhardii.* Phytochemistry 21:1195–1200
12. Guerrero MG, Vega JM, Losada M (1981) The assimilatory nitrate-reducing system and its regulation. Annu Rev Plant Physiol 32:169–240
13. Hewitt EJ, Notton BA (1980) Nitrate reductase systems in eukaryotic and prokaryotic organisms. In: Coughland M (ed) Molybdenum and molybdenum-containing enzymes. Pergamon Press, Oxford, pp 273–325
14. Johnson JL, Hainline BE, Rajagopalan KV (1980) Characterization of the molybdenum cofactor of sulfite oxidase, xanthine oxidase, and nitrate reductase. J Biol Chem 255:1783–1786
15. Losada M, Paneque A, Aparicio PJ, Vega JM, Cárdenas J, Herrera J (1970) Inactivation and repression by ammonium of the nitrate reducing system in *Chlorella.* Biochem Biophys Res Commun 38:1009–1015
16. Maldonado JM, Herrera J, Peneque A, Losada M (1973) Reversible inactivation by NADH and ADP of Chlorella fusca nitrate reductase. Biochem Biophys Res Commun 51:27–33
17. Maldonado JM, Notton BA, Hewitt EJ (1980) The reactivation of nitrate reductase from spinach (*Spinacia oleracia* L.) inactivated by NADH and cyanide, using trivalent manganese either generated by illuminated chloroplasts or as manganipyrophosphate. Planta 150:242–248
18. Maldonado JM, Vargas MA, Mauriño SG, Aparicio PJ (1981) Inactivation by acetylene of spinach nitrate reductase. Biochim Biophys Acta 661:112–119
19. Mauriño SG, Vargas MA, Aparicio PJ, Maldonado JM (1983) Blue-light reactivation of spinach nitrate reductase inactivated by acetylene or cyanide. Effects of flavins and oxygen. Physiol Plant 57:411–416
20. Moreno C, Aparicio PJ, Palacian E, Losada M (1972) Interconversion of the active and inactive forms of *Chlorella* nitrate reductase. FEBS Lett 26:11–14
21. Relimpio AM, Aparicio PJ, Paneque A, Losada M (1971) Specific protection against inhibitors of the NADH-nitrate reductase complex from spinach. FEBS Lett 17:226–230
22. Roldán JM, Butler WL (1980) Photoactivation of nitrate reductase from *Neurospora crassa.* Photochem Photobiol 32:375–381

23. Solomonson LP, Vennesland B (1972) Properties of a nitrate reductase of *Chlorella.* Biochim Biophys Acta 267:544–557
24. Syrett PJ (1981) Nitrogen metabolism of microalgae. In: Platt T (ed) Physiological bases of phytoplancton ecology. Can Bull Fish Aquat Sci 210:182–210
25. Vega JM, Herrera J, Relimpio JM, Aparicio PJ (1972) NADH-nitrate réductase de *Chlorella:* nouvelle contribution à l'étude de ses propriétés. Physiol Vég 10:637–652
26. Vennesland B (1981) HCN and the control of nitrate reduction. The regulation and the amount of active nitrate reductase present in *Chlorella* cells. In: Bothe H, Trebst A (eds) Biology of inorganic nitrogen and sulfur. Springer, Berlin Heidelberg New York, pp 233–243
27. Whitelam GC, Johnson CB, Smith H (1979) The control by phytochrome of nitrate reductase in the curd of light-grown cauliflower. Photochem Photobiol 30:589–594

Photoreceptor Mechanism of *Neurospora crassa*: Control Over the Electrophysiological Properties of Cell Membrane and Over the Level of Nucleotide Regulators

M. S. KRITSKY[1], T. P. AFANASIEVA[1], T. A. BELOZERSKAYA[1], L. M. CHAILAKHIAN[2], E. K. CHERNYSHEVA[1], S. YU. FILIPPOVICH[1], N. N. LEVINA[1], T. V. POTAPOVA[3], and V. YU. SOKOLOVSKY[1]

The fungus *Neurospora crassa* bears a physiological mechanism which provides for the control by light of the realization of its developmental program. When this organism is cultivated in the surface culture, light stimulates formation and maturation of conidia rich in carotenoid pigments [3, 14]. In some mutants conidiogenesis manifests itself with a circadian rhythm; the permanent illumination suppresses the rhytmicity [13]. In the submerged culture the conidia formation is hampered and carotenoids are accumulated in response to illumination directly in the vegetative hyphae. The action spectra of the physiological response – the photoinduction of carotenogenesis and the light suppression of circadian rhythm of conidiation – demonstrate the sensitivity of the photoreceptor to the short wave visible and to the near-ultraviolet radiation [3, 13, 14]. These action spectra, along with the results obtained in the experiments with riboflavin auxotrophic mutants [9], and with the data concerning photocatalytic redox reactions connected with the photoreception [5, 8] may indicate the possible role of the flavin chromophore in photoreception.

Functioning of the physiological mechanism of photoreception poses a question: what systems of metabolic regulation can be involved in the implementation of the physiological response? The regulation by light of the activity of the cell membrane is of special interest, due to its role in the control of cellular homeostasis. The other control mechanism which is studied in this paper is the regulation by light of some low molecular nucleotide regulators, including cyclic AMP.

To obtain the information concerning the relationship between the photoregulation mechanism and the function of the cell membrane, the effect of light on electrophysiological parameters, i.e., the membrane potential and the input resistance, was studied. The measurement of these parameters was carried out by using the standard intracellular microelectrode technique. The influence of illumination on the membrane potential was measured in two separate groups of experiments differing in conditions of illumination. In the first group the whole surface of the colony was illuminated for different periods of time and then compared with the nonilluminated

1 A.N. Bach Institute of Biochemistry, Academy of Sciences of the USSR, Moscow

2 Institute for Information Transmission Problems, Academcy of Sciences of the USSR, Moscow

3 A.N. Belozersky Laboratory of Molecular Biology and Bioorganic Chemistry, M.V. Lomonosov State University, Moscow, USSR

Blue Light Effects in Biological Systems
ed. by H. Senger

culture. In the second group the light spot about 200-300 μ in diameter was focused on the hyphae directly in the chamber for microelectrode measurement. Later the values of their membrane potential could be compared with the neighboring nonilluminated cells. The illumination was carried out by the spectral zone 350-500 nm with an intensity of 20,000 erg cm^2 s^{-1}. The visual microscopic control during cell puncturing was carried out with dim red light. In order to measure the input resistance of the membrane, the current pulses were passed through the second electrode put into the same cell. Input resistance was calculated as a ratio of the membrane potential shift to the value of the current pulse [12].

In both groups of experiments illumination of the cells by blue light led to hyperpolarization of the cell membrane. After 20-25 min illumination the cell membrane was hyperpolarized for 40-60 mV if compared with the dark control level. The increase of the input resistance was also observed in many cells when using local spot illumination. Interestingly, the time course of the development of the input resistance changes did not correspond to that of the membrane potential ones. The values of input resistance reached a maximum of 2-3 min (in some experiments 5-8 min) after the onset of illumination, that is during the period when the membrane potential values hardly changed. Moreover, after 30 min of illumination, i.e., at the maximum of the cell membrane hyperpolarization, input resistance in the light-exposed and dark cultures did not show distinct differences: 2.5 ± 0.4 Mohm after illumination and 3.1 ± 1.5 Mohm in darkness [12].

Thus, our work showed that the illumination of the *N. crassa* hyphae with blue light caused changes in the functions of the cell membranes. It is as yet difficult to realize what regulatory channels may couple the photoreceptor mechanism with the function of the cell membranes. In this connection two groups of data are noteworthy.

First, the possible connection between the light-activated redox functions in plasmalemma and the generation of the membrane potential. Second, the photoreceptor control over the level of cyclic AMP, which will be discussed later, the substance known to play an important role in the regulation of membrane ion currents. Since the most significant component of the membrane potential of *N. crassa* is generated by a proton electrogenic H^+-ATPase, it seems plausible to suspect that the hyperpolarization may occur as a result of its activation.

Underlying the photoinduced carotenogenesis is the control by the photoreceptor over a series of successive enzymatic reactions providing synthesis and dehydrogenation of phytoene. Since the dark-grown cells accumulate phytoene and no subsequent compounds of the biosynthetic pathway are formed, the dehydrogenation of phytoene can be regarded as a critical key step of the photoinduction mechanism. Light activates the dehydrogenation of phytoene as well as some preceding reactions: the formation of geranyl geranyl pyrophosphate and of phytoene. A prerequisite for the activation is stimulation of the expression of some nuclear genes: *albino 3, albino 2* and *albino 1*. These genes are responsible for the synthesis of carotenogenic enzymes prenyl transferase, phytoene synthetase and phytoene dehydrogenase (reviews: [4, 10]).

It was found that illumination of the *N. crassa* cells induces activation of phosphodiesterase and the transient fall of the cyclic AMP level during a lag period of photoinduction. The value of this fall fluctuates in the individual experiments and correlates with the value of the subsequent increment of carotenoids. The mutants with the impaired function of adenyl cyclase, such as *crisp I* or *frost,* demonstrate a diminished dark, i.e., constitutive level of cyclic AMP and increased content of the carotenoid pigments. On the other hand, inhibition of the phosphodiesterase activity by 1 methyl-3-isobuthyl-xanthine also inhibits the light-induced accumulation of carotenoids. These results may be summarized to indicate that a negative correlation exists between the level of cyclic AMP and the accumulation of carotenoids. Thus, regulation of the cyclic AMP level can be regarded as one of the mechanisms controlling the expression of carotenoid loci in the process of photoinduction [7].

This control mechanism constitutes, however, only one side of photoregulation process. In addition to enzyme proteins, the biosynthetic pathway needs some suitable low molecular co-factors for its free operation. For instance, the key reaction, dehydrogenation of phytoene, in addition to the enzyme phytoene dehydrogenase needs also $NADP^+$ as a cofactor [2, 11]. It is important to elucidate the mechanism of photoregulation of NAD^+kinase reaction which supplies $NADP^+$ to metabolic needs.

It was found that at different developmental stages light may activate NAD^+kinase via different mechanisms. In a vegetative mycelium grown in submerged culture, there prevails an activation mechanism based on the flavin-dependent photooxidation of NADH, the inhibitor of NAD^+kinase reaction [6]. The operation of this mechanism could be suspect from the changes in nicotinamide coenzymes level during the lag-period of photoinduction, when photooxidation of NADH and parallel to it increase of $NADP^+$/NAD ratio could be observed [5]. Further investigation of the NAD^+kinase properties allowed us to demonstrate the photoregulation mechanism in in vitro reconstitution experiments.

The other, more specific mechanism of regulation is active during the light-stimulated maturation of conidia. Illumination leads to a rapid activation of NAD^+kinase, which is insensitive to translation inhibitors, i.e., is not connected with the synthesis of new protein molecules. The increase of activity is accompanied by the activation of one of the molecular forms, the so-called form II having a molecular weight of 3.06×10^5 [1]. It is important that both activation mechanisms do not involve de novo protein synthesis. Thus they work in parallel to a mechanism inducing formation of carotenogenic enzymes.

When studying control mechanisms of the enzymes involved in photoinduction, cyclic AMP phosphodiesterase and NAD^+kinase, it should be kept in mind that control of their activity may be achieved by different molecular mechanisms. The results obtained in our group have shown, for instance, a different attitude of the enzymes to the metabolic control mechanism involving calcium ions. The activity of phosphodiesterase is under a negative control of calcium ion concentration. The experiments with halogenated phenothiazines, inhibitors of enzyme association with calmodulin, suggest a control mechanism based on interaction with this protein regulator. In contrast, NAD^+kinase activity does not depend on calcium ions in a broad range of concentration.

We may conclude that activation of photoreceptor leads to a parallel shift of very different parameters, including activity of some key enzymes, the level of nucleotide cofactors and regulators, selective ion transport and genome expression. The metabolic regulation mechanism under the control of photoreceptor is a complex, branched system. This could be illustrated by photocontrol of carotenogenesis, when both synthesis of new enzyme molecules and activation of already existing molecules should be considered. Formation of the ultimate physiological response in more complex systems, such as photomorphogenetic ones, can be expected to have a relatively more complex network of parallel and interacting metabolic events.

References

1. Afanasieva TP, Filippovich SYu, Sokolovsky VYu, Kritsky MS (1982) Developmental regulation of NAD⁺Kinase in Neurospora crassa. Arch Microbiol 133:307
2. Clarke I, Sandmann G, Bramley PM (1981) Cofactor requirements for the formation of carotenes and squalene in Phycomyces blakesleeanus. Abstr 6th Int Symp Carotenoids. July 1981, Liverpool, England
3. De Fabo EC, Harding RW, Shropshire W Jr (1976) Action spectrum between 260 and 800 nanometers for the photoinduction of carotenoid biosynthesis in Neurospora crassa. Plant Physiol 57:N2, 440
4. Harding RW, Shropshire W Jr (1980) Photocontrol of carotenoid biosynthesis. Annu Rev Plant Physiol 31:217
5. Kritsky MS, Chernysheva EK, Soboleva IS (1977) On the participance of nicotinamide coenzymes in the photoinduction of carotenoid pigment synthesis in the mycelium cells of Neurospora crassa. Prikladnaya Biokhim Mikrobiol/Appl Biochem Microbiol 13:N6, 901 (in Russian)
6. Kritsky MS, Chernysheva EK (1980) On the participance of nicotinamide coenzymes in the photoinduction of carotenoid pigment synthesis in the mycelium cells of Neurospora crassa. Dokl Akad Nauk SSSR 250:N2, 473
7. Kritsky MS, Sokolovsky VYu, Belozerskaya TA, Chernysheva EK (1982) Relationship between cyclic AMP level and accumulation of carotenoid pigments in Neurospora crassa. Arch Microbiol 133:206
8. Munoz V, Butler WL (1975) Photoreceptor pigment for blue light in Neurospora crassa. Plant Physiol 55:N2, 421
9. Paietta J, Sargent ML (1981) Photoreception in Neurospora crassa: Correlation of reduced light sensitivity with flavin deficiency. Proc Natl Acad Sci USA 78:5573
10. Perkins DD, Radford A, Newmeyer D, Björkman M (1982) Chromosomal loci of Neurospora crassa. Microbiol Rev 46:4, 426
11. Porter JW, Spurgeon SL (1979) Enzymatic synthesis of carotenes. Pure Appl Chem 151: N3, 609
12. Potapova TV, Levina NN, Belozerskaya TA, Chailakhian LM, Kritsky MS (1984) The investigation of electrophysiological responses of N. crassa to blue light. Arch Microbiol 137:262
13. Sargent ML, Briggs WR (1967) The effects of light on a circadian rhythm of conidiation in Neurospora. Plant Physiol 42:N6, 1504
14. Zalokar M (1955) Biosynthesis of carotenoids in Neurospora. Action spectrum of photoactivation. Arch Biochem Biophys 56:N2, 318

Genetic Analysis and Molecular Biology of Blue Light Responses

Genetic Analysis of Blue Light-Induced Responses in *Neurospora crassa*

F. DEGLI INNOCENTI and V. E. A. RUSSO[1]

Introduction

A genetic analysis of blue light sensory transduction chains has been done until now essentially only in *Phycomyces blakesleeanus*. In this organism it was found that two genes, *mad*A and *mad*B, are necessary for the photoinduction of five different effects. This pleiotropism has been considered an indication of the different photoresponses having their early steps in common, including the photoreceptor [12, 17, 18]. The disadvantage of *Phycomyces* lies in the fact that its genetics, as well as its biochemistry are still at a very primitive stage. On the contrary, *Neurospora crassa*, one of the most studied eukaryotic organisms, is well known from both a genetic [15] and from a biochemical point of view [6]. Furthermore, it has been studied also from a photobiological point of view (see below). The possibility of integrating information coming from genetic, biochemical, and photobiological studies makes this organism particularly suitable to approach the problem of photoreception at the molecular level.

The blue light effects known in *Neurospora* are: (1) inhibition of the circadian rhythm of conidiation of the band mutant; (2) phase shift of this circadian rhythm [13, 19]; (3) induction of carotenogenesis [8, 22]; (4) induction of protoperithecia formation [5]; (5) phototropic response of perithecial beaks [1, 7]; (6) light-promoted conidiation of starved mycelia of the double mutant albino-band [10]; Ninnemann personal communication. Contrary to *Phycomyces*, where only one photoreceptor has been postulated for five photoeffects, in *Neurospora* three different photoreceptors have been suggested: (a) a carotenoid for the photocarotenogenesis [4]; (b) a b-type cytochrome-flavin complex for the circadian rhythm [20]; (c) nitrate reductase for the light-promoted conidiation of starved mycelia [13]. It has recently been shown that nitrate reductase is not the photoreceptor for the effects on circadian rhythm [14] and for the photocarotenogenesis [2].

A genetic analysis of the blue light transduction processes in *Neurospora crassa* can be very useful (as it was in *Phycomyces*) in order to determine a possible intercorrelation between the different photoresponses. In *Neurospora crassa* mutants impaired in a process of photoinduction are already known; these mutants, called white collar (*wc*), display the normal constitutive carotenogenesis in the conidia (like wild type) but are defective in the photoinduced carotenogenesis in the mycelium [8, 9].

1 Max-Planck-Institut für molekulare Genetik, Abt. Trautner, Ihnestraße 63/73, D-1000 Berlin 33, FRG

Blue Light Effects in Biological Systems
ed. by H. Senger

Furthermore it has recently been shown that the phototropism of the perithecial beaks is blocked when *wc* strains are used as the protoperithecial parent [7]. We have asked ourselves three questions: (1) how many *wc* genes are there? (2) Are the *wc* mutants also "blind" in the photoinduction of protoperithecia? (3) Are the *wc* mutants "blind" in the blue light effects on the circadian rhythm? An answer to the first two questions is given here. We have as yet no answer to the third question.

Photoinduction of Carotenogenesis in White Collar Mutants

Up to now three *wc* mutations have been reported: two of them map in a locus (*wc*-1; allele P829; P4723) in linkage group VII, the other in a locus (*wc*-2; allele 234w) in linkage group I [15]. We have done a systematic study of new *wc* mutations. In this study eight mutants with typical *wc* phenotype have been isolated after UV mutagenesis of the WT; their photocarotenogenesis is affected but the morphology and linear growth rate are not altered (Degli Innocenti and Russo in preparation). Figure 1 shows the fluence-response curve of photocarotenogenesis of wild type (WT) at 34°C; by extrapolation the blue light fluence threshold has been determined to be about 8 J m^{-2}; a similar value was found for WT grown at 26°C. The WT curve is compared

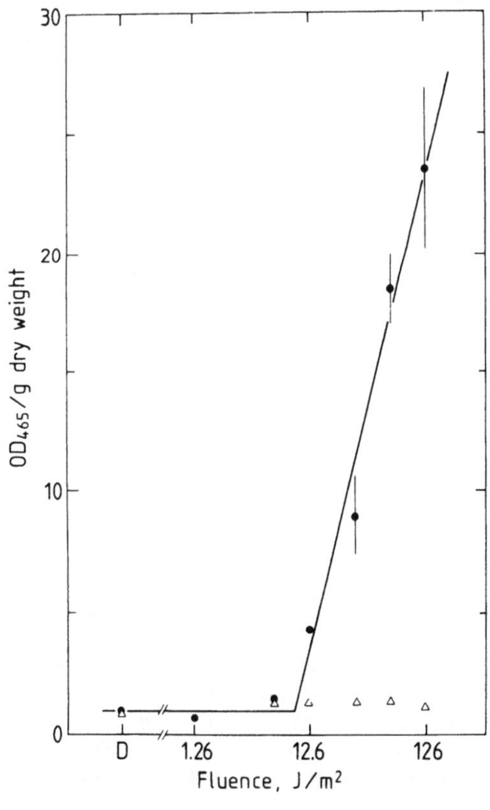

Fig. 1. Fluence-response curve of the photocarotenogenesis. The mycelia were grown for 46 h at 34°C on plates of minimal sorbose medium covered with dialysis membranes. In these conditions the formation of conidia is prevented. The mycelia were irradiated with blue light of different fluence rates for 60 s. The extraction of carotenoids was done after 90 min of further dark incubation. *D* dark; (●) wild type; average of four experiments; (△) white collar mutant (allele MK 1). *Bars* indicate the SEM (Russo and Degli Innocenti in preparation)

with the curve of a *wc* mutant. The same lack of response is displayed by all the other mutants but those carrying the mutations ER44 and ER24, when grown at 26°C. These mutations are temperature-sensitive: the photoinduction of carotenoids is blocked when the growth temperature is 34°C but it is only reduced at 26°C (Russo and Degli Innocenti in preparation). Complementation tests based on the analysis of photoinduction of carotenoids have indicated that all the *wc* mutations fall into two complementation groups corresponding to the already known *wc*-1 and *wc*-2 genes, in total seven mutations in the first group and four in the second (Russo and Degli Innocenti in preparation). Furthermore it has been shown with an inter-allelic recombination analysis that the mutations belonging to each of the two complementation groups are closely linked to the *wc* loci already mapped [16]. Therefore with our research we have isolated other *wc*-1 and *wc*-2 alleles but we have not found a new *wc* gene.

Photoinduction of Protoperithecia Formation in White Collar Mutants

We have recently shown that blue light induces the formation of protoperithecia (the female sexual structures) in *Neurospora crassa* [5]. The blue light fluence threshold of this effect is about 4.2 J m^{-2} (Fig. 2); this value is very similar to that determined by us for the photocarotenogenesis (about 8 J m^{-2}). The response of the mycelium depends only on the total fluence, independently of the time of exposition (Table 1). We wanted to determine whether the *wc* genes play a role also in the photoregulation

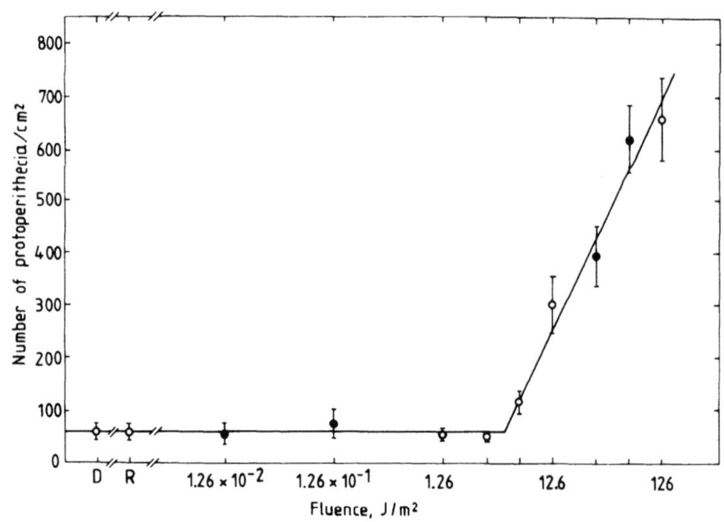

Fig. 2. Fluence-response curve of the photoinduction of protoperithecia formation. Wild-type mycelia, grown at 26°C for 72 h, were irradiated with blue light of different fluence rates for 60 s. The protoperithecia were counted after a further 24 h of dark incubation. *D* dark; *R* red light (mycelia irradiated with red light; 1620 J m^{-2}); (○) = average of four experiments with two replica; (●) = average of two experiments with two replica. *Bars* indicate the SEM. (After [5])

Table 1. Validity of the Bunsen-Roscoe reciprocity law

Fluence rate (W m^{-2})	Time (s)	Number of protoperithecia cm^{-2}
1.05	12	289 ± 39
5.25 x 10^{-1}	24	234 ± 36
2.1 x 10^{-1}	60	306 ± 28
1.05 x 10^{-1}	120	273 ± 34
5.25 x 10^{-2}	240	232 ± 23
Dark		53 ± 8

72-h-old wild-type mycelia were illuminated with blue light of different fluence rates and for different times so as to obtain the same fluence (12.6 J m^{-2}). The protoperithecia were counted after further 24 h of dark incubation. Data shown are the average of three experiments with three replica ± standard error of the mean. (After [5])

Table 2. Photoinduction of protoperithecia formation in *wc* and *al* mutants

Strain	C.G. (1)	Allele	Dark	Blue light
WT			100(2)	3091 ± 386
R 97	*wc*–1	P829	1.0 ± 0.6	10.7 ± 7.6
R 103	*wc*–1	ER57	3.5 ± 1.2	6.4 ± 1.8
R 110	*wc*–1	ER53	0.5 ± 0.5	1.2 ± 0.6
R 122	*wc*–1	MK1	2.1 ± 1.6	3.9 ± 0.8
R 126	*wc*–1	MK2	0.9 ± 0.6	1.7 ± 1.5
R 131	*wc*–1	P4723	8.1 ± 5	8.8 ± 3.1
R 145	*wc*–1	ER45	28 ± 11	26.7 ± 15.5
R 89	*wc*–2*ts*	ER44	200 ± 58	907 ± 153
R 100	*wc*–2	ER33	1.5 ± 0.2	1.0 ± 0.4
R 162	*wc*–2	234W	21 ± 7.4	19.0 ± 3.7
R 165	*wc*–2*ts*	ER24	160 ± 33	179 ± 28
R 184	*al*–2	ER7	38 ± 8	1839 ± 151
R 196	*al*–2	MK8	168 ± 38	4745 ± 618
R 200	*al*–1	ER2	261 ± 108	7548 ± 289

The formation of protoperithecia, both in the dark and after blue light irradiation (126 J m^{-2} given in 60 s, 24 h before counting), has been determined after 5 days of growth at 26°C. The number of protoperithecia per cm^2 yielded by the different strains was normalized in each experiment to the corresponding WT value in the dark considered equal to 100. Data shown are the average of two experiments with two replica ± SEM. (1) C.G. = complementation group; (2) 100 = 30.7 ± 6 protoperithecia cm^{-2} (Degli Innocenti and Russo in preparation)

of this other blue light effect. Table 2 shows the results obtained in a study of protoperithecial differentiation of different mutant strains (Degli Innocenti and Russo in preparation). These data indicate that: (1) the dark production of the *wc* mutants is lower than that of the WT; exceptions are the two *ts* (temperature-sensitive) *wc* strains; (2) the *wc* mutants do not show any significant increase in the production of protoperithecia after blue light irradiation; the only exception is the *ts wc* strain R89; (3) the albino (*al*) strains are photoinduced. The conclusion is that the *wc* mutations abolish this blue light effect and also cause a strong reduction of the protoperithecial differentiation in the dark. These effects are not due to a block in the "output", namely in the morphogenetic processes, because it is possible to induce the protoperithecial differentiation in the *wc* mutants using a different stimulus: the treatment of mycelia of each strain with suspensions of conidia of the same strain (Table 3; Degli Innocenti and Russo in preparation). The pleiotropic effect of the *wc* mutations suggests that the dark differentiation is controlled by the same processes of regulation which are responsible for the photoinduction.

Table 3. Induction of protoperithecia with Conidia

Strain	C.G.	Allele	No. of protoperithecia cm^{-2} ± SEM (A) with Conidia	(B) Control
WT			656 ± 173	92 ± 27
R 97	*wc*–1	P829	64 ± 14	4.3 ± 1.2
R 103	*wc*–1	ER57	821 ± 60	40 ± 7.5
R 110	*wc*–1	ER53	36.8 ± 8.4	2.3 ± 0.8
R 122	*wc*–1	MK1	106 ± 42	15 ± 6
R 126	*wc*–1	MK2	9.6 ± 3.4	0.9 ± 0.4
R 131	*wc*–1	P4723	101 ± 22	7.8 ± 1.5
R 145	*wc*–1	ER45	204 ± 68	16.4 ± 8.8
R 89	*wc*–2*ts*	ER44	242 ± 40	44 ± 13
R 100	*wc*–2	ER33	56 ± 10	15 ± 5.7
R 162	*wc*–2	234W	3.7 ± 1.1	1.8 ± 0.5
R 165	*wc*–2*ts*	ER24	125 ± 26	29 ± 6
R 184	*al*–2	ER7	72 ± 25	14.7 ± 4.8
R 196	*al*–2	MK8	272 ± 97	22 ± 5.8
R 200	*al*–1	ER2	599 ± 42	44 ± 26

After incubation in the dark for 5 days, each mycelium was spotted in marked areas with drops (5 μl) of: (A) a suspension of conidia having the same genotype of the mycelium (2×10^7 conidia ml^{-1}); (B) the conidia-free supernatant obtained after centrifugation of the same conidial suspension used in (A). This operation was performed under red light. The protoperithecia developed in the spotted areas were counted after further 3 days of incubation in the dark. Data shown are the average of at least two experiments with four replica ± SEM. (1) C.G. = complementation group (Degli Innocenti and Russo in preparation)

Discussion

If we assume that every gene of *Neurospora* is equally susceptible to UV mutagenesis [6], we can conclude on a statistical basis that there are probably only two *wc* genes (*wc*–1 and *wc*–2). Their gene products are necessary for photocarotenogenesis, for phototropism of perithecial beaks, for photoinduction of protoperithecia and for formation of protoperithecia in the dark. It would be very interesting to see if the *wc* mutants are also impaired in other photoresponses. At present we are investigating the effects of the *wc* mutations on the circadian rhythm of conidiation. It has been recently shown that white light transiently lowers the cAMP level [11] and that the blue-violet light causes a hyperpolarization of the membrane potential [3] in *Neurospora crassa*. *Wc* mutants should also be tested for these other light effects.

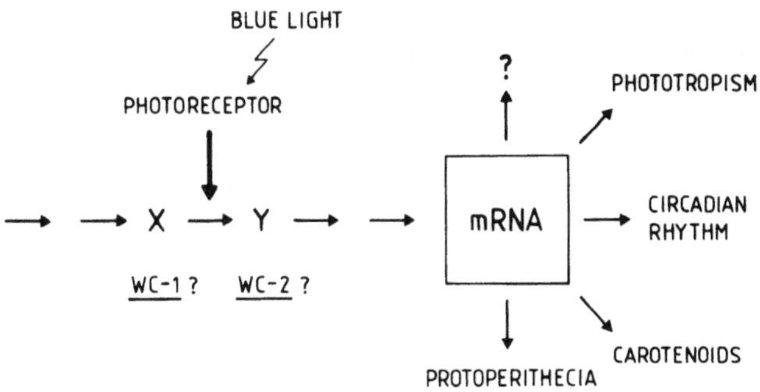

Fig. 3. Hypothetical model of the blue light sensory transduction chains in *N. crassa*. *Question marks* by *wc*–1? and *wc*–2? indicate that we do not know the sites of action of the *wc* gene products. See Discussion

Keeping in mind that in photocarotenogenesis a transcriptional control was claimed to operate [21; see also chapter by Mitzka-Schnabel et al. this vol.] and that Harding and Turner [9] have shown a photoinduction of two enzyme activities of the carotenoid pathway in WT but not in a *wc*–1 mutant (allele P829), then we can summarize all the data with the speculative model of Fig. 3. Our hypothesis is the following: (1) blue light modulates the activity of an enzyme directly (if the enzyme is the photoreceptor) or indirectly (if the photoreceptor does not have enzymatic activity); (2) this blue light-activated enzyme works in a pathway of processes which regulate the transcription of specific mRNA's; (3) the transcription and the following translation of those mRNA's cause the different photoresponses. The products of the *wc* genes control, in this model, two steps of the regulation pathway. We have no criteria to decide if the product of one of the two *wc* genes is the blue light photoreceptor. The constitutive carotenogenesis in the conidia and the formation of protoperithecia induced by conidia, which are both not affected by the *wc* mutations, should be controlled by processes downstream of the *wc* sites of action.

Acknowledgment. We wish to thank C. Ernsting and M. Götz for technical assistance, R. Bachman for secretarial assistance, and J.A.A. Chambers for reading the manuscript. This work was partially supported by the Deutsche Forschungsgemeinschaft.

References

1. Backus MP (1937) Phototropic response of perithecial necks in *Neurospora*. Mycologia 29: 383–386
2. Belozerskaya TA, Burikhanov SS, Chernyshova EK, Kritsky MS, Lvov NP (1982) Does nitrate reductase play a key role in photoinduction of carotenoid synthesis in *Neurospora crassa*? Neurospora Newslett 29:14–15
3. Belozerskaya TA, Kritsky MS, Potapova TV, Chailakhian LM (1982) Influence of light on membrane potential of *Neurospora crassa* cells. Biofizika 27:910–911
4. De Fabo EC, Harding RW, Shropshire W Jr (1976) Action spectrum between 260 and 800 nanometers for the photoinduction of carotenoid biosynthesis in *Neurospora crassa*. Plant Physiol 57:440–445
5. Degli Innocenti F, Pohl U, Russo VEA (1983) Photoinduction of protoperithecia in *Neurospora crassa* by blue light. Photochem Photobiol 37:49–51
6. Fincham JRS, Day PR, Radford A (1979) Fungal genetics. Blackwell, Oxford London Edinburgh Melbourne
7. Harding RW, Melles S (1983) Genetic analysis of phototropism of *Neurospora crassa* perithecial beaks using white collar and albino mutants. Plant Physiol 72:996–1000
8. Harding RW, Shropshire W Jr (1980) Photocontrol of carotenoid biosynthesis. Annu Rev Plant Physiol 31:217–238
9. Harding RW, Turner RV (1981) Photoregulation of the carotenoid biosynthetic pathway in albino and white collar mutants of *Neurospora crassa*. Plant Physiol 68:745–749
10. Klemm E, Ninnemann H (1978) Correlation between absorbance changes and a physiological response induced by blue light in *Neurospora*. Photochem Photobiol 28:227–230
11. Kritsky MS, Sokolovsky VY, Belozerskaya TA, Chernysheva EK (1982) Relationship between cyclic AMP level and accumulation of carotenoid pigments in *Neurospora crassa*. Arch Microbiol 133:206–208
12. Lipson ED (1980) Sensory transduction in *Phycomyces* photoresponses. In: Senger H (ed) The blue light syndrome. Springer, Berlin Heidelberg New York, pp 110–118
13. Ninnemann H (1979) Photoreceptors for circadian rhythms. In: Smith KC (ed) Photochemical and photobiological reviews, vol IV. Plenum Press, New York London, pp 207–266
14. Paietta J, Sargent ML (1982) Blue light responses in nitrate reductase mutants of *Neurospora crassa*. Photochem Photobiol 35:853–855
15. Perkins DD, Radford A, Newmeyer D, Björkman M (1982) Chromosomal loci of *Neurospora crassa*. Microbiol Rev 46:426–570
16. Russo VEA, Degli Innocenti F (1983) Recombination frequencies of mutations located in *wc*–1 and *wc*–2. Neurospora Newslett 30:11–12
17. Russo VEA, Galland P (1980) Sensory physiology of *Phycomyces blakesleeanus*. In: Hemmerich P (ed) Structure and bonding, vol 41. Springer, Berlin Heidelberg New York, pp 71–110
18. Russo VEA, Galland P, Toselli M, Volpi L (1980) Blue light induced differentiation in *Phycomyces blakesleeanus*. In: Senger H (ed) The blue light syndrome. Springer, Berlin Heidelberg New York, pp 563–569
19. Sargent ML, Briggs WR (1967) The effects of light on a circadian rhythm of conidiation in *Neurospora*. Plant Physiol 42:1504–1510
20. Senger H, Briggs WR (1981) The blue light receptor(s): primary reactions and subsequent metabolic changes. In: Smith KC (ed) Photochemical and photobiological reviews, vol VI. Plenum Press, New York London, pp 1–38
21. Subden RE, Bobowski G (1973) Evidence for actinomycin D inhibition of transcription of carotenoid loci in *Neurospora*. Experientia 29:965–967
22. Zalokar M (1955) Biosynthesis of carotenoids in *Neurospora*. Action spectrum of photoactivation. Arch Biochem Biophys 56:318–325

Genetic Determination of the Responses of Phycomyces to Light

E. CERDÁ-OLMEDO[1]

Introduction

The fungus *Phycomyces blakesleeanus* is modestly famous for the phototropism of its giant sporangiophores. But this is by no means the only effect of light: the metabolism and development of *Phycomyces* are affected by light in many other ways, many more, perhaps, than one would expect for a nonphotosynthetic organism.

Similar biochemical and morphological effects are brought about by other external stimuli. The cell must thus integrate the different informations received from the environment and use them to produce an advantageous response.

Genetic analysis allows us to determine the relations between the different regulatory effects and contributes to the understanding of their mechanisms.

Tropisms

The mutants with defective phototropism (*mad* mutants) provide an example of the use of genetic analysis in sensory physiology. Their study has been essential for our current views on phototropism (reviews: [2, 5, 17] and Lipson, this vol.).

Light is absorbed by a flavin photoreceptor [23], whose protein moiety is composed of the products of genes *madB* and *madC*. The product of gene *madA* intervenes soon thereafter, contributing to the adaptation to the prevailing light intensity. The action of other *mad* gene products (*madD* through *madH*) brings about the observed changes in growth velocity (mecism) and direction (tropism). These gene products modify the patterns of chitin breakage and synthesis (J. Ruiz-Herrera et al. 1983).

The growth of the sporangiophores is influenced by other environmental stimuli, including gravity (geotropism), wind (anemotropism), and the presence of obstacles in the vicinity of the growing zone (avoidance or fugotropism). The physical stimulus mediating avoidance is still unknown, but at least it has been clearly shown to be different from that of anemotropism [16].

Geotropism and fugotropism are grossly defective in mutants of genes *madD* through *madH*, but occur normally in mutants of genes *madA* through *madC*. This

1 Departamento de Genética, Facultad de Biología, Universidad de Sevilla, Sevilla, Spain

Blue Light Effects in Biological Systems
ed. by Senger

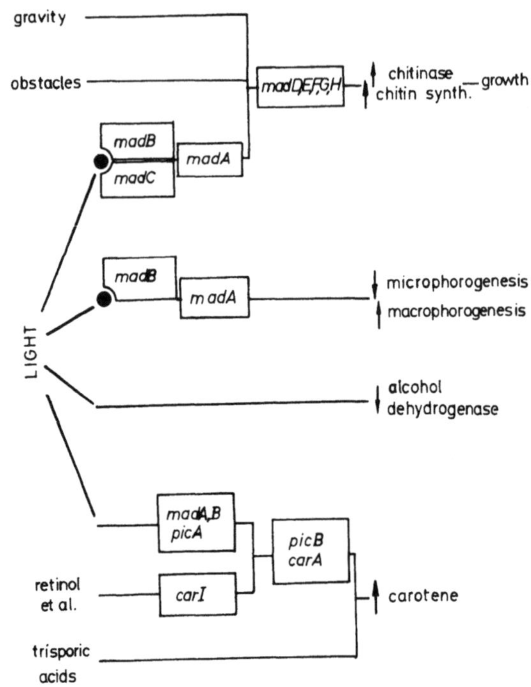

Fig. 1. Tentative overview of the sensory pathways of *Phycomyces.* The organism detects the environmental stimuli given in the *left side* of the graph, transfers the information as indicated by the *continuous lines,* and carries out the responses written in the *right side.* The necessary gene products are given in the *boxes. Black circles* represent the flavin in the photoreceptor

led to the concept of a branched sensory pathway, in which detectors for different environmental signals feed their information to a common effector mechanism regulating cell growth [3]. Genes *madA* through *madC* thus determine the light-specific elements of the sensory pathway governing the tropisms of the sporangiophores. The upper part of Fig. 1 summarizes these ideas.

In a similar way, to decide about the involvement of the *mad* gene products in other responses to light, it suffices to study such responses in the *mad* mutants, as shown in the following sections.

Photomorphogenesis

The velocity of growth and the overall look of *Phycomyces* under the usual culture conditions are unaffected by light, except for modest changes in the number of sporangiophores [27]. Various sets of experimental conditions bring the development of *Phycomyces* under stringent light control. The initiation of sporangiophores becomes light dependant on a special medium [1] or in a closed environment [25]. It is then possible to synchronize the growth of the sporangiophores and to induce bands of fruiting bodies by appropriately alternating light and darkness.

Normal sporangiophore photoinitiation requires the operation of the products of genes *madA* and *madB*, but not of the other *mad* genes [1, 11]. It thus likely that the corresponding mycelial photoreceptor is a flavoprotein having the *madB* gene product as the protein moiety; the product of gen *madA* must also participate in producing the internal signal governing this morphogenetic process. The normal supply of ß-carotene is essential for a normal response, but ß-carotene is believed to play a biochemical role other than that of photoreceptor [11].

The giant sporangiophores, or macrophores, of *Phycomyces* are up to several decimeters long and carry a sporangium containing about 100,000 spores. The size and the growth-guiding mechanisms of the macrophores make them suitable for spore dispersal, but wasteful under conditions rendering spore dispersal unlikely. It is thus not surprising that *Phycomyces* is able to develop another type of fruiting body, the microphores, measuring less than 3 mm in length and carrying a sporangium with only about 1000 spores.

Microphores are massively produced when *Phycomyces* is grown on certain very poor media in dark and cold environments. Under these conditions, blue light reverses the developmental program to the production of macrophores [30, 31]. This response has an extremely low light threshold and requires, like the other photomorphogenetic responses, the products of genes *madA* and *madB* [20]. The study of this effect of light would be extremely attractive if the experiments did not take several weeks and the results were not subject to relatively large quantitative fluctuations.

These difficulties are overcome with the experimental conditions recently used by Felix Gutiérrez-Corona in our laboratory. Microphores are produced en masse when cultures are grown on the usual agar media at 22° for about 56 h, then flooded with 12 ml of water and incubated further in the dark. Light given after the flood inhibits microphorogenesis and stimulates macrophorogenesis. These conditions are also very suitable for the study of a related problem, the chemical regulation of sporangiophore development.

Light does not have to act always through the *mad* photoreceptor. Thus, light strongly inhibits *Phycomyces* growth in the presence of abundant riboflavin or quinacrine [4], but this inhibition is independent of the *mad* genes [20]. It would not be surprising if light-activated riboflavin molecules disrupted essential cellular functions, for example respiration. The light-induced absorbance changes are likely to have a similar origin and be equally irrelevant for the adaptative behavior [18].

Photocarotenogenesis

The young mycelia, sporangiophores and sporangia of *Phycomyces* are yellow owing to the accumulation of ß-carotene (about 0.05 mg g^{-1} dry weight under the usual growth conditions in the dark). Light accentuates this pigmentation: the ß-carotene content is increased over tenfold by continuous exposure to 1 W m^{-2} blue light [3].

This photoinduction of ß-carotene synthesis is less marked in *madA* and *madB* mutants than in the wild type [3, 15, 19]. This suggested that the photoreceptor for this response might be the same mycelial photoreceptor implicated in photomorphogenesis.

The role of the *mad* genes on photocaroteneogenesis is, however, unclear. The behavior of the wild type at a given light intensity can be duplicated with *madA* and *madB* mutants by exposing them to a light intensity several times higher. This "displacement factor" is 10^4 to 10^5 times for phototropism or photomorphogenesis, but less than ten times for carotenogenesis. It is thus possible that genes *madA* and *madB* play only an indirect role in photocarotenogenesis. The question of the photoreceptor would then stay open.

The main argument against a carotene photoreceptor for phototropism is the normal behavior of the caroteneless mutants [3, 24]. A modified version of this test can be applied to the photoreceptor for increased carotenogenesis. If a mutant lacks ß-carotene, but contains a colorless intermediate in amounts much higher than the wild type, one may question whether the accumulation of this intermediate is affected by light.

Mutants unable to proceed beyond the synthesis of phytoene accumulate the same amount of phytoene in the light as in the dark [10, and Eduardo R. Bejarano unpublished]. Thus, strain C173, a *carB carR* double mutant unable to carry out the last six reactions of ß-carotene synthesis, contained 1.43 mg of phytoene g^{-1} dry weight when grown in the dark, and 1.57 when grown under continuous light (about 1 W m^{-2}). The two values are not significantly different (13 determinations each, standard deviations 0.12 and 0.13, respectively), but photoinduction to the same extent as in the wild type would have made them clearly different. The pathway is far from saturated at these phytoene levels. We conclude that photoinduction of the pathway requires the presence of ß-carotene.

The role of ß-carotene in photocarotenogenesis might be that of photoreceptor, as suggested from the detailed action spectrum in Neurospora [8] or a biochemical role unrelated to light absorption, as suggested for photomorphogenesis [11]. The possible photoreceptor role of the carotenes is a favorite matter for debate [7, 29, and this vol.].

Photocarotenogenesis should be placed in the wider context of carotene regulation. The accumulation of ß-carotene is increased by many chemicals, such as retinol, and by sexual activity, through the production of trisporic acids; the normal levels are maintained by a feedback control, determined by gene *carS* [6].

The products of other wild-type genes, such as *picA*, *picB*, and *carA*, are necessary for normal photocaroteneogenesis [19]. The *picA* mutant appears to be abnormal only for this response; the *picB* mutant is also defective for retinol stimulation. The wild-type *carA* product is required for the transfer of carotene precursors from an enzyme to another in the absence of chemical stimulation [22] and is essential for any photoinduction of the pathway.

Photoinhibition of the Carotenoprotein

The in vivo absorption spectrum of wild-type *Phycomyces* mycelia is similar to that of ß-carotene solutions. Fernando Parra, Amador de la Concha, and Francisco Murillo (unpublished) have isolated red strains of *Phycomyces* with in vivo absorption at 540 nm (Fig. 2). These strains contain ß-carotene in large amounts (about 100 times more than the wild type), but most of it is bound in a carotenoprotein complex, very similar to that of plants. A saline solution with 1% sodium deoxycholate extracts this complex and shows in vitro the characteristic 540-nm peak.

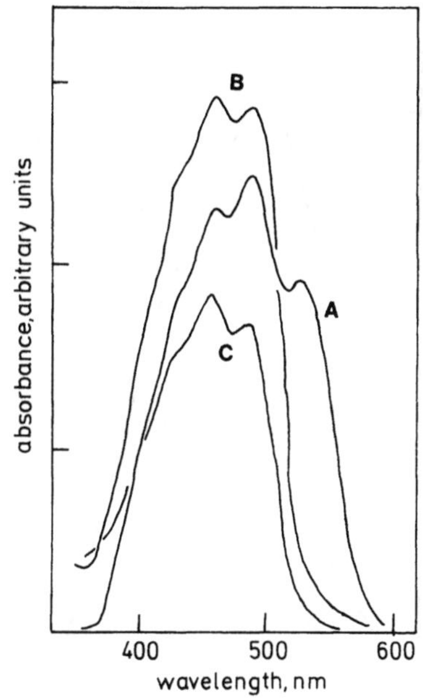

Fig. 2. In vivo absorption spectra of 4-day-old *Phycomyces* mycelia. *A* Strain S303R, producer of the carotenoprotein complex, grown in the dark. This mycelium contained 4.9 mg ß-carotene g^{-1} dry weight. *B* The same, grown under continuous white illumination (approx. 0.5 W m^{-2} white light). This mycelium contained 4.5 mg ß-carotene g^{-1} dry weight. *C* For comparison, strain C115, a *carS* ß-carotene-superproducing mutant. All spectra were read against a comparable mycelium of the albino strain C2

The carotenoprotein strains differ from the wild type in two hereditary factors. One of them is a *car*S mutation, removing the feedback control of carotene synthesis and leading to a large accumulation of ß-carotene [21]. The other is a cytoplasmic factor, which may be transferred from one strain to another through the formation and segregation of heterokaryons.

The carotenoprotein strains make the same amount of ß-carotene in the light and in the dark, but in the light all the pigment is free, while in the dark most of it is in the carotenoprotein complex. This effect of light is obvious to the naked eye by the change in mycelial color from bright red to bright yellow and is easily measured, even in vivo, by reading the absorption at 540 nm (Fig. 2). It should thus be an useful new experimental system in photobiology.

Photoinhibition of Alcohol Dehydrogenase

The central metabolism of *Phycomyces* is subtly influenced by light [14, 26–28]. The fixation of CO_2 via pyruvate carboxylase is increased and the accumulation of some citric-acid cycle intermediates is decreased in cultures grown in continuous light in comparison with the dark controls. The variations are usually small, rarely exceeding a factor of two.

A much larger effect was found by R. Garcés and J.R. Medina (unpublished) in Sevilla when they studied the NADPH-dependent alcohol dehydrogenase of *Phycomyces* [13]. *Phycomyces* produces small amounts of ethanol, but cannot utilize it as a carbon source. Mutants without alcohol dehydrogenase are perfectly viable; those which are recessive determine a single gene and the enzyme runs as a single band in

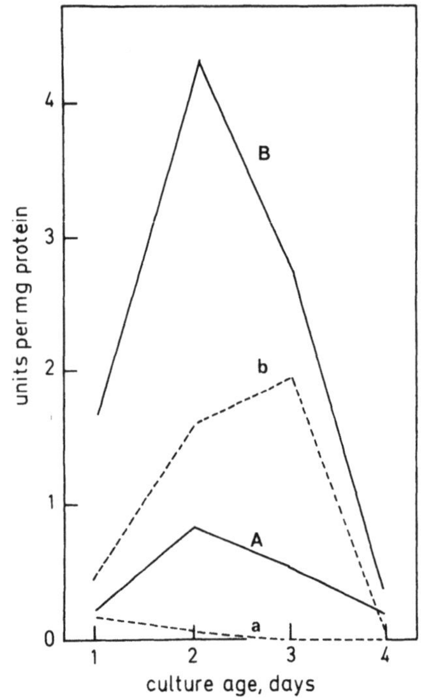

Fig. 3. Alcohol dehydrogenase activity of *Phycomyces mycelia* grown in liquid medium as a function of age. *A* Wild-type strain NRRL1555 grown in the dark. *a* Same grown under continuous white light (about 0.15 W m^{-2}). *B* Dark-grown strain C172, a *carA carB* double mutant producing no ß-carotene. *b* Same in the light

electrophoresis. Alcohol dehydrogenase is very scarce in germinating spores or in old mycelia, but quite appreciable in rapidly growing cultures. The striking finding is that light-grown cultures have much less alcohol dehydrogenase activity than those grown in the dark (Fig. 3). The thickness of the corresponding protein band follows the changes in activity, suggesting a different amount of enzyme, rather than a modification of its function. The enzyme itself is quite stable and insensitive to light.

Photoinhibition of alcohol dehydrogenase is independent of the *mad* photoreceptor: it occurs normally even in a triple *madA madB madC* mutant. The interruption of the carotene pathway, brought about either through mutations or through chemicals, results in vastly increased levels of alcohol dehydrogenase. These high levels are still subject to modulation by light.

The variations in alcohol dehydrogenase may be the amplified consequences of other metabolic alterations leading to an oversupply of NADPH and would serve just to regenerate NADP. They clearly indicate the existence in *Phycomyces* of a photoreceptor different from the one used for phototropism and photomorphogenesis. This new photoreceptor could be just an enzyme of central metabolism, rendered light sensitive by a flavin cofactor. The delicate coregulation of the metabolic reactions would then bring about variations in other pathways. A model for such as a photoreceptor would be the nitrate reductase of *Neurospora* (see Aparicio, Ninnemann this vol.), but the analogy should not be stretched further, because *Phycomyces* has no nitrate reductase.

Interconnections of the Sensory Pathways

The results described above may be tentatively summarized by the chart in Fig. 1, which indicates the different stimuli and responses, together with the corresponding receptors and the genes whose gene products are required for the elaboration of the responses.

It should be kept in mind that many kinds of possible mutants have never been looked for, that the behavioral investigation of the available mutant classes is far from complete and that the interrelations shown in Fig. 1 are therefore not always certain. I have excluded from the chart the responses incertae sedis, pending further investigation, such as the regulation of the carotenoprotein and the effects of red light, certain chemicals, and pressure on sporangiophore growth.

These limitations do not conceal that the complex behavior of *Phycomyces* is governed by a rather small number of genes, some of them put to work in several sensory pathways. The chart may be viewed as a guide for the biochemical investigation of the sensory transducers, because it indicates which elements are likely to be altered in the different mutants.

Acknowledgement. The author expresses his gratitude to the Alexander von Humboldt-Stiftung.

References

1. Bergman K (1972) Blue-light control of sporangiophore initiation in *Phycomyces.* Planta 107:53–67
2. Bergman K, Burke PV, Cerdá-Olmedo E, David CN, Delbrück M, Foster KW, Goodell EW, Heisenberg M, Meissner G, Zalokar M, Dennison DS, Shropshire W (1969) *Phycomyces.* Bacteriol Rev 33:99–157
3. Bergman K, Eslava AP, Cerdá-Olmedo E (1973) Mutants of *Phycomyces* with abnormal phototropism. Mol Gen Genet 123:1–16
4. Carlile MJ (1962) Evidence for a flavoprotein photoreceptor in *Phycomyces.* J Gen Microbiol 28:161–167
5. Cerdá-Olmedo E (1977) Behavioural genetics of *Phycomyces.* Annu Rev Microbiol 31: 535–547
6. Cerdá-Olmedo E, Torres-Martinez S (1979) Genetics and regulation of carotene biosynthesis. Pure Appl Chem 51:631–637
7. De Fabo E (1980) On the nature of the blue light photoreceptor: still an open question. In: Senger H (ed) The blue-light syndrome. Springer, Berlin Heidelberg New York, pp 187–197
8. De Fabo E, Harding RW, Shropshire W (1976) Action spectrum between 260 and 800 nanometers for the induction of carotenoid biosynthesis in *Neurospora crassa.* Plant Physiol 57:440–445
9. De la Concha A (1983) Biosintesis de carotenos y carotenoproteina en *Phycomyces blakesleeanus.* Tesis Doct, Univ Sevilla
10. Eslava AP, Alvarez MI, Cerdá-Olmedo E (1974) Regulation of carotene biosynthesis in *Phycomyces* by vitamin A and ß-ionone. Eur J Biochem 48:617–623
11. Galland P, Russo VEA (1979) Photoinitiation of sporangiophores in *Phycomyces* mutants deficient in phototropism and in mutants lacking ß-carotene. Photochem Photobiol 29: 1009–1014

12. Garcés R (1983) Genética de la deshidrogenasa del alcohol de ***Phycomyces*. Tesis Doct, Univ Sevilla**
13. Hartz TK, Houston MR, Lockwood LB (1978) Substrate specificity and kinetic properties of $NADP^+$-dependent alcohol dehydrogenase of *Phycomyces blakesleeanus*. Mycologia 70: 586–593
14. Hilgenberg W, Sandmann G (1977) Light-stimulated carbon dioxide fixation in *Phycomyces blakesleeanus* Bgff. Exp Mycol 1:265–270
15. Jayaram M, Leutwiler L, Delbrück M (1980) Light-induced carotene synthesis in mutants of *Phycomyces* with abnormal phototropism. Photochem Photobiol 32:241–245
16. Lafay JF, Matricon J (1982) Avoidance and wind responses of *Phycomyces blakesleeanus:* two distinct sensitivities. Exp Mycol 6:244–252
17. Lipson ED (1980) Sensory transduction in *Phycomyces* photoresponses. In: Senger H (ed) The blue-light syndrome. Springer, Berlin Heidelberg New York, pp 110–118
18. Lipson ED, Presti D (1977) Light-induced absorbance changes in *Phycomyces* photomutants. Photochem Photobiol 25:203–208
19. López-Díaz I, Cerdá-Olmedo E (1980) Relationship of photocarotenogenesis to other behavioural and regulatory responses in *Phycomyces* Planta 150:134–139
20. Lopéz-Díaz I, Cerdá-Olmedo E (1981) Light-controlled phorogenesis and mycelial growth in *Phycomyces* mutants. Curr Genet 3:23–26
21. Murillo FJ, Cerdá-Olmedo E (1976) Regulation of carotene synthesis in *Phycomyces* Mol Gen Genet 148:19–24
22. Murillo FJ, Torres-Martínez S, Aragón CMG, Cerdá-Olmedo E (1981) Substrate transfer in carotene biosynthesis in *Phycomyces*.Eur J Biochem 119:511–516
23. Otto MK, Jayaram M, Hamilton RM, Delbrück M (1981) Replacement of riboflavin by an analogue in the blue-light photoreceptor of *Phycomyces*. Proc Natl Acad Sci USA 78: 266–269
24. Presti D, Hsu WJ, Delbrück M (1977) Phototropism in *Phycomyces* mutants lacking ß-carotene. Photochem Photobiol 26:403–405
25. Russo VEA (1977) The role of blue light in synchronization of growth and inhibition of differentiation of stage I sporangiophore of *Phycomyces blakesleeanus*. Plant Sci Lett 10: 373–380
26. Sandmann G, Hilgenberg W (1978) Der lichtabhängige Intermediärstoffwechsel von *Phycomyces blakesleeanus* Bgff. I. Die CO_2-Fixierungsreaktion. Biochem Physiol Pflanz 173: 390–395
27. Sandmann G, Hilgenberg W (1979) Der lichtabhängige Intermediärstoffwechsel von *Phycomyces blakesleeanus* Bgff III. Der Einfluss von Asparagin an die Bildung der Sporangiophoren. Biochem Physiol Pflanz 174:794–801
28. Sandmann G, Hilgenberg W (1980) The light-dependent intermediary metabolism of *Phycomyces blakesleeanus* Bgff II. Formation and degradation of oxalacetic acid. Z Pflanzenphysiol 96:203–210
29. Shropshire W (1980) Carotenoids as primary photoreceptors in blue-light responses. In: Senger H (ed) The blue-light syndrome. Springer, Berlin Heidelberg New York, pp 172–186
30. Thornton RM (1973) New photoresponses of *Phycomyces*. Plant Physiol 51:570–576
31. Thornton RM (1975) Photoresponses of *Phycomyces blakesleeanus*: Initiation and development of sporangiophore primordia. Am J Bot 62:370–378

Blue Light Receptors in Phycomyces Investigated by Action Spectroscopy, Fluorescence Lifetime Spectroscopy, and Two-Dimensional Gel Electrophoresis

E. D. LIPSON, P. GALLAND, and J. A. POLLOCK [1]

Introduction

Phycomyces responses to blue light include phototropism, the light-growth response, light-induced carotene synthesis, and light-induced initiation of sporangiophores and sporangia (review: [7, 14]; Cerdá-Olmedo, this vol.). Action spectra for phototropism and the light-growth response [3] and several other lines of evidence (review: [7]; see also [12]) were compatible with the assumption of a single photoreceptor for most or all of these responses, probably a flavoprotein in the plasma membrane. However, it was found recently that more than one chromophore is involved in phototropism. Phototropic action spectra extended to a wavelength of 740 nm show a small peak at 630 nm (Löser and Schäfer, this vol., [5]). This peak probably corresponds to a new pigment since red light of wavelength 605 nm can partly revert phototropism elicited by low intensity blue light at 450 nm [10]. In the blue range, 380 to 500 nm, phototropism seems also to be mediated by more than one photoreceptor, since phototropic action spectra measured by two methods (photogeotropism and phototropic balance; see below) differ substantially [5].

In this chapter, we will first present phototropic action spectra for wild-type and mutant strains. Then we will show results on the wavelength dependence of adaptation kinetics for phototropism which suggest that adaptation is mediated at the level of the photoreceptors. Finally, we will indicate how approaches from biochemistry and fluorescence spectroscopy are being directed toward identification of molecules involved in photosensory transduction, in particular blue light receptors (cryptochromes).

Action Spectroscopy

There are two complementary ways to measure action spectra for phototropism in *Phycomyces.* The first involved photogeotropic equilibrium, i.e. the balance between phototropism, and geotropism evaluated after many hours of exposure to unilateral light. The measurements are performed near the phototropic threshold ($\sim 10^{-9}$ W m^{-2} for blue light). The other method employs phototropic balance between two lights of

1 Department of Physics, Syracuse University, Syracuse, NY 13210, USA

Blue Light Effects in Biological Systems
ed. by H. Senger

different quality and intensity. Application of these methods to wild-type and mutant strains permits investigation of whether or not a single receptor pigment is responsible for *Phycomyces* phototropism and helps in the identification of such pigment(s).

Photogeotropic Threshold

Figure 1 shows action spectra for photogeotropism for wild type and for representative night-blind mutants affected in the genes *mad*A, *mad*B, and *mad*C. The action spectra for wild type and *mad*A are identical to those published by Galland [5]; those for *mad*B and *mad*C have been redone (Galland and Lipson in preparation) to eliminate artifacts associated with convection. These action spectra are derived from fluence-response curves for photogeotropism (showing the equilibrium-bending angle versus the logarithm of the photon fluence rate; see [5]. These curves have varying

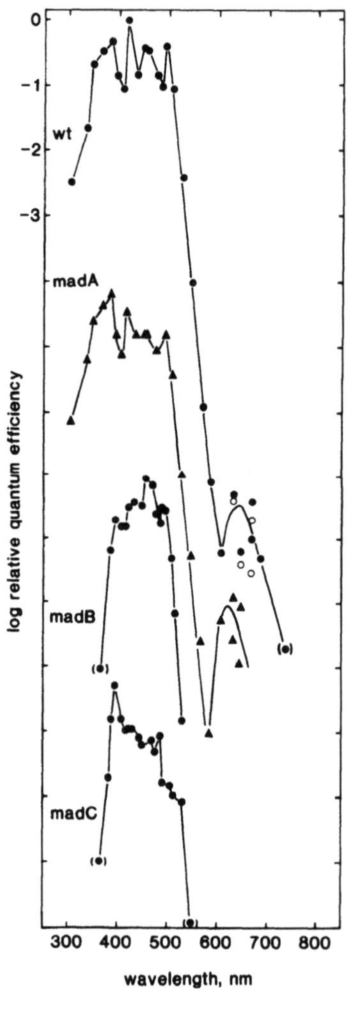

Fig. 1. Action spectra for photogeotropism in *Phycomyces*. Sporangiophores were exposed for 9 h to unilateral monochromatic light at various intensities in a "threshold box". The bending angles were measured from the vertical to determine the intensity for which the bending angle reached a half-maximal level of 40°; this intensity is a factor of 50 above the absolute threshold. The action spectra represent the relative quantum efficiency for achieving this angle as a function of wavelength. The spectra are shown for the wild-type strain, NRRL1555 (wt), and for the night-blind mutants, C47 (*mad*A), C109 (*mad*B), and L1 (*mad*C). The curves are displaced for clarity: the absolute efficiency in the blue region is reduced by a factor of about 10^3 for *mad*A, and about 10^5 for *mad*B and *mad*C. The data points shown in parentheses are upper limits

slopes and saturation levels – a result unexpected for a single receptor pigment. The wild-type action spectrum obtained from these curves (wild-type curve in Fig. 1) shows a complex shape with peaks at 414 and 491 nm that are uncharacteristic of flavin absorption. There are also peaks at 383 and 446 nm which are expected for flavin absorption; however the strength of the ultraviolet peak is uncharacteristically high. The peaks in the action spectrum at 414 and 491 nm appear too narrow to be attributed to simple absorption (by a single chromophore; see below). They suggest instead the interaction of two or more chromophores or perhaps the action of a single photochromic pigment.

The very large operating range of *Phycomyces* phototropism permits extension of the action spectrum out to wavelengths with very low efficiency. Figure 1 shows a peak at about 650 nm with relative efficiency of about 10^{-8}. The peak probably accounts for the tail up to 630 nm in the light-growth response action spectrum which Delbrück, Katzir, and Presti [4] had interpreted as direct triplet excitation; the mutant results below indicate that this peak is due to a different pigment (see [10]; also this vol.).

Figure 1 shows that the action spectrum for the *mad*A mutant is very similar in shape to that for wild type. However, the overall sensitivity is reduced by a factor of about 10^3. The similarity of the wild-type and *mad*A spectra indicate that the *mad*A mutant does not alter the photoreceptor system directly although the reduction in sensitivity could still be due to a reduction in concentration of receptor pigment. Alternatively, the reduction in sensitivity might occurs at a later step in the sensory transduction pathway. Thus the *mad*A mutation appears to affect either the concentration of receptor pigment or else the quantum efficiency for phototropism by way of subsequent dark reactions.

The spectra for *mad*B and *mad*C mutants are very different from those for wild type and *mad*A. In the *mad*B and *mad*C spectra, the prominent ultraviolet peak at 383 nm is missing as are the narrow peaks at 414 and 491 nm. The overall sensitivity is down for both by a factor of about 10^5. The *mad*B and *mad*C spectra differ in detail. In particular, around 450 nm, the *mad*B spectrum is about tenfold more sensitive than that of *mad*C. However, the cutoff of the *mad*B spectrum is shifted about 7 nm to shorter wavelengths in comparison to those of wild type and *mad*C.

Because the *mad*B and *mad*C action spectra are so abnormal, these mutants are probably altered directly in the photoreceptor complex. Such photoreceptor mutants can provide a genetic assay for the biochemical identification of the photoreceptor molecule(s) (see below). The reduction in sensitivity may be attributed to the loss of the abundant chromophore specialized for low intensity (perhaps acting as an antenna); the distinct shape (and comparably reduced sensitivity) of the *mad*B and *mad*C spectra may reflect the residual absorption spectrum of a chromophore for higher intensity; this latter pigment would presumably be present only at low concentration.

Phototropic Balance

Since the photogeotropism method must be performed near threshold, it does not permit investigation of the intensity dependence of the action spectra. This dependence can be studied instead by the phototropic balance method. This method in-

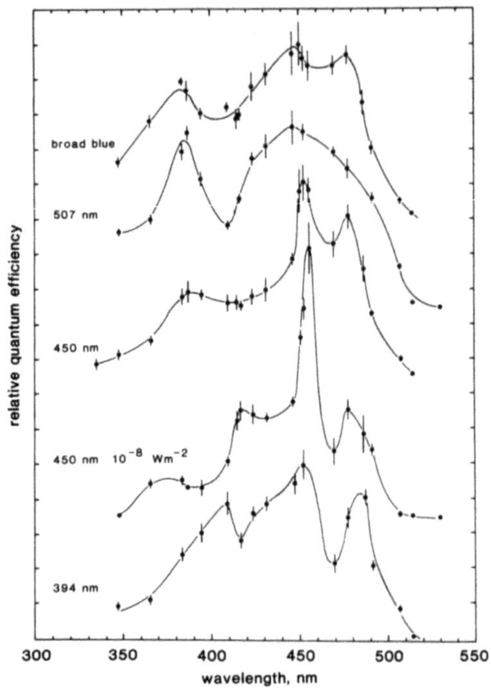

Fig. 2. Action spectra for phototropic balance for wild type. For each curve, sporangiophores were exposed to two opposed beams of light: a fixed "reference" beam, and a "test" beam with variable wavelength and intensity. The wavelength (or color) of the reference beam is given next to each curve; the broadband blue light came from a tungsten halogen lamp with a heat filter and a Corning 5-61 filter. The reference intensity was 10^{-4} W m^{-2} for the top three curves and for the bottom curve; for the remaining curve, the intensity was 10^{-8} W m^{-2} as indicated on the figure. The wavelength of the test beam is given on the abscissa. The ordinate represents the relative qunatum efficiency determined from the reciprocal of the test intensity (expressed as photon fluence rate) that balanced the effectiveness of the reference beam

volves dichromatic irradiation with two opposed beams: a reference beam and a test beam. Figure 2 shows wild-type action spectra for reference light consisting of broadband blue and wavelengths 507, 450, and 394 nm.

The spectrum for the broadband-blue reference beam shows peaks at 383, 450, and 477 nm and the typical shape associated with flavins. Similar spectra were measured previously by Curry and Gruen [1] and by Delbrück and Shropshire [3]. If there were a single receptor pigment (and no screening), then one would expect the shape of the spectra to be independent of the wavelength and intensity of the reference beam. However, Fig. 2 shows that the shape of the spectrum varies markedly with both of these parameters. In particular, the spectrum for 450 nm differs significantly from the one for broad blue. Further, for 450 nm we found that the shape of the spectrum depends significantly upon the reference intensity. At low intensity, a peak at 455 nm becomes very pronounced while the near ultraviolet peak practically disappears. With increasing intensity, the relative efficiency of 386 nm increases up to a factor of 1.4 at 10^{-1} W m^{-2}. Finally, the two other spectra with reference wavelengths of 507 and 394 nm are substantially different from each other and from the spectra with blue light.

As with the photogeotropic threshold action spectra, these phototropic balance spectra are incompatible with the hypothesis that a single receptor pigment mediates phototropism. The complex and variable structure is more compatible with a photoreceptor system involving antagonistically interacting chromophores and/or a photochromic pigment.

In retrospect, it is not so surprising that *Phycomyces* should need more than one pigment to handle the operating range spanning 10^{10}:1 in light intensity. In our visual systems, we have a duplex retina with rods and cones specialized respectively for scotopic and photopic vision. It seems that *Phycomyces* may have evolved with a similar strategy. As mentioned above, the *mad*B and *mad*C mutants may have their sensitivity limited to the upper half (logarithmic(of the operating range by lacking the low intensity photosystem (by analogy to night-blind humans who lack functional rods in their retinas).

Adaptation Kinetics

Phycomyces exhibits photosensory adaptation properties that are essential for the operation of phototropism and the light-growth response over the range of 10^{10} in light intensity [2, 8, 14]. These adaptation phenomena are similar kinetically to those in vision. Just as the kinetics of visual adaptation are tied closely to the photochemical kinetics of rhodopsin [13], we would expect that adaptation in *Phycomyces* would also involve the receptor pigment(s). It is hard to imagine that adaptation could be far removed from the process of photoreception, for then information on absolute intensity over many orders of magnitude would have to be carried through a number of metabolic steps. Although such connections between adaptation and photochemical kinetics in *Phycomyces* has been considered theoretically [5], direct experimental evidence has been lacking.

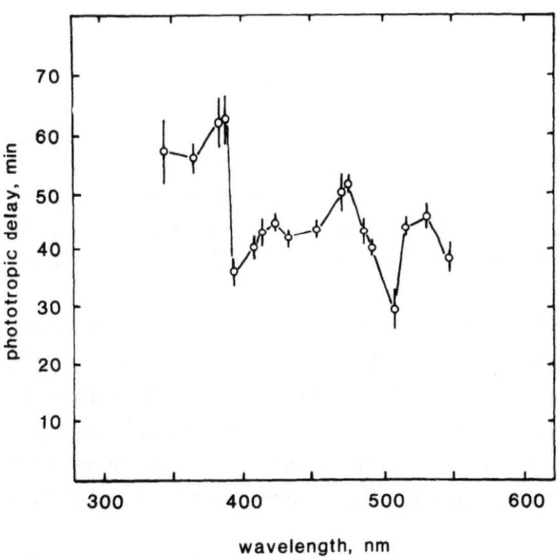

Fig. 3. Wavelength dependence of dark adaptation kinetics for phototropism. Sporangiophores were initially illuminated "symmetrically" by rotation at 6 r.p.m. with respect to a fixed horizontal light beam. After 45 min, the rotation was stopped (to produce a unilateral stimulus) and the intensity was reduced 500-fold. The delay time for the initiation for phototropic bending is represented on the ordinate. The abscissa represents the wavelength of the beam both before and after rotation was stopped. The initial intensity of the beam was 1.3 x 10^{-2} W m^{-2} for a wavelength of 491 nm. For other wavelengths, the intensity was adjusted for equal subjective intensity according to the photogeotropic action spectrum for wild type in Fig. 1

To investigate whether adaptation may occur at the photoreceptor(s), we have measured the wavelength dependence of adaptation kinetics by the phototropic delay method (Fig. 3); the details of the method are given in the figure caption. If the site of adaptation in the sensory transduction chain occurred downstream from the photoreceptor(s), one would expect no dependence of the adaptation kinetics upon wavelength. However, Fig. 3 shows that the kinetics do depend significantly upon wavelength. Therefore we conclude that adaptation and the photosystem(s) are intimately connected.

The dark adaptation kinetics for phototropism are biphasic [14]. For wild type, the decay is biexponential with time constants of 3 min and 9 min (for broadband blue light). Mutants in genes *mad*A and *mad*B show altered kinetics: time constants of 3 min and 15 min for *mad*A; 12 min alone for *mad*B. Thus these night-blind mutants are also adaptation mutants. The observation that certain adaptation mutants are also photoreceptor mutants (see above) reinforces the association between adaptation and photoreception.

The biphasic kinetics of adaptation vary significantly with wavelength (Galland and Lipson in preparation). In particular, our studies of the time course of dark adaptation at selected wavelengths show that the short time constant (~3 min) is independent of wavelength (at least it is similar for 383 nm and 409 nm), while the longer time constant depends critically on wavelength.

In this section, we have shown that the combination of action spectroscopy and genetics provides tools to help us study the components and processes of photosensory reception and adaptation. In the remainder of this chapter, we will discuss our biochemical studies of wild type and representative mutants aimed at finding the molecules involved in these processes –in particular the blue-light photoreceptors responsible for *Phycomyces* phototropism.

Biochemistry

Wild Type

To help identify the protein(s) and chromophores involved in phototropism, we have employed the technique of two-dimensional polyacrylamide gel electrophoresis [11]. Examples of such gels are shown in Fig. 4. The proteins are solubilized from a plasma membrane-enriched fraction obtained from short synchronous sporangiophores by differential centrifugation (Pollock, Sullivan, and Lipson in preparation). For fluorographic visualization of proteins, we have used two labels: ^{14}C-riboflavin as a specific label for proteins with covalently-bound flavins (Fig. 1a) and ^{35}S-sulfate as a general protein label (Fig. 1b,c). We have sought proteins with covalently bound flavin prosthetic groups, because, as is the case with other photoreceptors, the chromophore is expected to be covalently bound to the protein for optimal operation and energy transfer [16].

As indicated by the arrows in Fig. 4a, we have identified two flavoproteins, FP_1 (pI 6.6, 71 kdal) and FP_2 (pI 6.5, 59 kdal). These proteins also appear prominently in Fig. 4b with the general ^{35}S protein label. In separate studies (Pollock, Sullivan, and Lipson in preparation), we have found that the densities of both of these spots depend significantly upon illumination protocols during growth; specifically after 96 h of growth in the dark, a 1-min pulse of blue light (10^{-4} W m^{-2}) induced a seven-

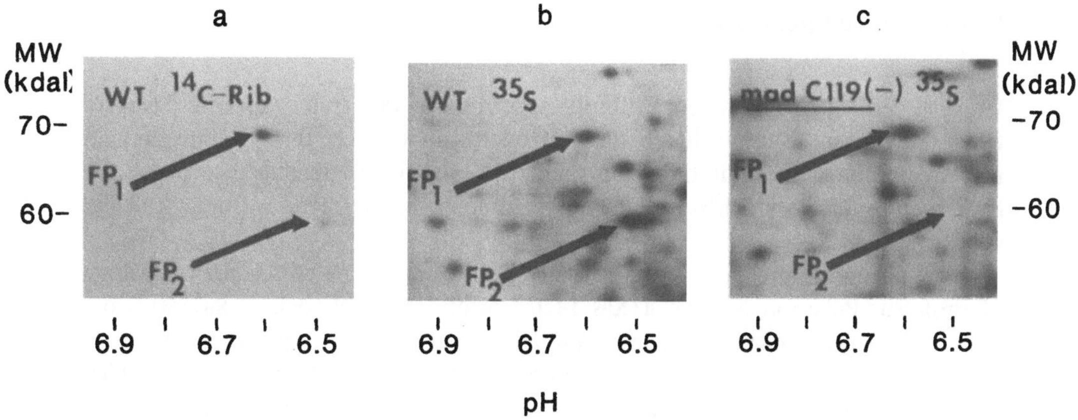

Fig. 4. Two-dimensional gels of proteins from a plasma-membrane-enriched fraction from synchronous sporangiophores. Only a small portion of the gels are shown here. The proteins were labeled by continous growth on media containing ^{35}S-sulfate or ^{14}C-riboflavin, and visualized by fluorography. *a* Proteins with covalently bound flavins revealed by labeling of wild-type strain with ^{14}C-riboflavin. *Arrows* indicate flavoproteins, FP_1 and FP_2. *b* Wild-type gel with ^{35}S label. Again flavoproteins FP_1 and FP_2 are indicated by *arrows*. *c* Gel of night-blind mutant, L1 (*mad*C) labeled with ^{35}S. The spot for flavoprotein FP_2 (*lower arrow*) is absent

fold increase in the density of FP_2 and a threefold decrease in FP_1. This effect may be due to a light-regulated synthesis of FP_1 and FP_2.

Mutants

Our studies with 2D gels have included certain night-blind mutants believed to be affected in the photoreceptor system (see above). We have found that certain spots appear or disappear in our comparative analyses of gels of different mutant strains. For the *mad*A strain, C21, two protein spots have appeared (pI 6.4, 59 kdal; and pI 6.45, 51 kdal). For the *mad*B strain, C111, one protein spot is absent (pI 6.6, 56 kdal).

The most dramatic difference we have noticed in mutant gels concerns the gene *mad*C. The spot corresponding to FP_2 protein is missing in the gels of certain *mad*C strains (Fig. 4c). However, its presence in gels from certain other strains carrying the same allele shows that the missing spot is not simply the *mad*C gene product. Regardless of whether or not FP_2 is associated with *mad*C, we believe that both of these covalent flavoproteins are candidates for components of the photoreceptor complex; the action spectra (Figs. 1 and 2) suggest that more than one flavoprotein may be involved in this complex.

Fluorescence Spectroscopy

We have performed fluorescence lifetime spectroscopy in search of specific flavoproteins with lifetimes shorter than the 5 ns characteristic of those flavoproteins that are fluorescent. Such short lifetimes are predicted [16] for flavin photoreceptors and have been reported similarly for purified plasma membrane fractions from corn coleoptiles [15].

We performed preliminary fluorescence lifetime measurements in the laboratory of Professor Pill-Soon Song at Texas Tech University in Lubbock, Texas (with the help of H. Sarker) on an SLM phase-modulation spectrofluorometer. Those measurements employed a plasma membrane-enriched fraction from liquid-grown mycelial cultures. A two-component heterogeneity analysis of the phase and modulation values at 10 MHz and 30 MHz indicated a major component with short lifetime (2 ns, 75%) and a longer lifetime component (7.4 ns, 25%).

Recently, we have performed fluorescence measurements on a similar instrument at the State University of New York Upstate Medical Center in the laboratory of Professor Anthony Martonosi (with the help of Dr. George Kracke) of the Biochemistry Department. These measurements were performed on purified samples from a plasma-membrane fraction from synchronous stage I sporangiophores. The purification in this case was done by Chromatofocusing (Pharmacia), an ion-exchange technique that separates proteins according to isoelectric point. Under the denaturing conditions we employed, this technique effectively eliminate free flavins from the sample while concentrating FP_1 and FP_2.

The phase and modulation results (at modulation frequencies of 16 MHz and 30 MHz) were fit as before with a two-component algorithm. We studied samples of both wild type and the *mad*C strain L1. In both cases, we found a short lifetime component; in the case of wild type, a predominant component of 1.5 ns (74%), and in L1 1.8 ns (83%). These results may indicate a specialized flavoprotein which serves as the photoreceptor molecule. The efficient depopulation of the excited state because of efficient photochemistry, expected for a photoreceptor molecule, would produce the shorter fluorescence lifetime [15, 16].

Conclusion

In summary, recent action spectroscopy results on *Phycomyces* sporangiophores indicate a photoreceptor complex with antagonistic chromophores or perhaps a photochrome. Adaptation is a compound process that occurs at the level of the receptor pigments and that involves more than one chromophore. With approaches from genetics, protein biochemistry, and fluorescence spectroscopy, we have begun to isolate and characterize components, in particular flavoproteins, of the photoreceptor complex.

Acknowledgments. We are indebted to P.S. Song, H. Sarkar, and G. Kracke for fluorescence lifetime analyses on our samples, and to D. Sullivan for his collaboration on the biochemistry work. This work was supported by grants to EDL from the National Science Foundation (PCM-8003915) and the National Institutes of Health (GM29707). EDL gratefully acknowledges a research fellowship from the Alfred P. Sloan Foundation.

References

1. Curry GM, Gruen HE (1959) Action spectra for the positive and negative phototropism of *Phycomyces* sporangiophores. Proc Natl Acad Sci USA 45:797–804
2. Delbrück M, Reichardt W (1956) System analysis for the light growth reaction in *Phycomyces.* In: Rudnick D (ed) Cellular mechanisms in differentiation and growth. Princeton Univ Press, Princeton, pp 3–44
3. Delbrück M, Shropshire W (1960) Action and transmission spectra of *Phycomyces.* Plant Physiol 35:194–204
4. Delbrück M, Katzir A, Presti D (1976) Responses of *Phycomyces* indicating optical excitation of the lowest triplet state of riboflavin. Proc Natl Acad Sci USA 73:1969–1973
5. Galland P (1983) Action spectra of photogeotropic equilibrium in *Phycomyces* wild type and three behavioral mutants. Photochem Photobiol 37:221–228
6. Lipson ED (1975) White noise analysis of *Phycomyces* light growth response system. I. Normal intensity range, II. Extended intensity ranges, III. Photomutants. Biophys J 15: 989–1045
7. Lipson ED (1980) Sensory transduction in *Phycomyces* photoresponses. In: Senger H (ed) The blue light syndrome. Springer, Berlin Heidelberg New York, pp 110–118
8. Lipson ED, Block SM (1983) Light and dark adaptation in *Phycomyces.* J Gen Physiol 81:845–851
9. Lipson ED (1983) Sensory Information Processing at the Cellular Level: The Light Response Systems of *Phycomyces.* In: Scott GP, Wahner HW (eds) Radiation and cellular response. Iowa State Univ Press, Ames, pp 133–152
10. Löser G, Schäfer E (1980) Phototropism in *Phycomyces*: a photochromic sensor pigment? In: Senger H (ed) The blue light syndrome. Springer, Berlin Heidelberg New York, pp 244–250
11. O'Farrell PH (1975) High resolution two-dimensional electrophoresis of proteins. J Biol Chem 250:4007–4021
12. Otto MK, Jayaram M, Hamilton RM, Delbrück M (1981) Replacement of riboflavin by an analogue in the blue-light photoreceptor of *Phycomyces.* Proc Natl Acad Sci USA 78: 266–269
13. Rushton WAH (1965) The Ferrier Lecture, 1962: Visual adaptation. Proc R Soc London Ser B 162:20–46
14. Russo VEA, Galland P (1980) Sensory physiology of *Phycomyces blakesleeanus.* Struct Bonding (Berlin) 41:71–110
15. Sarkar HK, Song PS, Leong TY, Briggs WR (1982) A fluorescence lifetime assay of a membrane-bound flavin from corn coleoptiles. Photochem Photobiol 35:593–595
16. Song PS (1980) Spectroscopic and photochemical characterization of flavoproteins and carotenoproteins as blue light photoreceptors. In: Senger H (ed) The blue light syndrome. Springer, Berlin Heidelberg New York, pp 157–171

The Quest for *Trichoderma* Cryptochrome

B. A. HORWITZ[1], J. GRESSEL[1], and S. MALKIN[2]

Introduction

The common soil fungus *Trichoderma harzianum* Rifai, (previous designation, *Trichoderma viride* Pers. ex Fries, ATCC 32173) grows as a vegetative mycelium in the dark, provided the nutrient supply is sufficient. A short pulse of blue light induces a local switch to sporulative development of asexual conidia. The conidiophores develop in a ring located where the perimeter of the colony had been when it received light [8]. The action spectrum for photoconidiation has maxima at 365, 430, 455, and 480 nm [9]. These peaks correspond to those of cryptochrome, a pigment(s) which, so far, has been identified as a working hypothesis but not as a molecule [7]. Kumagai and Oda [15] extended the action spectrum to the UV and found an additional peak at 310 nm which may not be present in all organisms having cryptochrome responses.

Blue light with this action spectrum modulates growth and tropism [10], physical properties of the cytoplasm [26], and morphogenesis in fungi [11] and other organisms [18]. It is not clear whether these effects have a common biochemical/physical explanation.

Trichoderma is a suitable organism for exploring how morphogenesis is triggered by light. The genetics of responses to blue light are being studied in *Phycomyces* and in *Neurospora* (see Cerdá-Olmedo and Innocenti, both this vol.). In *Trichoderma*, less is known about the genetics, but the photoresponse seems particularly suitable for study. Morphogenesis is triggered by a brief low energy light pulse; 6.6 μE m^{-2} half saturating the response.

Specific Biochemical Requirements for Conidiation

The metabolic requirements for conidiation are most probably different from those for vegetative growth. New specific mRNA's are probably required [1, 8], although these mRNA's could not yet be isolated with the techniques available 12 years ago [24].

1 Dept. of Plant Genetics, The Weizmann Institute of Science, 76100 Rehovot, Israel

2 Dept. of Biochemistry, The Weizmann Institute of Science, 76100 Rehovot, Israel

Blue Light Effects in Biological Systems
ed. by H. Senger

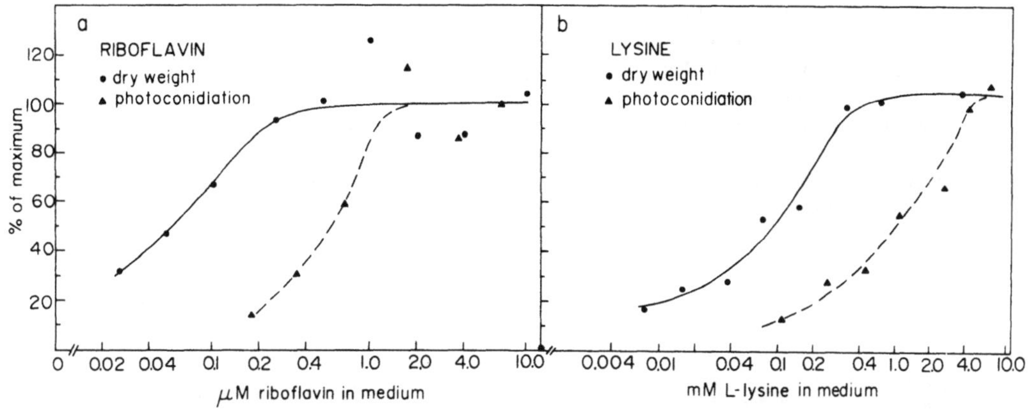

Fig. 1a,b. Two auxotrophic strains of *Trichoderma* needing more metabolites to conidiate than to grow. **a** riboflavin auxotroph; **b** lysine auxotroph. The mutants were isolated by D. Aviv (**a** redrawn from Horwitz and Gressel [12]; **b** unpublished results)

Two different auxotrophic strains that we studied needed supplementary nutrients to conidiate (Fig. 1); the genetic capability to respond exists, but was not expressed in the phenotype because the conditions were not right. We showed that the riboflavin auxotroph was actually photoinduced even at 0.2 μM riboflavin in the medium; it needed tenfold higher flavin levels for expression [12]. This does not imply that the cryptochrome chromophore is not a flavin; just that the added flavin was not needed for the photoresponse. In *Neurospora*, the effect of flavin deficiency could be counteracted by increasing the light level [19] in contrast to *Trichoderma*. Their data can be interpreted as an effect of flavin deficiency on dark or light reactions. They did not attempt to supplement with riboflavin after photoinduction to ascertain when the added flavin was needed. *Trichoderma* does not need to be growing to perceive light; it may be possible to photoinduce it at very low flavin levels using the auxotroph.

The Genetic Rationale

An initially indirect genetic rationale involves looking for mutants that do not respond to light. These should assist in finding candidates for cryptochrome, and help uncover the mechanism of action along with the chromophore. A single genetic defect can have many consequences. We designed criteria for isolation based on the postulated pathway shown in Fig. 2, to find mutants specific for the initial stages of the photoresponse. Mutants that respond normally to stress, but not to light, might lack cryptochrome or components involved in transducing photochemistry to the information that triggers morphogenesis. We isolated such mutants; two are shown in Fig. 3. We have named these *dim*, for "dimsighted," expressing that fact that the defects are often partial, and could be in different parts of the sensory pathway. We isolated 87 mutant strains that cannot be photoinduced at a very high fluence rate or require

Fig. 2. Postulated response pathways. *C* cryptochrome; *p* product of photoreaction(s); *J* start of the joint pathway. Possible sites of blockage by mutations are indicated. There are mutants for the joint pathway, and for the light-induced pathway; the individual steps affected are not yet identified

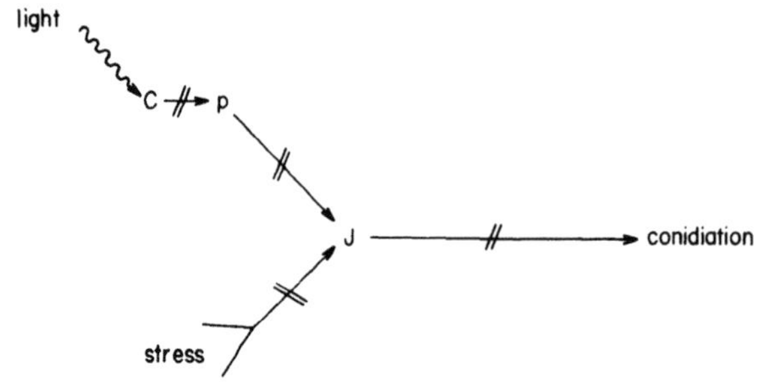

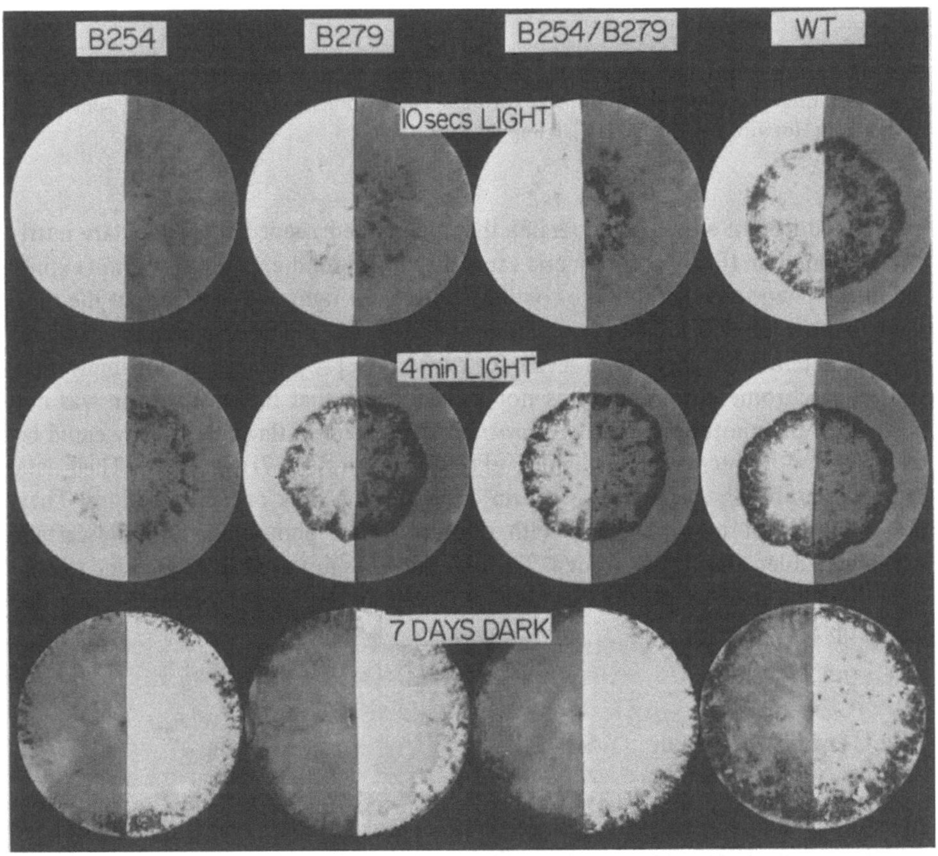

Fig. 3. Response of wild-type and *dim Trichoderma* strains to light and stress. Colonies were grown on filter paper soaked with liquid medium. Dark halves are stained to visualize mycelia. Note the partial complementation of B254 and B279. (Unpublished results)

more light to respond than the wild type. They can all be induced to conidiate by "end-of-dish" nutritional stress. Mutant strains of *Trichoderma* can be easily isolated after treatment of the spores with mutagens. The mycelium does not need to conidiate in order to survive: Greenshpan [6] isolated 38 mutants that do not respond to

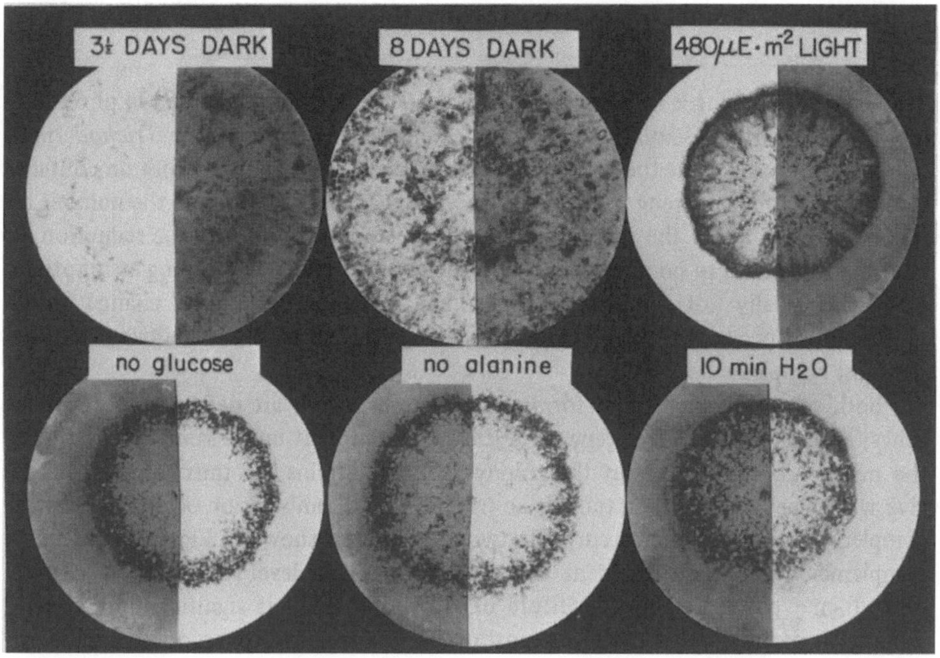

Fig. 4. Induction of conidiation by light and by various stresses. After 3.5 days in the dark, end-of-dish conidiation is just beginning. Light, given 1.5 days after planting, induced a discrete ring of conidia. End-of-dish conidiation is maximal after 8 days growth in the dark. Transferring the filter paper with the growing colony to medium lacking an immediately available carbon source (glucose) or nitrogen source (alanine) for 2 days leads to a discrete ring. Water-stressed colonies were kept in water for 10 min, then returned to the complete medium. All colonies were fixed after 3.5 days except for the 8-day treatment. (Unpublished results)

either light or stress. Bojňanská et al. [2] isolated 67 conidiation mutants; they apparently did not yet check whether these strains are equally defective in response to stress and light. Various compounds such as fluorouracil [8], or even D_2O [23] can inhibit conidiation without inhibiting growth.

Trichoderma has an alternate pathway for inducing conidiation: starvation (Figs. 3, 4) or some (but not all) other stresses. This may allow a unique opportunity to select mutants specific for the inductive part of morphogenesis. Probably the most formidable difficulty in this kind of study is selecting the mutants for the part of the pathway that is of interest. Many mutations prevent conidiation. Mutants that are able to stress conidiate, but do not respond to light would have defects in a set of genes clearly related to the photoresponse or its amplification.

Estimation of Numbers of Genes Involved

True genetic mapping is impossible because the *Trichoderma* strain used in photobiology lacks a perfect (sexual) stage. Complementation is quite easy in *Trichoderma*: one needs only to plant the two strains next to each other and they fuse on contact. Complementation mapping using heterokaryons was used to estimate the number of lesions that can affect the same function. Such estimates are rough; the reduction of eukaryotic genetics to combinatorial sampling requires many simplifying assumptions which are usually not completely correct; i.e., that there is an equal chance for all mutations to occur. Complementation indicated the parts of the pathway that are the most complex.

If both of the parent strains forming the heterokaryons are defective in different genes, normal and defective gene products will be present in the same cytoplasm, as the nuclei move freely about the colony. The two strains can thus complement to give wild type behavior even though no true genetic recombination occurs [3]. Partial complementation can also occur. The two *dim* mutants shown in Fig. 3 do not show complementation when tested at the standard low light level (20 μE m^{-2}). After a 480 μE m^{-2} pulse, the mixed culture of B254 and B279 is significantly different from both parents and is therefore most probably a true heterokaryon. A number of mechanisms could explain partial complementation: (a) Interallelic (i.e., within the same gene) complementation [25]. (b) The defective gene products in the cytoplasm prevent the normal ones from functioning perfectly. (c) A regulatory protein with a rapid turnover might not be able to reach defective nuclei. (d) The mutants may have multiple lesions. Multiple lesions seem relatively unlikely, as all 30 *dim* mutants derived from the wild type were without secondary spore color lesions and only one was an auxotroph. Still, certain groups of genes might co-mutate more frequently than others; without physical mapping we cannot resolve this question.

Additional markers are needed for pairs that do not complement to prove that heterokaryons were actually formed. Bojňanská et al. [2] found that some pairs of nonconidiating mutants would complement only following fusion of protoplasts from both strains. The best markers are auxotrophic: heterokaryons are tested on minimal medium that does not allow growth of either parent strain (forced heterokaryons). We used *dim* mutants isolated from riboflavin and lysine-requiring strains to construct a forced complementation table.

We used several methods to estimate the number of distinct classes of mutants isolated. The complementation data and the estimates are summarized in Table 1. The tests for the photoresponse mutants were done at a low light level (20 μE m^{-2}). Some cases of partial complementation (as in Fig. 3) were thus included in the list of noncomplementing pairs. Even at this light level, some pairs responded slightly more than the parent strains. These partially complementing pairs were also considered as noncomplementing. If partial complementation is due to the interaction between products of different genes rather than interallelic complementation, this assumption leads to an underestimate of the number of genes. Another order-of-magnitude estimate can be obtained from the frequency of isolation of the mutants. Thirty six *dim* mutants were obtained from 6225 mutagenized and viable colonies. The average viability of the mutagenized spores was 0.3%. If the mutation frequency was increased

Table 1. Estimates of the number of "photoresponse" genes

Type of mutants	Number of mutants sampled	Number complementation groups	Estimate of genes	Method
Conidiation[a]	38	31	90	MLE[b]
dim	17	12	24	MLE
dim (forced)	19	-	12	MLE
dim (forced)	19	-	12–30	Binomial[c]

[a] [6]

[b] MLE: Maximum Likelihood Estimator [16]; if there are n classes in a population of sample of size N, k classes will be represented in the sample. The expected value of k is $E(k) \cong n(1-e^{-N/n})$, assuming equal probability of selecting any class

[c] Hranueli et al. [14] assumed that the fraction of identical pairs follows a binomial distribution. We checked 14 "new" mutants against five "known" ones and found p = 0.29; the estimate is ± one standard deviation around p

from a spontaneous rate of about 10^{-6} by a factor of one over the viability, it should have increased to 3.33×10^{-4}. The corresponding number of genes is 17. Despite all the uncertainties involved in these estimations, all the methods suggest that 12-30 genes are available for selection by our procedure. The errors introduced by the separate assumptions might have cancelled out to some extent. Using similar estimation techniques, it was estimated that several hundred genes are apparently involved in the whole developmental process from aggregation to sporulation in *Dictyostelium* [17]. About 50 unique gene products seem to participate in the aggregation process in slime molds [27, 28]. We estimated from Greenshpan's [6] complementation table that about 90 *Trichoderma* gene products are required for conidiation (induced by either stress or light). Hranueli et al. [14] estimated 31 to 46 operons specific for sporulation in *Bacillus subtilis* which, as a prokaryote, should be less complex. The photoinductive part of the pathway in *Trichoderma* thus really does seem to be less complex than morphogenesis as a whole. There is reason to expect, though, that the number of genes actually involved in photoreception is even lower. In addition to one or a few genes for cryptochrome(s), we may have selected for many defects in the amplification or expression pathways. Only eight complementation groups were found for phototropism in *Phycomyces.* Of these, three have been associated with the photoreceptor (see Lipson this Vol.).

A strain that responds normally to stress should have normal dark reactions, if our strategy for selecting mutants is correct. In fact, many partially defective mutants were obtained. These may be "leaky" conidiation-defective strains, equally defective in light and stress-induced conidiation.

Wild-type colonies can be induced by transferring them to media lacking key nutrients for an extended period, or to distilled water for a 10-min period in the dark (Fig. 4). Only ten of the dim mutants respond normally to the distilled water treatment, but show no photoresponse or require more light than the wild type. The isolation of such mutants shows that a functioning light pathway is not necessary for induction by stress. Stress is thus a true alternate pathway.

There should be several steps between the primary photoact and entry into the joint pathway (J in Fig. 2). The 77 stress-inducible *dim* mutants that are still defective in response to a transient stress are clearly defective in reactions common to both pathways. The few mutants that respond to transient stress, the best photomimetic agent we have found so far, are the most likely candidates for photoreceptor (or early phototransduction) mutants.

Is Light a Stress?

It is possible to attempt to study the pathway of photomorphogenesis from the beginning rather than by going backwards from the joint pathway with stress. Ascertaining the first consequences of photoinduction is interesting from the physiological point of view. Combined with genetics, such data may help elucidate the response pathway and classify mutants as "early" or "late". Galun [5] and Greensphan [6] noticed that aerial hyphae develop at the growing edge of the colony, but soon fall over and fuse with the mycelial mat. After photoinduction, the aerial hyphae remain upright in a ring located where the perimeter was at photoinduction. These aerials eventually branch, forming conidiophores. New aerials also continue to develop from the mycelial mat and form conidiophores; these give rise to a normal ring even if existing aerials were removed before photoinduction [6].

The failure of the aerials to fall over is the earliest visible sign of morphogenesis. Turgor pressure in the vacuoles is thought to keep aerial hyphae upright. Actively growing fungal hyphae are not vacuolated all the way to the tip, but adverse conditions (stress) can increase vacuolation [29]. We found that transferring *Trichoderma* mycelia to water causes many hyphae to become vacuolated all the way to the tips. A photoinductive light pulse also increases vacuolation (Fig. 5). Blue light may act as a transient "stress" initiating changes leading to vacuolation and eventually conidiation [7].

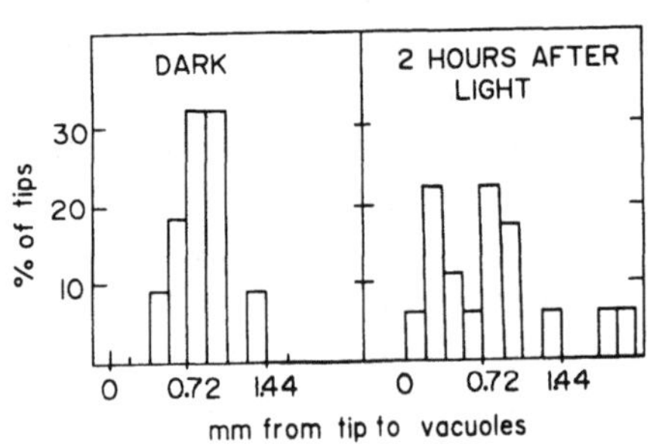

Fig. 5. Increase in vacuolation following photoinduction. The distance from the tip to the vacuolated region was microscopically measured for 20 tips each of dark control and photoinduced colonies. The colonies were grown on agar; the edge of the mycelial mat was observed by adding a drop of liquid medium and a cover glass. (Unpublished results)

The early biochemical and biophysical consequences of photoinduction in *Trichoderma* and other organisms are largely unknown. We are beginning to search for the synthesis of new proteins. Another possibility is that the earliest changes are electro-

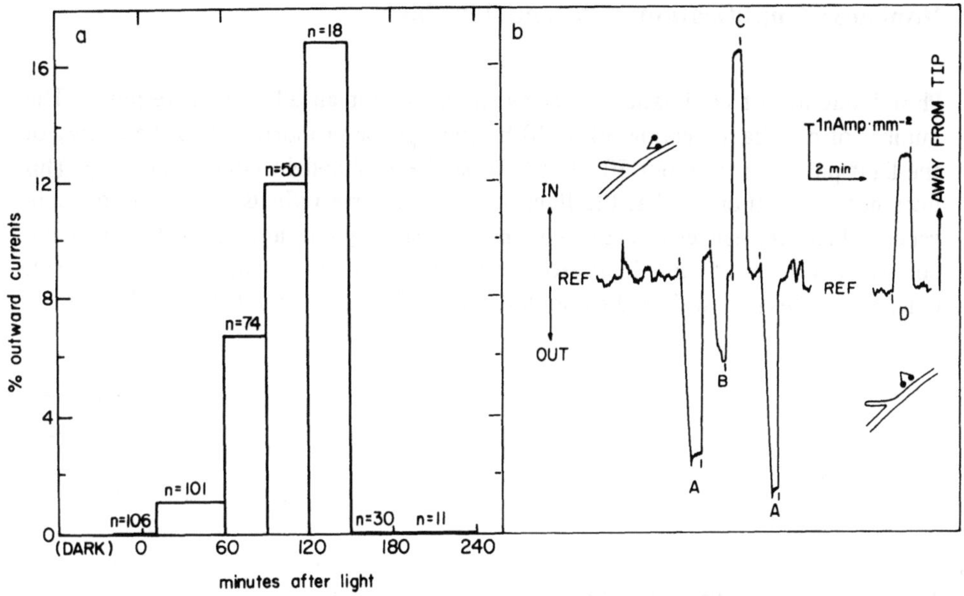

Fig. 6. Outward ionic currents specific to photoinduction around hyphae growing in liquid medium. **a** Percent of locations along the hyphae where the current, measured with an extracellular vibrating electrode, was outward rather than inward. No outward currents were found on dark control hyphae. *n*, the number of locations checked, is noted above each bar. **b** Recording of extracellular currents near a photoinduced hypha. The sketches indicate the direction of vibration of the probe. *A* and *B* are sites with inward current; *C* showed outward current. *B* was located between *A* and *C*, and the distance from *A* to *C* was 30 μm. The probe was first brought close to the hypha at *A*, moved to a reference site (*REF*) far from any hypha, then moved to position *B*. After recording at *C*, the current at *A* was measured again. Measurement *D*, with the probe vibrating parallel to the direction of growth, shows a complete current loop near the hypha. (Condensed from Horwitz et al. [13]

physiological. We measured electric currents around *Trichoderma* hyphae growing in liquid medium using an extracellular vibrating electrode [13]. A steady current enters growing hyphal tips as well as along the side of the youngest millimeter of the hyphae. The density of this current extrapolates to at least 1.5 n Amp·mm^{-2} at the surface of the tip and 7 n Amp·mm^{-2} along the sides of the hyphae. This pattern persisted during and after an irradiation with blue light. Photoinduction does not immediately change the currents, nor does it rearrange the currents near the tips. We detected some seemingly specific *outward* currents, 60 to 150 min after illumination, at about one ninth of the locations along the sides of the hyphae (Fig. 6). This sporadic, localized outward current pattern might be an early biophysical response to light. The appearance of the electrical changes approximately coincides with the appearance of increased vacuolation, and may somewhat precede it. Presumably there are biochemical changes that precede stabilization of aerials, vacuolization and the electrophysiological changes, occurring during the first hour after light. Any *dim* mutants having these changes must have a functioning photoreceptor. It will be interesting to see when new proteins are produced.

Nonlinear Amplification: a Potential Problem

Photoinduction of conidiation in *Trichoderma* is not an all-or-none response. The number of conidia appearing 30 to 40 h after light is proportional to the energy of the light pulse, up to about 50 μE m^{-2} broad-band blue-light (Figs. 7 and 8c). This does not mean, though, that the fluence-response curve reflects photoreceptor conversion. Light influences at least two aspects of subsequent development in the dark; colonies that received more light produce more conidiophores, and also develop faster (Fig. 8a,b). This may well lead to nonlinear amplification of the light signal.

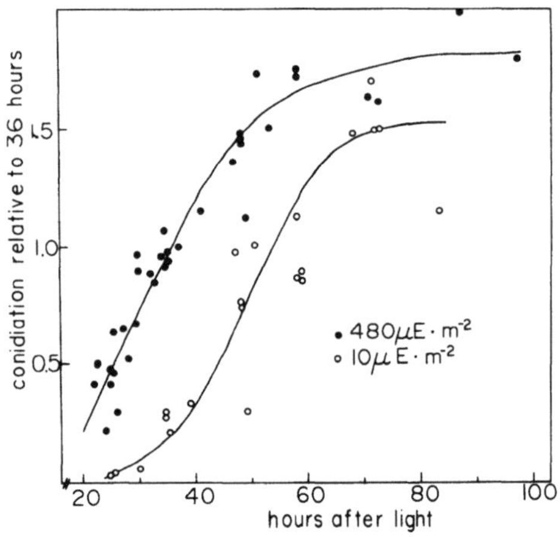

Fig. 7. Lag of conidiation after photoinduction at a low fluence. The data are from seven experiments, normalized to the value for 480 μE$\cdot$m^{-2}, a saturating pulse, 36 h after light. Even though the curves approach each other after 60 h, the colonies appear different: conidiation is more patchy and diffuse (though still limited to a discrete ring) for the lower fluence. (Unpublished results)

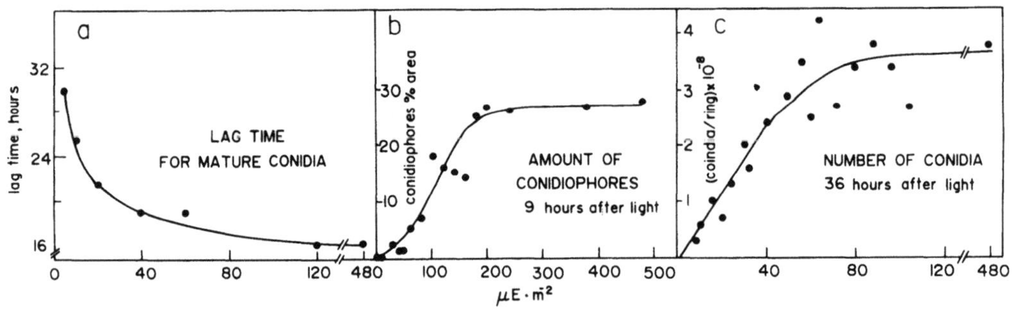

Fig. 8a-c. Different ways of measuring the effect of light. Conidiation was measured by the turbidity of spore suspensions; the amount of conidiophores was estimated by measuring the area covered by them on cotton blue stained colonies. The fluence changes the lag time (at very low fluences there appears to be an absolute threshold) as shown in **a** by fixing colonies on filter paper every 2 h. The amount of conidiophores follows a saturation curve **b**. Despite the nonlinearity shown by **a** and **b**, the number of conidia is proportional to the light level up to saturation at 36 h after light. (Unpublished results)

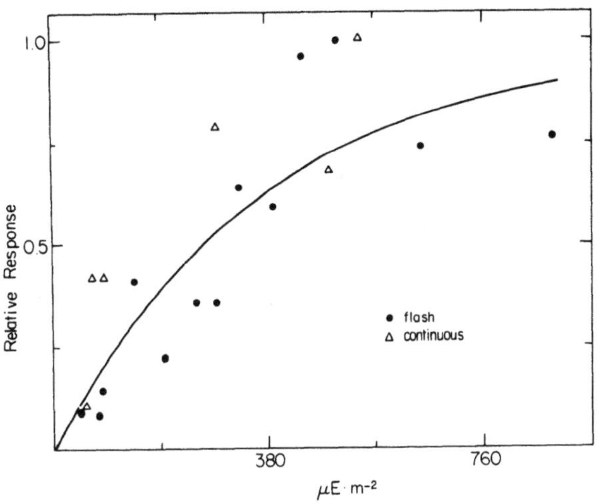

Fig. 9. The effect of light flashes. 6-mm-wide sections of dark grown colonies were irradiated with single 25 μs flashes from a Xenon flash lamp filtered through a Balzers K2 filter. Solid line is an exponential least-squares fit. Triangles are control values obtained by irradiating similar spots with continuous light. (Unpublished results)

At the blue peak in the action spectrum, 6.6 μE m^{-2} (447 nm) half-saturates the response [9]. If this normal photon exposure $F_{\perp}$ were not very different from the true photon fluence, F, reaching the photoreceptor, the following estimate can be made: for a simple first-order photoreaction, the fraction of photoacceptor converted is $1\text{-}e^{\phi \sigma F}$, where ϕ is the quantum yield, σ the absorption cross section and F the photon fluence. Substituting $F \simeq F_{\perp}$, $\phi = 1$ and $\sigma = 1.2 \times 10^3\ m^2\ mol^{-1}$ (the value for riboflavin at 450 nm), the fluence for half-saturation converts only 0.79% of the photoreceptor. The absorption cross section of cryptochrome is unlikely to have a value much larger than that of riboflavin. Optical corrections, while certainly important, probably cannot account for such a large discrepancy. Thus the initial concentration of cryptochrome does not seem to limit the response, and saturation is likely to depend on rate-limiting dark reactions, or on a steady state concentration of photoreceptor. We could not find any evidence for disagreement with Bunsen-Roscoe reciprocity, even with an exposure time of 25 μs (Fig. 9). Therefore, if photoreceptor regeneration is involved, it must be very rapid. There are other possibilities that should not be neglected, such as multiple photoreceptors, some of which could act as antenna pigments.

Separation of Light and Subsequent Dark Reactions

Colonies grown at 26°C, transferred to 3°C, illuminated with nonsaturating light in the cold, then put back at 26°C to conidiate respond normally at various fluences (Fig. 10). This suggests that the light reactions are likely to be purely photochemical, perhaps including charge transfer. Similar conclusions were reached with *Fusarium* [21]. The dark reactions amplify the molecules of photoproduct, p, to a visible and quantifiable conidiation response. There is a response level, R, at each time after light to each level of p which defines an amplification function, a(p). A photoreceptor mutation could change the light reactions. If this occurs, then after giving the same

Fig. 10. Photoinduction in the cold at nonsaturating fluences. Colonies were grown for 1.5 days at 26°C in the dark, then transferred to 3°C for 12 h. The 3°C colonies were induced at the end of the cold period, while the 26°C controls were induced before the time of the cold period. *Solid line* is a least-squares fit to data for controls that were not transferred to the cold. Conidiation was assayed 1.5 days after illumination. (Unpublished results)

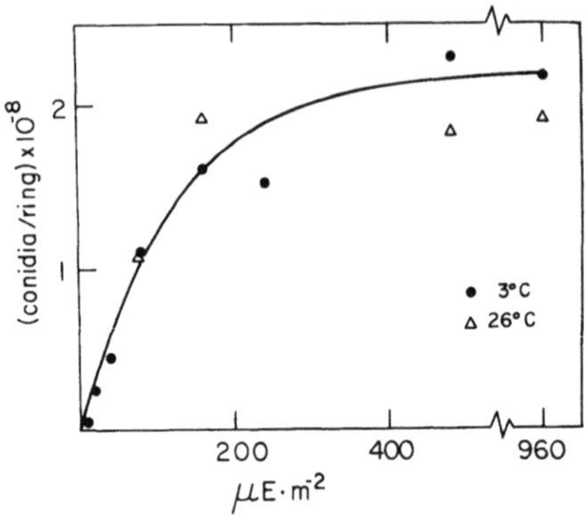

amount of light to wild type (wt) and mutant (m) colonies, $p_m \neq p_{wt}$ and the dark reactions are normal:

$$R_m = a(p_m) \neq a(p_{wt}) = R_{wt} \quad (1)$$

Giving more light to the mutant may compensate for the defect, $p_m(F_m) = p_{wt}(F_{wt})$. Equation (1) implies that the relative response will be the same at *all* sampling times, not just at the time for which the light fluences were adjusted. Mutants that fail this "identical development" criterion must be defective in the dark reactions, but they could be pleiotropic or have multiple defects, and be defective in the light reactions as well. Mathematically, this means that $a_m(p)$ and $a_{wt}(p)$ are different functions. This possibility has often been overlooked when considering mutants (e.g., [22]). The comparison of the wild type with strains B275 and B279 in Fig. 11 shows that both of these mutants must be defective (at least) in the dark reactions. Both of these mutants, particularly B275, have an abnormal pattern of stress conidiation. Neither responds to transient stress by transfer to water. This is also in agreement with defective amplifier reactions.

Fig. 11. Modified expression of conidiation suggesting that some mutants have defective dark reactions. Colonies of the wild type, B275 and B279 were grown on filter paper soaked with liquid medium supplemented with 0.09% deoxycholate to slow radial expansion rate; photoinduction was 1.5 days after inoculation. According to Eq. (1) , the three curves would fit each other (within variability) if the mutants were defective only in the light reactions. In contrast to this prediction, mutant colonies produced conidia more slowly, reaching lower values than the wild type, even though the light levels were adjusted to give the same values at 36 h.(B. Horwitz, A. Harari and J. Gressel, unpublished results)

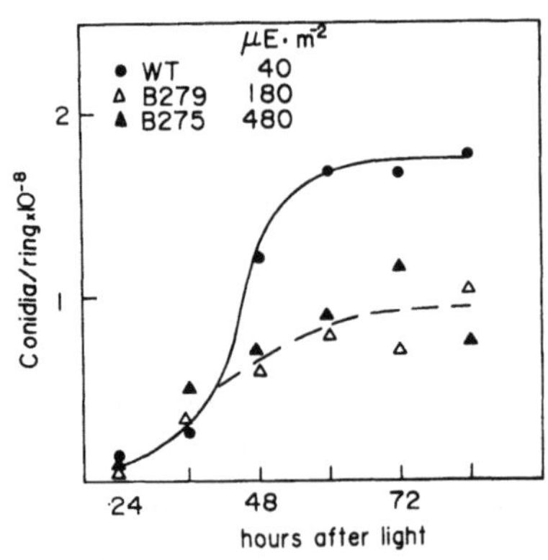

Mutants that *do* satisfy the "identical development" criterion need not be photoreceptor mutants. The functions p(F), and a(p) may have similar shapes: exponential and hyperbolic saturations, for example. It is not possible to distinguish experimentally between such curves because of variability in the data. This is a familiar problem in photomorphogenesis [4], and in any process with a "black box" amplifier.

Outlook

The genetic approach we are utilizing cannot unequivocally pinpoint a mutant lacking cryptochrome. Indeed, the lack of cryptochrome may be lethal and light perception may be a secondary function of the photoreceptor(s) [7, 20]. The genetic approach coupled with the physiological model (Fig. 2) reduced the number of possible mutants that need to be considered as cryptochrome defective. Mutants that respond to transient stress are much rarer than those that are defective in response to both light and stress. Biochemical and spectroscopic analysis of these strains should lead either to elucidation of cryptochrome, or to a perhaps less elusive but equally interesting protein involved in phototransduction.

Acknowledgments. We should like to thank Drs. D. Aviv (Weinman-Greenshpan), A. Dorn, E. Galun, and M.H. Weisenseel for useful discussions and collaborations. J.G. is the Gilbert de Bottom Professor of Plant Sciences.

References

1. Betina V, Spišiaková J (1976) Suppression of photo-induced sporulation in *Trichoderma viride* by inhibitors. Folia Microbiol 21:362–370
2. Bojňanská A, Sipiczki M, Ferenczy L (1980) Characterization of conidiation mutants in *Trichoderma viride* by hyphal anastomosis and protoplast fusion. Acta Microbiol Acad Sci Hung 27:305–307
3. Burnett JH (1975) Mycogenetics. Wiley, London, pp 85–99
4. Fukshansky L, Mohr H (1980) Boundary conditions for mathematical models in photomorphogenesis. In: De Greef J (ed) Photoreceptors and plant development. Univ Press, Antwerpen, pp 135–144
5. Galun E (1971) Scanning electron microscopy of intact *Trichoderma* colonies. J Bacteriol 108:938–940
6. Greenshpan D (1970) Morphogenetic studies on the fungus *Trichoderma viride*. Ph D thesis, Weizmann Inst Sci, Rehovot
7. Gressel J (1980) Photocontrol of transcription. In: Senger H (ed) The blue light syndrome. Springer, Berlin Heidelberg New York, pp 133–153
8. Gressel J, Galun E (1968) Morphogenesis in *Trichoderma*: Photoinduction and RNA. Dev Biol 15:575–598
9. Gressel J, Hartmann KM (1968) Morphogenesis in *Trichoderma*: Action spectrum of photoinduced sporulation. Planta 79:271–274
10. Gressel J, Horwitz B (1982) Gravitropism and phototropism. In: Smith H, Grierson D (eds) The molecular biology of plant development. Blackwell, Oxford, pp 405–433

11. Gressel J, Rau W (1983) Photocontrol of fungal development. In: Shropshire W, Mohr H (eds) Encyclopedia of plant physiology, vol 16. Springer, Berlin Heidelberg New York, pp 603–639
12. Horwitz BA, Gressel J (1983) Elevated riboflavin requirement for postphotoinductive events in sporulation of a *Trichoderma* auxotroph. Plant Physiol 71:200–204
13. Horwitz BA, Wiesenseel MH, Dorn A, Gressel J (1984) Electric currents around growing *Trichoderma* hyphae, before and after photoinduction of conidiation. Plant Physiol 74: 912–916
14. Hranueli D, Piggot PJ, Mandelstam J (1974) Statistical estimate of the total number of operons specific for *Bacillus subtilis* sporulation. J Bacteriol 119:684–690
15. Kumagai T, Oda Y (1969) An action spectrum for photoinduced sporulation in the fungus *Trichoderma viride*. Plant Cell Physiol 10:387–392
16. Lewontin RC, Prout T (1956) Estimation of the number of different classes in a population. Biometrics 12:211–223
17. Loomis WF (1980) Genetic analysis of development in *Dictyostelium*. In: Leighton T, Loomis WF (eds) The molecular genetics of development. Academic, London New York, pp 179–209
18. Mohr H (1980) Interaction between blue light and phytochrome in photomorphogenesis. In: Senger H (ed) The blue light syndrome. Springer, Berlin Heidelberg New York, pp 97–109
19. Paietta J, Sargent ML (1978) Photoreception in *Neurospora crassa*: Correlation of reduced light sensitivity with flavin deficiency. Proc Natl Acad Sci USA 78:5573–5577
20. Presti D, Delbrück M (1978) Photoreceptors for biosynthesis, energy storage and vision. Plant Cell Environ 1:81–100
21. Rau W (1983) Photoregulation of carotenoid biosynthesis. In: Porter JW, Spurgeon SL (eds) Biosynthesis of isoprenoid compounds. Wiley, London, pp 123–157
22. Russo VEA, Galland P (1980) Sensory physiology of *Phycomyces blakesleeanus*. Struct Bonding 41:71–110
23. Siegel M, Galun E (1978) Differential effects of deuterium oxide on the growth and morphogenesis of *Trichoderma viride*. Plant Cell Physiol 19:851–856
24. Stavy (Rodeh) R, Galun E, Gressel J (1972) Morphogenesis in *Trichoderma*: RNA-DNA hybridization studies. Biochim Biophys Acta 259:321–329
25. Suzuki DT, Griffeths AJF, Lewontin RC (1981) An introduction to genetic analysis. Freeman, San Francisco, p 408
26. Virgin H (1952) An action spectrum for the light-induced changes in viscosity of plant protoplasm. Physiol Plant 5:575–582
27. Warren AJ, Warren WD, Cox EC (1975) Genetic complexity of aggregation in the cellular slime mold *Polysphondylium violaceum*. Proc Natl Acad Sci USA 72:1041–1042
28. Williams KL, Newell PC (1976) Genetic study of aggregation in the cellular slime mold *Dictyostelium discoideum* using complementation analysis. Genetics 82:287–307
29. Zalokar M (1965) Integration of cellular metabolism. In: Ainsworth GC, Sussman AS (eds) The fungi, vol I. Academic, London New York, pp 403–408

Enzyme Regulation and Carbohydrate Metabolism

Blue Light Effects on the Level of Translation and Transcription

G. RICHTER [1]

Introduction

Recent findings suggest a relationship between light absorbed by a suitable photoreceptor system and the synthesis of discrete RNA species respective and proteins in developmental processes. Many of these are controlled by phytochrome. As to its mode of operation a mechanism of "differential gene activation and gene repression" has been proposed [14]. Indeed, recent studies on light-induced chloroplast differentiation seem to support this hypothesis. When dark-grown plants had been treated with red light the level of translatable mRNAs for several nuclear-coded proteins like light-harvesting chlorophyll a,b protein (LHCP; [1, 5]), the small subunit (SSU) and the large subunit (LSU) of ribulosebisphosphate carboxylase/oxygenase (RuBPCase; [19, 21]), and the precursor of a plastid-encoded polypeptide of 32,000 MW [12] increases drastically. The light-induced changes also include a decrease or disappearance of mRNA's respective of the corresponding plastid proteins which is well documented for NADPH-protochlorophyllide oxidoreductase (PCR; [1, 13]).

Do we have equally convincing evidence for the action of blue quanta on translation and transcription?

To answer this question it is necessary to examine systems with the following features: Upon irradiation for a certain period of time with blue light of low energy fluence rate (EFR) activation of gene(s) occurs within a minimum of time leading to the transcription of new and different discrete species of mRNA; their appearance and amounts are assessable by hybridization analysis; concomitantly, new and different translation products appear and can be visualized both in-vivo and in-vitro.

Thus the task of this treatise is to evaluate whether we have meanwhile successfully reached this "Utopia" – as Dr. Gressel put it so nicely at the first meeting in Marburg [6] – or at least come near to it. I would like to concentrate on examples which seem suitable for pursuing the action of blue light (in this term I include near-ultraviolet radiation) under the auspices outlined above. Since several of these systems will be reviewed in detail by other authors in this volume I shall dwell only on points which I feel are relevant for the topic of this treatise. It is divided into four major parts, each describing a system with features requiring translation and transcription products: slime molds where blue light induces an important developmental step,

1 Institut für Botanik, Universität Hannover, Herrenhäuser Straße 2, D–3000 Hannover 1, FRG

Blue Light Effects in Biological Systems
ed. by H. Senger

fungi with a photoinduced formation of carotenoids, plant cell cultures with blue light-dependent pathways of secondary metabolism, and cultured plant cells which develop chloroplasts exclusively in blue light.

Influence of Blue Light on RNA and Protein Synthesis in Nonphotosynthetic Organisms

Slime molds and fungi are very useful to elucidate blue light effects on translation and transcription since they lack chlorophyll and generally phytochrome. Thus interference by these photoreceptor systems do not occur, and the analysis of the blue light effect is perhaps less complicated.

Photoinduced Development of Slime Molds

Schreckenbach and coworkers [16] have chosen the slime mold *Physarum polycephalum* (Myxomycete) to study the effect of blue quanta on translation and transcription during an important developmental process in the life cycle of this organisms: the starvation-induced transition of microplasmodia to microsclerotia (= spherules), i.e., encysted dormant structures, which is completely blocked by exposure of plasmodial suspensions to blue light in a reversible fashion (= photoinhibition of spherulation). This phenomenon and the receptor system involved will be covered in a separate chapter of this volume; therefore, I shall restrict myself to the molecular events.

During the period of spherule formation in dark incubated plasmodia two time periods were chosen to follow the appearance of spherulation-specific proteins: 8 h after transfer and 16–32 h, respectively. The protein patterns obtained were compared with those from plasmodia kept in continuous blue light over the same lengths of time. During the period from 16–32 h of starvation in darkness which is the essential period of spherule formation, 20 major proteins are synthesized (= differentiation-specific proteins); concomitantly, 26 prominent mRNA species appear which are translatable in the rabbit reticulocyte system. The picture changes drastically upon illumination with blue light over this period: with the complete inhibition of spherulation as the major effect also the formation of spherule-specific proteins and mRNA species ceases. On the other hand, only three polypeptide components are synthesized de novo in blue light, indicating that no novel differentiated plasmodial state is induced. Blue light seems mainly to conserve the mRNA activities characteristic for the starvation state rather than to induce new prominent mRNA species.

The photoregulation of distinct mRNA's mediated by a photoreceptor of the cryptochrome type as found by Schreckenbach and coworkers is effective only up to ~ 10 h of starvation; beyond this time point a decreasing photosensitivity of the starved microplasmodia is observed.

Generally, the results indicate that spherulation of *Physarum* plasmodia is regulated at a level below translation, probably at that of transcription. One could term this phenomenon as "negative photocontrol" of a developmental step by blue light.

Induction of Carotenogenesis in Fungi by Blue Light

The photoinduced synthesis of carotenoids in fungi [17] offers a good chance to study transcriptional and translational changes. As these pigments do not seem to exhibit a rapid turnover a transient increase in the mRNA's for the required enzymes is to be expected. There is, indeed, a blue light-induced de novo synthesis of carotenogenic enzymes [15]. By labeling in vivo with [^{35}S] methionine (90 min after photoinduction) the specific activity of at least four polypeptides is markedly increased. They are constituents of a membrane fraction containing about 80% of the enzyme activities catalyzing the formation of phytoene respective of carotenoids in *Neurospora.* After photoinduction the amount of poly(A)$^+$-mRNA as well as the translational activity was enhanced in two species ; moreover, the translation products after separation by two-dimensional gel electrophoresis reveal that about 15 prominent mRNA species are synthesized or at least markedly augmented as a result of blue light action. Some of the polypeptides translated in vivo compare well with their molecular masses to in vivo proteins (for details see separate article of the authors in this vol.). These results support earlier findings which indicate that blue light exerts its influence in carotenogenesis at a level below translation.

Developmental Changes of Plant Cells Induced by Blue Light

Transcription Following Photoinduction of Pathways of the Secondary Metabolism

The induction of flavonoid glycoside synthesis in suspension cultured parsley cells (*Petroselinum hortense*) by light (~ 290 nm) under the control of phytochrome has been subject of several excellent reviews [7, 8]. Therefore, I shall limit my presentation to those points which will contribute to our topic.

Parsley cells synthesize glycosides of three flavones and three flavonols when irradiated with visible or long-wavelength ultraviolet light by a common reaction sequence. The first three reactions are catalyzed by enzymes of the general phenylpropanoid metabolism, and give rise to 4-coumaroyl-CoA from phenylalanine. During the subsequent formation of the flavonoid glycosides a group of 10–12 enzymes is involved. Significant differences exists between the two groups of enzymes in respect to the mode of induction: Group I, without light just by subculturing (= "dilution effect"); group II, strictly by light and with a different timing as compared with the enzyme activities of group I.

For the enzymes of each group a very selective mode of induction has been found, particularly under the conditions of short-term irradiation: those of group I with phenylalanine ammonia lyase (PAL) as representative reach their maximal activity significantly earlier than the enzymes of group II represented by flavonone synthase (FLS) and UDP-apiose synthase (UAS); the same holds true for the formation of corresponding proteins (Fig. 1). The rates of synthesis for the three enzymes as measured in vivo and in vitro (using the mRNA activities) increase rapidly upon irradiation with a maximum after 6–12 h, then decline exponentially. PAL synthesis

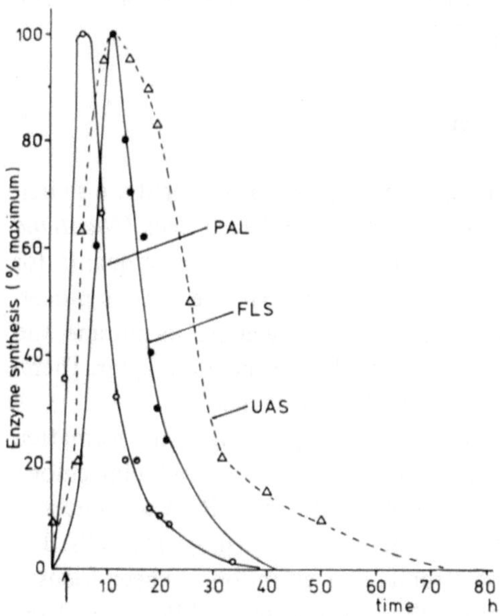

Fig. 1

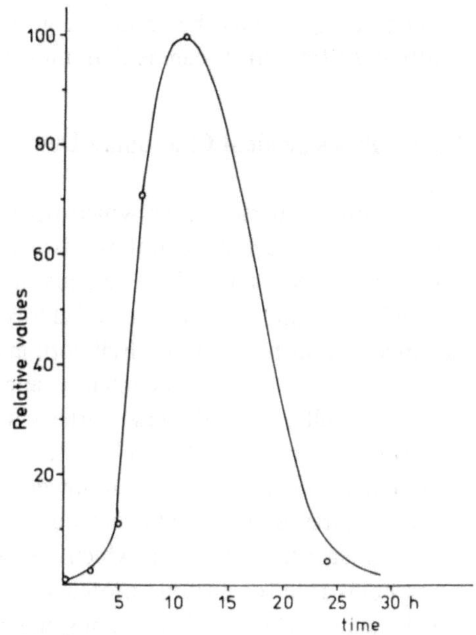

Fig. 2

Fig. 1. Light-induced changes in the rates of subunit synthesis for phenylalanine ammonialyase (PAL), flavonone synthase (FLS) and UDP-apiose synthase (UAS) after short-term irradiation for 2,5 h (*arrow*) of cultured parsley cells [20]

Fig. 2. Time course of light-induced changes in the amount of chalcone synthase (CHS) mRNA in parsely cells. A culture was irradiated for 2.5 h, total RNA was extracted from samples taken at the times indicated, blotted and hybridized to ^{32}P-labeled plasmid DNA carrying the specific insert of cDNA. Relative amounts of ^{32}P-cDNA hybridized by CHS mRNA (o) were obtained by scanning the gel strips. The curve reflects the relative changes in the CHS mRNA after the conditions of short-term induction. (After [10])

reaches its maximal value several hours earlier than the synthesis of FLS and UAS. From the close relationship between mRNA and enzyme activities the authors concluded that light-induced effects on activity and amount of the enzymes mainly or exclusively originate from changes in the level of the respective mRNA species.

One major question could not be answered in these studies: do the observed changes in mRNA activities reflect increased transcription of the genes or change in the turnover of the mRNA's? An answer was sought by Kreuzaler and coworkers [10] by following changes in the amount of chalcone synthase (CHS) mRNA in response to the irradiation of parsley cells. The quantitative determination of the mRNA sequence was achieved by using its cloned cDNA as a hybridization probe (Fig. 2). The results demonstrate that light-induced changes in activity and quantity of CHS are reflected in the amount of the active specific mRNA species isolated from polyribosomes. The data suggest that light causes a true induction of the synthesis of CHS mRNA; i.e., a transient increase in the rate of transcription. Changes in the rates of mRNA processing or degradation would be another explanation of the observed effects. For an ultimate answer to the problem how light regulates enzyme synthesis other techniques must be applied, e.g., work with isolated nuclei where only newly synthesized transcripts can be determined [4, 22].

Blue Light-Dependent Chloroplast Development

From cultured parsley cells which do not form chloroplasts in the light we switch over to their greening counterparts: suspension cultured tobacco cells (*Nicotiana tabacum* var. Samsun). Here we encounter a system which requires blue quanta for the induction and maintenance of chloroplast development [3]. The various processes involved such as chlorophyll formation, synthesis of thylakoid membrane and stroma proteins are well documented and will be covered in a separate chapter. Therefore, they will only be discussed briefly in the context under review.

We have shown that in cell cultures of tobacco blue light triggers the formation of polypeptides and pigments followed by their insertion into the developing chloroplast membranes [18]. Following treatment with blue light the mRNA sequences of several main proteins appear which are almost undetectable in dark-grown cells. The same holds true for enzyme proteins of the stroma. Our attention has been focused on the synthesis of ribulosebisphosphat carboxylase/oxygenase (RuBPCase) and the specific mRNA's for the small and the large subunit. If dark-grown tobacco cells are transfered to continuous blue light increasing amounts of RuBPCase are detectable from day 3 onwards (Fig. 3). A lag-period of about 3 d precedes the massive accumu-

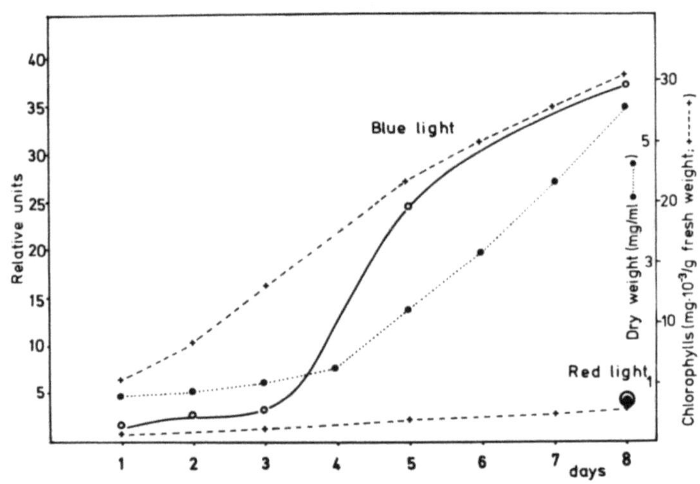

Fig. 3. Total amount of the protein of ribulose - 1,5 - bisphosphate and carboxylase/oxygenase (RuBPCase) under different light conditions. Dark-grown tobacco cells were transfered into either blue light or red light of identical EFR. Relative amounts were obtained by scanning of the Coomassie brilliant blue stained band of the large subunit (*LSU*) in the SDS-polyacrylamide gels by which the enzyme preparations from cells after various times of irradiation were separated. The increase of chlorophyll and cell mass (as dry weight) in blue light was also followed

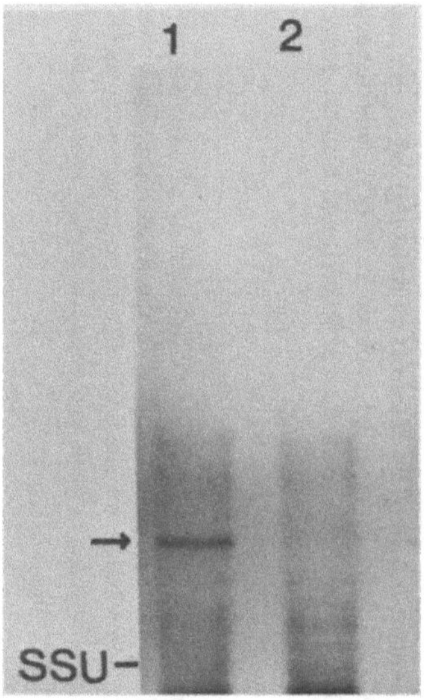

Fig. 4. In vitro translation products of polysomes from blue light-treated (*1*) and dark-grown (*2*) tobacco cells immunoprecipitated with antibody raised against RuBPCase, separated by SDS-polyacrylamide gel electrophoresis and visualized by autoradiography. The precursor polypeptide of SSU appears in the region of $M_r \approx 20{,}000$ (*arrow*)

lation of the enzyme in blue light leading to an 8-10-fold increase within 8-10 d. In cells treated with red light, however, the enzyme quantity remained unchanged throughout the experiment; in this respect red light cells compare well with dark-grown ones.

We have extended our studies to the level of the mRNA coding for SSU which is of the poly$(A)^+$-type. Polysomes from irradiated and from dark-grown cells were introduced into a read-through translational system followed by immunoprecipitation of the specific product. Only polysomes from blue light-treated cells yielded the specific precursor polypeptide of SSU (= pSSU; $M_r \approx 20{,}000$) as the prominent ^{35}S-labeled product (Fig. 4). Accordingly, in darkness the amount of translatable mRNA for SSU is insignificant. From this data it is reasonable to conclude that the stimulatory effect of blue light on the level of polysomal mRNA for SSU of RuBPCase could account for the massive accumulation of the complete enzyme – provided that both subunits are formed and assumbled coordinately.

With the methods used it was not possible to distinguish between a blue light-dependent activation or inactivation of stored mRNA sequences and a blue light-induced change in the steady-state concentration of the mRNA sequences. The quantitative determination of a particular mRNA sequence becomes possible if a complementary DNA to this mRNA is available and can be used as a hybridization probe. Two structural sequences which encode LSU of RuBPCase and the precursor polypeptide ($M_r \approx 35{,}000$) of the 32,000 MW plastid membrane protein and were isolated from chloroplast DNA of mustard leaves (*Sinapsis alba*) by Link [11] and cloned in *E. coli* served as the probes. As these clones (pSA 204, pSA 452) share sequence homology with the maize genes an appropriate selection of the corresponding mRNAs of tobacco cells could be expected. A modified "dot hybridization" [23] was applied to assess the levels of the mRNA's for LSU and the precursor polypeptide of 35,000 MW in blue light-irradiated tobacco cells. Samples of total RNA or of plas-

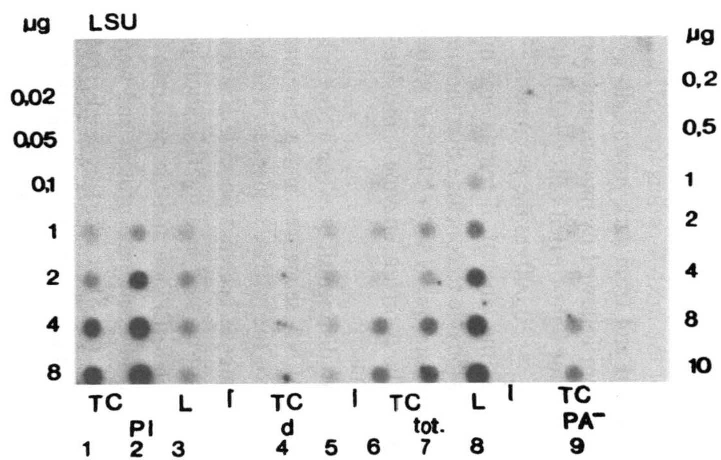

Fig. 5

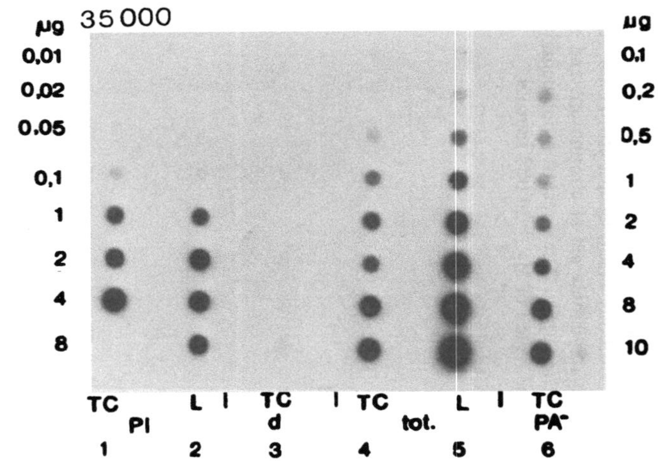

Fig. 6

Fig. 5. Blue light-induced changes after 12 d of illumination on hybridizable mRNA for the large subunit (*LSU*) of RuBPCase in tobacco cells (*TC*). Dot hybridization on nitrocellulose to the ^{32}P-labeled plasmid containing the sequence for *LSU*. The amounts of RNA dotted in the rows *1* and *2* are given on the *left*, for the *other rows* on the *right*. Rows *1–3* plastid RNA *(Pl)* from illuminated tobacco cells (*TC*) and tobacco leaves (*L*); rows *4* and *5* total RNA from dark-grown tobacco cells (*d*); rows *6–8*, total RNA (*tot.*) from greened tobacco cells and tobacco leaves, respectively; row *9*, poly(A)-minus-mRNA (PA^-) from greened tobacco cells

Fig. 6. As described for Fig. 5 except hybridizable mRNA for the precursor polypeptide of the 32,000 MW plastid membrane protein was dotted. Hybridization to the ^{32}P-labeled plasmid containing the corresponding sequence. The amounts of RNA dotted in the rows *1* and *2* are given on the *left*, those for the *other rows* on the *right*. Rows *1* and *2* plastid RNA *(Pl)* from illuminated tobacco cells (*TC*) and tobacco leaves (*L*); row *3* total RNA from dark-grown (*d*) tobacco cells; row *4* and *5* total RNA (*tot.*) from greened tobacco cells and tobacco leaves; row *6* poly(A)-minus-mRNA (PA^-) from greened tobacco cells

tid RNA fractions were denatured and dotted onto nitrocellulose filters and hybridized to the ^{32}P-labeled probes. A high steady-state level of LSU exists in blue light-treated cells and in their plastids, respectively (Fig. 5) as indicated by the appearance of hybrids which were visualized by autoradiography. These are practically undetectable in corresponding preparations from dark-grown cells. Similarly, a pronounced accumulation of the mRNA coding for the 35,000 MW polypeptide is found in blue light-grown tobacco cells which is lacking in dark-grown ones (Fig. 6).

In another experimental approach the amount of the mRNA for the 35,000 MW precursor polypeptide was followed from the onset of blue light illumination until day 4. During this period of time the intensity of radioactivity bound to the dots increased rapidly indicating that blue light affects drastically the steady-state concentration of the specific mRNA in the tobacco cells (Fig. 7).

These results, together with those obtained for the LSU mRNA, constitute evidence that blue light exerts its influence on the actual sequence level of the plastid mRNA's encoding plastid membrane and stroma proteins. Thus a parallel seems to exist to the action of red light mediated by phytochrome in pea seedlings: the levels of the mRNA's for both subunits of RuBPCase are likewise photocontrolled [19]. These data support the notion that light may strongly influence the transcription of specific genes. On the other hand, it cannot be excluded that the stability of transcripts is affected. An ultimate answer to the problem how blue quanta regulate chloroplast development on the level of transcription will come from studies employing suitable systems like isolated nuclei [4] or a plastid complex consisting of DNA, RNA polymerase and protein factors which is transcriptionally active [9; Link unpublished].

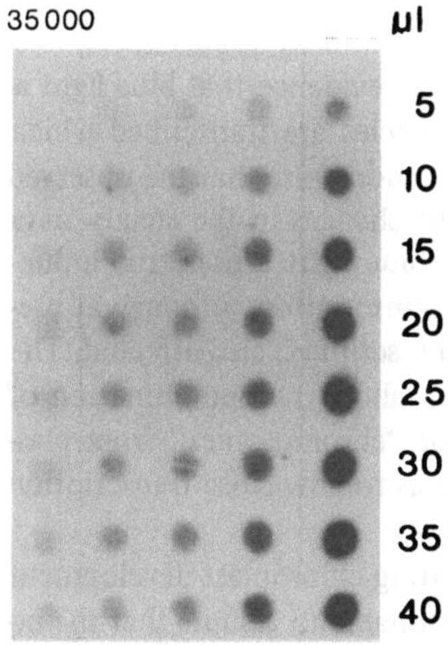

Fig. 7. Effect of blue light on the steady-state concentration of mRNA encoding the 35,000 MW precursor polypeptide of the 32,000 MW plastid membrane protein in tobacco cells. Total RNA in homogenates from 1-g samples of cells harvested on the days indicated were dotted onto nitrocellulose (5–40 µl) after serial dilution of each sample. ^{32}P-labeled plasmid with the specific insert served as hybridization probe

Pursuing this goal we succeeded in isolating and partially purifying a plastid transcription complex from greening tobacco cells which catalyzed the synthesis of RNA when supplied with the four ribonucleoside triphosphates as substrate and incu-

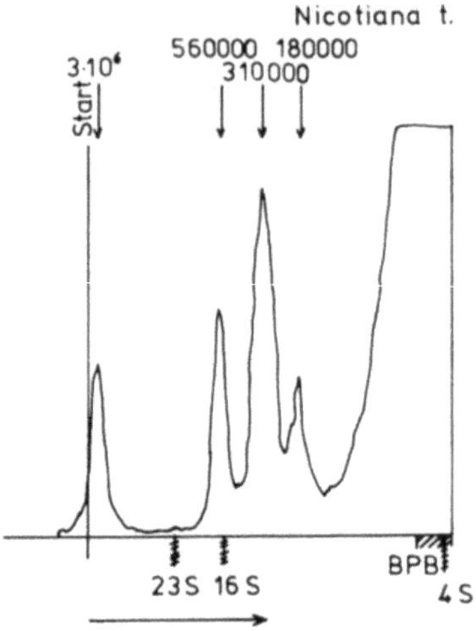

Fig. 8. In vitro synthesis of RNA's by a transcriptionally active complex of DNA and RNA polymerase isolated from plastids of greening tobacco cells. ATP, GTP, CTP and [α–^{32}P] UTP in equimolar amounts served as substrates. After deproteinization and ethanol precipitation the transcripts were separated in agarose gel (1.8%) and visualized by autoradiography. The positions of the plastid ribosomal RNA species and of the marker dye (bromphenolblue *BPB*) are indicated (Kretschmer unpublished)

bated for 20–120 min. Several abundance classes of transcripts are detectable the proportion of which varies with the stages of chloroplast development (Fig. 8). Preliminary hybridization studies employing plastid gene sequences (see above) indicate that selective transcripts of plastid DNA are among the RNA species synthesized in vitro.

Concluding Remarks

The analysis of the systems presented here provide ample evidence that blue light is implicated in affecting transcription; distinct mRNA species are transcribed which lead to specific translational products. Several findings demonstrate that the observed light-induced changes in mRNA activities correspond to changes in the steady-state concentration of mRNA. Moreover, they suggest that blue light causes true induction of mRNA, i.e., a transient increase in the rate of transcription, although at present it cannot be excluded that the stability of those transcripts is also affected. The answer to this problem may come from systems which allow the determination of only newly transcribed sequences like isolated nuclei or "chromosomes" from plastids. With the data available no statement can be made as to how close transcription is to the photoact and what processes are affected first.

As the numerous proteins which are synthesized during chloroplast development are encoded both by nuclear and plastid genes it is reasonable to assume that the induction of the two mRNA levels is coordinately regulated by blue light. RuBPCase offers a choice system to support this hypothesis.

Acknowledgments. We thank Dr. G. Link (Freiburg) for the generous gift of the plasmids pSA 204 and pSA 452 and Prof. K. Zetsche (Gießen) for providing us with the antibody to RuBPCase.

References

1. Apel K (1979) Phytochrome-induced appearance of mRNA activity for the apoprotein of the light-harvesting chlorophyll a,b protein of barley (*Hordeum vulgare L.*). Eur J Biochem 97:183–188
2. Apel K (1981) The protochlorophyllide holochrome of barley (*Hordeum vulgare L.*). Phytochrome induced decrease of translatable mRNA coding for the NADPH-oxidoreductase. Eur J Biochem 120:89–93
3. Bergmann L, Berger Ch (1966) Farblicht und Plastidendifferenzierung von Zellkulturen von *Nicotiana tabacum* var. *"Samsun"*. Planta 69:58–69
4. Galagher TF, Ellis RJ (1982) Light-stimulated transcription of genes for two chloroplast polypeptides in isolated pea leaf nuclei. EMBO J 1:1483–1498
5. Gollmer I, Apel K (1983) The phytochrome-controlled accumulation of mRNA sequences encoding the light-harvesting chlorophyll a,b protein of barley (*Hordeum vulgare L.*) Eur J Biochem 133:309–313
6. Gressel J (1980) Blue light and transcription. In: Senger H (ed) The blue light syndrome. Springer, Berlin Heidelberg New York, pp 133–153
7. Hahlbrock K, Grisebach H (1979) Enzymic control in the biosynthesis of lignin and flavonoids. Annu Rev Plant Physiol 30:105–130
8. Hahlbrock K. Gardiner SE, Matern U, Ragg H, Schröder J (1979) The role of light in the induction of mRNAs for phenylalanine ammonia lyase and related enzmes in plant cell cultures. In: Leaver CJ (ed) Genome organization and expression in plants. NATO Adv Study Inst Series: Ser Life Sci. Plenum Press, New York London, pp 187–194
9. Hallick RB, Lipper C (1976) Selective transcription of the *Euglena gracilis* chloroplast chromosome in vitro. In: Nierlich DP, Rutter WJ, Fox CF (eds) Molecular mechanisms in the control of gene expression. INC-UCLA Symp Mol Cell Biol, vol V. Academic, London New York, pp 355–360
10. Kreuzaler F, Ragg H, Fautz E, Kuhn DN, Hahlbrock K (1983) UV-induction of chalcone synthase mRNA in cell suspension cultures of *Petroselinum hortense.* Proc Natl Acad Sci USA 80:2591–2593
11. Link G (1981) Cloning and mapping of the chloroplast DNA sequences for two messenger RNAs from mustard (*Sinapis alba L.*) Nucleic Acids Res 9:3681–3694
12. Link G (1982) Phytochrome control of plastid mRNA in mustard (*Sinapis alba L.*) Planta 154:81–86
13. Meyer G, Bliedung H, Kloppstech K (1983) NADPH-protochlorophyllide oxidoreductase: Reciprocal regulation in mono- and dicotyledonean plants. Plant Cell Rep 1:26–29
14. Mohr H (1972) Lectures on photomorphogenesis. Springer, Berlin Heidelberg New York
15. Mitzka-Schnabel U, Rau W (1981) Subcellular site of carotenoid synthesis in *Neurospora crassa.* Phytochemistry 20:63–69
16. Putzer H, Werenskiod K, Verfuerth C, Schreckenbach Th (1983) Blue light inhibits slime mold differentiation at the mRNA level. EMBO J 2:261–267
17. Rau W (1980) Blue light-induced carotenoid biosynthesis in microorganisms. In: Senger H (ed) The blue light syndrome. Springer, Berlin Heidelberg New York, pp 283–298
18. Richter G, Beckmann J, Groß M, Hundrieser J, Schneider Ch (1982) Blue light-induced synthesis of chloroplast proteins in cultured plant cells. In: Proc Clin Biol Res, vol 102 B. Cell function and differentiation, part B: Biogenesis of energy transducing membranes and membrane and protein energetics (FEBS vol 65). AR Liss, New York, pp 267–276
19. Sasaki Y, Sakihama T, Kamikubo T, Shinozaki K (1983) Phytochrome-mediated regulation of two mRNAs, encoded by nuclei and chloroplasts of ribulose-1,5-bisphosphate carboxylase/oxygenase. Eur J Biochem 113:617–620
20. Schröder J, Kreuzaler F, Schäfer E, Hahlbrock K (1979) Concomitant induction of phenylalanine ammonia lyase and flavanone synthase mRNAs in irradiated plant cells. J Biol Chem 254:57–65

21. Tobin EM (1981) Phytochrome-mediated regulation of messenger RNAs for the small subunit of ribulose-1,5-bisphosphate carboxylase and the light-harvesting chlorophyll a,b protein in *Lemna gibba.* Plant Mol Biol 1:35–51
22. Tobin EM, Silverthorne J, Stiekema WJ, Wimpee CF (1983) Chloroplast biogenesis. Cambridge Univ Press, in press
23. White BA, Bancroft FC (1982) Cytoplasmic dot hybridization. J Biol Chem 257:8569–8572

Light-Induced Changes in the Protein Pattern Translated in Vivo and in Vitro Accompanying Carotenogenesis in *Neurospora crassa* and *Fusarium aquaeductuum* *

U. MITZKA-SCHNABEL, E. WARM, and W. RAU [1]

Introduction

Photocontrol of carotenogenesis in some fungi, which is one of the intensively studied responses to blue light [9] may serve as a model system for other blue light-induced phenomena during development in eukaryotic organisms.

According to current models derived from inhibitor studies and RNA labeling, the expression of the light signal during photoinduction of carotenogenesis is assumed to be realized by the means of transcriptional events [2, 6, 9, 10]. Even though, the involvement of light-induced gene expression remains speculative unless specific photoregulated mRNA's have been identified.

For a more direct analysis of a transcriptional control two approaches have been attempted: (1) In vivo labeled membrane polypeptides that contain the bulk of carotenogenic enzyme activity and (2) in vitro translated cellular mRNA's were separated by two-dimensional electrophoresis and analyzed for light-induced changes.

Materials and Methods

Biological Material, Growth Condition and Illumination. N. crassa and *F. aquaeductuum* were grown as submerse cultures for 20 h respectively 18 h in the dark, harvested by filter suction, resuspended in buffer-glucose and accomodated for 2 h to these step-down conditions. Samples were kept in the dark or illuminated with banks of white light fluorescent tubes (incident fluence rate 60 W m^{-2}) as a source for blue light. Otherwise the experimental conditions were as described earlier [5, 6].

Labeling, Isolation and Extraction of Membrane Proteins. For pulse labeling of *N. crassa*, mycelia were incubated with 25 μCi L–(^{35}S)–methionine in the dark or light for 90 min. Uptake of methionine was stopped by adding ice-cold deionized water. The mycelia were then collected onto glass fibre discs and homogenized by sonication. The membrane fraction was pelleted by fractionated centrifugation at 115,000 g_{av} for 2 h and tested for carotenogenic activity as described previously [6].

* Dedicated to Roswitha Dengler †

1 Botanisches Institut der Universität München, Menzinger Straße 67, 8000 München 19, FRG

Blue Light Effects in Biological Systems
ed. by H. Senger

RNA Extraction and Fractionation of Poly(A)-RNA. Total RNA was extracted by the SDS-proteinase K method according to Hall et al. [3]. Prior to the separation of poly(A)-RNA by oligo-dT-cellulose affinity chromatography (Sigma, Heidelberg) as described by Aviv and Leder [1], contaminating polysaccharides were removed by cellulose chromatography [7].

In Vitro Translation of Poly(A)-RNA. Usually 1 μg poly(A)-RNA was translated in a rabbit reticulocyte lysate in the presence of 25 μCi (925 x 10^3 Bq) L-(^{35}S)-methionine following the manufacturer's protocol (NEN, Dreieich, FRG).

Two-Dimensional Polyacrylamide Gel Electrophoresis. (^{35}S)-labeled membrane proteins and aliquots from the in vitro translation products were lyophilized and resuspended in O'Farrell's lysis buffer [8] with Servalytes and 0.2% SDS (w/v). For two-dimensional electrophoresis standard methods were applied [8] with the isoelectric focusing in tubes resulting in a pH range from 4 to 7 and the second dimension in 10% to 16% SDS-polyacrylamide slab gels (1.5 mm thick). After fixation and Coomassie staining the gels were processed for fluorography [4] and exposed to X-ray film at -70°C. Exposure was repeated and optimized for equal density.

Results and Discussion

Photoregulation of in Vivo Translated Proteins

Carotenogenic enzymes are localized in the membrane fraction. Therefore the membrane proteins from dark-grown and illuminated mycelia were separated by two-dimensional gel electrophoresis. About 270 polypeptides were visibly separated in the stained gels (not shown), at least two of which were markedly intensified in the membranes from illuminated mycelia. This expectedly low percentage of light regulated proteins has been tried to selectively intensify by pulse labeling with (^{35}S)-methionine. The labeling period was for 80 min starting from 10 to 90 min after onset of light according to the appearance of carotenogenic enzymes as deduced from the "lag phase" for carotenoid accumulation, inhibitor studies (for review [9]) and a biphasic fluence response curve [11].

The two-dimensional separation of the labeled membrane proteins exhibit at least 4 polypeptides which were conspicuously increased in response to photoinduction of carotenogenesis (Fig. 1). These polypeptides might represent a multi-enzyme complex as postulated by De La Guardia [12]. At present, however, these changes in the protein pattern are not proved to be unequivocally linked to carotenogenesis.

Photoregulation of mRNA's

A de novo synthesis of enzyme proteins is preceded by the transcription and processing of the corresponding mRNA's. Cellular poly(A)-RNA, extracted from dark and illuminated mycelia of *N. crassa* and *F. aquaeductuum* revealed high translational

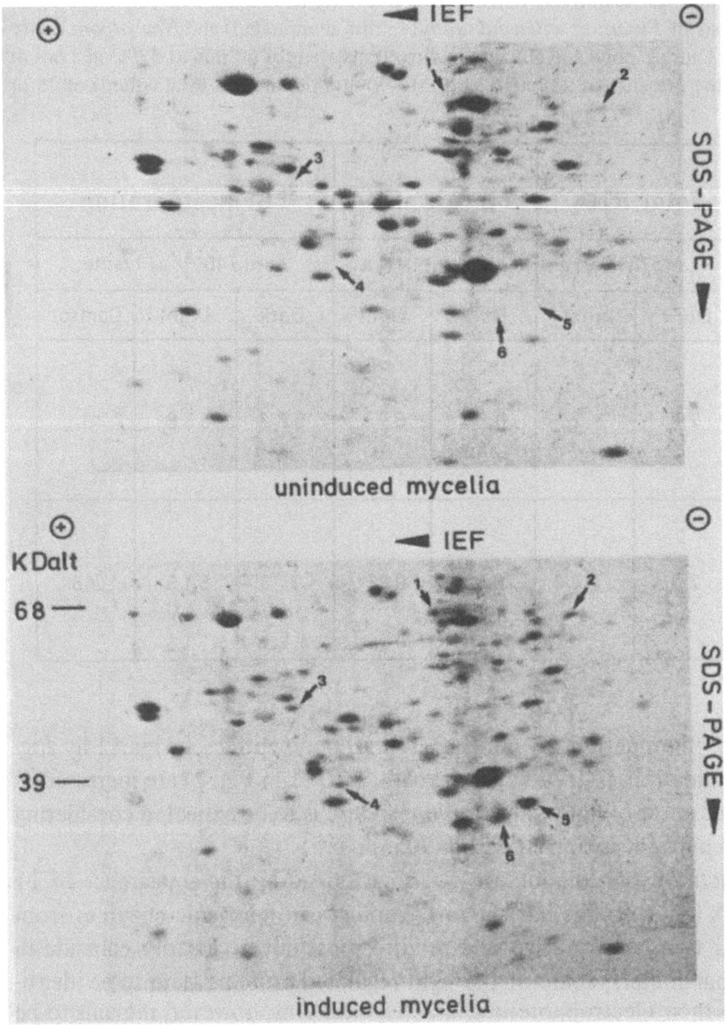

Fig. 1. Two-dimensional electrophoresis fluorographs (enlarged sections) of (^{35}S)-methionine labeled membrane proteins extracted from dark-grown (uninduced) and illuminated (induced) mycelia of *N. crassa*. The arrows indicate polypeptides increasing in response to photoinduction. Mycelia were labeled for 80 min in the presence of 25 μCi (925 x 10^3 Bq) L-(^{35}S)-methionine (1250 Ci/mmol = 4,6 x 10^3 Bq/mmol) and 1 μmol unlabeled L-methionine added 10 min after start of incubation respectively onset of light

capactiy in vitro reflected by the stimulation of (^{35}S)-methionine incorporation (Table 1). Remarkably the amount of poly(A)-RNA is increased in both illuminated fungi and might indicate the higher biosynthetic activity during carotenogenesis (Table 1). While translation of the blank lysate results in just one labeled polypeptide (not shown) about 400 polypeptides are electrophoretically separated upon the addition of *Neurospora* mRNA (Fig. 2). No polypeptides were found to disappear

Table 1. Amount of total RNA, poly(A)-RNA and its translational activity with respect to light-induced carotenogenesis of *Fusarium aquaeductuum* (45 min illuminated) and *Neurospora crassa* (30 min illuminated). One μg poly(A)RNA was in vitro translated for 60 min at 37°C in a rabbit reticulocyte lysate in the presence of 925 x 10^3 Bq L-(^{35}S)-methionine in a total volume of 25 μl

	Total RNA		Poly(A)$^+$-RNA		^{35}S-incorporation		
	mg/g f.wt.		% of tot. RNA		cpm x 10^{-3}/μl lysate		
	Dark	Light	Dark	Light	Dark	Light	Control
Fusarium	2.8	3.2	0.54	0.72	52.3	53.8	11.1
Neurospora	2.3	2.4	0.49	0.97	31.3	59.5	7.8

during the course of illumination. By contrast, several polypeptides are markedly augmented during illumination, four of which (spot 4, 7, 8, 11 in Fig. 2) are increased as early as 10 min after onset of light. Such rapid response is to be expected considering the first carotenoid pigment to appear within 30 min.

In *F. aquaeductuum* photoinduction is also attended by the appearance of 11 prominent mRNA's (fluorograms not shown). Though carotenogenic enzymes from different fungi and translated in vivo and in vitro must not necessarily coincide in their biochemical parameters, some of the light-regulated proteins seem to be identical with respect to their electrophoretic characteristics. This, however, remains to be confirmed by immunologic tests.

Conclusions

The results obtained from the separation of in vivo labeled membrane proteins and translation products of cellular mRNA's support previous findings that photoinduction of carotenogenesis – as a model for positive response to blue light – is regulated at the level of transcription.

Acknowledgments. This work was supported by grants of the Deutsche Forschungsgemeinschaft to Dr. W. Rau. The skilful technical assistance of Mrs. T. Kydd and Miss B. Herrmann is gratefully acknowledged.

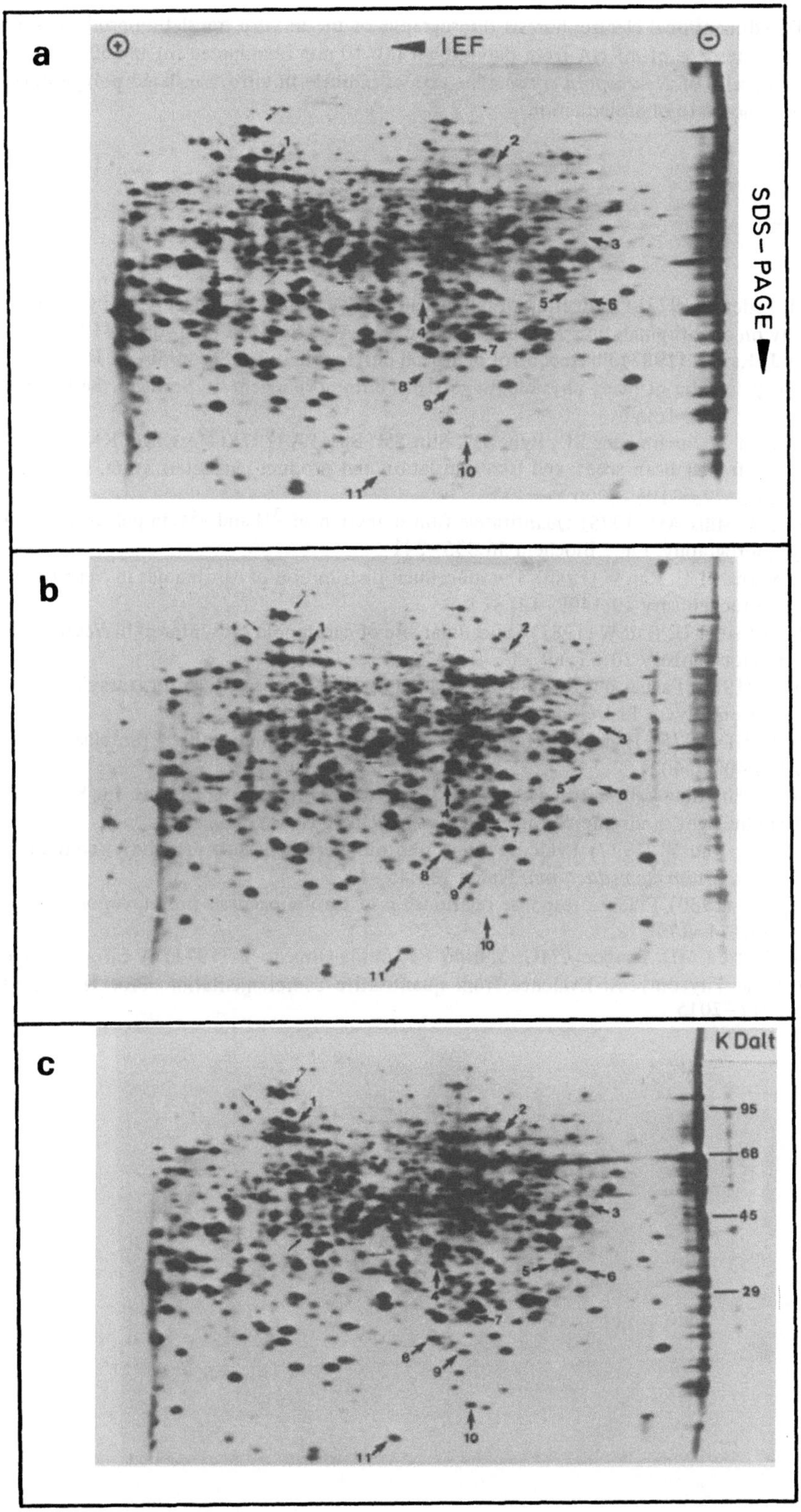

Fig. 2a-c (caption see opposite page)

Fig. 2a-c. Two-dimensional electrophoresis fluorographs of the in vitro translation products synthesized in the presence of mRNA from dark-grown (a), 10 min illuminated (b) and 30 min illuminated (c) mycelia of *Neurospora crassa*. The arrows indicate in vitro translated polypeptides increasing in response to photoinduction

References

1. Aviv H, Leder P (1972) Purification of biologically active globin messenger RNA by chromatography on oligothymidylic acid cellulose. Proc Natl Acad Sci USA 69:1408–1417
2. Gressel J, Rau W (1983) Photocontrol of fungal development. In: Shropshire W Jr, Mohr H (eds) Encyclopedia of plant physiology, vol 16. Photomorphogenesis. Springer, Berlin Heidelberg New York Tokyo
3. Hall TC, Ma Y, Buchbinder BU, Pyne JW, Sun SM, Bliss PA (1978) Messenger RNA for G1 protein of french bean seeds; cell free translation and product characterization. Proc Natl Acad Sci USA 75:3196–3200
4. Laskey RA, Mills AD (1975) Quantitative film detection of ^{3}H and ^{14}C in polyacrylamide gels by fluorography. Eur J Biochem 56:335–341
5. Mitzka-Schnabel U, Rau W (1980) The subcellular distribution of carotenoids in *Neurospora crassa*. Phytochemistry 19:1409–1413
6. Mitzka-Schnabel U, Rau W (1981) Subcellular site of carotenoid biosynthesis in *Neurospora crassa*. Phytochemistry 20:63–69
7. Mozer TJ (1980) Partial purification and characterization of the mRNA for α-amylase from barley aleurone layers. Plant Physiol 65:835–837
8. O'Farrell PH (1975) High resolution two-dimensional electrophoresis of proteins. J Biol Chem 250:4007–4021
9. Rau W (1980) Blue light-induced carotenoid biosynthesis in microorganisms. In: Senger H (ed) The blue light syndrome. Springer, Berlin Heidelberg New York
10. Schrott EL, Rau W (1977) Evidence for a photoinduced synthesis of poly(A) containing mRNA in *Fusarium aquaeductuum*. Planta 136:45–48
11. Schrott EL (1980) Fluence response relationship of carotenogenesis in *Neurospora crassa*. Planta 150:174–179
12. De La Guardia MD, Aragon CMG, Murillo FJ, Cerda-Olmedo E (1971) A carotenogenic aggregate in Phycomyces: Evidence from quantitative complementation. Proc Natl Acad Sci 68:2012–2015

Light-Dependent Greening of the *Chlorella* Mutant G-2: Spectral Dependency and Chloroplast Translation Effects

G. GALLING [1]

Introduction

Numerous microalgae are capable of producing chlorophyll and active thylakoid membranes in the dark. The genetic loci of this light-independent greening system are easily attacked by mutagenesis with physical or chemical mutagens, since from most of these algae mutants are obtainable which have lost their ability to green in the dark. In higher plants, dark-grown leaves mostly contain proplastids or etioplasts whereas in mutant strains of *Chlorella* [9], or *Chlamydomonas* [15] thylakoid structures can be found which apparently are free of chlorophyll, see also [4].

The process of development of an active chloroplast can be followed in algal mutants which have lost their greening capability in the dark. In many cases, blue light induces chloroplast development [2, 17], whereas in other organisms, blue plus red or even red plus yellow green light are needed [16]. This indicates that various light receptors are to be found in the algal world. Higher plants, however, seem to be mainly influenced by the phytochrome system [14].

We describe here a mutant of *Chlorella fusca* 211/8b which requires blue and red light for greening. From antibiotic experiments, the life time of the chloroplast-translated prerequisites for greening can be deduced.

Materials and Methods

From the unicellular green alga *Chlorella fusca,* strain 211/8b of the algal collection, University of Göttingen, FRG (the former Marburg strain of *C. pyrenoidosa*), a mutant has been derived by UV light mutagenesis and screening for light-dependent greening on dark-incubated polytrophic agar slabs [7]. The stability of the mutation was maintained for 6 years after culture on agar and in liquid media [6].

Liquid cultures were grown in a mineral medium supplied with 1% of glucose according to Lorenzen and Kuhl [12], and aerated with air or with stirring and kept at 25°C or 32°C.

To determine the spectral dependency of greening, the algae were plated on agar slabs with the medium cited above or they were embedded in smooth agar on the sur-

1 Botanical Institute, Technical University, Braunschweig, FRG

Blue Light Effects in Biological Systems
ed. by H. Senger

face of normal agar slabs. From the light of a projector equipped with a lamp of 250 W/24 V, a slit was projected on a triangular prism, and the spectrum was shined through a silver mirror directly onto the surface of the agar slab. The regions corresponding to the blue, green, yellow, red and far-red parts were indicated on the Petri dish. Greening was monitored after 24 h illumination by microscopy. In liquid cultures, greening was induced either by illumination with blue or red light, obtained with glass filters SCHOTT BG 28, VG 9, BR 665, RG 9, (all obtained from SCHOTT, Mainz, FRG) or with white fluorescent light.

The antibiotic cleocin (= clindamycin = 7-Cl-lincomycin, kindly supplied by The UPJOHN Cie, Kalamazoo, USA) was applied in concentrations of 2-5 $\mu g\ ml^{-1}$, either continuously or at the onset of illumination.

Growth was monitored by OD measurement at 600 nm and by cell counting in a THOMA chamber.

Chlorophyll was extracted from whole cells with hot methanol according to McKinney [13].

Results and Discussion

In analogy to Engelmann's classical experiment on the spectral dependency of photosynthesis, we applied the projection of prismatically split light on layers of dark-grown mutant cells. As shown in Fig. 1, greening can be observed in the region both of blue and red light. According to the energy characteristics of the electrical bulb, in the red light region the energy flow is 30 times higher than in the blue light. Furthermore, the red

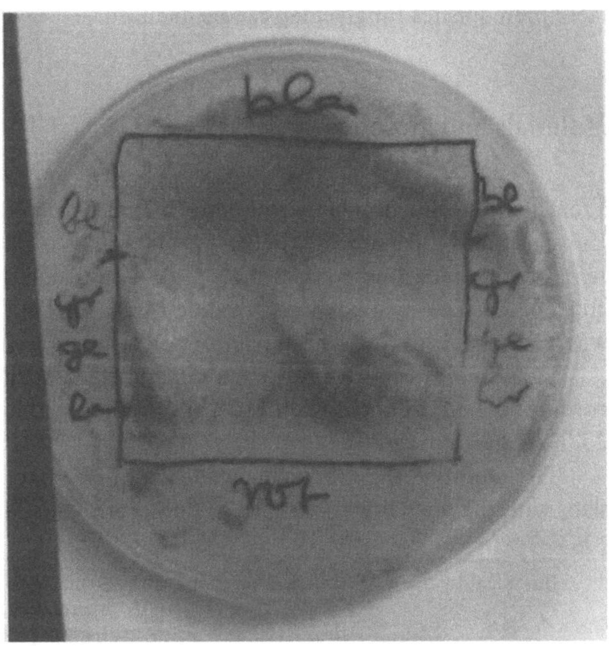

Fig. 1. Chlorella fusca g-2 plated on agar and illuminated with prismatically split light for 24 h after preculture in the dark. Print taken from a color print (Ectachrome) on black and white Ilford HP 4 with orange filter

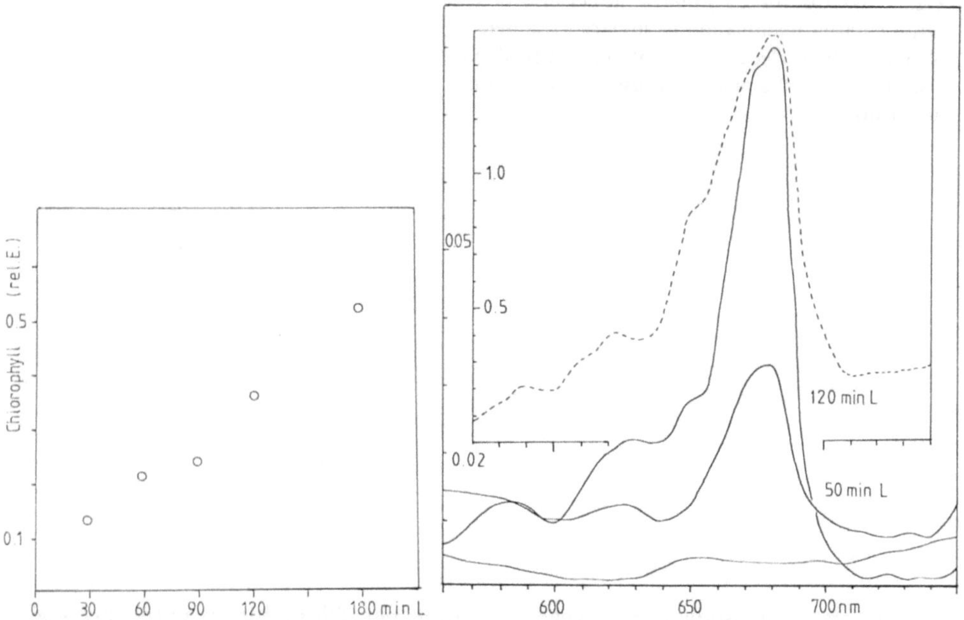

Fig. 2a,b. Chlorophyll synthesis in *Chlorella fusca* g-2. Increase of chlorophyll as a function of illumination time (**a**). Dark-grown cells were illuminated for various time intervals, and chlorophyll was extracted and determined according to McKinney [13]. In vivo spectra of dark-grown and illuminated cultures of *C. fusca* g-2 (**b**). After short- or long-term illumination spectra were registered in a double beam photometer with dark-grown cells as reference. To reduce the light scattering, the cuvettes were turned by 90°. *Dotted line* long-term illumination (24 h)

light region is narrower than that of the blue light. This explains the much sharper greening band and the greater intensity in the red light area.

In *Euglena,* the need of blue plus red light for greening has been shown [17], and the autocatalytic involvement of protochlorophyllide on the greening process has been postulated. In *Chlorella* and *Scenedesmus,* blue and red light were shown to have antagonistic effects in the protein metabolism, in respiration, in carbohydrate and in membrane function [10, 11, 18].

In another set of experiments, the dependency of greening from the time of illumination was investigated. As shown in Fig. 2a, the amount of chlorophyll found in the cells after illumination and further dark incubation follows a linear function to the time of light treatment in the first 3 h of greening. This indicates that greening of the g-2 mutant is not an all-or-nothing-mechanism. In order to demonstrate the in vivo spectra of greening cells, difference spectra of illuminated and dark-grown cells were written in a double beam photometer after turning the cuvettes by 90°. In Fig. 2b, it can be seen that after short-term illumination, only a double peak in the 674 and 682 nm region is observed and, after longer illumination, the shoulder at 654 nm appears. This indicates that the synthesis of chlorophyll a slightly precedes that of chlorophyll b. Both pigments seem to be present in the thylakoids as functional complexes even after very short illumination periods [3].

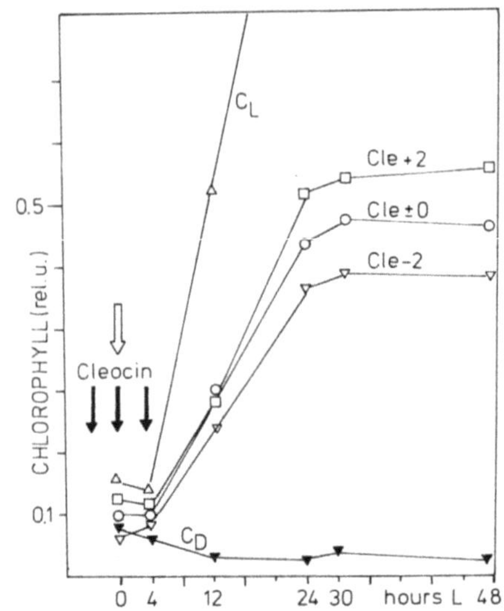

Fig. 3. Increase of chlorophyll in cultures of *C. fusca* g-2 after illumination in absence or presence of 5 $\mu g\ ml^{-1}$ of cleocin. The antibiotic was added just prior to, at the onset and just afer onset of illumination

As previously shown [8], oxygen evolution per chlorophyll is extremely high during the first stages of greening. This underlines the rapid formation of photosynthetic units during the first hours of greening. In order to study the involvement of the translation systems during greening, cells were treated with cleocin, a chlorosubstituated derivative of lincomycin. This drug has been shown to be the only anti-70 S ribosome agent which inhibits greening of cells without interfering with growth and cell division during long experimental procedures [4]. From data obtained with *Chlamydomonas reinhardii* and with cell-free translation systems, we deduce that the possible site of action is on the ribosomal protein responsible for the binding of 50 S subunits to the chloroplast membrane.

Continuous treatment of *Chlorella* cells, either wild type or the g-2 mutant, prevents any greening in light or darkness. We applied the drug prior to, during, and shortly after onset of illumination of dark-grown g-2 cultures in liquid medium.

As shown in Fig. 3, the greening of cells starts with half the rate as in untreated controls but, after about 24 h, it levels off. Cell growth, however, continues for longer periods. In the untreated culture, the chlorophyll formation continues to a final level, about five times higher than with cleocin.

The duration of chlorophyll synthesis is not dependent on the time of cleocin treatment, since in all three cultures the leveling off is at the same time point. Only the amount of chlorophyll varies with the application time of the antibiotic. Possibly, in darkness, the continuous synthesis of one prerequisite of greening takes place on 70S ribosomes. Only after illumination, does this substance induce chlorophyll synthesis (or chlorophyll charging of membranes respectively). One or more chloroplastic proteins, involved in this process, have a life-time of about 24 h. We do not yet know if it is coincidence or correlation that the life cycle of the algal strain we use is within the same margin.

References

1. Bingham S, Schiff JA (1979) Biochem Biophys Acta 547:531–543
2. Brinkmann G, Senger H (1978) Plant Cell Physiol 19:1427–1437
3. French CS, Brown JS, Lawrence MC (1972) Plant Physiol 49:421–429
4. Galling G (1976) In: Bücher Th et al. (eds) Genetics and biogenesis of chloroplasts and mitochondria. Elsevier/North Holland Biomed Press, Amsterdam, pp 53–60
5. Galling G (1978) In: Akoyunoglou G et al (eds) Chloroplast development. Elsevier/North Holland Biomed Press, Amsterdam, pp 439–444
6. Galling G (1981) In: Akoyunoglou G (ed) Photosynthesis, vol V. Chloroplast development. Balaban Int Sci Serv, Philadelphia, pp 465–472
7. Galling G (1982) In: Edelman et al. (eds) Methods in chloroplast molecular biology. Elsevier/North Holland Biomed Press, Amsterdam, pp 65–72
8. Galling G (1982) In: Jahrestag Dtsch Bot Ges 1982, Poster 054
9. Gergis MS (1969) Arch Mikrobiol 68:187–190
10. Kowallik W (1966) In: Energy conversion by the photosynthetic apparatus. Brookh Symp Quant Biol 19:467–477
11. Kowallik W (1966) Planta 69:292–295
12. Lorenzen H, Kuhl A (1964) In: Prescott DM (ed) Methods in cell physiology. Academic, London New York, pp 159
13. McKinney G (1941) J Biol Chem 140:315–322
14. Mohr H (1981) In:Akoyunoglou G (ed) Photosynthesis, vol V. Chloroplast development. Balaban Int Sci Serv, Philadelphia, pp 869–883
15. Ohad I, Siekevitz P, Palade G (1967) J Cell Biol 35:521–552
16. Schiff JA (1974) Les cycles cellulaires. Coll Int CNRS 240:79–93
17. Schiff JA (1980) In: Senger H (ed) The blue light syndrome. Springer, Berlin Heidelberg New York, pp 269–280
18. Senger H, Schoser G (1966) Z Pflanzenphysiol 54:308–320

Increase in RNA by Blue Light in a Chlorophyll-Free *Chlorella* Mutant

W. KOWALLIK and D. BARTLING [1]

Introduction

When irradiated with blue light of low intensity (~ 300 μW cm^{-2}) cells of a chlorophyll-free carotenoid-containing mutant of the unicellular green alga *Chlorella vulgaris* produce additional protein on the expence of reserve carbohydrates [8, 18]. In crude extracts of such illuminated organisms the specific activities of several enzymes with rather low turnover rates, like the carbohydrate degrading pyruvate kinase [17] and the amino acid producing glutamine-2-oxoglutarate-aminotransferase (Kowallik and Neuert this vol.), are enhanced by up to 100%. Partial suppression of this light effect by inhibitors of protein biosynthesis such as cycloheximide and actinomycin nourishes the assumption that blue irradiation leads to a de novo synthesis of the enzyme proteins. Such a synthesis might then be the consequence of an increased production of specific mRNA's, somehow provoked by the short wavelength visible quanta. An increased accumulation of total RNA by blue light was described for wild-type *Chlorella* several years ago [12], and the wavelength dependence of this phenomenon [28] has been found to correspond with that of the increased protein production [13] and enhanced enzyme activity as well ([23], Kowallik and Neuert this vol.). But also blue light effects on transcription have been proposed already. They are derived mainly from data on greening of *Euglena* [6, 24], on carotenoid formation in the mold *Fusarium* [26], and on conidiation of *Trichoderma* [3, 9]. Most recently they have gained special support by the demonstration of an enhanced level of translatable mRNA for the small subunit of ribulosebisphosphate carboxylase in cultured tobacco cells in which the activity of that enzyme increased drastically after blue light exposure [11] and by the appearance of several translatable mRNA species after photoinduction of carotenogenesis in *Fusarium aquaeductuum* and *Neurospora crassa* (Mitzka-Schnabel et al. this vol.).

With all this information available and in spite of its complexity, we still wanted to look closer for changes by blue irradiation in our mutant cell's nucleic acid system, especially in its mRNA. The blue-enhanced catabolic carbohydrate metabolism of this alga has been examined most extensively [14, 15, 18], so that additional information on early events following blue light absorption of just this organism appears especially desirable.

1 Fakultät für Biologie, Universität Bielefeld, 4800 Bielefeld 1, FRG

Blue Light Effects in Biological Systems
ed. by H. Senger

Materials and Methods

The chlorophyll-free mutant strain 211-11 h/20 of the green alga *Chlorella vulgaris* [27] was cultured heterotrophically in nutrition solution with 1% of glucose in the dark as described elsewhere [12].

For the experiments cells of 10-day-old static batch cultures were transferred to fresh nutrition solution without glucose and adjusted to a density of 100 μl packed cell volume (pcv) ml^{-1}. Illumination with blue or red light was performed in a set-up used by Dresbach and Kowallik [7], passing the beam of a Xenon high pressure arc (Osram XBO 450 W) through interference filters with maximal transmission at 457 nm or 729 nm, respectively.

Nucleic acids were isolated using a modified method of Brandt and Ingversen [4]: 1000 μl pcv were ground in a mortar under liquid nitrogen cooling and the nucleic acids extracted with 10 ml buffer (0.1 M TRIS, 0.1 M NaCl, 0.01 M EDTA pH 8.5) plus 1% SDS by shaking for 15 min at room temperature. After centrifugation, the pellet was ground and extracted twice again. The resulting supernatants were combined, mixed with diethylpyrocarbonate and deproteinized by shaking with an equal volume of phenolchloroform:isoamylalkohol (50:50:1) for 20 min. Thereafter they were cooled to 4°C and spun at 10,000 *g*. The lower, organic phase was reextracted twice again with equal volumes of buffer before the combined aqueous phases were washed with an equal volume of chloroform:isoamylalkohol (24:1). From these purified aqueous phases the nucleic acids were precipitated by the addition of two volumes of ethanol overnight at -20°C, spun down at 12,000 *g* at -8°C, washed twice with cool ethanol and finally dissolved in 0.05 M TRIS, 0.01 M NaCl, 0.01 M EDTA buffer of pH 7.5. The purity of this preparation was controlled by the ratio of its UV absorption at 260 and 280 nm. The content of the nucleic acids was judged by the A_{260} unit, which means an optical density at 260 nm of 1.0.

For poly(A^+)RNA isolation we first separated the DNA in an isopycnic CsCl gradient (modified after [5] consisting of a 2 ml 1.85 g cm^{-3} cushion and a 5 ml, 1.58 *g* cm^{-3} the nucleic acids containing supernant. Gradients were run for 18 h at 120,000 *g* at 22°C in a Sorvall TI-865 rotor.

From the resulting RNA pellet poly(A^+)RNA was purified by oligo(dT)cellulose (BRL Catalog No. 5940) chromatography (modified procedure of [2]): 5 ml of buffer (10 mM TRIS, 1 mM EDTA, 0.1% SDS, 0.5 M NaCl pH 7.5) containing up to 60 A_{260} units of RNA were elaborately shaken with 250 mg of oligo(dT)cellulose in sterile glass tubes to bind poly(A^+)mRNA to the affinity matrix. The tube was then centrifuged for 3 min at 3000 *g* and the affinity matrix washed extensively four times with binding buffer. Poly(A^+)RNA was finally eluted from the matrix in four fractions of 2 ml elution buffer (10 mM TRIS, 1 mM EDTA, 0.1% SDS pH 7.5) at 45°C; the highest amount being in the first two fractions. For further experiments poly(A^+) RNA was precipitated by the addition of 0.3 M Na-acetate and 2.5 volumes of ethanol at -20°C overnight. After centrifugation the pellet was dissolved in bidest. water.

Polyacrylamide gel electrophoresis of the nucleic acids was either performed in 2.4 and 7.5% (w/v) "tandem" gels according to [10] with the modification of [20] or in 3.4% (w/v) uniform gels as described by Loening [19]. Gels in quartz tubes (100 x

6 mm) were scanned at 254 nm in an ISCO 1310/UA 5 gel scanner or, when prepared as horizontal slaps, stained with ethidium bromide.

Results

From blue illuminated cells of a chlorophyll-free *Chlorella* mutant more total nucleic acids ($OD_{260/280}$ = 2.1) can be buffer-extracted than from dark-kept controls. The additional amount increases gradually with the time of light exposure: Although the data obtained allow the assumption of an almost immediate effect, it takes 16-20 h before a final enhanced level is reached (Fig. 1). Investigations of the intensity dependence of this effect reveal an influence of very small amounts of blue light already but saturation at about 600 $\mu W\ cm^{-2}$ (Fig. 2). Red light proves to be ineffective.

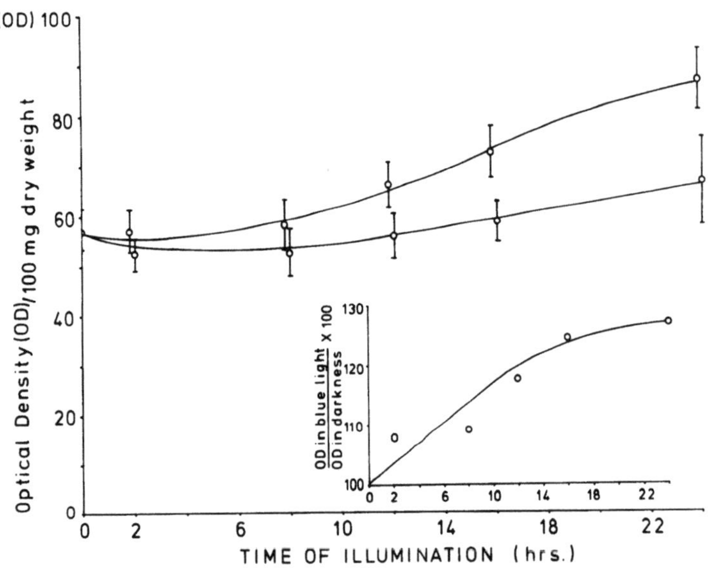

Fig. 1. Dependence on the duration of illumination of the enhancing effect of blue light (λ 457 nm, 600 $\mu W\ cm^{-2}$) on total nucleic acids (OD_{260}) of "resting" cells of *Chlorella*

In trying to find out which of the different nucleic acids of the cell responds to blue illumination we first separated them by electrophoresis on polyacrylamide gels. For better separation of most of the species expected these gels were prepared as "tandem gels" of 7.5% and 2.4%.

Figure 3 shows the resulting UV absorption profile from scanning at 254 nm after electrophoresis of 3 h. There are five distinct peaks, which from the literature [30, 31] represent DNA, 25/23S rRNA, 18/16S rRNA, 5S rRNA and tRNA. mRNA is not expected to be seen by this procedure because of its small portion. Comparable separation of the nucleic acids of blue illuminated cells reveals an identical peak dis-

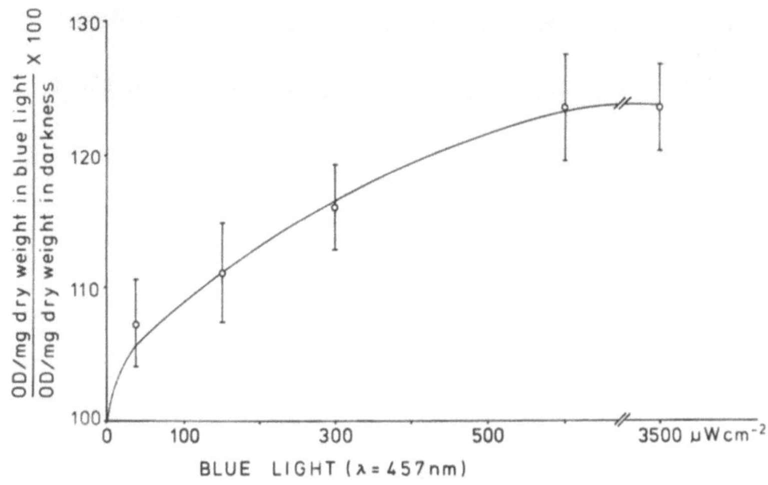

Fig. 2. Dependence on light intensity of the enhancing effect of blue illumination (λ 457 nm) on total nucleic acids (OD_{260}) of "resting" cells of *Chlorella*. Time of illumination was 16 h

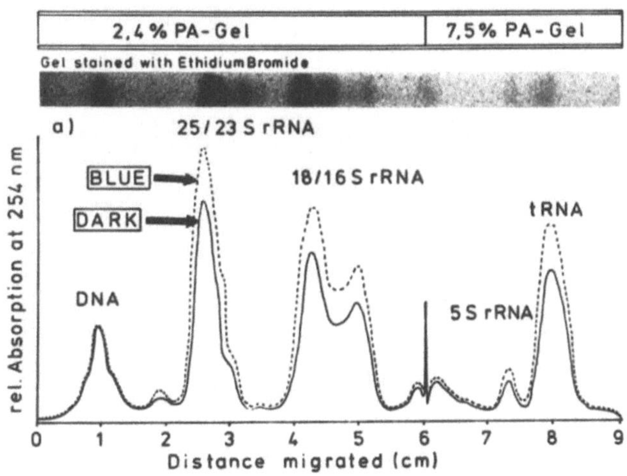

Fig. 3. Separation by polyacrylamide gel (2.4 and 7.5%) electrophoresis of total nucleic acids from dark-kept or blue-illuminated (λ 457 nm, 600 μW cm^{-2}, 16 h) *Chlorella* cells

tribution. In assuming that in the nongrowing cells used there has not been any DNA propagation the DNA-normalized curve exhibits all other peaks enhanced by an equal amount of 25-30% (Fig. 3). This must be taken to indicate an equal enhancement of all – cytoplasmic and plastidic – ribosomal and transfer RNA species of the cell on blue illumination.

To obtain the still missing information on the behavior of messenger RNA we prepared its poly(A^+) fraction. After separating DNA and parts of the low molecular RNA by CsCl centrifugation, the remaining RNA was shaken with oligo(dT)cellulose to separate its poly(A^+) fraction by binding to this matrix. Subsequent elution finally led to a fraction which, on testing in an in vitro translation experiment with a

Fig. 4. Polyacrylamide gel (3.4%) electrophoresis of isolated poly(A^+)mRNA from *Chlorella* cells. *Vertical lines* indicate the positions of the other RNA species

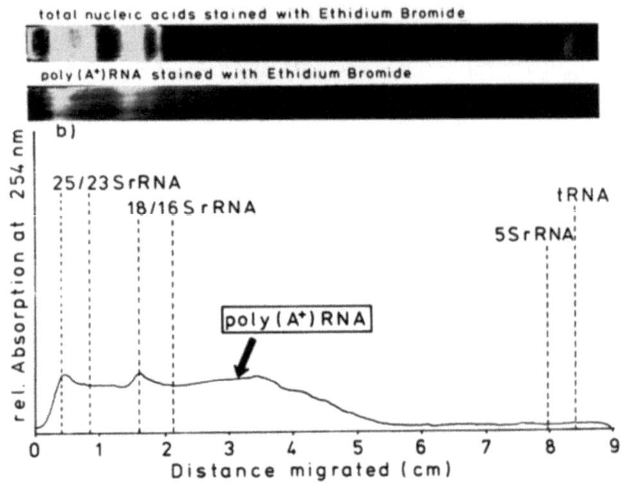

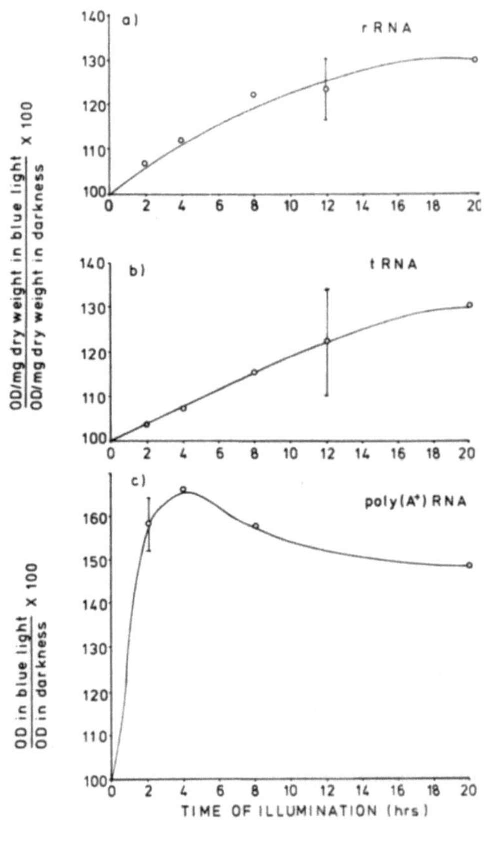

Fig. 5. Dependence on the duration of illumination of the enhancing effect of blue light (λ 457 nm, 600 μW cm^{-2}) on rRNA, tRNA, and poly(A^+)mRNA (OD_{260}) of "resting" cells of *Chlorella*

commercial wheat germ system, proved to be the mRNA wanted. Its additionally performed polyacrylamide gel electrophoresis (3.4%) results in a broad band in the 23 to 12S region, confirming the well-known difference in the size of mRNAs (Fig. 4). Also the amount of poly(A^+)mRNA thus obtained is found enhanced after blue light treatment of the algal cells.

Although thus all RNA species examined are found to be blue light increased, there is a slight difference in the extent of this effect: while rRNA and tRNA show a maximum enhancement of about 30%, poly(A^+)mRNA increases up to 60%. Besides this there is, however, a more profound difference in the time course of the response. While it takes about 16 h until rRNA and tRNA are maximally enhanced, about 2 h are sufficient to elevate the level of poly(A^+)mRNA to its final hight (Fig. 5). Our data do not allow an exact answer to the question of the effect of still shorter times of illumination, but it appears to be rather immediate.

Discussion

It should first be pointed out that in our experiments – to give only information on comparatively long-term effects of blue illumination on the nucleic acid system of the yellow *Chlorella* mutant – the observed increases indicate enhanced levels of the substance in question. This is different from the results of short-term labeling experiments of Steup [29, 30], Steup et al. [31] and Steup and Ssymank [32] done with wild-type and mutant cells of the same organism which mainly prove an enhancement of the turnover rates by blue light.

The coincidence between the intensity and the time dependencies of the blue-enhanced ribosomal and transfer RNA formation reported in this paper and protein production and increased carbohydrate breakdown described earlier [8] probably simply indicates the obvious interdependence of these processes. The much faster appearance of increased mRNA, however, must be interpreted differently. Being ahead for hours it seems to be closer to the primary effect of blue radiation. On investigating such an effect, several possibilities can be imagined. There could be the activation of a specific RNA polymerase generally described as occurring on illumination of corn by Apel and Bogorad [1], either direct or indirect by a change of the environment. As a first alternative, the model experiments with isolated enzymes of Schmid [25], (see also [22]), as a second, the effects of blue radiation on intracellular permeability barriers discussed elsewhere [33] must be mentioned. It could, however, also consist in releasing some specific block, previously being responsible for slowing down transcription. In that case no random increase in mRNA was to be expected, since in the mutant cells examined only some enzyme proteins appear to be additionally synthesized, others to be degraded and still others to be uninfluenced by blue light [14, 22]. To obtain an answer on the question if blue light somehow brings about an enhancement in specific mRNA species we have begun in vitro translation experiments with the isolated poly(A^+)mRNA from dark-kept and blue-irradiated *Chlorella* cells.

Acknowledgment. The authors are indebted to Miss Petra Gayk for typing the manuscript.

References

1. Apel K, Bogorad L (1976) Light-induced increase in the activity of maize plastid DNA-dependent RNA polymerase. Eur J Biochem 67:615–620
2. Avin H, Leder P (1972) Purification of biologically active globin messenger RNA by chromatography on oligothymidylic acid-cellulose. Proc Natl Acad Sci USA 69:1408–1412
3. Betina V, Zajakova J (1978) Inhibition of photo-induced Trichoderma viride conidiation by inhibitors of RNA synthesis. Folia Microbiol 23:460–464
4. Brandt A, Ingversen J (1978) Isolation and translation of horein messenger RNA from wild type and mutant endosperms in barley. Carlsberg Res Commun 43:451–470
5. Brunk CF, Leick V (1969) Rapid equilibrium isopycnic CsCl gradients. Biochim Biophys Acta 179:136–144
6. Cohen D, Schiff JA (1976) Events surrounding the early development of Euglena chloroplasts. 10. Photoregulation of the transcription of chloroplastic and cytoplasmic ribosomal RNAs. Arch Biochem Biophys 177:201–216
7. Dresbach C, Kowallik W (1974) Eine fördernde Wirkung von Blaulicht auf die Carotinoidbildung einer gelben Chlorella-Mutante. Planta 120:291–297
8. Georgi M (1974) Über blaulichtbedingte Veränderungen im Kohlenhydrat- und Proteingehalt einer chlorophyllfreien Mutante von Chlorella vulgaris. PhD thesis, Univ Köln, FRG, 77 pp
9. Gressel J (1980) Blue light and transcription. In: Senger H (ed) Blue light syndrome. Springer, Berlin Heidelberg New York, pp 133–153
10. Grossbach U, Weinstein IB (1968) Separation of ribonucleic acids by polyacrylamide gel-electrophoresis. Anal Biochem 22:311–320
11. Hundrieser J, Richter G (1982) Blue light-induced synthesis of ribulosebisphosphate carboxylase in cultured plant cells. Plant Cell Rep 1:115–118
12. Kowallik W (1962) Über die Wirkung des blauen und roten Spektralbereichs auf die Zusammensetzung und Zellteilung synchronisierter Chlorellen. Planta 58:337–365
13. Kowallik W (1965) Die Proteinproduktion von Chlorella im Licht verschiedener Wellenlängen. Planta 64:191–200
14. Kowallik W (1982) Blue light effects on respiration. Annu Rev Plant Physiol 33:51–72
15. Kowallik W, Gaffron H (1966) Respiration induced by blue light. Planta 69:92–95
16. Kowallik W, Neuert G (1984) Enhancement of GOGAT activity by blue light in Chlorella. This volume
17. Kowallik W, Ruyters G (1976) Über Aktivitätssteigerungen der Pyruvatkinase durch Blaulicht oder Glucose bei einer chlorophyllfreien Chlorella-Mutante. Planta 128:11–14
18. Kowallik W, Schätzle S (1980) Enhancement of carbohydrate degradation by blue light. In: Senger H (ed) Blue light syndrome. Springer, Berlin Heidelberg New York, pp 344–360
19. Loening UE (1967) The fractionation of high-molecularweight ribonucleic acid by polyacrylamide gel-electrophoresis. Biochem J 102:251–257
20. Mirault ME, Scherrer K (1971) Isolation of preribosomes from hela cells and their characterization by electrophoresis on uniform and exponential-gradient-polyacrylamide gels. Eur J Biochem 23:372–386
21. Mitzka-Schnabel U, Warm E, Rau W (1984) Light-induced changes in the protein pattern translated in vivo and in vitro accompanying carotenogenesis in Neurospora crassa and Fusarium aquaeductuum. This volume
22. Ruyters G (1984) Effects of blue light on enzymes. This volume
23. Ruyters G, Kowallik W (1980) Further studies of the light-mediated change in the activity of pyruvate kinase of a chlorophyll-free Chlorella mutant. Z Pflanzenphysiol 96:29–34
24. Schiff JA (1980) Blue light and the photocontrol of chloroplast development in Euglena. In: Senger H (ed) Blue light syndrome. Springer, Berlin Heidelberg New York, pp 495–511

25. Schmid GH (1980) Conformational changes caused by blue light. In: Senger H (ed) Blue light syndrome. Springer, Berlin Heidelberg New York, pp 198–204
26. Schrott EL, Rau W (1977) Evidence for a photoinduced synthesis of poly(A) containing mRNA in Fusarium aquaeductuum. Planta 136:45–48
27. Schwarze P, Frandsen NO (1960) Herstellung von Chlorella Farbmutanten mit Hilfe von radioaktiven Isotopen. Naturwissenschaften 47:47
28. Senger H, Bishop NI (1968) An action spectrum for nucleic acid formation in an achlorophyllous mutant of Chlorella pyrenoidosa. Biochim Biophys Acta 157:417–419
29. Steup M (1975) Die Wirkung von blauem und rotem Licht auf die Synthese ribosomaler RNA bei Chlorella. Arch Microbiol 105:143–151
30. Steup M (1977) Blue light-dependent regulation of cytoplasmic ribosomal RNA synthesis in Chlorella. Arch Microbiol 112:277–282
31. Steup M, Ssymank V, Winkler U, Glock H (1977) Photoregulation of transfer and 5S ribosomal RNA synthesis in Chlorella. Planta 137:139–144
32. Steup M, Ssymank V (1978) Photoregulation of ribosomal RNA synthesis in a mutant of Chlamydomonas reinhardii. Ber Dtsch Bot Ges 91:243–250
33. Widell S, Larsson C (1984) The role of membranes in blue light effects. This volume

Enzyme Regulation and Carbohydrate Metabolism

Effects of Blue Light on Enzymes

G. Ruyters [1]

Introduction

It is generally believed that enzymes are probable candidates for regulation within the sequence light perception – transduction – metabolic response. Therefore it is not surprising that in recent years several reviews have dealt with the effect of light on enzymes [2, 8, 23, 24, 25, 41, 42, 68, 73, 89]. However, strangely enough, the specific influence of the blue light (BL) photoreceptors has been neglected to some extent. In some cases this was certainly due to the fact that the reviews had been intentionally restricted to specific problems, e.g., to the action of phytochrome [68] or to chloroplast enzymes [8]. Only general reviews on BL responses contain short chapters on the action of BL on enzymes [31, 70, 71, 84, 85], and, of course, some information about this topic is included in the proceedings of the first Blue Light Conference [69]. It might be of some value, therefore, to present a review specifically on the effect of BL on enzymes. In this review, the present status of knowledge will be summarized not only to inform, but also to encourage further work in this interesting and important field.

General Aspects

The Range of Enzymes Affected by BL

Table 1 tries to summarize the present information of BL effects on enzymes reported for various organisms. It should be stressed that the organisms, not the isolated enzymes have been irradiated. The table may emphasize the variety of enzymes and organisms; altogether the number of enzymes affected is around 27, the complex enzyme systems of carotenoid biosynthesis not included. In this number of 27, BL effects mediated via phytochrome or chlorophyll should not be included, although the involvement of only the BL photoreceptor(s) has not been demonstrated in many cases. In fact, as far as I know, up to now a respective wavelength dependence has existed only for pyruvate kinase of the chlorophyll-free *Chlorella* mutant no. 20 [60], and two more wavelength dependences are presented in this volume for glut-

1 Faculty of Biology IV, University of Bielefeld, 4800 Bielefeld, FRG

Blue Light Effects in Biological Systems
ed. by H. Senger

Table 1. Enzymes reported to be under blue light control

Enzyme	Plant	Effect	Reference
ALA dehydratase	*Chlorella* M 20	+	[13]
	Scenedesmus C-2A'	+	[72]
ALA synthetase	*Scenedesmus* C-2A'	+	[72]
Aldolase	*Euglena gracilis*	+	[67]
c-AMP phosphodiesterase	*Phycomyces blakesleeanus*	+	[12]
Catalase	*Triticum aestivum*	+	[17]
DOVA-dehydrogenase, DOVA-transaminase	*Scenedesmus* C-2A'	+	[72]
Fumarase	*Euglena gracilis*	+	[22]
Glucose-6-phosphate dehydrogenase	*Acrochaetium daviesii,*	+	[81]
	Wolffia arrhiza	+	[16]
Glyceraldehyde phosphate dehydrogenase (NAD-dep.)	*Chlorella* M 20	–	[13]
	Chlorogonium elongatum	+	[74]
	Scenedesmus obliquus	+	[37]
Glyceraldehyde phosphate dehydrogenase (NADP-dep.)	*Chlamydomonas reinhardii*	+	[75]
	Chlorella M 20	+	[13]
	Chlorogonium elongatum	+	[74]
	Euglena gracilis	+	[67]
	Scenedesmus obliquus	–	[37]
	Vicia faba	+	[47]
	Zea mays	+	
Glycolate oxidase	*Dolichos lab lab*	–	[19]
	Pisum	+	[86]
	Triticum aestivum	+	[17]
	Phaseolus	+	[87]
	Zea mays	+	
Glyoxylate aminotransferase	*Phaseolus*	+	[87]
	Zea mays	+	
Hydroxypyruvate reductase	*Triticum aestivum*	+	[17]
Isocitratase	*Chlorogonium elongatum*	–	[74]
NAD kinase	*Neurospora crassa*	+	[1]
NADP protochlorophyllide oxidoreductase	*Nicotiana tabacum* (callus)	–	[52]
Nitrate reductase	*Agmenellum quadruplicatum*	–	[76]
	Chlamydomonas reinhardii	+	[7]
	Pisum arvense	+	[27]
	Zea mays	+	
	Sinapis alba	+	[88]
	Triticum	+	[77]
	Zea mays	+	[49]
Nitrite reductase	*Chlorella pyrenoidosa*	+	[78]
Phenylalanine ammonia-lyase	*Cucumis sativus*	+	[6]
	Zea mays	+	[15]

Table 1 (cont.)

Enzyme	Plant	Effect	Reference
Phosphoenolpyruvate carboxylase	*Chlamydomonas reinhardii*	+	[75]
	Chlorella ellipsoidea	+	[45]
	Chlorella M 20	+	[13, 55]
	Chlorella M 125	+	[28]
	Chorogonium elongatum	+	[74]
	Dolichos lab lab	+	[19]
	Scenedesmus obliquus	+	[37]
	Vicia faba	+	[47]
	Zea mays	+	
6-Phosphogluconate dehydrogenase	*Acrochaetium daviesii*	+	[81]
Phosphoglycerate kinase	*Euglena gracilis*	+	[67]
Pyruvate kinase	*Acetabularia mediterranea*	+	[66]
	Chlorella M 20	+	[34]
	Chlorella M 125	+	Ruyters (unpubl.)
	Scenedesmus C-2A'	+	[58]
Ribulose bisphosphate carboxylase	*Anabaena flos-aquae*	–	[11]
	Chlamydomonas reinhardii	+	[75]
	Chlorella M 20	+	[13]
	Chlorogonium elongatum	+	[74]
	Euglena gracilis	+	[67]
	Microcystis aeruginosa	–	[11]
	Nicotiana tabacum (callus)	+	[26]
	Vicia faba	+	[47]
	Zea mays	+	
Succinate dehydrogenase	*Euglena gracilis*	+	[22]
UDPG-pyrophosphorylase	*Acetabularia mediterranea*	+	[66, 83]

amate synthase (GOGAT) of the yellow *Chlorella* mutant (Kowallik and Neuert this Vol.) and for nitrate reductase of *Chlamydomonas*, respectively (Aparicio and Azuara this Vol.). In order to give a comprehensive view, however, the results cited here have not been rigorously selected according to this criterion.

The large number of enzymes reported to be influenced by BL can be divided mainly into two groups, photosynthetic enzymes including those of pigment synthesis and photorespiration, and carbohydrate degrading enzymes. Figures 1 and 2 may help to elucidate this; Fig. 1 shows the carbohydrate breakdown, Fig. 2 the photosynthetic carbon metabolism including photorespiration and chlorophyll synthesis in a somewhat simplified form, without regarding compartmentation. Blue light-controlled enzymes are given in frames.

Of the 27 enzymes listed 21 can be associated either with autotrophic or heterotrophic carbon metabolism, indicating a special role of these two processes in BL enhanced metabolism.

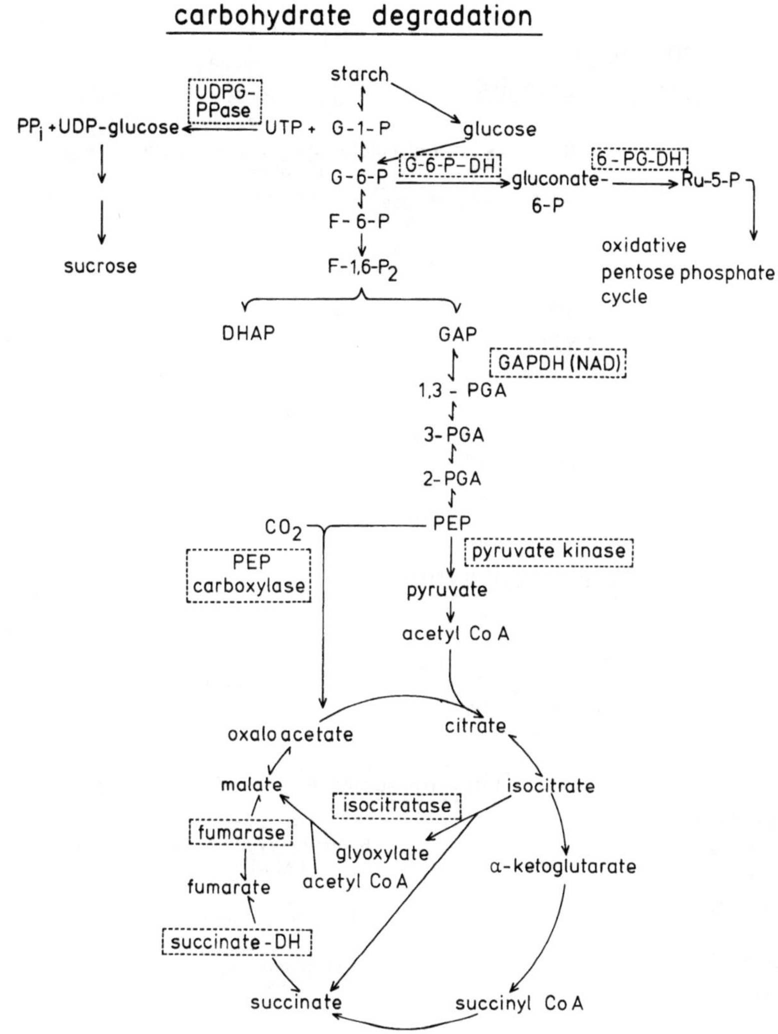

Fig. 1. Pathways of carbohydrate degradation in simplified form; blue light-affected enzymes in frames (for blue light-affected enzymes of the chlorophyll-free *Chlorella* mutant no. 20, see Fig. 4 and Table 6)

Control Mechanisms of Enzyme Activity

After demonstrating that quite a number of enzymes have been reported to be under BL control, the mechanisms involved should be discussed. At first the question has to be answered, how in general enzymes can be regulated, before we can find out in a second step which possibilities are realized in the case of BL.

Control of enzyme activity can be achieved mainly in two ways (Fig. 3):
1. In the so-called "coarse control" the amount of active enzyme, the enzyme capacity [48], is regulated mostly by changing the rate of synthesis or degradation. In some

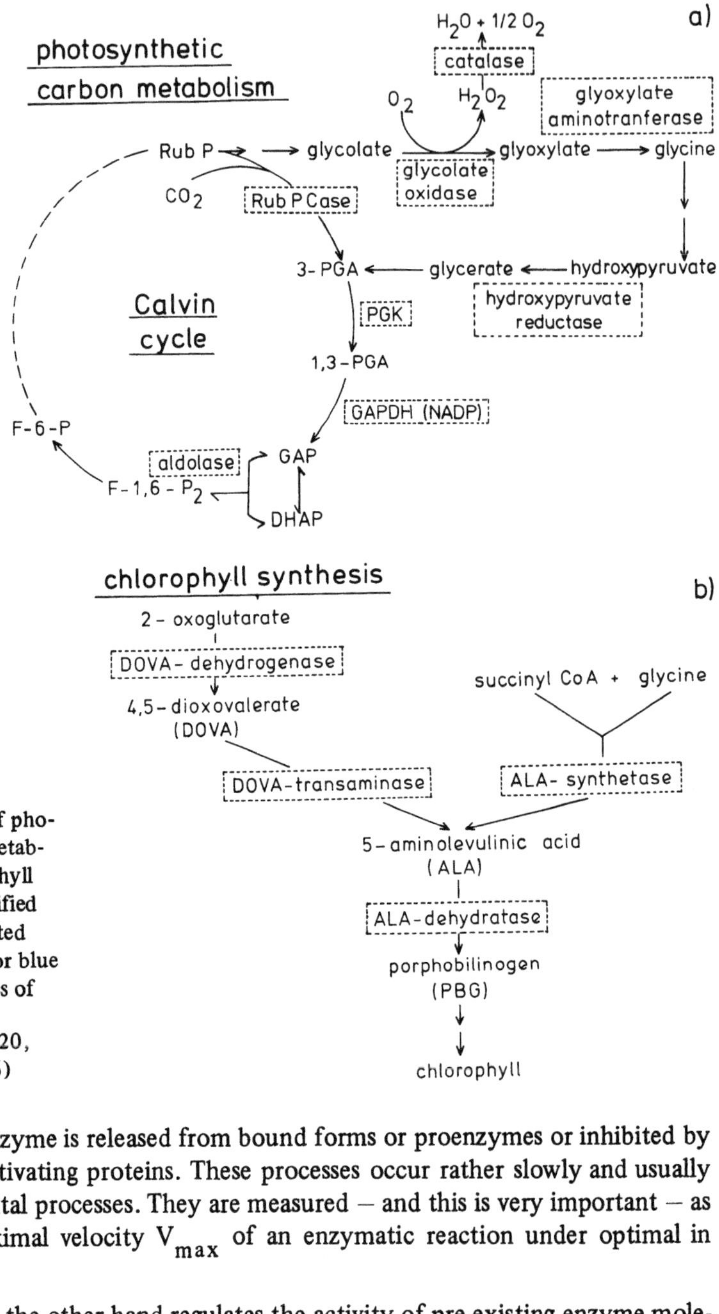

Fig. 2a,b. Pathways of photosynthetic carbon metabolism (a) and chlorophyll synthesis (b) in simplified form; blue light-affected enzymes in frames (for blue light-affected enzymes of the chlorophyll-free *Chlorella* mutant no. 20, see Fig. 4 and Table 6)

special cases the enzyme is released from bound forms or proenzymes or inhibited by production of inactivating proteins. These processes occur rather slowly and usually during developmental processes. They are measured – and this is very important – as change in the maximal velocity V_{max} of an enzymatic reaction under optimal in vitro conditions.

2. Fine control on the other hand regulates the activity of pre-existing enzyme molecules. This can be achieved by changing the concentration of effectors – including intermediates, substrates and ions, by chemical modifications (e.g., redox reactions, cf. Krauß et al. this Vol.; phosphorylation) or by physical interactions (e.g., dissociation – association). Since no protein synthesis is involved, fine control can be achieved rather quickly. The question now is, which mechanisms are used by BL to exert its effects on enzymes?

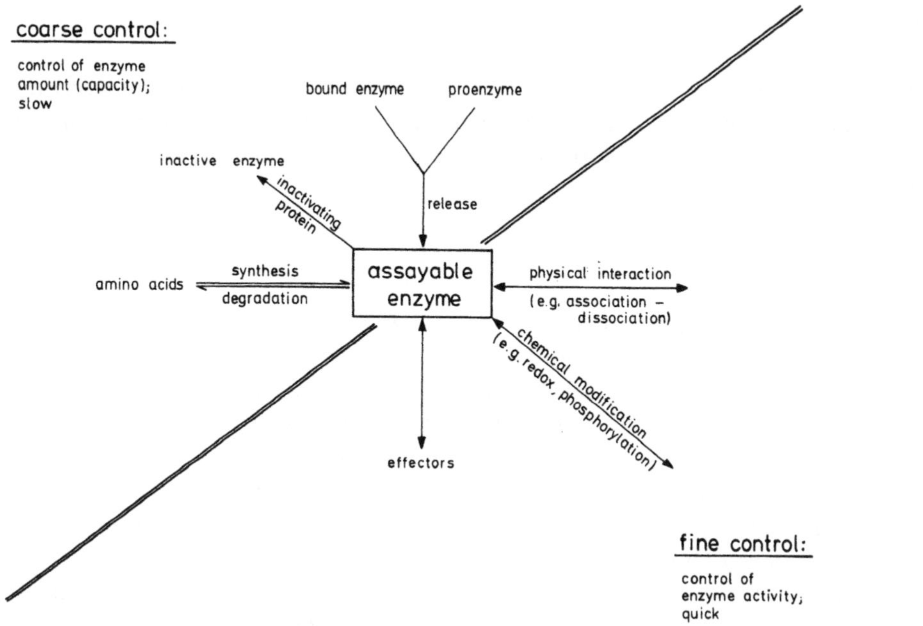

Fig. 3. Control mechanisms of enzyme activity

Control Mechanisms Realized in BL

Coarse Control by BL

The question which control mechanisms are realized in BL is not easy to answer. What has been done in most studies is to measure enzyme activity with an in vitro assay under, if at all, optimized conditions. The obtained maximal velocity V_{max} is a measure for the total amount of active enzyme, the enzyme capacity. In most cases BL-mediated changes in V_{max} are considered to be indicative of protein synthesis. In addition, some experiments with protein synthesis inhibitors have been reported (e.g., [28, 34]). Both results together strongly suggest the involvement of protein synthesis in the BL control of enzymes, but proof can only come from somewhat more sophisticated techniques like immunological procedures or comparative density labeling. In fact, very recently Hundrieser and Richter [26] (Richter this Vol.) were able to prove blue light-induced de novo synthesis of RubPCase in cultured tobacco cells. The increase in enzyme activity was reflected in the level of translatable messenger RNA specific for its small subunit, as shown by the isolation of the in vitro translation products by immunoprecipitation. Thus BL is able to influence enzyme synthesis, in other words, it can exert coarse control.

Fine Control by BL

After demonstrating the existence of coarse control, fine control of enzyme activity by BL, i.e., modification of the activity of already existing enzyme molecules, should be regarded next. Two possibilities can be distinguished: direct or indirect action of BL.

Direct Action of BL

In the case of direct action of BL, the blue radiation has to be absorbed by the enzyme molecule itself or its prosthetic group. Not the whole organisms were irradiated in these experiments, but the isolated and more or less purified enzymes. Since most enzymes and their coenzymes do not absorb visible light, however, this mechanism is restricted to very few cases. The best studied example is nitrate reductase; containing FAD, cytochrome b_{557} and molybdenum as cofactors, the inactivated enzyme from fungi, green algae and higher plants can be reactivated by BL, indicating that FAD is the primary photoreceptor (review: Ninnemann this Vol.).

Some other enzymes have been reported to become BL-sensitive upon the addition of FMN in vitro. Table 2 summarizes these results mainly obtained by Schmid and Codd. In most cases the enzyme activity is reduced by BL in the presence of FMN. Only the activity of glycine oxidase of the colorless *Chlorella* mutant no. 125 is enhanced by BL under such conditions, and the activity of a FAD containing glucose oxidase – suppressed by removing the FAD – can be restored with blue light in the presence of FMN. The physiological significance of all these results, however, is hardly understood.

Table 2. Blue light effects on enzymes in vitro upon addition of FMN (1)

Enzyme	Material	Effect	Reference
Amino acid oxidase	Hog kidney	–	[63]
Glucose oxidase	*Aspergillus*	+	[64]
Glycine oxidase	*Chlorella* M 125	+	[65]
Glycolate oxidase	*Nicotiana*	–	[62]
Lactate dehydrogenase	Yeast	–	[63]
Malate dehydrogenase	*Euglena* *Nicotiana* *Zea mays* *Pig heart*	–	[9]
Nitrate reductase	*Spinacia*	–	[82]
Ribulose bisphosphate carboxylase	*Anabaena* *Microcystis* *Scenedesmus* *Sphaeocystis*	–	[11]
Transketolase	*Nicotiana*	–	[10]
Uricase	Hog liver	–	[63]
Xanthine oxidase	Bovine milk	–	[63]

Table 3. Blue light effects on enzymes in vitro upon addition of FMN (2) (Kowallik and Stein pers commun)

Enzyme	Material	Effect
Glyceraldehyde phosphate dehydrogenase (NAD-dep.)	Rabbit muscle	–
Hexokinase	Yeast	–
Phosphofructokinase	Rabbit muscle	–
	Chlorella M 20	–
Pyruvate kinase	Rabbit muscle	–
	Chlorella M 20	–

Most recently Kowallik and Stein again investigated the influence of BL on enzymes in the presence of FMN in connection with the work on enzymes of the carbohydrate metabolism of *Chlorella* (Table 3). Hexokinase from yeast, phosphofructokinase (PFK), NAD-dependent glyceraldehyde 3-phosphate dehydrogenase and pyruvate kinase (PK) from rabbit muscle as well as PFK and PK from extracts of the yellow *Chlorella* mutant no. 20 all were inhibited under such conditions. Since the latter result is different from the results for these enzymes from irradiated *Chlorella* cells [3, 34], this finding raises new questions as to the physiological meaning of this effect.

Indirect Action of BL

Indirect action of BL means in this context that BL does not act on the enzyme molecule itself, but affects its activity, e.g., via changes in the level of effectors or substrates. As mentioned above, measurement of enzyme activity under optimal in vitro conditions can only give the amount of active enzyme, but does not provide insight into the in vivo activity. The latter is very difficult if not impossible to measure, since the complete physiological situation has to be elucidated, i.e., the in vivo concentration of substrates and products, of cofactors, effectors, ions, the enzyme protein itself and all other possible influencing parameters have to be considered. But in order to get at least some idea of fine control of enzyme activity, the identification of nonequilibrium reactions in metabolic pathways such as carbohydrate breakdown often provides valuable evidence for regulatory reactions. The term "nonequilibrium" is used for those reactions that are far from thermodynamic equilibrium as determined by estimation of substrates and products. Some more attributes are generally associated with regulatory enzymes [39]. So the enzyme capacity is usually rather low, its dependence on substrate concentration often deviates from Michaelis-Menten kinetics, as shown by nonlinear Lineweaver-Burk plots, indicating cooperativity. Also the enzyme activity is often allosterically regulated, and many reactions catalyzed by regulatory enzymes have high free energy changes and are practically irreversible. Often regulatory enzymes are located near branchpoints in metabolic pathways.

So for our problem it might be useful to try to find out regulatory enzymes with the help of these criteria and then study possible BL-mediated changes in substrates and effectors and their influence on these regulatory enzymes in vitro.

This was to some extent done by us in recent years for pyruvate kinase (PK) of the chlorophyll-free mutant no. 20 of *Chlorella vulgaris.* Besides phosphofructokinase, pyruvate kinase is considered an important regulatory enzyme of the glycolytic pathway in plants [80]. We were able to prove this for our yellow *Chlorella* mutant by comparing the mass action ratio of the PK reaction with its equilibrium constant. From the data of Kowallik and Scheil [35] for ADP and ATP and our own preliminary results for pyruvate and PEP, we calculate a mass action ratio of 0.10 in the dark and of 0.58 in blue light. These results allow the following conclusions: (1) The mass action ratio for PK of the yellow *Chlorella* mutant is in both conditions far from the equilibrium constant of 11,000 [43], i.e., this PK, catalyzing a nonequilibrium reaction, seems to be a regulatory enzyme. (2) The difference in the mass action ratios of dark and blue light, arising from the increase in pyruvate and ATP and the decrease in PEP and ADP, respectively, indicates that the PK reaction in the cells is really stimulated by blue light.

After proving that PK plays a regulatory role also in the yellow *Chlorella* mutant no. 20, kinetic and regulatory properties of this enzyme had to be investigated. At first our further studies revealed that *Chlorella* M 20 contains two isoenzymes of PK, possibly located in the cytosol and the plastid, respectively; their ratio is influenced by blue light, which seems to convert one isoenzyme into the other [57, 59]. The isoenzymes were partially purified in order to study the influence of several effectors. Table 4 shows the effect of AMP, ATP, P_i, citrate, F-6-P and FbP on the maximal velocity V_{max}. While AMP, P_i, F-6-P and FbP exert no marked influence, ATP and citrate inhibit V_{max} of both isoenzymes. Table 5 demonstrates the influence of the same effectors on $S_{0.5}$ (PEP), i.e., on substrate affinity. Again P_i, F-6-P and FbP have only minor influence. ATP and citrate increase the $S_{0.5}$ values, i.e., they decrease the affinity; on the contrary, AMP decreases the $S_{0.5}$ values, i.e., increases the affinity of both isoenzymes toward PEP. In total, these two tables indicate that both PK isoenzymes have major potential for regulation in vitro. But this may be of physiological significance as well. For the yellow *Chlorella* mutant an enhanced ATP and decreased ADP and AMP level after blue light irradiation was demonstrated [35]. These results indicate that indeed the in vivo activity of PK might be controlled by blue light-mediated changes in the concentration of substrates – ADP – and effectors – AMP, ATP.

Table 4. Effect of metabolites (final concentration 12 mM) on the maximal velocity (V_{max}) of pyruvate kinase isoenzymes of the yellow *Chlorella* mutant no. 20 (V_{max} of the control is set 1). (After [59])

Addition	PK_1	PK_2
None	1.00	1.00
AMP	1.00	1.09
ATP	0.40	0.71
P_i	0.96	1.19
Citrate	0.42	0.62
F-6-P	0.91	0.96
FbP	0.95	1.10

Table 5. Effect of metabolites (final concentration 12 mM) on PEP affinity ($S_{0.5}$) of pyruvate kinase isoenzymes of the yellow *Chlorella* mutant no. 20 ($S_{0.5}$ of the control is set 1). (After [59])

Addition	PK_1	PK_2
None	1.00	1.00
AMP	0.55	0.44
ATP	1.82	3.23
P_i	1.23	1.65
Citrate	1.57	5.47
F-6-P	1.12	1.04
FbP	1.18	1.43

Also PEPCase has to be mentioned when discussing blue light effects on enzyme activity. Some more indirect evidence for BL-enhanced in vivo activity of this enzyme came from $^{14}CO_2$ fixation studies by Miyachi and coworkers [28, 40, 45]. Blue light stimulates dark $^{14}CO_2$ fixation; immediately after the onset of the irradiation much of the label is found in aspartate, derived via oxaloacetate from carboxylation of PEP. This indicates that in fact PEPCase activity increases rapidly under blue light. So, BL really seems to be able to modulate enzyme activity by changing the supply of substrates or effectors.

Physiological Significance

Now, after summarizing the range of enzymes affected by blue light and discussing the regulatory mechanisms, an attempt should be made to elucidate the physiological meaning of the blue light-mediated enzyme control. As mentioned above, the blue light affected enzymes can be divided, with only very few exceptions, into two groups, photosynthetic enzymes and carbohydrate degrading enzymes.

Photosynthesis

It is well known that BL is necessary for the development and maintenance of a functioning photosynthetic apparatus [85]. Chloroplasts photosynthesizing under prolonged red light age progressively and can be restored by addition of blue light [53, 79] (Schmid this Vol.). Chloroplast development is very carefully regulated; at least three photoreceptors are involved: phytochrome, the BL photoreceptor(s), and the chlorophylls themselves [29]. It is not always easy to elucidate the respective contribution of these photoreceptors, but generally speaking phytochrome seems to be more important for higher plants, the blue light photoreceptor(s) for lower plants, especially for green algae. In this context, enzymes of photosynthesis, especially RubPCase and NADP-dependent GAPDH, have repeatedly been reported to be under BL control in a range of organisms (see Table 1). In addition, some enzymes of chlorophyll synthesis in *Scenedesmus* mutant C-2A' [72] and those of carotenoid synthesis in various organisms are also influenced by blue light [20, 50].

Interestingly also our yellow *Chlorella* mutant cells obviously possess some remaining BL responses, which are to be seen in connection with chloroplast development [13, 56]. So this organism does not only contain photosynthetic enzymes like RubPCase, NADP-dependent GAPDH and δ-ALAD, but their capacity can also be enhanced by blue light [13, 14].

It should be emphasized in this context that not only photosynthetic enzymes are formed during chloroplast development, but also enzymes of the carbohydrate breakdown or at least their plastidic isoenzyme. We were able to demonstrate this recently for PK in greening pea plants [61]. This leads to the second group of blue light-influenced enzymes, those of carbohydrate breakdown.

Carbohydrate Degradation

The most comprehensive work on blue light effects on enzymes of the carbohydrate breakdown and related pathways has been done by Kowallik and his coworkers. Using cells of the chlorophyll-free *Chlorella* mutant no. 20 they measured nearly all enzymes of carbohydrate degradation, as shown in Fig. 4. It is well known that BL stimulates carbohydrate degradation and respiration in higher plants and green algae [31]. In our opinion this stimulation seems to play a crucial role in blue light-regulated carbon metabolism, and it is, in addition to enhanced CO_2 fixation, one of the earliest measurable blue light responses. At first these enzyme studies were undertaken to get some idea about the mechanism of the BL-enhanced starch breakdown and respiration. But it soon turned out that the obtained BL-induced changes in enzyme capacity occurred much too slowly for rapid regulation of starch breakdown [56]. In the case of PK and PEPCase, for instance, more than 12 h of blue illumination are needed to yield maximal stimulation of their capacity [60, 55]. These long time courses, characteristic for coarse control, rather point to a developmental process and we have some indications that, at least in the case of PK, the changes in enzyme capacity have to be seen in connection with chloroplast development [56, 58].

Table 6 summarizes the results for enzymes of the yellow *Chlorella* mutant no. 20. Altogether the picture is rather complex. Enzymes are stimulated by blue light, e.g., pyruvate kinase and PEPCase, others, e.g., enolase, are inhibited or not affected like PFK. Is there any plausible explanation for these results? First it should be emphasized again that these results do not reflect the fast changes of in vivo enzyme activity, which certainly occur during the blue light-enhanced starch breakdown; they rather represent developmental changes. So it is possible that enzymes which increase in capacity are synthesized in greater amounts during a developmental change, while those which decrease undergo metabolic turnover to provide the amino acids for the synthesis of other enzymes. The latter seems reasonable, since there is no net protein synthesis in these resting *Chlorella* cells suspended in buffer [18].

This attempt at explanation separates the quick stimulation of carbohydrate degradation and the later changes in enzyme capacity to some extent by connecting these with developmental processes like chloroplast development; thus it completely

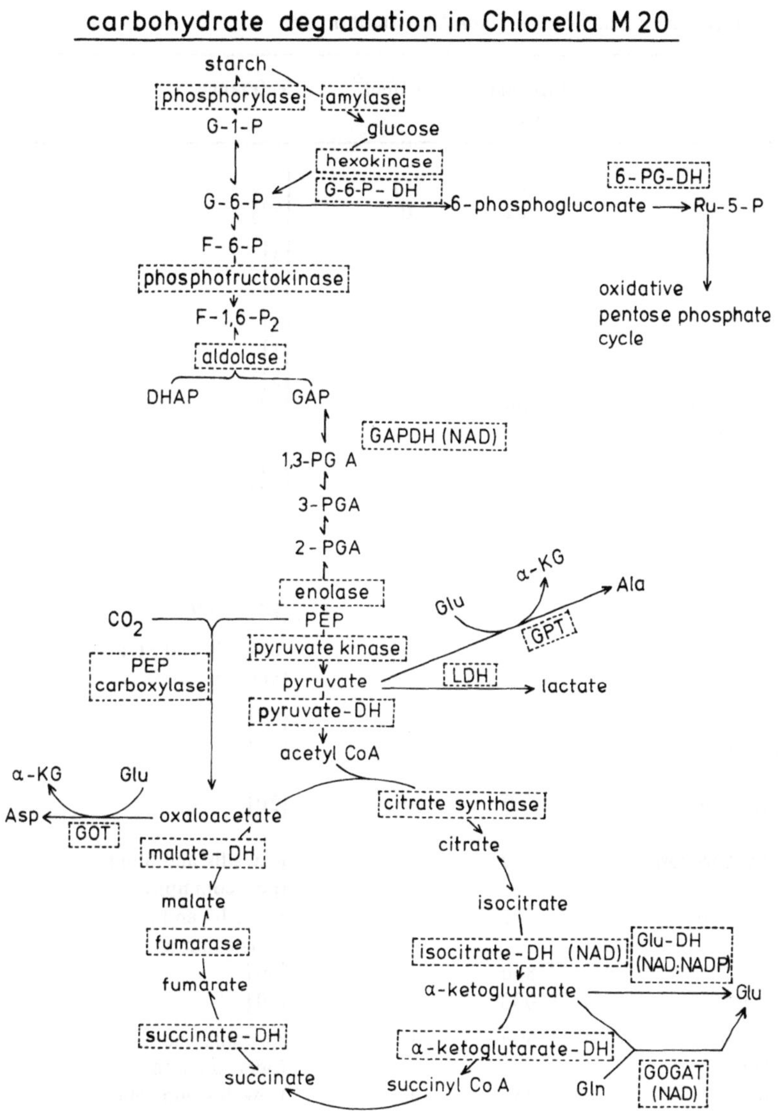

Fig. 4. Pathways of carbohydrate degradation and of related amino acid metabolism in simplified form; enzymes measured in the chlorophyll-free *Chlorella* mutant no. 20 in frames

separates fine control and coarse control. A second explanation concentrates solely on the enhancement of carbohydrate breakdown. Namely, it is also conceivable that the demonstrated changes in enzyme capacity of carbohydrate-degrading enzymes represent some kind of readjustment of the metabolism. The flux through a pathway like carbohydrate degradation could be quickly regulated by fine control, but then this could be followed by changes in enzyme capacities – coarse control – so that some optimum balance between flux and activity is re-established [5]. These might be possible explanations to the very complicated results obtained for BL-mediated changes in enzyme capacity of *Chlorella* M 20.

Table 6. Blue light effects on enzymes of the *Chlorella* mutant no. 20

Enzyme	Enzyme capacity[a]	Effect[b]	Reference
Amylase	70	(+)	[54]
Phosphorylase	15	0	[54]
G-6-P Dehydrogenase	70	0	[3]
6-PG Dehydrogenase	200	(–)	[3]
Hexokinase	140	(–)	[3]
Phosphofructokinase	80	0	[3]
Aldolase*	160	+	[3]
Glyceraldehyde phosphate dehydrogenase (NAD-dep.)	1400	(–)	[13]
Enolase	400	–	[3]
Pyruvate kinase	100	+	[34]
PEP carboxylase	20	+	[55]
Pyruvate dehydrogenase	10	?	Kowallik and Stein (pers commun)
Lactate dehydrogenase	30	(+)	Kowallik and Stein (pers commun)
Citrate synthase	300	(–)	[30]
Isocitrate dehydrogenase (NAD-dep.)	40	+	[30]
Isocitrate dehydrogenase (NADP-dep.)	80	(+)	[30]
2-oxoglutarate dehydrogenase	30	–	Kowallik and Stein (pers commun)
Succinate dehydrogenase	15	–	Kowallik and Stein (pers commun)
Fumarase	800	0	[30]
Malate dehydrogenase	12000	(–)	[30]
Glutamate oxaloacetate transaminase (GOT)	1500	(–)	Kowallik and Neuert (pers commun)
Glutamate pyruvate transaminase (GPT)	250	(–)	Kowallik and Neuert (pers commun)
Glutamate dehydrogenase (NAD-dep.)	150	0	Kowallik and Neuert (pers commun)
Glutamate dehydrogenase (NADP-dep.)	30	0	Kowallik and Neuert (pers commun)
Glutamate synthase (GOGAT)	5	+	[33]

[a]Enzyme capacity given in nmol/min mg protein.

[b]Effects smaller than 15% are indicated with brackets.

Asterisk at aldolase indicates that it could not be distinguished between glycolytic and photosynthetic enzyme activity.

Certainly much more information is necessary on fine control of carbohydrate-degrading enzymes. Only for PK do some data exist, presented above, and some additional results are given in our ^{31}P NMR study of *Scenedesmus* C-2A' (Oh-hama et al. this Vol.). However, for a complete picture all intermediates and effectors of the carbohydrate breakdown and a time course for their changes in darkness and blue light have to be measured. Much work remains to be done especially in elucidating the mechanism of the quick stimulation of carbohydrate degradation and respiration by blue light, but also in completely clarifying the physiological meaning of the changes in enzyme capacity presented in some detail for the yellow *Chlorella* mutant no. 20.

Conclusion

Figure 5 tries to summarize the discussed results. Enzymes are certainly influenced by blue light, although it has only been proved in a few cases that the obtained response is solely due to the action of the BL photoreceptor(s). In most studies the maximal velocity V_{max} of a reaction has been measured under optimal conditions, and thereby the amount of active enzyme, the enzyme capacity. Blue light-mediated changes in V_{max}, together with protein synthesis inhibitor studies, strongly indicate an effect of blue light on protein synthesis. This conclusion is supported by the demonstrated enhancement of mRNA for a few special cases.

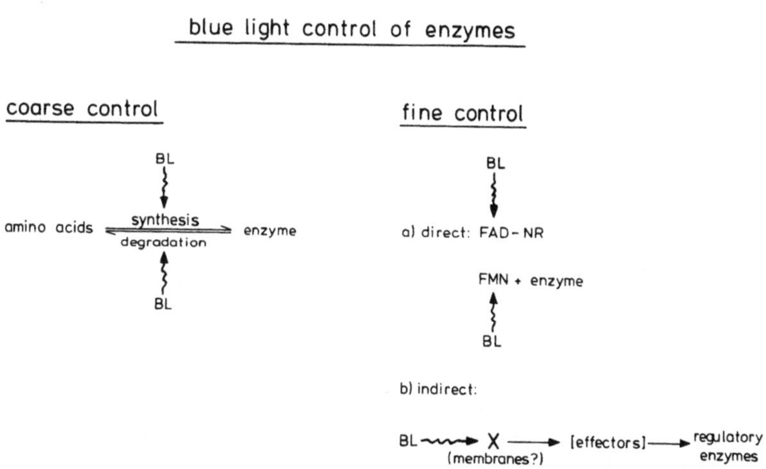

Fig. 5. Control mechanisms of enzyme activity exerted by blue light (NR = nitrate reductase)

These slow changes in enzyme capacity represent coarse control of metabolism and are characteristic for developmental processes; above all chloroplast development has to be considered.

But blue light can also influence the activity of already existing enzyme molecules and thus exert fine control. The enzyme molecule can either be influenced directly via its prosthetic group as in the case of nitrate reductase, or made sensitive to blue light by the addition of FMN. BL might also change the level of substrates, cofactors and effectors, thereby controlling the activity of regulatory enzymes, as shown for pyruvate kinase. However, not much information is available on this latter mechanism.

It is this quick regulation of pathways, however, which is certainly involved in the sequence light perception – transduction → metabolic response. Thus the main task for the future should be to obtain more information about quick changes in the level of intermediates, cofactors and effectors with special regard to the compartmentation of these compounds. Since the first steps of starch breakdown occur in the chloroplast, the following in the cytoplasm and in the mitochondria, membrane transport is certainly involved. A recent review tries to answer the question whether plant photoreceptors act at the membrane level [51]. Blue light might thus also bring about its effects via altered membrane permeability or altered membrane transport, e.g., via the phosphate translocator of the chloroplast envelope. Similar speculations have been expressed repeatedly [21, 32, 38]; no direct data exist, however. Thus, besides studies on the effect of BL on enzymes, those on membranes should be intensified in order to gain deeper insight into the primary mode of action of shortwavelength light.

Acknowledgments. The author wants to thank Prof. Dr. W. Kowallik for providing unpublished data and for valuable discussions. Thanks are also due to Drs. H.J. Newbury and J.A. Raven for sending manuscripts in press and to Mr. R. Church for reading the English. I am also indebted to the Deutsche Forschungsgemeinschaft for covering the costs to attend this meeting.

References

1. Afanasieva TP, Filippovich SY, Sokolovsky VY, Kritsky MS (1982) Developmental regulation of NAD^+ kinase in Neurospora crassa. Arch Microbiol 133:307–311
2. Anderson LE (1979) Interaction between photochemistry and activity of enzymes. In: Gibbs M, Latzko E (eds) Encyclopedia of plant physiology, New Ser, vol VI. Springer, Berlin Heidelberg New York, pp 271–281
3. Angele S (1981) Über die Wirkung von Blaulicht auf Kohlenhydrat abbauende Enzyme bei Chlorella. PhD thesis, Univ Bielefeld, FRG
4. Aparicio PJ, Azuara MP (1984) Metabolic and spectral characterization of nitrate-dependent nitrite release induced by blue light in Chlamydomonas reinhardii. This volume
5. Ap Rees T (1977) Conservation of carbohydrates by the nonphotosynthetic cells of higher plants. Symp Soc Exp Biol 31:7–32
6. Attridge TH, Smith H (1974) Density-labeling evidence for the blue-light-mediated activation of phenylalanine ammonium lyase in Cucumis sativus seedlings. Biochim Biophys Acta 343:452–464
7. Azuara MP, Aparicio PJ (1983) In vivo blue-light activation of Chlamydomonas reinhardii nitrate reductase. Plant Physiol 71:286–290
8. Buchanan BB (1980) Role of light in the regulation of chloroplast enzymes. Annu Rev Plant Physiol 31:341–374
9. Codd GA (1972) The photoinhibition of malate dehydrogenase. FEBS Lett 20:211–214
10. Codd GA (1972) The photoinactivation of tobacco transketolase in the presence of flavin mononucleotide. Z Naturforsch 27b:701–704

11. Codd GA, Stewart R (1980) The photoinactivation of microalgal ribulose bisphosphate carboxylase. In: Senger H (ed) The blue light syndrome. Springer, Berlin Heidelberg New York, pp 392–400
12. Cohen RJ, Atkinson MM (1978) Activation of Phycomyces adenosine 3',5'-monophosphate phosphodiesterase by blue light. Biochem Biophys Res Commun 83:616–621
13. Conradt W (1980) Über die Wirkung kurzwellig-sichtbarer Strahlung auf plastidäre und cytoplasmatische Enzyme einer chlorophyllfreien Chlorellamutante. PhD thesis, Univ Bielefeld, FRG
14. Conradt W, Ruyters G (1980) Blue light-effects on enzymes of the carbohydrate metabolism in Chlorella. 2. Glyceraldehyde 3-phosphate dehydrogenase (NADP-dependent). In: Senger H (ed) The blue light syndrome. Springer, Berlin Heidelberg New York, pp 368–371
15. Duke SO, Naylor AW (1976) Light control of anthocyanin biosynthesis in Zea seedlings. Physiol Plant 37:62–68
16. Eichhorn M, Augsten H (1977) Die Wirkung von Blau- und Rotlicht auf die Aktivität der Glucose-6-phosphat-Dehydrogenase und das Adenylatsystem bei Wolffia arrhiza unter steady state Bedingungen. Z Pflanzenphysiol 85:147–152
17. Feierabend J (1975) Developmental studies on microbodies in wheat leaves. III. On the photocontrol of microbody development. Planta 123:63–77
18. Georgi M (1974) Über blaulichtbedingte Veränderungen im Kohlenhydrat- und Proteingehalt einer chlorophyllfreien Mutante von Chlorella vulgaris. Ph D thesis, Univ Köln, FRG
19. Gnanam A, Habib Mohamed A, Seetha R (1980) Comparative studies on the effect of ammonia and blue light on the regulation of photosynthetic carbon metabolism in higher plants. In: Senger H (ed) The blue light syndrome. Springer, Berlin Heidelberg New York, pp 435–443
20. Harding RW, Shropshire W (1980) Photocontrol of carotenoid biosynthesis. Annu Rev Plant Physiol 31:217–238
21. Hase E (1980) Effects of blue light on greening in microalgae. In: Senger H (ed) The blue light syndrome. Springer, Berlin Heidelberg New York, pp 512–625
22. Horrum MA, Schwartzbach SD (1980) Nutritional regulation of organelle biogenesis in Euglena: Photo- and metabolite induction of mitochondria. Planta 149:376–383
23. Hug DH (1978) The activation of enzymes with light. Photochem Photobiol Rev 3:1–33
24. Hug DH (1981) Photoactivation of enzymes. Photochem Photobiol Rev 6:87–138
25. Hug DH, O'Donnell PS, Hunter JK (1980) Light activation of enzymes. Photochem Photobiol 32:841–848
26. Hundrieser J, Richter G (1982) Blue light-induced synthesis of ribulosebisphosphate carboxylase in cultured plant cells. Plant Cell Rep 1:115–118
27. Jones RW, Sheard RW (1977) Effect of blue and red light on nitrate reductase level in leaves of maize and pea seedlings. Plant Sci Lett 8:305–311
28. Kamiya A, Miyachi S (1975) Blue light-induced formation of phosphoenolpyruvate carboxylase in colorless Chlorella mutant cells. Plant Cell Physiol 16:729–736
29. Kasemir H (1979) Control of chloroplast formation by light. Cell Biol Int Rep 3:197–214
30. Kirsch M (1981) Nachweis von Enzymen des Tricarbonsäurecyclus bei einer chlorophyllfreien Chlorella-Mutante und ihre Beeinflussung durch Außenfaktoren. PhD thesis, Univ Bielefeld, FRG
31. Kowallik W (1982) Blue light effects on respiration. Annu Rev Plant Physiol 33:51–72
32. Kowallik W, Kirst R (1975) Über unterschiedliche Temperaturabhängigkeiten des Atmungsgaswechsels einer chlorophyllfreien Chlorella-Mutante im Dunkel und im Licht. Planta 124: 261–266
33. Kowallik W, Neuert G (1984) Enhancement of GOGAT activity by blue light in Chlorella. This volume
34. Kowallik W, Ruyters G (1976) Über Aktivitätssteigerungen der Pyruvat-Kinase durch Blaulicht oder Glucose bei einer chlorophyllfreien Chlorella-Mutante. Planta 128:11–14
35. Kowallik W, Scheil I (1976) Lichtbedingte Veränderungen des ATP-Spiegels einer chlorophyllfreien Chlorella-Mutante. Planta 131:105–108
36. Krauß W, Biber W, Stransky H, Hager A (1984) Blue light mediated reversible inhibition of ATP-dependent proton accumulation in microsomal vesicles from Zea mays. This volume

37. Kulandaivelu G, Sarojini G (1980) Blue light-induced enhancement in activity of certain enzymes in heterotrophically grown cultures of Scenedesmus obliquus. In: Senger H (ed) The blue light syndrome. Springer, Berlin Heidelberg New York, pp 372–380
38. Laudenbach B, Pirson A (1969) Über den Kohlenhydratumsatz in Chlorella unter dem Einfluß von blauem und rotem Licht. Arch Microbiol 67:226–242
39. Latzko E, Kelly GJ (1979) Enzymes of the reductive pentose phosphate cycle. In: Gibbs M, Latzko E (eds) Encyclopedia of plant physiology, New Ser, vol VI. Springer, Berlin Heidelberg New York, pp 239–250
40. Miyachi S, Miyachi S, Kamiya A (1978) Wavelength effects on photosynthetic carbon metabolism in Chlorella. Plant Cell Physiol 19:277–288
41. Montagnoli G (1977) Biological effects of light on proteins: Enzyme activity modulations. Photochem Photobiol 26:679–683
42. Newbury HJ (1983) Photocontrol of enzyme activation and inactivation. Philos Trans R Soc Ser B 303:433–441
43. Newsholme EA, Start C (1973) Regulation in metabolism. Wiley, New York
44. Ninnemann H (1984) The nitrate reductase system. This volume
45. Ogasawara N, Miyachi S (1970) Regulation of CO_2 fixation in Chlorella by light of varied wavelengths and intensities. Plant Cell Physiol 11:1–14
46. Oh-hama T, Ruyters G, Furihata K, Seto H, Miyachi S (1984) ^{31}P NMR studies in Scenedesmus C-2A' in darkness and blue light. This volume
47. Poyarkova NM, Drozdova IS, Voskresenskaja NP (1973) Effects of blue light on the activity of carboxylating enzymes and $NADP^+$-dependent glyceraldehyde 3-phosphate dehydrogenase in bean and maize plants. Photosynthetica 7:58–66
48. Queiroz O (1979) Rhythms of enzyme capacity and activity as adaptive mechanisms. In: Gibbs M, Latzko E (eds) Encyclopedia of plant physiology, New Ser, vol VI. Springer, Berlin Heidelberg New York, pp 126–139
49. Rao LVM, Datta N, Guha-Mukherjee S, Sopory SK (1982) The effect of blue light on the induction of nitrate reductase in etiolated excised maize leaves. Plant Sci Lett 28:39–47
50. Rau W, Schrott EL (1979) Light-mediated biosynthesis in plants. Photochem Photobiol 30:755–765
51. Raven JA (1983) Do plant photoreceptors act at the membrane level? Philos Trans R Soc Ser B 303:403–417
52. Richter G (1984) Blue light effects on the level of translation and transcription. This volume
53. Ried A (1969) Über die Wirkung blauen Lichts auf den photosynthetischen O_2-Austausch von Chlorella. Planta 87:333–346
54. Ruyters G (1977) Über Aktivitätsänderungen Kohlenhydrat-abbauender Enzyme von Chlorella im Dunkel und im Blaulicht. PhD thesis, Univ Köln, FRG
55. Ruyters G (1980) Blue light-enhanced phosphoenolpyruvate carboxylase activity in a chlorophyll-free Chlorella mutant. Z Pflanzenphysiol 100:107–112
56. Ruyters G (1981) Effects of blue light on respiration and enzyme activity in a yellow Chlorella mutant. In: Akoyunoglou G (ed) Proc 5th Int Congr Photosynth, vol V. Balaban Int Sci Serv, Philadelphia, pp 905–914
57. Ruyters G (1981) Isoenzymes of pyruvate kinase in a chlorophyll-free Chlorella mutant and their blue-light-mediated interconversion. Z Pflanzenphysiol 103:109–113
58. Ruyters G (1982) Effects of blue light on pyruvate kinase activity during chloroplast development of unicellular green algae. Photochem Photobiol 35:229–231
59. Ruyters G (1982) Regulatory properties of pyruvate kinase isoenzymes from a chlorophyll-free Chlorella mutant in darkness or blue-light. Z Pflanzenphysiol 108:207–214
60. Ruyters G, Kowallik W (1980) Further studies of the light-mediated change in the activity of pyruvate kinase of a chlorophyll-free Chlorella mutant. Z Pflanzenphysiol 96:29–34
61. Ruyters G, Miyachi S (1983) Changes in plastidic and cytoplasmic pyruvate kinase activity during chloroplast development of pea in blue and red light. Plant Cell Physiol 24:863–871
62. Schmid GH (1969) The effect of blue light on glycolate oxidase of tobacco. Hoppe-Seyler's Z Physiol Chem 350:1035–1046
63. Schmid GH (1970) The effect of blue light on some flavin enzymes. Hoppe-Seyler's Z Physiol Chem 351:575–578

64. Schmid GH (1971) Photoregulation of β-D-glucose oxidase by blue light. Phytochemistry 10:2041–2042
65. Schmid GH, Schwarze P (1969) Blue light enhanced respiration in a colorless Chlorella mutant. Hoppe-Seyler's Z Physiol Chem 350:1513–1520
66. Schmid R (1984) Blue light effects in specific organisms: Acetabularia. This volume
67. Schmidt G, Lyman H (1974) Photocontrol of chloroplast enzyme synthesis in mutant and wild-type Euglena gracilis. In: Avron M (ed) Proc 3rd Int Congr Photosynth. Elsevier, Amsterdam, pp 1755–1764
68. Schopfer P (1977) Phytochrome control of enzymes. Annu Rev Plant Physiol 28:223–252
69. Senger H (ed) (1980) The blue light syndrome. Springer, Berlin Heidelberg New York
70. Senger H (1982) The effect of blue light on plants and microorganisms. Photochem Photobiol 35:911–920
71. Senger H, Briggs WR (1981) The blue light receptor(s): Primary reactions and subsequent metabolic changes. Photochem Photobiol Rev 6:1–38
72. Senger H, Klein O, Dörnemann D, Porra RJ (1980) The action of blue light on 5-aminolevulinic acid formation. In: Senger H (ed) The blue light syndrome. Springer, Berlin Heidelberg New York, pp 541–551
73. Smith H, Billett EE, Giles AB (1977) The photocontrol of gene expression in higher plants. In: Smith H (ed) Regulation of enzyme synthesis and activity in higher plants. Academic, London New York, pp 93–127
74. Stabenau H (1972) Aktivitätsänderungen von Enzymen bei Chlorogonium elongatum unter dem Einfluß von rotem und blauem Licht. Z Pflanzenphysiol 67:105–112
75. Steup M (1972) Über Beziehungen zwischen Kohlenhydrat-, Protein- und Phosphatstoffwechsel bei Chlamydomonas reinhardii Dangeard im blauen und roten Spektralbereich. PhD thesis, Univ Göttingen, FRG
76. Stevens SE, van Baalen C (1974) Control of nitrate reductase in a blue-green alga. Arch Biochem Biophys 161:146–152
77. Stoy V (1955) Action of different light qualities on simultaneous photosynthesis and nitrate assimilation in wheat leaves. Physiol Plant 8:963–985
78. Strotmann H (1967) Blaulichteffekte auf die Nitritreduktion von Chlorella. Planta 73: 376–380
79. Terborgh J (1966) Potentiation of photosynthetic oxygen evolution in red light by small quantities of monochromatic blue light. Plant Physiol 41:1401–1410
80. Turner JF, Turner DH (1980) The regulation of glycolysis and pentose phosphate pathway. In: Davies DD (ed) Biochemistry of plants, vol II. Academic, London New York, pp 279–316
81. van der Velde HH, Guiking P, van der Wulp D (1975) Glucose-6-phosphate dehydrogenase and 6-phosphogluconate dehydrogenase in Acrochaetium daviesii cultured under red, white and blue light. Z Pflanzenphysiol 76:95–108
82. Vargas MA, Maurino SG, Maldonado JM, Aparicio PJ (1982) Photoinactivation of spinach nitrate reductase sensitized by flavin mononucleotide. Evidence for the involvement of singlet oxygen. Photochem Photobiol 36:223–228
83. Vettermann W (1973) Mechanism of the light-dependent accumulation of starch in chloroplasts of Acetabularia, and its regulation. Protoplasma 76:261–278
84. Voskresenskaja NP (1972) Blue light and carbon metabolism. Annu Rev Plant Physiol 23:219–234
85. Voskresenskaja NP (1979) Effect of light quality on carbon metabolism. In: Gibbs M, Latzko E (eds) Encyclopedia of plant physiology, New Ser, vol VI. Springer, Berlin Heidelberg New York, pp 174–180
86. Voskresenskaja NP, Grishina GS, Sechenska M, Drozdova IS (1970) After-effect of blue and red light on oxidation of glycolic acid by pea chloroplasts and homogenates. Sov Plant Physiol 17:859–865

87. Voskresenskaja NP, Khodzhiev AK (1973) After-effect of red and blue light on activity of glycolate oxidase and glyoxylate aminotransferases in plants. Sov Plant Physiol 20:254–259

88. Wild A, Holzapfel A (1980) The effect of blue and red light on the content of chlorophyll, cytochrome f, soluble reducing sugars, soluble proteins and the nitrate reductase activity during growth of the primary leaves of Sinapis alba. In: Senger H (ed) The blue light syndrome. Springer, Berlin Heidelberg New York, pp 444–451

89. Zucker M (1972) Light and enzymes. Annu Rev Plant Physiol 23:133–156

Influence of Blue Light on NADP-Dependent Glyceraldehyde 3-Phosphate Dehydrogenase Activity in *Chlorella*

W. KOWALLIK and N. GROTJOHANN[1]

Introduction

NADP-dependent glyceraldehyde 3-phosphate dehydrogenase, a key enzyme of photosynthesis, is subject to multifarious light regulations. While in experiments with higher plant seedlings its synthesis turned out to be phytochrome controlled [3, 5, 6, 17, 18], its activity in fully developed green organisms has been found to be regulated by photosynthesis. The low turnover rate measurable in extracts from dark-kept plants increases drastically on illumination. This light effect can be mimicked by application to the isolated enzyme of reducing agents such as dithiothreitol, from which it is supposed to derive in vivo from a reduction of inactive enzyme present in darkness or reductive enzyme interconversion [1, 19]. A special thioredoxin system reduceable by ferredoxin of the photosynthetic electron transport chain is considered responsible for this activation (literature: [4]).

Besides these, however, still another influence of visible radiation on this enzyme might exist. For several organisms its activity has been reported greater in crude extracts of samples previously exposed to blue instead of red light of equal fluence rates. Found in greening *Chlorogonium* [23] as well as after prolonged exposure of leaves or fully grown bean, corn and other plants [20, 24], some specific connection of this effect with photosynthesis is not definitely excluded, but by no means proved or even easily explainable at present.

In the past we have examined several enzymes on a response to blue illumination of a chlorophyllfree *Chlorella* mutant. Among those being enhanced was also the NADP-dependent glyceraldehyde 3-phosphate dehydrogenase. Its rather low turnover rate in crude extracts from dark-kept cells was found enhanced by maximally 60% after blue light treatment [7]. This and the above information in the literature made it appear rewarding to look for the behavior of that enzyme in wild-type cells autotrophically grown in blue or in red light of equal dry matter production.

1 Fakultät für Biologie, Universität Bielefeld, 4800 Bielefeld, FRG

Blue Light Effects in Biological Systems
ed. by H. Senger

Material and Methods

All experiments were performed with the unicellular green alga *Chlorella vulgaris* strain 211-11 h obtained from the Algae Culture Collection at the University Göttingen, FRG.

The organism was grown autotrophically at 30°C in inorganic nutrient solution [9] in a light-thermostate described by Kuhl and Lorenzen [15]. Blue light for growth was produced by passing the radiation of fluorescent tubes (Osram L 115 W/20 Sa cool white) through a blue Plexiglas filter, red light by using Osram L 40 W/32 warm white DeLuxe and a red Plexiglas filter. Equal quantum fluence rates of 22 x 10^{18} Q m^{-2} s^{-1} (QSM 11-2500; Techtum Instruments), each leading to equal dry matter production were applied. The cultures were kept in the respective light quality for several months with periodic dilution with fresh nutrition solution every 4 days.

For the irradiation experiments the cells were spun down, washed twice with dist. water, resuspended in inorganic nutrient solution to give a packed cell volume of 6.2 μl ml^{-1}, and treated in a set-up after Dresbach and Kowallik [8]. Xenon high pressure arcs (Osram 150 W/XBO) were used as light sources. At fluence rates greater than 4 x 10^{18} quanta m^{-2} s^{-1} their radiation again was passed through Plexiglas filters, at fluence rates smaller than 4 x 10^{18} quanta m^{-2} s^{-1} interference filters (λ 459 nm and 677 nm) with halfwidths of ~20 nm (Schott & Gen. Mainz) could be used.

The specifically light-treated cells were harvested, washed twice with dist. water on a centrifuge and resuspended in phosphate buffer (0.1 mol l^{-1}). One volume of this suspension was mixed with 1.6 vol of glass beads (0.5 mm Ø) and the mixture shaken in a Vibrogen cell mill (Bühler, Tübingen) for 2 x 5 min at 2°–4°C. The resulting homogenate, containing 95% broken cells [13], was centrifuged in a Sorval RC-S Superspeed Refrigerated Centrifuge for 30 min at 22,000 *g* at 4°C and the supernatant (= crude extract) used for enzyme activity determination.

The enzyme activity was measured according to Bergmeyer [2]: 3 ml cuvettes contained: 50 mmol TEAE pH 8.6; 2.6 mmol EDTA; 7.6 mmol $MgSO_4$; 0.44 mmol NADPH; 3.2 mmol ATP; 2.5 U PGK and 10 μl crude cell extract. The reaction was started by adding 18 mmol glycerate 3-phosphate. Soluble protein was determined according to Lowry et al. [16].

Results

When cells of the unicellular green alga *Chlorella vulgaris* are grown autotrophically in white light of 1300 lx, the activity of NADP-dependent glyceraldehyde 3-phosphate dehydrogenase measurable in their crude extracts amounts to about 1600 nmol mg^{-1} protein min^{-1}. This is in rather good accordance with comparable data from the literature [25]. Exposure of such cells to the blue or the red end of the visible spectrum (Fig. 1), each applied at quantum fluence rates resulting in equal dry matter productions (22 x 10^{18} quanta m^{-2} s^{-1}), leads, however, to smaller in vitro turnover rates of that photosynthetic enzyme. In the respective crude cell extracts

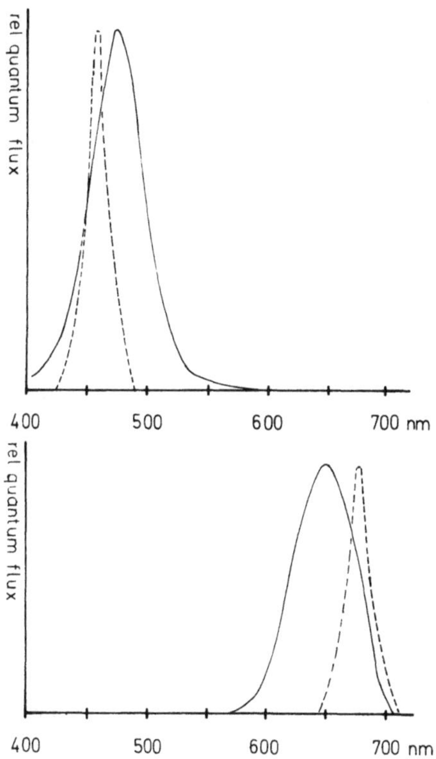

Fig. 1. Spectral distribution of the red and blue light fields used. *Solid line* when more, *broken line* when less than 4 x 10^{18} quanta m^{-2} s^{-1} were applied

only about 1200 nmol mg^{-1} protein min^{-1} of glycerate 3-phosphate are converted to glyceraldehyde 3-phosphate. Obviously both light qualities must work together to yield a maximum enzyme activity. Confirming this generally by applying the same amount of total quanta as a mixture of red and blue light, we found, however, that the extent of the rise of the turnover rate was pronouncedly dependent on the ratio of both light qualities. Combining 90% of red with only 10% of blue wavelengths resulted in the greatest in vitro activity measured. The then converted 1800 nmol mg^{-1} protein min^{-1} exceeded even those found to be metabolized in the extracts of white-grown cells (Fig. 2).

Since for pyruvate kinase and for glutamine-2-oxoglutarate aminotransferase of a chlorophyll-free mutant of our organism blue light-dependent enhancements of the activities have been considered to depend at least partly on a de novo synthesis of the enzymes [12, 13], we examined the influence of chloramphenicol and cycloheximide on the enhancing effect of the wavelength combination. As seen in Fig. 3C there is a slight decrease in the turnover rates, but still the rate for cells from red + blue irradiation is greater by about 45%. This is somewhat less than with unpoisoned cells, leaving the possibility that a small portion of the total enhancement depends on de novo synthesis of the enzyme, but indicating that most of it must derive from some other effect.

Thinking of the above-mentioned involvement of the photosynthetic electron transport, we added DCMU to the cell samples. Again the rise in in vitro activity of red + blue – compared with red-grown cells – was evident (Fig. 3B). Since at the

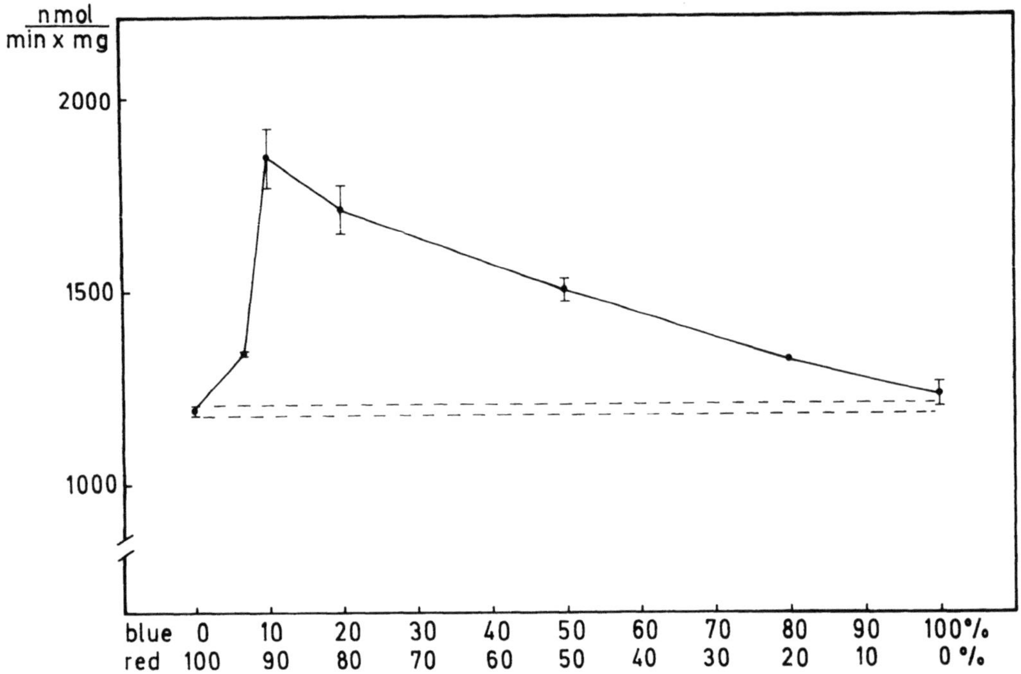

Fig. 2 ▲ ▼ Fig. 3

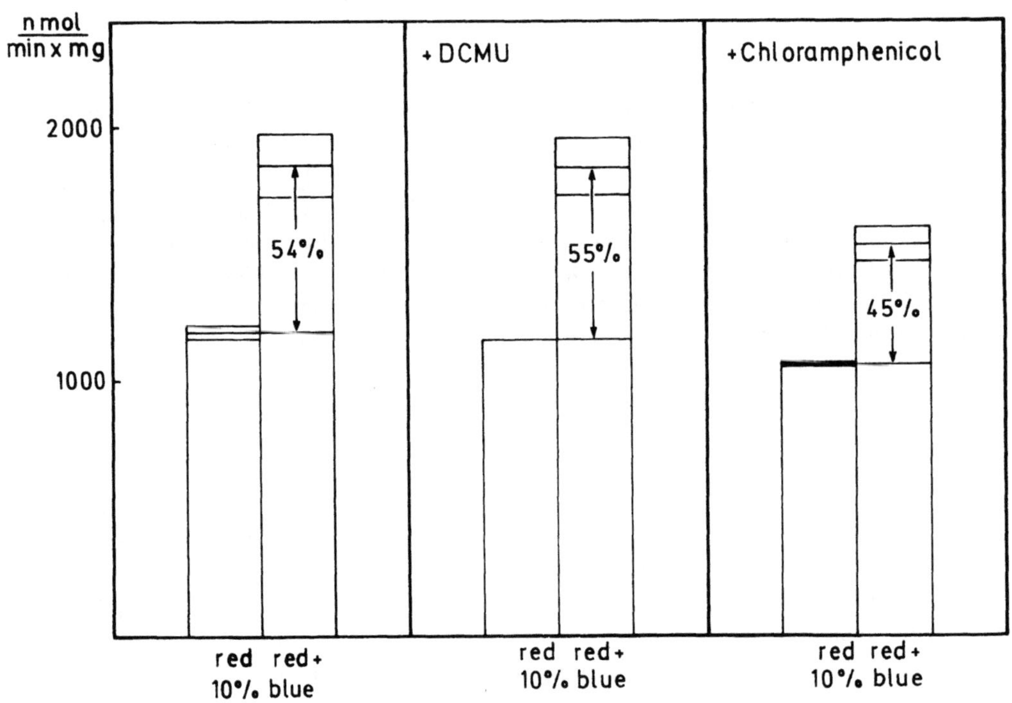

Fig. 2. Activity of NADP-dependent glyceraldehyde 30phosphate dehydrogenase in crude extracts of *Chlorella vulgaris* grown at 22 x 10^{18} quanta m^{-2} s^{-1} of blue + red light of different ratios

Fig. 3. Activity of NADP-dependent glyceraldehyde 3-phosphate dehydrogenase in crude cell extracts of *Chlorella vulgaris* grown in red or red + blue (90/10) light in absence or presence of DCMU (10^{-5} mol l^{-1}) or chloramphenicol (3 x 10^{-4} mol l^{-1})

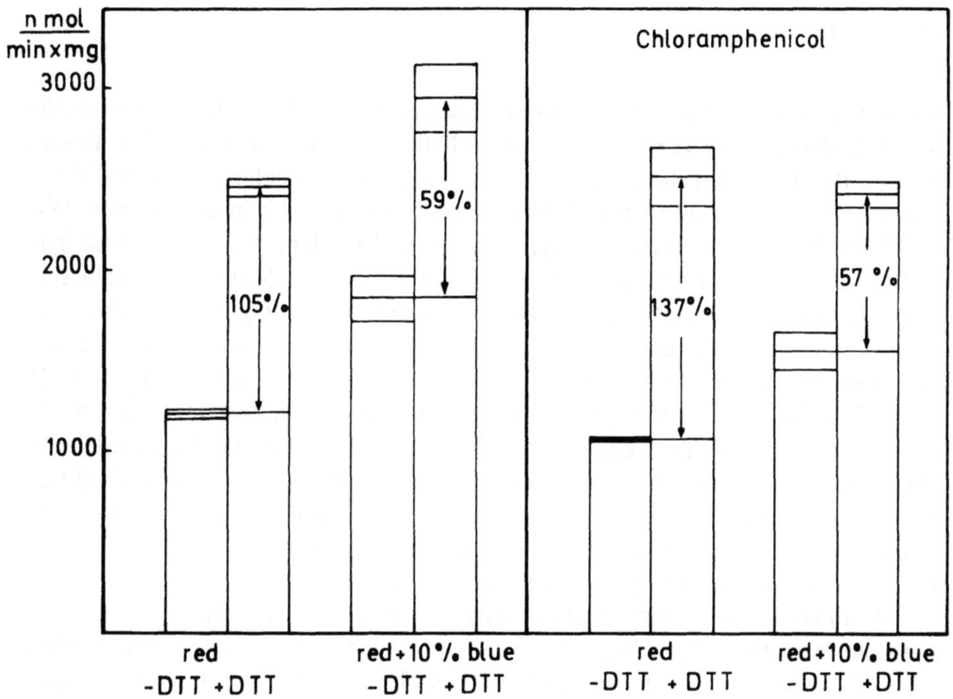

Fig. 4. Activity of NADP-dependent glyceraldehyde 3-phosphate dehydrogenase in crude cell extracts of *Chlorella vulgaris* grown in red or red + blue (90/10) light before and after treatment with 50 mmol dithiothreitol

concentration used O_2-production was found to be completely suppressed, a light effect via the ordinary known photosynthetic reactions – especially reduced ferredoxin – is not easily to be seen.

For optimal functioning of NADP-dependent glyceraldehyde 3-phosphate dehydrogenase an involvement of several reduced sulfhydryl groups is known [21]. We therefore had to take into account the possibility of not measuring the true capacity of the enzyme present in the crude cell extracts, since partial oxidations might have developed variably during their preparation. To obtain information on this we tested the influence of the artificial reducing agent dithiothreitol. Addition of 50 mmol to crude cell extracts kept at 0°C for 60 min resulted in the ensuing test in a pronounced increase in the measurable activity. Exhibiting only small variations in equally treated samples, the enhancement was profoundly different for extracts of red + blue- or of red-irradiated cells. While in the latter there was an increase in activity of about 105%, it was only about 60% in the former (Fig. 4). But still the turnover rate in the sample of red-exposed cells remained by about 20% lower (2950/2450). This latter difference finally was absent on DTT treatment of extracts of chloramphenicol poisoned cells. In that case an average activity of the enzyme of ~ 2450 nmol mg^{-1} protein min^{-1} was determined likewise in extracts from red- and from red + blue-illuminated cells (Fig. 4).

Discussion

There is an important difference between the influence of blue light on the activity of NADP-dependent glyceraldehyde 3-phosphate dehydrogenase reported in this paper and that found for the activation of other enzymes of a chlorophyll-free *Chlorella* mutant [10] or for an increased formation of chlorophyll b in our organism (Kowallik and Schürmann this Vol. [14]). While in all the latter cases traces of blue light sufficed to provoke the respective effect – unchanged achievable at higher light intensities – in this case a specific ratio of blue to red radiation only yields a maximum response. For those effects following a saturation curve the activation of a specific, blue-absorbing pigment is usually considered, the nature of which being largely undecided. The significance of a given ratio of two wavelengths, however, might be explained better by the involvement of a pigment system acting like phytochrome. Including two interconvertable forms, both – but differently – blue and red light-absorbing, such a system at given ratios of blue to red radiation would produce specific steady states with supposedly only one being most potent in bringing forth a specific reaction. We shall have to try to confirm or disprove this idea.

Whatever blue-absorbing pigment might be involved, its primary action is still quite unclear. The attempt can therefore only be made to point out ideas possibly worth following up. From the studies with inhibitors of protein biosynthesis some de novo synthesis of the enzyme protein might be assumed, which from other discussions (see Kowallik and Bartling this Vol. [11]) might derive from a somehow specifically enhanced transcription. But most of the difference in enzyme activity measurable in the extracts from red- or from red + blue-grown algae does not depend on different amounts of enzyme, since almost the same – enhanced – turnover rates can be obtained in both samples on treatment with the artificial reducing agent dithiothreitol. The explanation at hand is that the enyzme – activated by reduction – is present in a more reduced state in extracts from blue + red-grown cells than in that from red-grown algae. If so, the question is in which way blue light might cause such a situation.

Since it is well known that NADP-dependent glyceraldehyde 3-phosphate dehydrogenase underlies an activation by reduction by the thioredoxin system [4], the idea of a special blue light effect on that system suggests itself. However, greater turnover rates in extracts of blue illuminated cells were also present when their photosynthetic electron transport had previously been inhibited by DCMU, thus excluding thioredoxin reduction required by ferredoxin of the photosynthetic electron transport chain. Whether any other electron source might become acceptable by blue light action cannot be judged at present.

Quite another speculation is that blue light somehow might cause the formation of an enzyme species, more – or at least more permanently – reduced. Blue-dependent changes in enzymes present have been described as occurring in pyruvate kinase, the substrate affinity of which has been found to be increased after the proper illumination [22]. Also the sometimes reported conversion of an NAD- into an NADP-dependent enzyme species, which is assumed to be connected with reductive processes [19], should be mentioned.

The effect observed, however, does not necessarily demand a more permanently reduced enzyme species. It could also depend on a greater reduction potential of the crude cell extract, which would keep the enzyme in a more reduced state during isolation and subsequent testing. Again no source for such a potential can easily be seen at the moment.

Doubtless all these ideas are highly speculative, but still their study might help clarify the peculiar action of blue light on living plant cells.

Summary

In crude extracts of autotrophically grown *Chlorella vulgaris* the activity of NADP-dependent glyceraldehyde 3-phosphate dehydrogenase is 1600 nmol mg^{-1} protein min^{-1}, when white, but only 1200 nmol mg^{-1} protein min^{-1} when red or blue light is used for equal dry matter production. Simultaneous application of both these light qualities results in increased activities, the extent of which differs markedly with the ratio of both wavelength regions. At 90% red plus 10% blue light a maximum turnover rate of about 1800 nmol mg^{-1} protein min^{-1} can be measured in vitro. Poisoning with DCMU does not prevent this effect, making it appear independent of the photosynthetic electron transport. Application of chloramphenicol or of cycloheximide does lead to a small recuction of the enhancement, but still allows 80% of it to appear, thus indicating only minor involvement of de novo enzyme synthesis. The difference in the in vitro activity of red- or red + blue-grown cells shrinks or – in the case of chloramphenicol poisoning – even disappears by treating the crude extracts with dithiothreitol. From this it is concluded that at a proper ratio of red to blue light either a more permanently reduced NADP-glyceraldehyde 3-phosphate dehydrogenase species or an environment with greater reducing capacity are produced.

Acknowledgments. We thank Miss Petra Gayk for typing this manuscript.

References

1. Anderson LE, Lim T-C (1972) Chloroplast glyceraldehyde 3-phosphate dehydrogenase: light-dependent change in the enzyme. FEBS Lett 27:188–191
2. Bergmeyer HV (1970) Methoden der enzymatischen Analyse. Verlag Chemie, Weinheim
3. Brüning K, Drumm H, Mohr H (1975) On the role of phytochrome in controlling enzyme levels in plastids. Biochem Physiol Pflanz 168:141–156
4. Buchanan BB (1980) Role of light in the regulation of chloroplast enzymes. Annu Rev Plant Physiol 31:341–374
5. Cerff R (1973) Glyceraldehyde 3-phosphate dehydrogenase and glyoxylate reductase. I. Their regulation under continuous red and far red light in the cotyledons of Sinapis alba. Plant Physiol 51:76–81
6. Cerff R (1974) Inhibitor dependent, reciprocal changes in the activities of glyceraldehyde 3-phosphate dehydrogenase in Sinapis alba cotyledons. Z Pflanzenphysiol 73:109–118

7. Conradt W, Ruyters G (1980) Blue light-effects on enzymes of the carbohydrate metabolism in Chlorella. 2: Glyceraldehyde 3-phosphate dehydrogenase (NADP-dependent). In: Senger H (ed) Blue light syndrome. Springer, Berlin Heidelberg New York, pp 368–371
8. Dresbach Ch, Kowallik W (1974) Eine fördernde Wirkung von Blaulicht auf die Carotinoidbildung einer gelben Chlorella-Mutante. Planta 120:291–297
9. Kowallik W (1962) Über die Wirkung des blauen und roten Spektralbereichs auf die Zusammensetzung und Zellteilung synchronisierter Chlorellen. Planta 58:337–365
10. Kowallik W (1982) Blue light effects on respiration. Annu Rev Plant Physiol 33:51–72
11. Kowallik W, Bartling D (1984) Increase in RNA by blue light in a chlorophyll-free Chlorella mutant. This volume
12. Kowallik W, Neuert G (1984) Enhancement of GOGAT activity by blue light in Chlorella. This volume
13. Kowallik W, Ruyters G (1976) Über Aktivitätssteigerungen der Pyruvat-Kinase durch Blaulicht oder Glucose bei einer chlorophyllfreien Chlorella-Mutante. Planta 128:11–14
14. Kowallik W, Schürmann R (1984) Chlorophyll a/chlorophyll b ratios of Chlorella vulgaris in blue or red light. This volume
15. Kuhl A, Lorenzen H (1964) Handling and culturing of Chlorella. In: Prescott DM (ed) Methods in cell physiology, vol I. Academic, London New York, pp 159–187
16. Lowry OH, Rosebrough NJ, Farr AL, Randall RJ (1951) Protein measurement with the folin phenol reagent. J Biol Chem 193:265–275
17. Marcus A (1959) Photocontrol of formation of red kidney bean leaf triphosphopyridine nucleotide linked triosephosphate dehydrogenase. Plant Physiol 34:126–129
18. Margulies M (1965) Relationship between red light mediated glyceraldehyde 3-phosphate dehydrogenase formation and light dependent development of photosynthesis. Plant Physiol 40:57–61
19. O'Brien MJ, Easterby JS, Powls R (1977) Glyceraldehyde 3-phosphate dehydrogenase of Scenedesmus obliquus. Effects of dithiothreitol and nucleotide on enzyme specificity. Biochim Biophys Acta 481:348–358
20. Poyarkova NM, Drozdova IS, Voskresenskaya NP (1973) Effect of blue light on the activity of carboxylating enzymes and $NADP^+$ dependent glyceraldehyde 3-phosphate dehydrogenase in bean and maize plants. Photosynthetica 7:58–66
21. Rossmann MG, Liljas A, Brändén C-J, Banaszak LJ (1975) Evolutionary and structural relationships among dehydrogenases. In: Boyer PD (ed) The enzymes, vol XI/A. Academic, London New York, pp 61–102
22. Ruyters G (1981) Isoenzymes of pyruvate kinase in a chlorophyll-free Chlorella mutant and their blue-light-mediated interconversion. Z Pflanzenphysiol 103:109–113
23. Stabenau H (1972) Aktivitätsänderungen von Enzymen bei Chlorogonium elongatum unter dem Einfluß von rotem und blauem Licht. Z Pflanzenphysiol 67:105–112
24. Ziegler H, Ziegler I (1965) Der Einfluß der Belichtung auf die $NADP^+$ abhängige Glycerinaldehyd-3-Phosphat Dehydrogenase. Planta 65:369–380
25. Ziegler H, Ziegler I, Schmidt-Clausen HJ, Müller B, Dörr I (1969) Activation of $NADP^+$-dependent glyceraldehyde-3-phosphate dehydrogenase in relation to photosynthetic electron transport. Prog Photosynth Res III:1636–1645

Enhancement by Blue Light of GOGAT Activity in *Chlorella*

W. KOWALLIK and G. NEUERT[1]

Introduction

When autotrophically grown in light of different spectral regions wild-type cells of *Chlorella* exhibit pronounced differences in their basic composition, as they contain more protein at the exposure to blue, more carbohydrates, however, on application of red wavelengths [4]. This effect of different light quanta is not derived from an influence on photosynthesis proper. It is equally present in DCMU-poisoned [5] or in suitably mutated [2] cells, and thus appears located in the further pathway of the primary photosynthetic products. Because of nondisturbing chlorophyll absorption, its analysis has largely been tried with a chlorophyll-free carotenoid-containing mutant of *Chlorella vulgaris.*

In this organism a blue-dependent increase in the breakdown of reserve carbohydrates has been reported [6], which in part might depend on the enhanced capacities of carbohydrate degrading enzymes like pyruvate kinase [8].

Whether such a regulation of the carbon skeleton-producing pathway is accompanied by corresponding changes in amino acid-forming enzymes has been begun to be tested in this publication. Two glutamate producing enzymes – glutamate dehydrogenase (GlDH) and glutamine-2-oxoglutarate-aminotransferase (GOGAT) – and two glutamate-consuming enzymes – glutamate-oxaloacetate-transaminase (GOT) and glutamate-pyruvate-transaminase (GPT) – have been chosen.

Material and Methods

For the experiments a chlorophyll-free mutant of *Chlorella vulgaris,* No. 211-11 h/20 of the Algae Collection, Institute of Plant Physiology, University Göttingen, was used. The organism was grown in a mineral nutrition solution plus 1% glucose at 30°C in the dark [7].

For irradiation cell samples were washed twice with distilled water on a centrifuge and resuspended in 0.1 M phosphate buffer pH 7.5 to yield a final density of 80-100 μl packed cell volume ml^{-1}. Blue light was obtained either by passing the radiation of four high pressure mercury lamps (Philips HPI, 250 W) through a Plexiglas

1 Fakultät für Biologie, Universität Bielefeld, 4800 Bielefeld, FRG

Blue Light Effects in Biological Systems
ed. by H. Senger

filter cutting off all wavelengths longer than 550 nm, or by using a Xenon high pressure arc (XBO, 150 W) and interference filters with halfband widths of around 10 nm. Further details of the set-up used have been described earlier [8, 16].

Enzyme activities were measured in crude cell extracts which were prepared by shaking cell samples, mixed with 1.6 volumes of glass beads (Ø 0.3 mm) at 0°-3°C for 5 min in a cell mill (Bühler, Tübingen) and spinning the resulting homogenate at 20,000 *g* for 10 min at 4°C.

The activities were determined spectrophotometrically by following the oxidation of NAD(P)H at 334 nm. The final volumes of 1.5-2.735 ml of the reaction mixtures contained:

– glutamate dehydrogenase:
a) NADH-dependent: 50 mM HEPES buffer pH 8.0; 500 mM NH_4Cl: 1mM $CaCl_2$; 5 mM 2-oxoglutarate; 0.4 mM NADH; 0.025 ml extract; Vol.: 2.725 ml
b) NADPH-dependent: 50 mM HEPES buffer pH 7.6; 500 mM NH_4Cl; 2 mM $CaCl_2$; 5 mM 2-oxoglutarate; 0.4 mM NADPH; 0.1 ml extract; Vol.: 1.5 ml

– glutamine-2-oxoglutarate-aminotransferase:
100 mM phosphate buffer pH 6.9; 50 mM 1-glutamine;
1 mM 2-oxoglutarate; 0.4 mM NADH; 0.1 ml extract; Vol.: 2.3 ml

– glutamate-oxaloacetate-transaminase:
100 mM aspartate; 50 mM glycylglycine buffer pH 8.0; 20 mM 2-oxoglutarate; 0.2 mM NADH; 0.05 ml malate dehydrogenase (1.375 U/ml); 0.01 ml extract; Vol.: 2.615 ml

– glutamate-pyruvate-transaminase:
250 mM alanine; 100 mM phosphate buffer pH 7.5; 20 mM 2-oxoglutarate; 0.2 mM NADH; 0.01 ml lactate dehydrogenase (0.7 U/ml); 0.025 ml extract; Vol.: 2.735 ml

Soluble protein was determined according to Lowry et al. [9]. To secure additionally the data on glutamine-2-oxoglutarate-aminotransferase activity, the glutamate produced was proved by thin layer chromatography. After stopping the enzyme reaction with ethanol 20 μl aliquots were applied to cellulose thin layers (250 μm), chromatographed in n-butanol-acetic acid-water (80:20:20, vol:vol:vol:), treated with ninhydrine and developed for 10 min at 110°C.

Results

In crude extracts of a chlorophyll-free mutant of *Chlorella vulgaris* activities of the glutamate-producing enzymes GlDH and GOGAT and of glutamate-consuming transaminases GPT and GOT were demonstrated. Consistent with reports on other organisms, in this alga two species of GlDH were also shown to be present: one using NADH the other NADPH as an electron donor. For GOGAT, however, there was only NADH, but no NADPH-dependent activity detectable, whereas the additionally known ferredoxin-dependent species of this enzyme, also described as occurring in *Chlorella* [13], has not been tested for in this investigation. Its indirect involvement

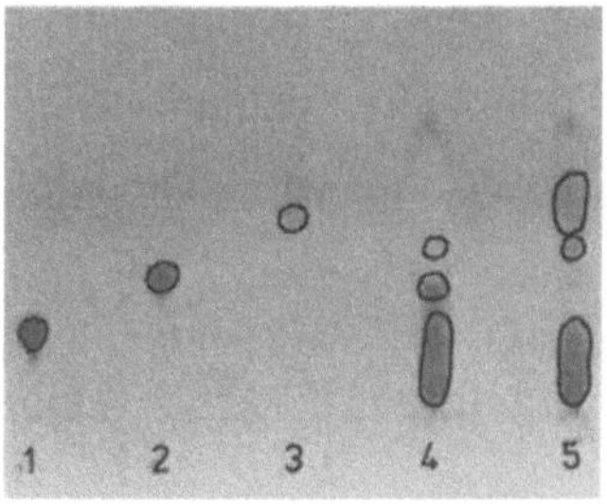

Fig. 1. TLC of the components of the reaction mixture for GOGAT determination *4*. In *5* the enzyme has been inhibited by 1 mM azaserine

has been excluded, however, by applying an antibody against ferredoxin to the reaction mixture. Like the control serum, this preparation did not alter the glutamine/2-oxoglutarate-dependent oxidation of NADH. The latter's dependence on GOGAT activity was further proved by the inhibiting effect (= 90%) of 1 mM azaserine and finally by demonstrating the reaction product glutamate by thin layer chromatography (Fig. 1).

In all above cases the optimized in vitro turnover rates were markedly dependent on the nutritional state of the algae. In well-supplied, growing cells they were 6-8 times higher for GOGAT, 2 to 3 times greater for G1DH and GPT, but – on the contrary – by about 50% smaller for GOT than in nondividing algae. This is shown in Table 1 which in addition indicates that the activities of the different enzymes differ by orders of magnitude. The turnover rate of the NADH-dependent GOGAT is extremely low, making this enzyme a favorite candidate for coarse regulation.

Table 1. Activities of glutamate producing or consuming enzymes (nmol mg^{-1} protein min^{-1}) in crude extracts of growing or "resting" (= nondividing) cells of a chlorophyll-free *Chlorella* mutant kept in darkness or illuminated with blue light ($\lambda < 550$ nm, 300 $\mu W\ cm^{-2}$, 18 h)

enzymes	GlDH (NADH)	GlDH (NADPH)	GOGAT (NADH)	GOT	GPT
growing dark	485	23	40	865	445
resting dark	150	30	5	1 560	265
resting blue	145	31	10	1 412	209

Correspondingly, blue illumination of the intact cells resulted either in no significant changes or in small decreases of the activities of both species of G1DH, GPT and of GOT; it led, however, to an increased turnover rate in the NADH-dependent GOGAT. This light effect was rather small in growing (~ 10%), but reasonably pronounced (~ 100%) in "resting" cells (Table 1). It developed within 8-10 h (Fig. 2) and saturated at rather low intensities already (Fig. 3). Its wavelength dependence yielded greatest effects around λ 460 nm. There was a steep drop to shorter as well as to longer wavelengths, resulting in about 15% of the maximum effect at λ 400 nm

Fig. 2. Dependence on the duration of illumination of the enhancement of GOGAT activity by blue light ($\lambda < 550$ nm, 300 $\mu W\ cm^{-2}$) in "resting" cells of a chlorophyll-free *Chlorella* mutant

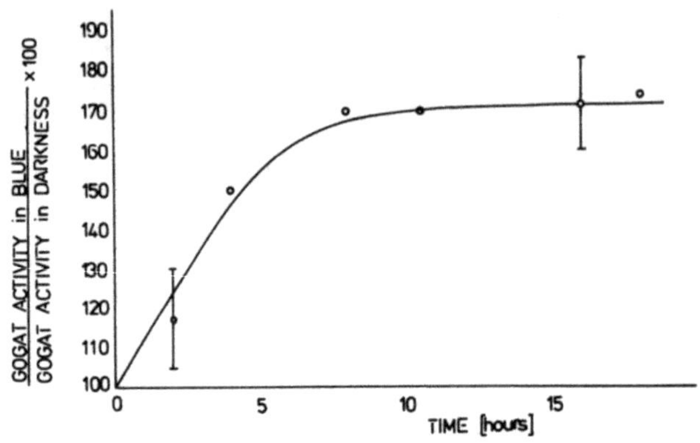

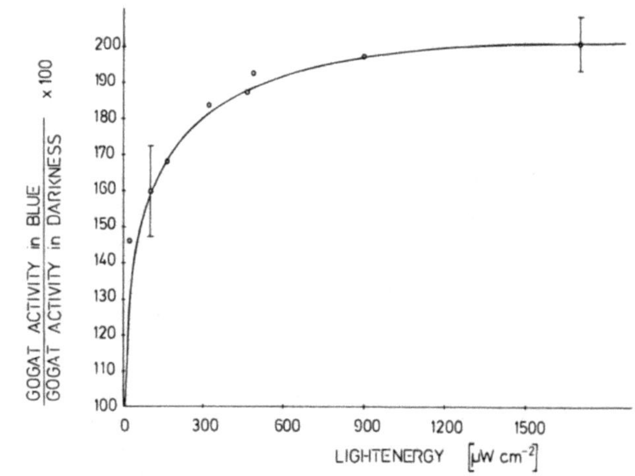

Fig. 3. Intensity dependence of the enhancement of GOGAT activity by blue light (λ 457 nm) in "resting" cells of a chlorophyll-free *Chlorella* mutant. Time of illumination was 18 h

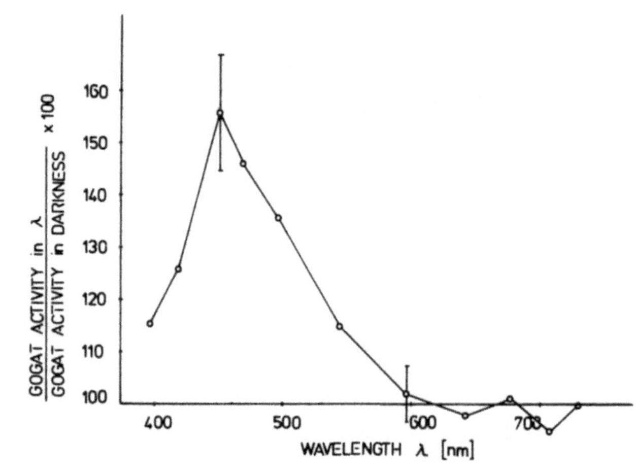

Fig. 4. Wavelength dependence (0.85×10^{18} quanta $m^{-2}\ s^{-1}$) of the enhancement of GOGAT activity by light in "resting" cells of a chlorophyll-free *Chlorella* mutant. Time of illumination was 18 h

Table 2. GOGAT (NADH) activity (nmol mg^{-1} protein min^{-1}) of *Chlorella* cells kept in darkness or irradiated with blue light ($\lambda < 580$ nm, 600 $\mu W\ cm^{-2}$, 8 h) in absence or presence of cycloheximide (5 $\mu g\ ml^{-1}$)

	- cycloheximide	+ cycloheximide
dark	5.00	4.00
blue	10.60	3.70

only but in no increase whatsoever in green, yellow, and red light (Fig. 4). Whether there is a greater effect of quanta of the short wavelength blue/UV-A region again – as might be expected from the comparable spectra of increased pyruvate kinase activity [16] and of enhanced carbohydrate breakdown and protein production [2] – remains open, since technical reasons prevented us from measuring.

Looking for the point of attack of blue light we found its effect abolished completely on poisoning the cells with cycloheximide (Table 2). This indicates a light-dependent de novo synthesis of the enzyme protein, a conclusion which has also been drawn from comparable results with phosphoenolpyruvate carboxylase [3, 15], pyruvate kinase [8] and carotenogenesis in fungi [14].

Discussion

There is increasing evidence that the glutamine synthetase/glutamine-2-oxoglutarate aminotransferase pathway is a major reaction to produce glutamate in plant cells. Three species of GOGAT have been described as occurring, distinguishable by the different electron donors used. NADH, NADPH and reduced ferredoxin have all been found suitable to drive this reaction [17]. In heterotrophic cells – such as the chlorophyll-free *Chlorella* mutant used in this investigation – the NADH-dependent species appears to be preferentially active, while the ferredoxin-dependent species exhibits no or only minor activity. Beside, for nongreen plant organs this has been reported for dark-kept algae such as *Chlamydomonas* [1] and also for etiolated tissues of higher plants [10, 11, 20, 21]. Nevertheless we shall have to check also for the activity of the ferredoxin-dependent species in our yellow mutant.

Changing from darkness to light leads to a pronounced increase in the ferredoxin-dependent enzyme; but to our knowledge no changes in the activity of the NADH-dependent species have been reported thus far. The enhancing influence of visible radiation appears to depend strictly on the photosynthetic electron transport, since it can be prevented by DCMU [10, 12, 18] and, from a most recent report, is connected with the well-known enzyme regulatory thioredoxin system [19]. This light effect is thus clearly different from what we present in this publication.

Because of its low activity the NADH-dependent enzyme appears preferably suited for coarse control, which could be brought about by an increased synthesis or some manufacturing of the enzyme protein. This conclusion is consistent with the finding of a depressing effect of the protein biosynthesis inhibitor cycloheximide. Since several sulfhydryl groups have been found necessary for an optimal activity of the enzyme [17], from the observation of its greater stability in extracts from dark-kept cells compared with that of blue-illuminated samples, one speculation for such an action might be the formation of a more permanently reduced species.

Comparing the above results with the enhancing effect of blue light on carbohydrate breakdown, on pyruvate kinase activity, and on the accumulation of protein, the conspicuous similarities in energy and wavelength dependencies, as well as in the time courses of the effects are striking. These might be explained by a central metabolic regulation in green organisms taking place on transfer from darkness to light, perhaps performed at the level of transcription as principally discussed elsewhere in this volume. If so, the effect is also expected to occur in the green wild-type cell, which still has to be checked. But the existing interdependencies of the carbohydrate-degrading and the protein-producing metabolism, also, could yield these results. An increased supply of 2-oxoglutarate, derived from an enhanced breakdown of carbohydrate reserves, for instance, might initiate the provision of greater 2-oxoglutarate-consuming enzyme activities, just as its increased consumption might also cause increases in the activities of the producing enzymes. Certainly fine control of already existing enzyme molecules appears more likely by such events; however, data on pool sizes or short-term responses are too scarce at present to allow more than mere speculations.

Acknowledgments. We thank Prof. Dr. G.H. Schmid for providing us with ferredoxin and its antisera and Miss P. Gayk for drawing the figures and typing the text.

References

1. Cullimore JV, Sims AP (1981) Pathway of ammonia assimilation in illuminated and darkened Chlamydomonas reinhardii. Phytochemistry 20:933–940
2. Georgi M (1974) Über blaulichtbedingte Veränderungen im Kohlenhydrat- und Proteingehalt einer chlorophyllfreien Mutante von Chlorella vulgaris. PhD thesis, Univ Köln, Germany, 75 pp
3. Kamiya A, Miyachi S (1975) Blue light-induced formation of phosphoenolpyruvate carboxylase in colorless Chlorella mutant cells. Plant Cell Physiol 16:729–736
4. Kowallik W (1962) Über die Wirkung des blauen und roten Spektralbereichs auf die Zusammensetzung und Zellteilung synchronisierter Chlorellen. Planta 58:337–365
5. Kowallik W (1969) Der Einfluss von Licht auf die Atmung von Chlorella bei gehemmter Photosynthese. Planta 86:50–62
6. Kowallik W (1982) Blue light effects on respiration. Annu Rev Plant Physiol 33:51–72
7. Kowallik W, Kirst R (1975) Über unterschiedliche Temperaturabhängigkeiten des Atmungsgaswechsels einer chlorophyllfreien Chlorella-Mutante im Dunkel und im Licht. Planta 124: 261–266
8. Kowallik W, Ruyters G (1976) Über Aktivitätssteigerungen der Pyruvatkinase durch Blaulicht oder Glucose bei einer chlorophyllfreien Chlorella-Mutante. Planta 128:11–14
9. Lowry OH, Rosebrough NJ, Farr AL, Randall RJ (1951) Protein measurement with the folin phenol reagent. J Biol Chem 193:265–275
10. Matoh T, Takahashi E (1981) Glutamate synthase in greening pea shoots. Plant Cell Physiol 22:727–731
11. Matoh T, Takahashi E (1982) Changes in the activities of ferredoxin- and NADH-glutamate synthase during seedling development of peas. Planta 154:289–294

12. McKenzie GH, Ch'ng AL, Gayler KR (1979) Glutamine synthetase/glutamine:α-ketoglutarate aminotransferase in chloroplasts from the marine alga Caulerpa simpliciuscula. Plant Physiol 63:578–582
13. Miflin BJ, Lea PJ (1982) Ammonia assimilation and amino acid metabolism. In: Boulter D Partier B (eds) Encyclopedia of plant physiology. New Ser. vol 14 A. Nucleic acids and proteins in plants I, structure, biochemistry and physiology of proteins. Springer, Berlin Heidelberg New York, pp 5–64
14. Rau W (1980) Blue light-induced carotenoid biosynthesis in microorganisms. In: Senger H (ed) The blue light syndrome. Springer, Berlin Heidelberg New York, pp 283–298
15. Ruyters G (1980) Blue light-enhanced phosphoenolpyruvate carboxylase activity in chlorophyll-free Chlorella mutant. Z Pflanzenphysiol 100:107–112
16. Ruyters G, Kowallik W (1980) Further studies of the lightmediated change in the activity of pyruvate kinase of a chlorophyll-free Chlorella mutant. Z Pflanzenphysiol 96:29–34
17. Stewart GR, Mann AF, Fentem PA (1980) Enzymes of glutamate formation: glutamate dehydrogenase, glutamine synthetase, and glutamate synthase. In: Miflin BJ (ed) The biochemistry of plants, vol V. Amino acids and their derivatives. Academic, London New York, pp 271–327
18. Tischner R, Hüttermann A (1980) Regulation of glutamine synthetase by light and during nitrogen deficiency in synchronous Chlorella sorokiniana. Plant Physiol 66:805–808
19. Tischner R, Schmidt A (1982) A thioredoxin-mediated activation of glutamine synthetase and glutamate synthase in synchronous Chlorella sorokiniana. Plant Physiol 70:113–116
20. Wallsgrove RM, Harel E, Lea PJ, Miflin BJ(1977) Studies on glutamate synthase from the leaves of higher plants. J Exp Bot 28:588–596
21. Wallsgrove RM, Lea PJ, Miflin BJ (1982) The development of NAD(P)H-dependent and ferredoxin-dependent glutamate synthase in greening barley and pea leaves. Planta 154: 473–476

Blue Light Effects on Carbon Metabolism in *Dunaliella*

G. RUYTERS[1,2], T. HIROSAWA[2], and S. MIYACHI[2]

Introduction

Blue light exerts regulatory effects on carbon metabolism of green algae and higher plants (review: [17]). Especially the carbohydrate breakdown and respiration appear to be markedly stimulated by short wavelength light under certain conditions [13, 16].

However, the underlying mechanisms of these and other blue light responses are not well understood. In particular no intracellular localization of certain responses has been made; this is because most of the previous work has been done with *Chlorella*, *Scenedesmus* and other green microalgae, from which no intact chloroplasts have so far been isolated.

Dunaliella tertiolecta was chosen as object for the present investigation because of two major advantages:

1. In *Dunaliella*, carbohydrate degradation is known to be induced by salt stress (review: [4]). If we were able to show that blue light enhanced carbohydrate breakdown, some analogous conclusions about the mechanism might be possible.
2. *Dunaliella* lacks a rigid cell wall. Therefore isolated chloroplasts should be easier to obtain. In fact, a few reports dealing with chloroplast isolation from this organism exist already [12, 15], although the degree of intactness is still a matter of discussion [5, 8].

Before trying to obtain intact chloroplasts from *Dunaliella* for the localization of certain blue light responses, we had to demonstrate these responses first to prove the usefulness of this alga for studying blue light effects. In the following we therefore report on light effects on carbohydrate breakdown and respiration in this alga.

Material and Methods

Dunaliella tertiolecta (Butcher) was kindly supplied by Prof. H. Takada of Tezukayama College, Nara, Japan. The cells were grown autotrophically at 25°C and aera-

---

1 Present address: Faculty of Biology IV, University of Bielefeld, 4800 Bielefeld 1, FRG

2 Institute of Applied Microbiology, University of Tokyo, Bunkyo-ku, Tokyo 113, Japan

Blue Light Effects in Biological Systems
ed. by H. Senger

ted with air enriched with 1.5% CO_2. The growth medium developed by Johnson et al. [10] was modified by replacing its micronutrients with Arnon's A-5 solution [1]. The cells were continuously illuminated with 5 x 40 W fluorescent lamps (Toshiba fish-looks) at an intensity of 800 $\mu W\ cm^{-2}$.

Cells of the (late) logarithmic growth phase were used for experiments. The cells were harvested by centrifugation (5 min at approx. 650 *g*) and resuspended in fresh growth medium at a density of about 1 μl packed cells ml^{-1}. The packed cell volume was estimated with haematokrit by centrifugation at approx. 2000 *g* for 20 min. Photosynthesis was blocked by the addition of 2×10^{-5} M DCMU at the beginning of each experiment. Light of different intensities and qualities was provided by Xenon lamps (500 W, Ushio Electric Co., Ltd. Tokyo, Japan) equipped with interference (half-band width 11 to 20 nm, average 13.9 nm) and, if necessary, neutral density filters.

Carbohydrate was estimated according to Dubois et al. [6]. In order to determine the amount of starch, low molecular weight carbohydrates were removed with 80% methanol before the carbohydrate assay. O_2 uptake as a measure for respiration was followed with a Clark type oxygen electrode (Yellow Springs Instruments, Yellow Springs, Ohio, USA).

Results and Discussion

Preliminary experiments showed that blue light was indeed able to stimulate carbohydrate degradation and O_2 uptake in *Dunaliella.* In the following we tried to find the conditions for pronounced blue light responses. Since it is known, that dark starvation is often very effective [7, 11, 14], we first investigated the influence of dark pre-incubation on blue light-enhanced O_2 uptake and carbohydrate degradation. Figure 1 shows the results, a for O_2 uptake, b for starch breakdown. In the case of *Dunaliella,* dark starvation for 24 h is not a good means for obtaining pronounced blue light effects on O_2 uptake, which is different from previous results for *Chlorella* and *Scenedesmus.* A 2- to 3-h dark incubation, however, proved to be sufficient to obtain significant and nearly constant effects (Fig. 1a). Also for the blue light-mediated enhancement of starch degradation a 3-h dark incubation seems to be useful (Fig. 1b). No marked differences were found between cells suspended in the culture medium and in various buffers containing 0.5 M NaCl.

In order to examine whether respiration and starch breakdown of *Dunaliella* are specifically enhanced by blue light, cells were irradiated with blue or red light of different intensities. Figure 2a, which shows the dose responses for light-stimulated O_2 uptake, indicates that not only blue but also red light is effective. For both light qualities, intensities as low as approx. 45 $\mu W\ cm^{-2}$ (= 1×10^{14} quanta $cm^{-2}\ s^{-1}$) brought about complete saturation and approx. 9 $\mu W\ cm^{-2}$ (= 2×10^{13} quanta $cm^{-2}\ s^{-1}$) half-saturation. Similar low light requirements are known for other green algae (summary: [18]).

Starch degradation of *Dunaliella* was also found to be enhanced by red light, though much less than by blue light. As a representative example Fig. 2b compares the stimulating effect of these light qualities on starch breakdown.

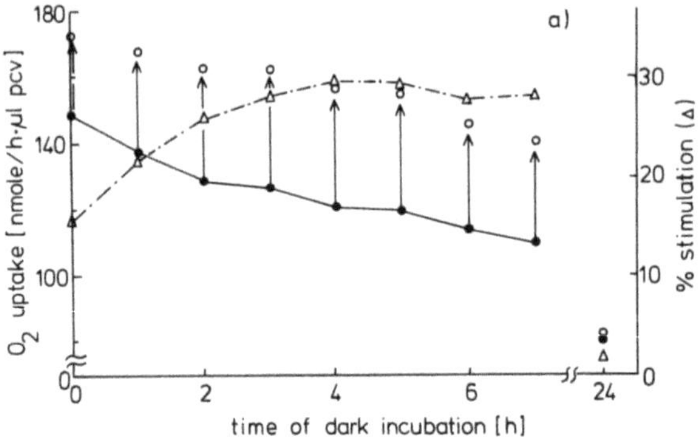

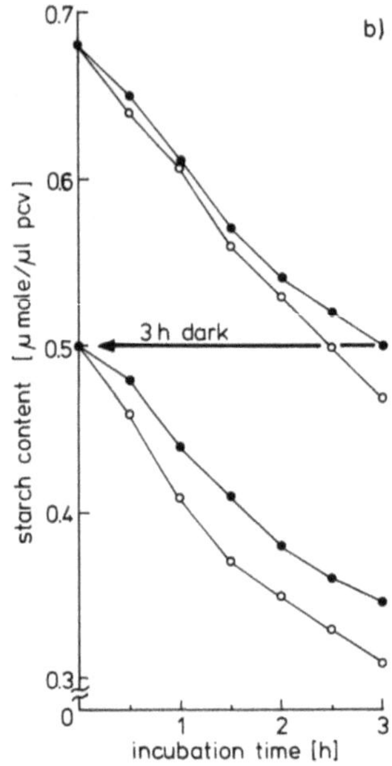

Fig. 1a. Influence of dark pre-incubation on blue light effects in *Dunaliella*: respiration. O_2 uptake of cells kept in the dark for the times indicated was measured in the dark (*closed circles*) or in blue light (*open circles;* 450 nm; 540 μW cm^{-2}). *Triangles* show the resulting enhancement (B-D/D x 100); **b** Influence of dark pre-incubation on blue light effects in *Dunaliella:* starch breakdown. Decrease in starch content was measured in darkness (*closed circles*) or blue light (*open circles*; 450 nm; 540 μW cm^{-2}) without or after 3 h dark incubation

In order to get some idea on the photoreceptor(s) involved, a wavelength dependence for the stimulation of O_2 uptake was measured. The fluence rate applied was 2 x 10^{13} quanta cm^{-2} s^{-1}, which half-saturated the blue and red light effects as shown above. The resulting wavelength dependence has three peaks at approx. 370, 450 and 650 nm (Fig. 3). The appearance of the broad maximum around 650 nm is quite different from the reported action spectra for the enhancement of O_2 uptake in other green algae. This wavelength dependence, however, resembles that obtained for chlorophyll synthesis and other light responses in pre-illuminated *Scenedesmus* mutant C-2A' [3]. From their wavelength dependence these authors assumed that the light responses of pre-illuminated cells were mediated via the blue light photoreceptor and via protochlorophyll(ide) (or a heme pigment). A similar explanation could be true for our results. Since our *Dunaliella* cells were grown in white light, they have, of course, chlorophyll; therefore our action spectrum may show a red light maximum at around 650 nm and not at 620 nm, and may thus be a result of a cooperation of chlorophyll and the blue light photoreceptor. Of course, our results also do not rule out the involvement of phytochrome.

Two preliminary results may support the idea that two photoreceptors are involved in the light responses in *Dunaliella.* A different influence of dark incubation between the blue and red light effect was found, the latter not changing greatly in dependence on dark incubation. In addition, the lag time after which the light-stimulated O_2 uptake rate is reached seems to be shorter in red than in blue light. Of course, much more work is necessary to prove the action of two photoreceptors and their identity with the blue light photoreceptor and chlorophyll, respectively.

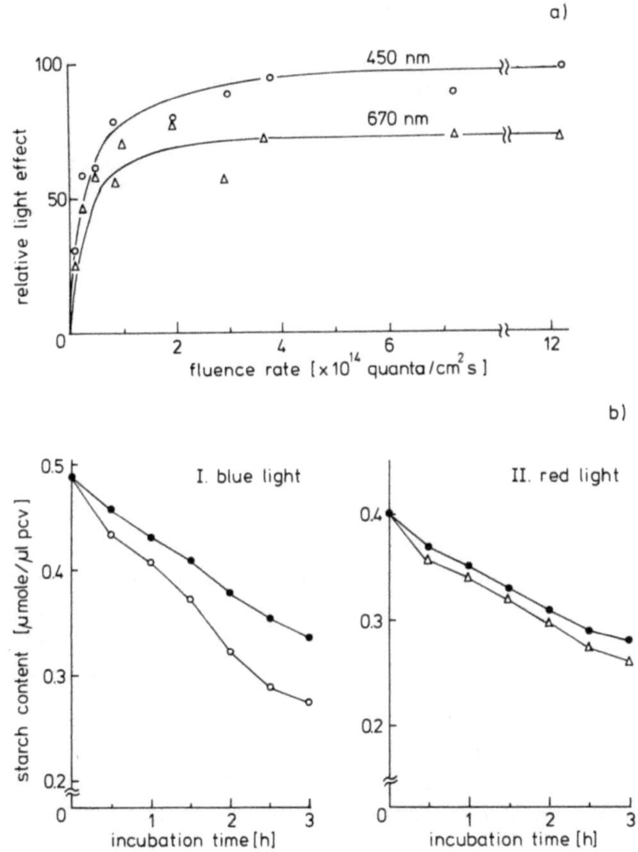

Fig. 2a,b (figure caption see opposite page)

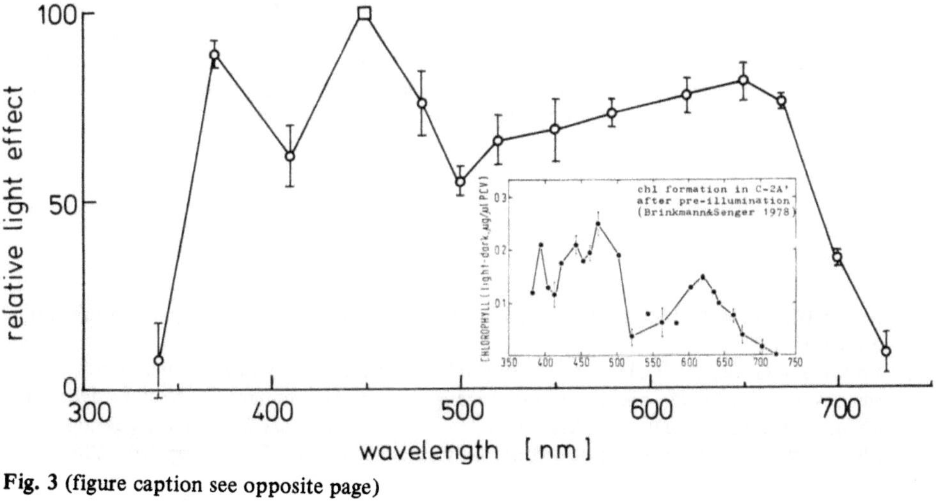

Fig. 3 (figure caption see opposite page)

Fig. 2a. Dose response for enhanced respiration of *Dunaliella*; O_2 uptake was measured in the dark and in blue or red light of different intensities; *circles*: relative blue light effects (450 nm); *triangles*: relative red light effects (670 nm); effect of 12 x 10^{14} quanta cm^{-2} s^{-1} of blue light is set 100; **b** Comparison of the effect of blue and red light on starch breakdown in *Dunaliella*. Decrease in starch content was measured in darkness (*closed circles*), blue light (*open circles*: 450 nm; 540 μW cm^{-2}) or red light (*triangles*; 670 nm; 360 μW cm^{-2})

Fig. 3. Wavelength dependence of light-enhanced O_2 uptake; relative effects with standard deviation (2 x 10^{13} quanta cm^{-2} s^{-1}; effect of 450 nm is set 100); for comparison the *inset* shows the wavelength dependence for chlorophyll formation in pre-illuminated *Scenedesmus* C-2A' after [2]

Conclusions

The results presented here demonstrate that blue light is able to stimulate starch breakdown and O_2 uptake in *Dunaliella*. Besides some similarities to other investigated algae, e.g., the very low light requirement for saturation of the effects (Fig. 2), some differences exist. Prolonged dark starvation does not lead to more pronounced effects; 2- or 3-h dark incubation is rather effective instead (Fig. 1). It was also found that not only blue but also red light is effective in bringing about those stimulations (Fig. 3). Red light-stimulated respiration has not been reported for green algae, and there are only a few reports on the effect of red light on respiration for higher plants (review: [9]). In these cases they either were assumed to be mediated through the photosynthetic metabolism or shown to be due to phytochrome action.

Wavelength effects in *Dunaliella* are somewhat more complicated than those in other green algae because of the possible involvement of two photoreceptors. However, we were able to demonstrate basic light responses of carbon metabolism in *Dunaliella*, thus proving the usefulness of this alga for studying light responses. In the future it may be possible to isolate chloroplasts and to localize light responses in different cellular compartments.

Acknowledgments. G.R. is indebted to the Japan Society for the Promotion of Science and to the Alexander von Humboldt Foundation (Feodor Lynen Program) for a postdoctoral fellowship. S.M. and T.H. would like to thank the Deutsche Forschungsgemeinschaft for covering the travel costs to attend this meeting. The work presented was supported by a grant from the Nissan Science Foundation. Thanks are also due to Mr. Robert Church for reading the English.

References

1. Arnon DI, Wessel G (1953) Vanadium as an essential element for green plants. Nature (London) 172:1039
2. Brinkmann G, Senger H (1978) Light-dependent formation of thylakoid membranes during the development of the photosynthesis apparatus in pigment mutant C-2A' of *Scenedesmus obliquus*. In: Akoyunoglou G, Argyroudi-Akoyunoglou JH (eds) Chloroplast development. Elsevier/North-Holland Biomedical Press, Amsterdam New York Oxford, pp 201–206

3. Brinkmann G, Senger H (1980) Blue light regulation of chloroplast development in Scenedesmus mutant C-2A'. In: Senger H (ed) The blue light syndrome. Springer, Berlin Heidelberg New York, pp 526–540
4. Brown AD, Borowitzka LJ (1979) Halotolerance of Dunaliella. In: Levandowsky M, Hunter SH (eds) Biochemistry and physiology of protozoa, vol I. Academic, London New York, pp 139–190
5. Brown AD, McCLilley R, Marengo T (1982) Osmoregulation in Dunaliella. Intracellular distribution of enzymes of glycerol metabolism. Z Naturforsch 37c:1115–1123
6. Dubois M, Gilles KA, Hamilton JK, Rebers PA, Smith F (1956) Colorimetric method for determination of sugars and related substances. Anal Chem 28:350–356
7. Georgi M (1974) Über blaulichtbedingte Veränderungen im Kohlenhydrat- und Proteingehalt einer chlorophyllfreien Mutante von Chlorella vulgaris. PhD thesis, Univ Köln
8. Gimmler H, Lotter G (1982) The intracellular distribution of enzymes of the glycerol cycle in the unicellular green alga Dunaliella parva. Z Naturforsch 37c:1107–1114
9. Graham D (1980) Effects of light on "dark" respiration. In: Davies DD (ed) The Biochemistry of plants, vol II. Academic, London New York, pp 525–579
10. Johnson MK, Johnson EJ, MacElroy RD, Speer HL, Bruff BS (1968) Effects of salts on the halophilic alga Dunaliella viridis. J Bacteriol 95:1461–1468
11. Kamiya A, Miyachi S (1974) Effects of blue light on respiration and carbon dioxide fixation in colorless Chlorella mutant cells. Plant Cell Physiol 15:927–937
12. Kombrink E, Wöber G (1980) Preparation of intact chloroplasts by chemically induced lysis of the green alga Dunaliella marina. Planta 149:123–129
13. Kowallik W (1982) Blue light effects on respiration. Annu Rev Plant Physiol 33:51–72
14. Kowallik W, Gaffron H (1966) Respiration induced by blue light. Planta 69:92–95
15. Lynch DV, Thompson GA (1982) Low temperature-induced alterations in the chloroplast and microsomal membranes of Dunaliella salina. Plant Physiol 69:1369–1375
16. Miyachi S, Kamiya A, Miyachi S (1977) Wavelength-effects of incident light on carbon metabolism in Chlorella cells. In: Mitsui A, Miyachi S, San Pietro A, Tamura S (eds) Biological solar energy conversion. Academic, London New York, pp 167–182
17. Senger H (ed) (1980) The blue light syndrome. Springer, Berlin Heidelberg New York
18. Senger H, Briggs WR (1981) The blue light receptor(s): Primary reactions and subsequent metabolic changes. Photochem Photobiol Rev 6:1–38

^{31}P-NMR Studies in *Scenedesmus* C-2A′ in Darkness and Blue Light

T. OH-HAMA[2], G. RUYTERS[1], K. FURIHATA[2], H. SETO[2] and S. MIYACHI[2]

Introduction

Blue light stimulates carbohydrate degradation and respiration of green algae (review: [8, 10, 12]); the underlying mechanism, however, is not well understood. Since all reactions are enzyme-catalyzed, enzymes are certainly candidates for regulation. Indeed several enzymes have been reported to be blue light-stimulated (see Ruyters this Vol.). The finding of a blue light-mediated enhancement of PEP carboxylase and pyruvate kinase activity [6, 9], together with the reported blue light-enhanced carbohydrate breakdown of a chlorophyll-free *Chlorella* mutant under anaerobic conditions [7] strongly indicates an important role of the glycolytic pathway in these blue light phenomena. Since the demonstrated changes in maximal catalytic activities of pyruvate kinase and PEPCase occur too slowly to account for the quick stimulation of respiration and carbohydrate breakdown [19], these processes might be regulated by changes in the concentration of effectors, cofactors and substrate of regulatory enzymes. However, except for adenosine phosphates [11], no data on the level of these compounds and their possible changes in blue light are available.

Since the glycolytic intermediates – except the final product pyruvate – and many important cofactors and effectors are phosphate compounds, they are detectable by ^{31}P NMR. ^{31}P-NMR has become a well-established method for the study of the bioenergetics and metabolism of bacterial and animal systems. It is a useful tool to monitor changes in the level of organic phosphate compounds. Inorganic phosphate, which is important as one of the controlling factors in metabolism, can be measured as well. Moreover, ^{31}P-NMR is the only means to study phosphorylated metabolites in vivo without destroying the cells. By use of intact cells, information about the cellular environment of the phosphate compounds, intracellular compartmentation and the internal pH values can also be obtained. Since blue light might cause changes in intracellular pH values, thereby modulating rates of enzyme-catalyzed reactions, knowledge of these data would be very valuable.

Very few reports, however, describe the application of ^{31}P-NMR to higher plants, and to our knowledge only one study exists for green algae, which is mainly concerned with polyphosphates in intact cells [3]. Therefore, the aim of our present paper is to report the first steps of the adaptation of ^{31}P-NMR to the study of ex-

1 Present address: Faculty of Biology IV, University of Bielefeld, 4800 Bielefeld, FRG

2 Institute of Applied Microbiology, University of Tokyo, Bunkyo-ku, Tokyo 113, Japan

Blue Light Effects in Biological Systems
ed. by H. Senger

tracts and whole cells of green algae and to describe some preliminary results about their phosphate compounds and cellular pH values in darkness and under blue light.

As object for our present investigation we chose the *Scenedesmus* mutant C-2A', heterotrophically grown in the dark. This mutant has been widely used for the study of blue light effects on carbon metabolism as well as on chloroplast development [1, 24].

Material and Methods

Algal Culture. The strain of *Scenedesmus obliquus* mutant C-2A' and the growth medium were the same as those reported previously [17, 24]. The cells were cultured by shaking in the dark for 4-6 days at 29°C, and harvested at the stationary growth phase. For experiments the algae were suspended in 1.3 mM phosphate buffer (pH 7.3) and shaken in the dark for further 20 h. After centrifugation the cells were washed with 8 mM NaCl and suspended in 4 mM TES buffer (pH 7.3) at a concentration of 100 μl packed cell volum ml^{-1} (unless otherwise indicated in the text); 2.5 ml packed cells or their perchloric acid (PCA) extract, respectively, was used for each ^{31}P-NMR measurement.

Irradiation of Algal Suspensions. A 5 KW Xe-lamp was used as the light source. Blue (400-500 nm) and red light (600-700 nm) were obtained by the use of broad band interference filters (Vacuum Optics Corp. Japan). Light intensity was measured as described previously [18].

Preparation of Samples for NMR Measurement. At the end of the experiment a handful of grated ice was added to the algal suspension to preserve its physiological state. After centrifugation the cells were resuspended in 4 mM TES buffer containing 1 mM EDTA and 10% D_2O at a concentration of 40% packed cell volume ml^{-1} suspension (for measurement of intact cells), or treated with 1/3 volume of 6 N PCA for extraction. Combined PCA extracts (final concentration of PCA: 1 N) were neutralized with NaOH to pH 8.2, then EDTA and D_2O were added as described above, unless stated otherwise in text.

^{31}P NMR Spectra. The ^{31}P-NMR spectra were recorded with a JEOL FX- or GX-400 spectrometer working at 162.2 MHz under conditions of broad-band proton decoupling at room temperature. Chemical shifts were expressed in ppm relative to 85% phosphoric acid and were measured mostly relative to the external 20 mM phosphoric acid dissolved in 0.1 M HCl containing D_2O (*R* in figures). This solution was filled into a capillary tube, which was inserted at the center of the sample tube used for spinning (15 mm diameter). Repetition time and pulse number of the measurements are indicated in each legend.

Results

Phosphate Compounds Detectable in *Scenedesmus* Extracts

Figure 1 shows typical ^{31}P-NMR spectra of extracts from dark-grown *Scenedesmus* mutant cells; the extract was prepared 45 min after addition of 6 mM glucose to the cells. The spectra should demonstrate the phosphate compounds which were detectable in *Scenedesmus* cells with our method. Without the addition of EDTA to the PCA extract, the spectrum totally lacked resolution and had high noise levels (Fig. 1a). This broadening of the signals in ^{31}P NMR had been presumed to be the effect of paramagnetic ions such as Mn^{2+} [16] and was eliminated by chelating those with 1 mM EDTA (Fig. 1b). Peak assignments from left to right in Fig. 1b are sugar phosphates including G-6-P and FbP, orthophosphate (P_i), NTP_γ and NDP_β, NTP_α and NDP_α (and probably also NADH and NAD), and NTP_β. These assignments were made by comparing the chemical shifts obtained with available literature data (e.g., [4]). In the case of G-6-P and FbP the assignment was confirmed by adding small quantities of both phosphate compounds (purchased from commercial sources) to the extract (Fig. 1c). In addition ^{31}P-NMR spectra of several more phosphate compounds were recorded under our pH conditions for comparison (Fig. 2a-e). The peak at 18.4 ppm in Fig. 2b coincides with the β-phosphate of Mg-bound ATP [2]. By adding 6 mM EDTA or in the absence of extracts the peak shifted to 20.5 ppm (Fig. 2a), corresponding with the value in an environment free of divalent metal ions; this indicates that the content of divalent cations (probably mainly magnesium) in PCA extracts of the algal cells was too large to be chelated by 1 mM EDTA.

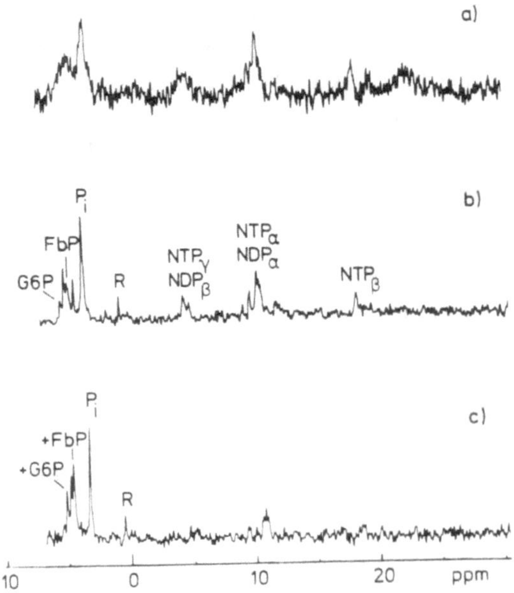

Fig. 1a-c. ^{31}P-NMR spectra of neutralized PCA extracts (pH 8.2) of *Scenedesmus* C-2A' cells; 6 mM glucose added to the cells 45 min before extraction; spectrum measured without (a) or with 1 mM EDTA (b); in (c) 0.3 mM G6P and FbP were added; measurement conditions: 1000 pulses, repetition time 1.1 s; *R* references (20 mM H_3PO_4 in 0.1 M HCl)

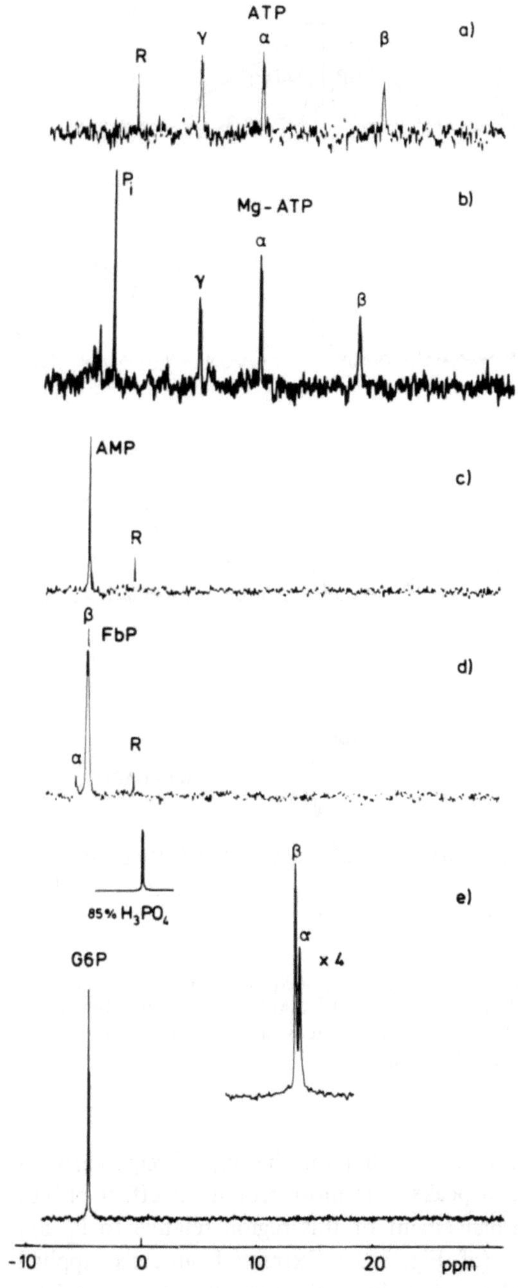

Fig. 2a-e. ^{31}P-NMR spectra of authentic substances (1 mM) at pH 8.2. a ATP; b ATP + extract (= Mg-ATP); c AMP; d FbP; e G6P; measurement conditions: 350 pulses, repetition time 2.0 s

Changes in the Level of Phosphate Compounds in *Scenedesmus* C-2A' Cells After Irradiation with Blue Light or After Feeding Glucose

Figures 3 and 4 show ^{31}P-NMR spectra of PCA extracts of *Scenedesmus* cells which had been illuminated with blue light, or kept in the dark with or without exogenous

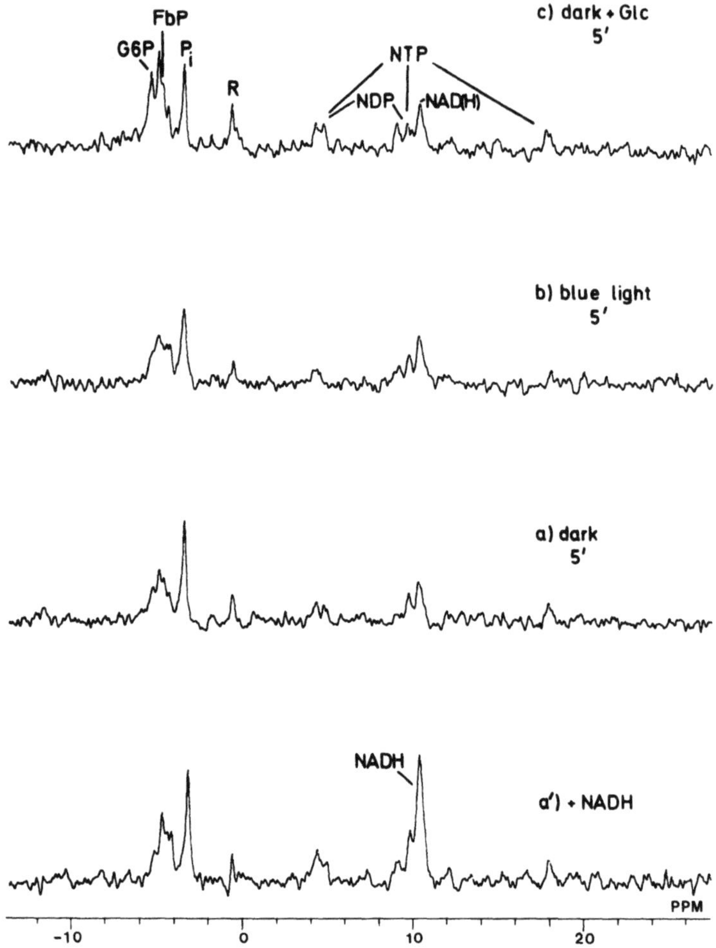

Fig. 3a-c. ^{31}P-NMR spectra of neutralized PCA extracts (pH 8.2) of *Scenedesmus* C-2A' cells held under different conditions. **a** dark control (5 min); **a'** dark + 0.5 mM NADH; **b** 5 min blue light-irradiated cells (400-500 nm; approx. 600 μW cm^{-2}); **c** 6 mM glucose added 5 min before extraction; measurement conditions: 1000 pulses, repetition time 1.1 s

glucose for either 5 or 15 minutes. A rather broad band in the sugar-P region around −4 to −5 ppm might result from several peaks. The most prominent effect of blue light irradiation for 5 min was the enhancement of this region relative to P_i and broadening of the band on the left side (cf. Fig. 3a,b). Extracts from cells supplied with glucose for 5 min also showed enhanced peaks in the sugar-P region; especially G-6-P and FbP levels seem markedly elevated (Fig. 3c). However, in regions where a number of closely spaced resonances are formed, such as the sugar-P region, the assignment should be verified by treatment with specific enzymes and by measuring the pH dependences of the chemical shifts. In Fig. 3a' NADH has been added to the extract from dark-kept cells to prove the agreement with the respective peaks in the spectra of the extracts.

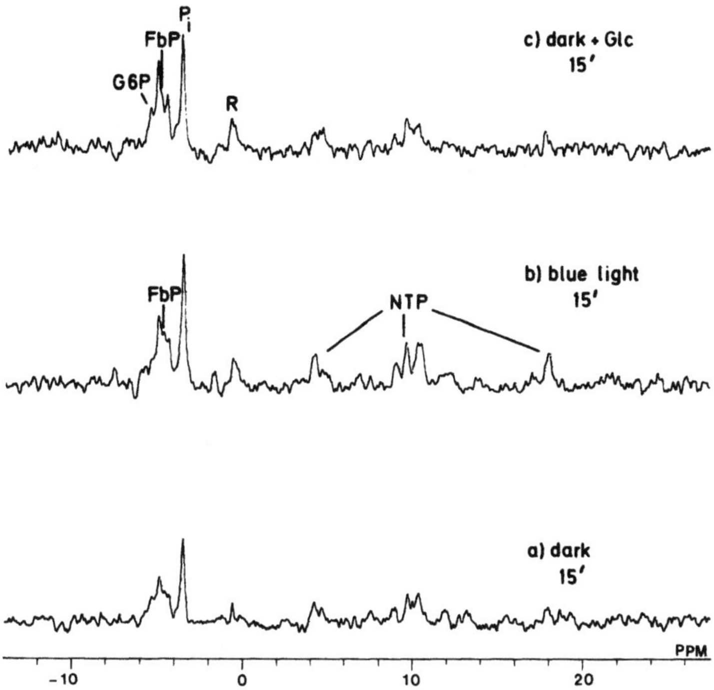

Fig. 4a-c. ^{31}P-NMR spectra of neutralized PCA extracts (pH 8.2) of *Scenedesmus* C-2A' cells held under different conditions. **a** dark control (15 min); **b** 15 min blue light-irradiated cells (400-500 nm; approx. 600 μW cm^{-2}); **c** 6 mM glucose added 15 min before extraction; measurement conditions: 1000 pulses, repetition time 1.1s

In addition to changes in the sugar-P region some increases in the nucleotides (probably mainly ATP) occurred after 15 min of blue light irradiation or application of exogenous glucose (Fig. 4.).

^{31}P-NMR Spectra of Intact *Scenedesmus* Cells Irradiated with Blue or Red Light

Figure 5 shows ^{31}P-NMR spectra of intact cells of *Scenedesmus obliquus* C-2A' which had been irradiated with blue or red light or incubated in the dark for 15 min. A comparison of the three spectra reveals differences especially between the blue light-irradiated and the dark-kept cells, while red light is less effective. Blue light enhanced the level of sugar-P (among them probably G-6-P and FbP), and of nucleotide phosphates, especially NTP as seen from the NTP peak. A peak of UDPG also appears in this light quality. Figure 5 also indicates the occurrence of large amounts of inorganic polyphosphates in algal cells, which are known to be enhanced by blue light in *Chlorella*, though after long illumination times [22].

Previous studies have shown that the intracellular pH can be measured by the chemical shift of the intracellular P_i [14]; the value of P_i is less affected by divalent metal ions, salt, etc. in the medium. A rather broad peak at −2.05 ppm arose from P_i

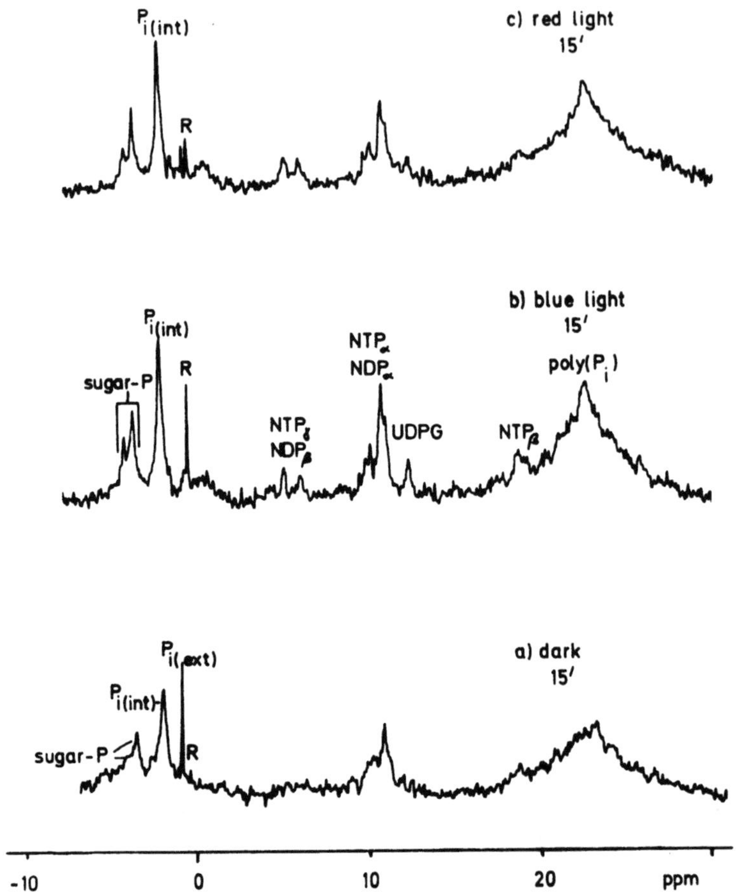

Fig. 5a-c. ^{31}P-NMR spectra of intact cells of *Scenedesmus* C-2A', suspended in 4 mM TES buffer pH 7.3 containing D_2O and 1 mM EDTA; cell density about 400 µl packed cells ml^{-1}. **a** dark control; **b** 15-min blue light-irradiated cells (400-500 nm; approx. 600 µW cm^{-2}); **c** 15-min red light-irradiated cells (600-700 nm; approx. 400 µW cm^{-2}); measurement conditions; 350 pulses, repetition time 2.0 s; $P_{i(int)}$: intracellular P_i; $P_{i(ext)}$: P_i in the medium

inside the cells. This chemical shift of P_i indicates an intracellular pH of the mutant cells of around 6.8, which was not shifted after irradiation with blue or red light. The sharp peak at 0.98 ppm in Fig.5a arose from extracellular P_i.

Discussion

Levels of phosphorous metabolites were followed with ^{31}P-NMR in PCA extracts of *Scenedesmus obliquus* mutant C-2A', which had been grown heterotrophically in the dark. Upon irradiation with blue light and also upon addition of glucose in the dark, the level of sugar-P, including G-6-P and most probably FbP, was enhanced relative to that of P_i (Figs. 3, 4). These results suggest an immediate enhancement of starch degradation by blue light or of the exogenously supplied gluocse, respectively, via increases in vivo activities of hexokinase and phosphofructokinase. These enzymes, together with pyruvate kinase, are generally considered the regulatory enzymes of glycolysis in plants [23]. (Since pyruvate, the product of the PK reaction, is not a phosphorylated compound, its possible changes cannot be measured with ^{31}P NMR; the level of its substrate PEP is usually very low and can probably not be detected by our method.)

There are a few reports on ^{31}P-NMR of inorganic polyphosphates (poly-P) from intact cells, e.g., from yeast [15, 21], a blue-green alga [5], and from a green alga [3]. Since only a part of the poly-P in *Chlorella* cells can be extracted with PCA [13], its concentration in PCA extracts of *Scenedesmus* cells may be too low to be detected with ^{31}P-NMR. In ^{31}P NMR spectra of intact cells, however, large peaks of poly-P appear in addition to the peaks of sugarphosphates, nucleotides and P_i (Fig. 5). Like the spectra of PCA extracts, the spectra of intact cells also show that especially the levels of sugar-P and nucleotides are enhanced by blue light.

The results with intact cells also demonstrate the possibility of estimating the intracellular pH of green algae with ^{31}P-NMR. The pH value was found to be around 6.8 in *Scenedesmus* mutant C-2A' and was not affected by irradiation of the cells for 15 min. Also no different pH values of different cellular compartments could be seen, which is not surprising, however, because the mutant cells are not able to perform photosynthesis and also do not contain vacuoles.

Although it is not possible to reach more definitive conclusions at this early stage of our investigation, we think that the present results already demonstrate the usefulness of ^{31}P-NMR for the study of the metabolism of green algae. Thus it might be possible to gain further insight into the mechanism of the well-known blue light-enhanced carbohydrate degradation and respiration of green algae by this method. Since it usually takes several minutes before the blue light-enhanced respiration rate is achieved, especially the very early phases of blue light action on carbon metabolism are interesting to be followed.

Summary

^{31}P-NMR spectra obtained at 162.2 MHz are presented for intact cells of *Scenedesmus obliquus* mutant C-2A' and their perchloric acid extracts. Algal cells were incubated under blue or red light or in darkness with or without glucose. Effects of these treatments on the levels of phosphorous metabolites in the algal cells were followed. In intact cells and their extracts, nucleotide di- and tri-phosphate, NAD, orthophosphate, fructose bisphosphate, glucose 6-phosphate and inorganic polyphosphate could be detected.

The chemical shift of inorganic phosphate in intact cells indicates an intracellular pH of the *Scenedesmus* mutant of around 6.8, which is not changed upon irradiation with blue or red light.

Acknowledgments. G.R. is indebted to the Japan Society for the Promotion of Science and to the Alexander von Humboldt Foundation (Feodor Lynen Program) for a postdoctoral fellowship. This work was supported by a grant from the Ministry of Education, Science and Culture of Japan. Thanks are also due to Mr. Robert Church for reading the English.

References

1. Brinkmann G, Senger H (1980) Blue light regulation of chloroplast development in Scenedesmus mutant C-2A'. In: Senger H (ed) The blue light syndrome. Springer, Berlin Heidelberg New York, pp 526–540
2. Colman A, Gadian DG (1976) ^{31}P nuclear-magnetic-resonance studies on the developing embryos of Xenopus laevis. Eur J Biochem 61:387–396
3. Elgavish A, Elgavish GA, Halmann M (1980) Intracellular phosphorus pools in intact algal cells. FEBS Lett 117:137–142
4. Gadian DG, Radda GK, Richards RE, Seely PJ (1979) ^{31}P NMR in living tissue: The road from a promising to an important tool in biology. In: Shulman RG (ed) Biological application of magnetic resonance. Academic, London New York, pp 463–535
5. Kallas T, Dahlquist FW (1981) Phosphorous-31 nuclear magnetic resonance analysis of internal pH during photosynthesis in the Cyanobacterium Synechococcus. Biochemistry 20: 5900–5907
6. Kamiya A, Miyachi S (1975) Blue light-induced formation of phosphoenolpyruvate carboxylase in colorless Chlorella mutant cells. Plant Cell Physiol 16:729–736
7. Kowallik W (1969) Eine fördernde Wirkung von Blaulicht auf die Säureproduktion anaerob gehaltener Chlorellen. Planta 87:372–384
8. Kowallik W (1982) Blue light effects on respiration. Annu Rev Plant Physiol 33:51–72
9. Kowallik W, Ruyters G (1976) Über Aktivitätssteigerungen der Pyruvat-Kinase durch Blaulicht oder Glucose bei einer chlorophyllfreien Chlorella-Mutante. Planta 128:11–14
10. Kowallik W, Schätzle S (1980) Enhancement of carbohydrate degradation of blue light. In: Senger H (ed) The blue light syndrome. Springer, Berlin Heidelberg New York, pp 344–360
11. Kowallik W, Scheil I (1976) Lichtbedingte Veränderungen des ATP-Spiegels bei einer chlorophyllfreien Chlorella-Mutante. Planta 131:105–108
12. Miyachi S, Kamiya A, Miyachi S (eds) (1977) Wavelength effects of incident light on carbon metabolism in Chlorella cells. In: Biological solar energy conversion. Academic, London New York, pp 167–182
13. Miyachi S, Kanai R, Mihara S, Miyachi Sh, Aoki S (1964) Metabolic roles of inorganic polyphosphates in Chlorella cells. Biochim Biophys Acta 93:625–634
14. Moon RB, Richards JH (1973) Determination of intracellular pH by ^{31}P magnetic resonance. J Biol Chem 248:7276–7278
15. Nicolay K, Scheffers WA, Bruineberg PM, Kaptein R (1982) Phosphorus-31 nuclear magnetic resonance studies of intracellular pH, phosphate compartmentation and phosphate transport in yeasts. Arch Microbiol 133:83–89
16. Ogawa S, Rottenberg H, Brown TR, Shulman RG, Castillo CL, Glynn P (1978) High resolution ^{31}P nuclear magnetic resonance study of rat liver mitochondria. Proc Natl Acad Sci USA 75:1796–1800
17. Oh-hama T, Senger H (1975) The development of structure and function in chloroplast of greening mutants of Scenedesmus III. Biosynthesis of δ-aminolevulinic acid. Plant Cell Physiol 16:395–405
18. Oh-hama T, Hase E (1978) Blue light effect on chlorophyll formation in Chlorella protothecoides. Photochem Photobiol 27:199–202
19. Ruyters G (1981) Effects of blue light on respiration and enzyme activity in a yellow Chlorella mutant. In: Akoyunoglou G (ed) Proc 5th Int Congr Photosynth, vol V. Balaban Int Sci Serv, Philadelphia, pp 905–914

20. Ruyters G (1983) Effects of blue light on enzymes. This volume
21. Salhany JM, Yamane T, Shulman RG, Ogawa S (1975) High resolution ^{31}P nuclear magnetic resonance studies of intact yeast cells. Proc Natl Acad Sci USA 72:4966–4970
22. Steup M, Pirson A (1974) Über den Einfluß des blauen und roten Spektralbereichs auf Phosphatfraktionen, besondere Polyphosphate, bei Grünalgen. Biochem Physiol Pflanz 166:447–459
23. Turner JF, Turner DH (1980) The regulation of glycolysis and pentose phosphate pathway. In: Davies DD (ed) The biochemistry of plants, vol II. Academic, London New York pp 279–316
24. Watanabe M, Oh-hama T, Miyachi S (1980) Light-induced carbon metabolism in an early stage of greening in wild type and mutant C-2A' cells of Scenedesmus obliquus. In: Senger H (ed) The blue light syndrome. Springer, Berlin Heidelberg New York, pp 332–343

Pigment Biosynthesis

Effects of Blue Light on the Formation od 5-Aminolevulinic Acid and Chlorophyll in Cultured Tobacco Cells

A. KAMIYA[1], I. IKEGAMI,[1] and E. HASE[2]

Introduction

Chlorophyll (Chl) biosynthesis is regulated at the step of 5-aminolevulinic acid (ALA) synthesis (cf. [3]). ALA formation in higher plants is controlled by two light systems, Pchl and phytochrome, which act as photoreceptors. In the first, the inhibition of ALA-synthesizing activity by Pchl accumulated in dark-grown seedlings is removed by light as a result of photoconversion of Pchl into Chl [7, 13]. In the other, ALA-synthesizing activity is potentiated by red light; formation of ALA-synthesizing enzyme is probably induced [7, 9, 13–15] through the mediation of phytochrome as a photoreceptor. In some algae which require light for Chl formation, blue light induces ALA-forming activity [17, 18] in dark-grown cells, while red light as well as blue light enhances ALA formation in greening algal cells [19, 22].

Chlorophyll formation in tissues like tobacco callus [5], potato tubers [4] and wheat roots [6] has been reported to proceed only under blue light. Our previous studies [11] showed that marked accumulation of chlorophyll occurred in callus cells of *Nicotiana glutinosa* when they were grown under blue light, while under strong red light no chlorophyll accumulated. Furthermore weak red light as well as blue light was effective for the phototransformation of Pchl into Chl, and strong red light caused photodestruction of Pchl. Strong white light was found to be effective for the photoreduction of Pchl to Chl with only slight destruction of Pchl, suggesting that blue light counteracts the destructive effect of red light. It was inferred that blue light enhances the supply of a reductant required for the photoreduction of Pchl to Chl.

In this paper, effects of blue and red light on the ALA formation in tobacco callus cells were described in correlation with chlorophyll synthesis.

1 Laboratory of Chemistry, Faculty of Pharmaceutical Sciences, Teikyo University, Sagamiko, Kanagawa 199–01, Japan

2 Laboratory of Chemistry, Faculty of Medicine, Teikyo University, Hachioji, Tokyo 192–03, Japan

Blue Light Effects in Biological Systems
ed. by H. Senger

Materials and Methods

Culture Conditions. Callus cells of *Nicotiana glutinosa* were grown on an agar medium in the dark at 25°C as described previously [11].

Light Conditions. The source of light was a bank of 40 W cool white fluorescent tubes (Fishlux; Toshiba Co. Ltd., Tokyo). The blue light ($370 < \lambda < 560$ nm, max. 470 nm) and red light (> 580 nm) were obtained with a blue filter (Paraglass 315K, Kyowa Gas Chemical Industry Co. Ltd., Tokyo) and a red filter (Paraglass 102K), respectively. The spectral properties of the blue and red filters and the light source were described previously [11].

Measurement of ALA Content. The ALA formation in dark-incubated or greening callus cells was followed employing levulinic acid (LA) as a competitive inhibitor of ALA dehydratase [12].

Measurements of Chlorophyll Content and Fluorescence Intensity. Chlorophyll was extracted with 80% acetone from the callus cells frozen and thawed in the dark. The amount of chlorophyll was determined spectrophotometrically by the method of Arnon [1]. The fluorescence emitted from the callus cells was measured at room temperature (ca. 20°C) with an optical set-up described previously [11].

Measurement of Oxygen Uptake. Callus cells (0.5g) were suspended in 50 mM phosphate buffer (pH 7.5) and the rate of oxygen uptake was determined with a Clark-type electrode (Rank Brothers, London) at 25°C. To measure the rate of oxygen uptake under blue light, blue light (1,400-90,000 mW m^{-2}, $360 < \lambda < 520$ nm, max. 430 nm) was obtained with a blue filter (Corning 5-60); light source was a 500 W Xenon lamp.

Results

Effects of Blue and Red Light on the Formation of 5-Amino Levulinic Acid in Callus Cells

Figure 1 shows the formation of chlorophyll and the ALA-synthesizing activity in callus cells of *Nicotiana glutinosa* illuminated with blue and red light for 4 days at various intensities. The enhancing effects of blue light on the formation of chlorophyll and the ALA-synthesizing activity were saturated at 600-800 mW m^{-2}, which were similar to the saturating intensities for the blue light effects on the photoconversion of Pchl to Chl in callus cells reported previously [11]. As seen in Fig. 1a, red light of low intensity (ca. 200 mW m^{-2}) was slightly effective for the accumulation of chlorophyll. However, when its intensity was increased, the chlorophyll content decreased. These results suggest that Pchl was destroyed by red light of high intensity,

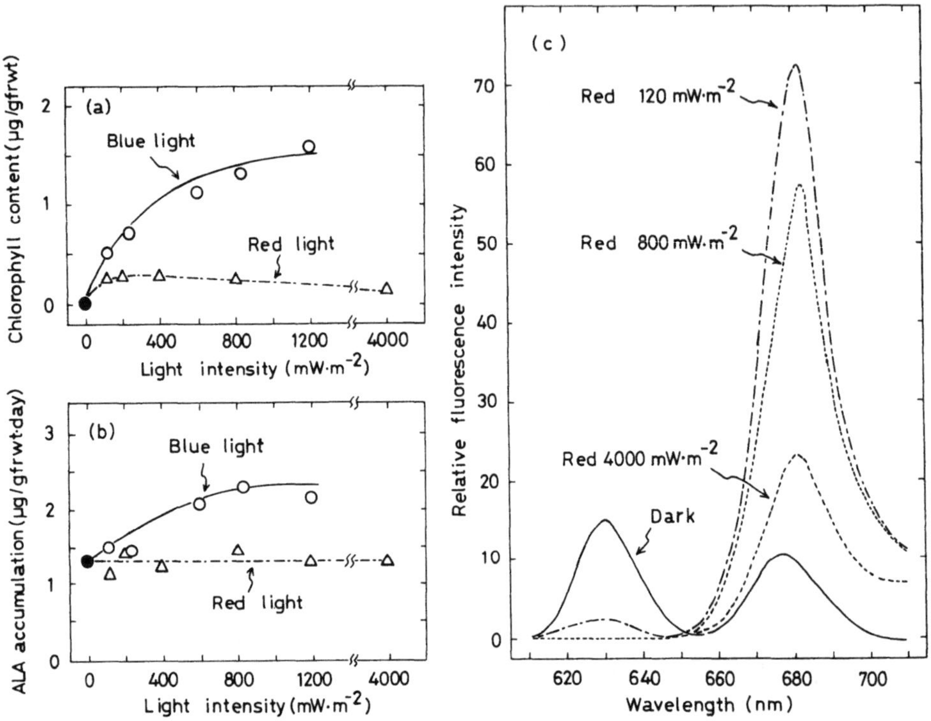

Fig. 1a-c. Effects of blue and red light intensity on the synthesis of Chl and ALA-synthesizing activity in callus cells. **a** Dark-grown callus cells were illuminated with blue and red light of different intensities for 4 days. The amount of chlorophyll was determined with 80% acetone extract; **b** The callus cells pre-illuminated with blue and red light of various intensities for 4 days were incubated for 1 day in the presence of LA (70 mM) under the same light conditions. *Closed circles* in **a** and **b** show the results in the dark. ALA content is given as the mean of three experiments; **c** Dark-grown callus cells were illuminated with red light of different intensities for 4 days, and fluorescence spectra emitted from the callus cells were measured

as demonstrated previously [11]. However, red light, either strong or weak, did not increase the rate of ALA formation (Fig. 1b). These results indicate that the ALA-synthesizing activity in callus cells is enhanced by blue but not red light. The slight Chl formation under weak red light seems to be due to the circumstance that Pchl was phototransformed without destruction, into Chl. As shown in Fig. 1c, when dark-grown callus cells were illuminated with weak red light (120 mW m^{-2}) for 4 days, a fluorescence peak at 682 nm, which is attributable to Chl-678 (absorption maximum; 678 nm, see [12]), appeared and the intensity of fluorescence at 631 nm, which is attributable to Pchl, decreased to one-fourth of that of dark-grown callus cells. When the intensity of the red light was increased, the fluorescence intensity at 682 nm markedly decreased and Pchl disappeared. The results that strong red light caused complete disappearance of Pchl, while showing no enhancement of ALA formation (Fig. 1b) support the inference that the Pchl level in dark-grown callus cells was too low to inhibit the ALA synthesis [12]. As shown in Fig. 1c, dark-grown callus cells showed a fairly high peak around 676 nm, besides a 631 nm peak. This peak seems to be attributable to dark-accumulated Chl. Details will be presented elsewhere.

Time Course of Blue Light Enhancement of ALA Formation

Figure 2 shows a time course of enhancement of the ALA-synthesizing activity of callus cells during the greening process under blue light. Although the time course varied somewhat among different batches of callus, the figure shows that a fairly long time was needed for full enhancement of the ALA-synthesizing activity by blue light. Though not shown in the figure, the ALA-synthesizing activity of the cells, which had been illuminated with blue light for 6 days, declined very slowly during the subsequent dark incubation [12].

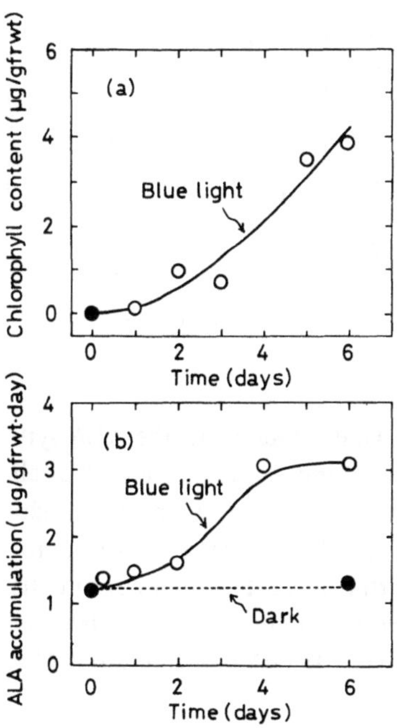

Fig. 2a,b. Time courses of Chl accumulation and enhancement of ALA-synthesizing activity in callus cells under blue light. **a** Time course of Chl accumulation under blue light. The amount of chlorophyll was determined with 80% acetone extract; **b** Dark-grown callus cells were illuminated with blue light for 6 h, 1, 2, 4, or 6 days, then incubated under blue light for another day in the presence of LA (70 mM). The dark sample (*closed circles*) was incubated in the dark for 1 day in the presence of LA (70 mM). Light intensity, 1200 mW m^{-2}

Effects of Blue Light on ALA-Synthesizing Activity and Respiration in Callus Cells Incubated on Media Containing Various Concentrations of Sucrose

Callus cells were inoculated onto sucrose-free agar medium and incubated for 12 days in the dark. The carbon-starved callus cells thus obtained were inoculated onto agar media containing various concentrations of sucrose and incubated for 8 days under blue light and in the dark. Figure 3 shows the effects of blue light on the ALA-synthesizing activity and the rate of O_2-uptake as a function of the initial concentration of sucrose. With increasing concentrations of sucrose in the medium, both the ALA-synthesizing activity and the rate of respiration in callus cells incubated in the dark

Fig. 3a,b. Effects of blue light on the ALA-synthesizing activity and the activity of respiration of callus cells as a function of the concentration of sucrose added. Callus cells grown in the sucrose-free agar medium for 12 days in the dark were incubated on agar media containing various concentrations of sucrose under blue light and in the dark for 8 days. To study the ALA-synthesizing activity **a)** callus cells were incubated under the same light conditions for another day in the presence of LA (100 mM). Light intensity, 1200 mW m^{-2}. Oxygen uptake was determined in the dark at 25°C **(b)**

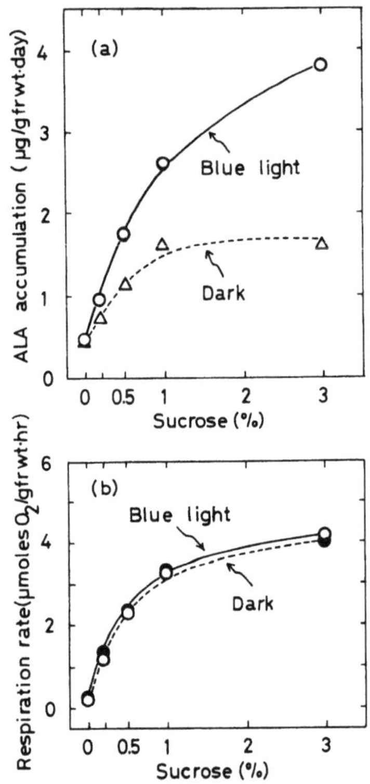

increased and reached a constant level at 1% sucrose. Under blue light, the ALA-synthesizing activity in the cells was always higher than that incubated in the dark at all sucrose concentrations used. However, respiration was not enhanced by blue light when the respiratory activity was measured in the dark. The enhancement was not observed either when the O_2 uptake was measured with callus cells under continuous illumination with blue light [12]. These results indicate that the enhancing effect of blue light on the ALA-synthesizing activity in callus cells is due to stimulation of a process other than respiration.

Effects of Blue Light on Chl Formation and Respiration in Carbon-Starved Callus Cells Incubated on Media Containing Various Concentrations of Glutamate

After incubation in sucrose-free medium in the dark for 12 days, carbon-starved callus cells were illuminated with strong red light for one day to destroy Pchl. The Pchl-free callus cells were inoculated onto agar media containing various concentrations of glutamate, and incubated for 2 days under blue light and in the dark. Table 1 shows the effects of blue light on Chl formation and respiration of the carbon-starved callus cells. The activity of carbon-starved callus cells for Pchl formation and respiration showed only a slight increase in the dark during 2 days of incubation with glutamate. Under blue light, on the other hand, the activity of Chl formation, but not respiration, strikingly increased, depending on the glutamate concentration: blue light en-

Table 1. Effects of blue light on Chl formation and respiration in carbon-starved callus cells after provision of glutamate

Light condition	Glutamate (mM)	Pchl[a] (F631)	Chl[b] (F683)	Respiration (μmol O_2 g^{-1} fr. wt. h^{-1})
Dark	0	1.8		1.6
	10	2.0		2.1
	30	2.4		2.0
Blue light	0		2.8	1.7
	10		6.4	2.1
	30		8.2	2.1

[a]Relative fluorescence intensity at 631 nm emitted by callus cells
[b]Relative fluorescence intensity at 683 nm emitted by callus cells

After incubation in sucrose-free agar medium for 12 days in the dark, callus cells were illuminated with red light (4000 mW m^{-2}) for one day to destroy Pchl, and inoculated onto agar media containing different concentrations of glutamate under blue light (1200 mW m^{-2}) and in the dark for 2 days, respectively. Respiration was measured at 25°C in the dark

hancement of Chl formation was observed at high as well as low concentrations of glutamate.

Discussion

As reported previously [11, 12], dark-grown callus cells contained Pchl-628 (absorption maximum, 628 nm), which was converted to Chl-672 under blue or weak red light and subsequently transformed into Chl-678 independent of light. The over-all rate of transformation of Pchl-628 to Chl-672 under weak red light was virtually the same as that under blue light. However, strong red light, but not blue light, caused the photodestruction of Pchl. Strong white light was found to be effective for the photoreduction of Pchl to Chl with only slight destruction of Pchl, suggesting that blue light counteracts the destructive effect of red light. From these results, we inferred that blue light enhances a certain step(s) involved in the production of a reductant required for the photoreduction of Pchl to Chl (Fig. 4).

As shown in Fig. 1a, the chlorophyll accumulation under the conditions used was much more pronounced under blue light than under red light at all intensities examined. This blue light effect was saturated at the same intensity (600-800 mW m^{-2}) as the saturating intensity of the blue light enhancing the ALA-synthesizing activity. These results indicate that the marked accumulation of Chl under blue light was due to the blue light enhancement of ALA-synthesizing activity (see Fig. 4).

The ALA formation in dark-grown tobacco callus cells was clearly enhanced by blue light, but not at all by red light (Fig. 1b). This excludes the involvement of either Pchl or phytochrome as the photoreceptor in the light enhancement of ALA formation in callus cells. Girnth et al. [8] showed that the formation of ALA-synthesiz-

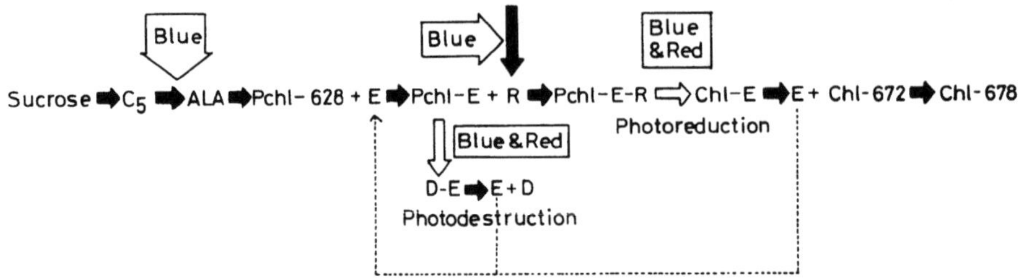

Fig. 4. A model for the formation of ALA and chlorophyll in tobacco callus cells. *White arrows* light processes; *black arrows* dark processes. *R* a reductant; *E* a photoenzyme; *D* a product of photodestruction

ing enzyme in dark-grown seedlings of barley could be induced by several flashes of either blue or red light. More recent studies [9] demonstrated that red light flashes were more effective than blue ones, and suggested phytochrome as a likely photoreceptor.

ALA formation in an early phase of greening of dark-grown microalgae is dependent on blue light (cf. [10]). In this case, carbon metabolism, including degradation of reserve carbohydrates and respiration, are also stimulated by blue light [21]. This blue light effect seems to enhance the supply of substrates and cofactors, which appears to limit the ALA synthesis at early stages of light-induced greening of algal cells (cf. [10]). In callus cells, however, blue light did not enhance respiration, as can be seen from Fig. 3. Callus cells were incubated on an agar medium rich in sucrose in most experiments, which probably provided sufficient substrates and cofactors for ALA synthesis. As seen from Fig. 3 and Table 1, ALA formation was significantly enhanced by blue light when callus cells were incubated with sucrose or glutamate of widely varing concentrations.

Glutamate has been shown to be a substrate for ALA synthesis in higher plants [2] and algae [20]. Thus, the blue light effect on ALA formation in callus cells does not seem to bc due to the enhancement of the production of substrates and cofactors.

Wider de Xifra et al. [23] detected ALA synthetase in extracts of cultured soybean cell, and found that the enzyme activity increases when dark-grown callus cells are exposed to light. Senger et al. [22] showed that the function of key enzymes for ALA synthesis is dependent on light, probably blue light, in dark-grown cells of the mutant C-2A' of *Scenedesmus obliquus.* Although further studies are needed to come to any conclusion, the enhancing effect of blue light on ALA formation in dark-grown tobacco callus cells seems to be mainly due to induction of additional formation of ALA-synthesizing enzyme, which occurs when active Chl synthesis is started in light, demanding rapid provision of ALA. Such effects of blue light on "anaplerotic" processes are known in the carbon metabolism of dark-incubated algal cells [16]. The result that long continuous irradiation of blue light was required for full enhancement of ALA formation in callus cells is in sharp contrast to the rapid and full induction of ALA-synthesizing enzyme by only several flashes of red or blue light, probably mediated by phytochrome, in dark-grown seedlings of barley [9]. However, a blue light receptor, instead of phytochrome, might play the regulatory role in the induction of ALA-synthesizing enzyme in unorganized callus cells, or the blue light effect in highly organized plant tissues, though existing, might be hidden by the striking red light effect mediated by phytochrome.

References

1. Arnon MD (1949) Copper enzymes in isolated chloroplasts. Polyphenoloxidase in *Beta vulgaris.* Plant Physiol 24:1–15
2. Beale SI, Gough SP, Granick S (1975) Biosynthesis of σ-aminolevulinic acid from the intact carbon skeleton of glutamic acid in greening barley. Proc Natl Acad Sci USA 72:2719–2723
3. Beale SI (1978) σ-Aminolevulinic acid in plants: Its biosynthesis, regulation, and role in plastid development. Annu Rev Plant Physiol 29:95–120
4. Berger Ch, Bergmann L (1967) Farblicht und Plastidendifferenzierung im Speichergewebe von *Solanum tuberosum* L. Z Pflanzenphysiol 56:439–445
5. Bergmann L, Berger Ch (1966) Farblicht und Plastidendifferenzierung in Zellkulturen von *Nicotiana tabacum* "Sumsun". Planta 69:58–69
6. Björn LO (1965) Chlorophyll formation in excised wheat roots. Physiol Plant 18:1130–1142
7. Ford MJ, Kasemir H (1980) Correlation between 5-aminolaevulinate accumulation and protochlorophyll photoconversion. Planta 150:206–210
8. Girnth C, Gough SP, Kannangara CG (1980) Induction of σ-aminolevulinate synthesizing activity by light. In: De Greef J (ed) Photoreceptors and plant development. Univ Press, Antwerpen, pp 261–267
9. Gough SP, Girnth C, Kannangara CG (1981) σ-Aminolevulinate synthesis in greening barley. 1. Regulation. In: Akoyunoglou G (ed) Photosynthesis, V. Chloroplast development. Balaban Int Sci Serv, Philadelphia, pp 107–116
10. Hase E (1980) Effects of blue light on greening in microalgae. In: Senger H (ed) The blue light syndrome. Springer, Berlin Heidelberg New York, pp 512–525
11. Kamiya A, Ikegami I, Hase E (1981) Effects of light on chlorophyll formation in cultured tobacco cells I. Chlorophyll accumulation and phototransformation of protochlorophyll (ide) in callus cells under blue and red light. Plant Cell Physiol 22:1385–1396
12. Kamiya A, Ikegami I, Hase E (1983) Effects of light on chlorophyll formation in cultured tobacco cells II. Blue light effect on 5-aminolevulinic acid formation. Plant Cell Physiol 24:799–809
13. Klein S, Katz E, Neeman E (1977) Induction of σ-aminolevulinic acid formation in etiolated maize leaves controlled by two light systems. Plant Physiol 60:335–338
14. Masoner M, Kasemir H (1975) Control of chlorophyll synthesis by phytochrome I. The effect of phytochrome on the formation of 5-aminolevulinate in mustard seedlings. Planta 126:111–117
15. Miller GW, Denney A, Wood JK, Welkie GW (1979) Light-induced delta-aminolevulinic acid in dark-grown barley seedlings. Plant Cell Physiol 20:131–143
16. Miyachi S, Kamiya A, Miyachi Sh (1977) Wavelength effects of incident light on carbon metabolism in *Chlorella* cells. In: Mitsui A et al. (eds) Biological solar energy conversion. Academic, London New York, pp 167–182
17. Oh-hama T, Hase E (1978) Blue light effect on chlorophyll formation in *Chlorella protothecoides.* Photochem Photobiol 27:199–202

18. Oh-hama T, Senger H (1978) Spectral effectiveness in chlorophyll and 5-aminolevulinic acid formation during regrenning of glucose-bleached cells of *Chlorella protothecoides*. Plant Cell Physiol 19:1295–1299

19. Oh-hama T, Hase E (1981) Role of light in 5-aminolevulinic acid formation in wild strain and mutant C-2A' cells of *Scenedesmus obliquus*. Plant Cell Physiol 22:747–757

20. Oh-hama T, Seto H, Otake N, Miyachi S (1982) ^{13}C-NMR evidence for the pathway of chlorophyll biosynthesis in green algae. Biochem Biophys Res Commun 105:647–652

21. Senger H, Bishop NI (1972) The development of structure and function in chloroplasts of greening mutants of *Scenedesmus* I. Formation of chlorophyll. Plant Cell Physiol 13: 633–649

22. Senger H, Klein O, Dörnemann D, Porra RJ (1980) The action of blue light on 5-aminolevulinic acid formation. In: Senger H (ed) The blue light syndrome. Springer, Berlin Heidelberg New York, pp 541–551

23. Wider de Xifra EA, Batle AM del C, Tigier HA (1971) σ-Aminolaevulinate synthetase in extracts of cultured soybean cells. Biochim Biophys Acta 235:511–517

The Blue Light Factor in Sun and Shade Plant Adaptation

K. HUMBECK and H. SENGER [1]

Introduction

Sun and shade plants are adapted to specific light conditions. They can live in extremely different environments. Shade plants can grow in the deep shade under the canopy of a rainforest and sun plants in the full sunlight of open fields.

The question arises: on the one hand, how are the cells adapted to allow the shade plants to survive in this specific light environment found in the shade of other plants? and on the other, what is the structure of the photosynthetic apparatus in the sun plants which enables them to use the high light intensities of their natural environment for high photosynthetic capacities?

The differences between sun and shade plants are best documented in the light saturation curves of photosynthesis (Fig. 1). Sun plants such as *Tidestromia oblongifolia* are characterized by a high respiration rate, a compensation point at higher light intensities and high photosynthetic capacities. Some sun plants do not even reach saturation of photosynthesis at a very high light intensity. On the contrary, shade plants such as *Alocasia macrorrhiza* reach saturation of photosynthesis and their compensation point at low light intensities and have low respiration rates [1, 2]. Surprisingly, both sun and shade plants work with the same quantum efficiency at low intensities (cf. [18]).

In the plant kingdom one can distinguish between plants which are genetically determined to be sun- or shade-type plants and those which can adapt to a sun environment as well as to a shade environment. The genetically fixed plants will not survive or will grow much more slowly after transfer to the opposite light conditions. All algae tested belong to the adaptable type. Adaptable plants seem to respond not only to the intensities of the light applied, but also to its quality. This is to be expected, since the spectral distribution of light is different in various biotops.

Adaptation to Light Intensity

Figure 2 shows light saturation curves of photosynthetic oxygen evolution of *Scenedesmus obliquus* grown under low (5 W m^{-2}) and high (28 W m^{-2}) intensities of white

1 Fachbereich Biologie/Botanik, Philipps-Universität, 3550 Marburg, FRG

Blue Light Effects in Biological Systems
ed. by H. Senger

Fig. 1. Light dependence of net CO_2 uptake of the sun plant *Tidestromia* and the shade plant *Alocasia*. (After [1])

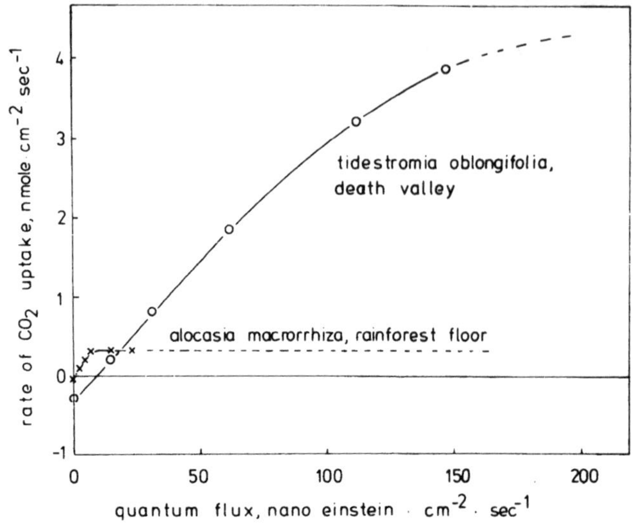

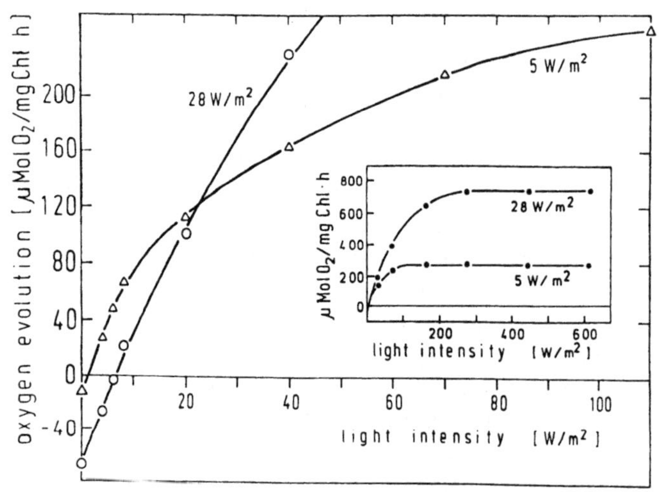

Fig. 2. Photosynthetic oxygen evolution of cultures of *Scenedesmus obliquus* grown under high or low light intensities of white light. Cultures were grown homocontinuously at 30°C and aerated with 3% CO_2 in air under 28 W m^{-2} or 5 W m^{-2}. Oxygen evolution was determined polarographically in an oxygraph under various intensities of red light (> 620 nm)

light [18]. The light saturation curves for *Scenedesmus* cells grown under high or low intensities show the same characteristics for capacity, compensation point and dark respiration as sun and shade plants respectively.

Another parameter different in low and high light-adapted plants is the chlorophyll content [18]. The curve in Fig. 3 shows the dependence of the chlorophyll concentration in *Scenedesmus* cells upon the light intensity of white light provided during growth. By adaptation to high intensities, the chlorophyll content per cell mass is less than in cultures grown under low conditions. These results found in the green alga

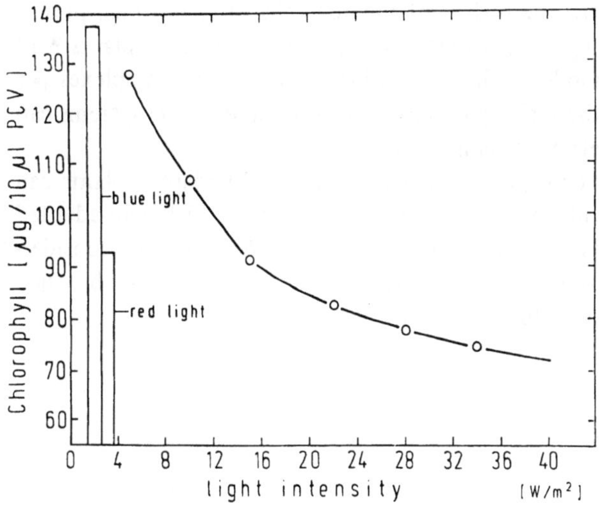

Fig. 3. Chlorophyll content of homocontinuous cultures of *Scenedesmus obliquus* grown under various light intensities of white light (o – o) or under different light qualities (red light: filter Röhm & Haas, 502; blue light: filter combination Schott DAL 444, DAL 424 and AL 423; both: 2.5 W m^{-2}). Other growth conditions as in Fig. 2

Scenedesmus are in agreement with those reported for higher plants which were adapted to low and high light conditions [19].

Besides the differences in photosynthetic capacity and chlorophyll content, variations in the leaf anatomy [9, 17], chloroplast structure [3], enzyme activities [10, 20] and electron transport system [19] are reported for higher plants.

Table 1. Cytochrome, plastoquinone (PQ) and P-700 content of cells from *Scenedesmus obliquus* grown under low or high light conditions. For further growth conditions see Fig. 2

Conditions	n mol mg^{-1} Chl cyt f	cyt b-563	cyt b-559 HP	PQ	mol Chl mol^{-1} P-700
Low light (5 W m^{-2})	1.3	1.6	2.8	6.7	197
High light (28 W m^{-2})	3.3	3.1	3.9	25.4	211

As in higher plants, in the photosynthetic electron transport chain of *Scenedesmus* the pools of cytochromes and plastoquinone determined on a chlorophyll basis are larger in high light than in low light-adapted cells [7]. Computations of molar concentrations of electron carriers per chlorophyll for *Scenedesmus* grown under different light intensities (see Table 1) compare very well with the data reported for higher plants adapted to low or high light conditions [3, 19]. Differences in the ratio of P-700: chlorophyll in higher plants [3, 19] and *Scenedesmus* [7] adapted to low and high light conditions are not considered significant. They reach values up to 10% measured in chloroplast particles prepared from the cells of *Scenedesmus*.

Based on ratios of redox carriers/chl and chl/P-700 Wild [19] proposed that in high light plants the structure of the photosynthetic apparatus can be described by the tripartite model of Butler and Kitajima [5] with one light-harvesting complex and the two reaction centers (each with one antenna) per one electron transport chain.

This model serves to explain how the high light-adapted cells carry out high photosynthetic efficiency under high light intensities. For the low light-adapted plants Wild [19] proposed a multipartite model with more chlorophyll protein complexes per electron carriers compared to the high light cells. Thus several antennae systems can function together to serve one electron transport chain.

This multipartite model shows quite well that the low light-adapted plants can catch all available light better with their antennae systems than the high light-adapted ones. This model (Fig. 5) serves well the conditions in *Scenedesmus*. Even considering a small increase in chl b in the low light-adapted cells, this change would be small in comparison to the change in the chl/redox carrier ratios.

Adaptation to Light Quality

Considerable alterations in the spectral distribution of light can be observed in different habitats. Holmes and Smith [11] measured the spectral energy distribution above a wheat canopy in the sunlight and in the shade of the wheat plants. They found that plants living in the shade of a vegetation have an extremely different spectral energy influx than plants exposed directly to sunlight. In the shade of the wheat plants the green and the far-red part of the light is present in a higher percentage, since the blue and the red part of the spectrum is filtered away by the green leaves above.

Blue and Red Light Irradiation of Higher Plants

Lichtenthaler et al. [13] irradiated etiolated seedlings of barley continuously with broad band blue and red light. They found that the content of chl a as well as of chl b per leaf is less after blue light irradiation. Furthermore, there were differences in the Hill activity of these blue and red light-exposed chloroplasts. The higher activity per chlorophyll could be found in the blue light-adapted cells. In the ultrastructure, they could find great differences: there are fewer grana in the cells grown in blue than in the red light-adapted [4]. These differences in photosynthetic activity, pigment content and chloroplast structure in the blue and red light-adapted plants are similar to those found in sun and shade respectively.

The most possible photoreceptor to mediate this adaptation to the light quality is the phytochrome system [11, 19].

Blue and Red Light Irradiation of the Green Alga *Scenedesmus obliquus*

Scenedesmus cultures adapted to blue or red light conditions demonstrate characteristic differences. Figure 4 shows the light saturation curves of photosynthetic oxygen evolution of *Scenedesmus* cells grown a long time under red or blue light. Blue light-adapted cells exhibit a lower photosynthetic capacity, dark respiration and compensation point and reach saturation at lower light intensities than those cells adapted to red light. In these asepcts the blue and red light-adapted cells resemble low and high light-grown cells respectively.

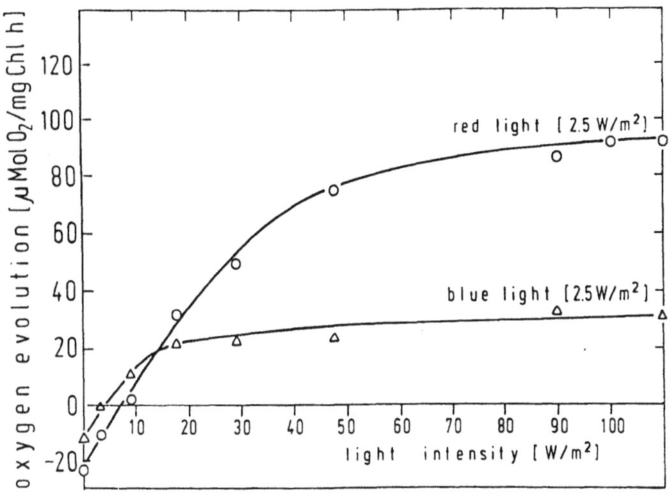

Fig. 4. Photosynthetic oxygen evolution of cultures of *Scenedemus obliquus* grown under different light qualities. The filters are the same as described in Fig. 3. Other conditions as in Fig. 2

The chlorophyll content per cell mass of cells grown under different light qualities is shown in Fig. 3. For this parameter again the blue light mimics the low intensity. A different behavior can be observed for the chl a/chl b ratio. This value is 5.5 for the red light and 3.1 for the blue light cells. This is a most significant effect and was therefore studied in more detail. The distribution of chlorophyll among the different chlorophyll-protein complexes, analyzed by LDS-PAGE according to Delepelaire and Chua [6] support this result (Humbeck et al. this Vol.). The amount of LHCP relative to the reaction center chlorophyll-protein complexes is higher in the blue light-adapted *Scenedesmus* cultures than in the red light-adapted. Similar results, which indicate again a larger photosynthetic unit size (Chl/RC) after blue light irradiation of *Scenedesmus,* are obtained by greening of a pigment mutant in blue or red light (Humbeck et al. this Vol.).

The fact that for the algae blue light causes the same low intensity adaptation phenomena as red light does for higher plants might be explained by their phylogenetical adaptation to the different environments. The natural spectral environments of the algae and of shade plants demonstrate great differences. With increasing depth of water the red part of the spectrum is filtered away and the best transmission is restricted to the blue-green region [8].

Conclusions

Earlier investigations on higher plants considered the photosynthetic apparatus of sun plants to be identical with that of high intensity-adapted plants and the same identity for shade with low intensity-adapted plants. This seems not to be true on the molecular level.

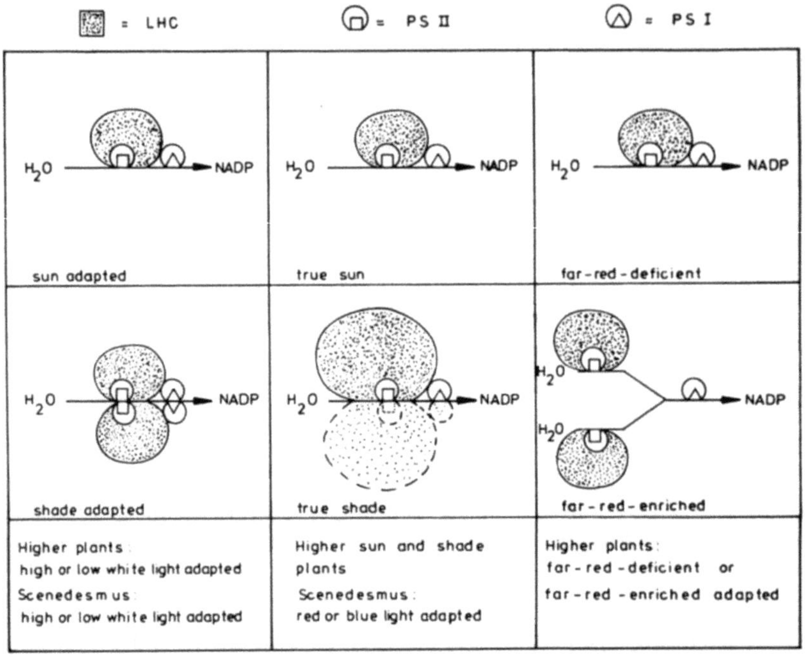

Fig. 5. Preliminary model of the organization of the photosynthetic apparatus in plants exposed to various environmental conditions. LCH = light harvesting chlorophyll; PSI/II = photosystem I/II, each with the reaction center and the close connected antenna system. This model includes the results of [15, 16, 19]

Whereas Wild [19] discusses the same photosynthetic unit size (chl/RC) for the same plant grown under high or low intensity, Malkin and Fork [15] found variations in the photosynthetic unit size (chl/RCII) between sun and shade plants from the natural environment. This was measured by the method of fluorescence induction [14]. They showed clearly that all shade plants investigated had a larger photosynthetic unit size than the sun plants. Two types of adaptation seem to exist: the low intensity-adapted cells have more photosynthetic units of the same size as high intensity-adapted cells per electron transport chain, whereas the true shade plant has larger photosynthetic units with a lower chl a/b ratio than the sun plant. This situation is described by the model shown in Fig. 5.

Investigations on *Scenedesmus* grown under high or low light intensities clearly follow the scheme proposing multiplying of the photosystems of equal sizes [7, 18].

Besides the effects of low or high intensities of white light, the photosynthetic apparatus adapts to various light qualities. Higher plants and algae showed a different behavior, i.e., blue light irradiation mimicked sun conditions in the higher plants and shade conditions in the algae. This might be explained by differences in their natural spectral environment.

Whereas the photosynthetic reactions of *Scenedesmus* grown under blue light are very similar to those of cells grown under low intensities of white light, on the molecular level the adaptation seems to be achieved by enlargement of the unit size (Humbeck et al. this Vol.).

The measurements with higher plants adapted to red or blue light demonstrate correspondingly that red light mimics the low intensity in aspect of photosynthetic reactions [13, 19]. The PS unit size is not discussed under these conditions. A few determinations on higher plants adapted with white light enriched or deficient of the far red portion of the spectrum are concerned with the molecular level of the photosynthetic apparatus [16]. They indicate a third type of adaptation, i.e., the increase of the ratio of PSII/PSI.

The experiments with photosynthetic organisms adapted to various intensities or qualities of light document again the flexibility of the photosynthetic apparatus to respond on the molecular level to environmental factors to ensure optimal function.

Abbreviations

LHCP = light harvesting chlorophyll-protein complex, LDS = lithium dodecyl sulfate, PAGE = polyacrylamide gel electrophoresis.

Acknowledgments. We thank the Deutsche Forschungsgemeinschaft (SFB 103) for generous financial support, Mrs. I. Koss for technical assistance and Mrs. I. Krieger, Mrs. P. Schreiber, and Mr. H. Becker for their help during the preparation of the manuscript.

References

1. Berry JA (1975) Adaptation of photosynthetic processes to stress. Science 188:644–650
2. Boardman NK (1977) Comparative photosynthesis of sun and shade plants. Annu Rev Plant Physiol 28:355–377
3. Boardman NK, Björkman O, Anderson JM, Goodchild DJ, Thorne SW (1975) Photosynthetic adaptation of higher plants to light intensity; relationship between chloroplast structure, composition of the photosystems and photosynthetic rates. In: Avron M (ed) Proc 3rd Int Congr Photosynth III. Elsevier, Amsterdam, pp 1809–1827
4. Buschmann G, Meier D, Kleudgen HK, Lichtenthaler HK (1978) Regulation of chloroplast development by red and blue light. Photochem Photobiol 27:195–198
5. Butler WL, Kitajima M (1975) A tripartite model for chloroplast fluorescence. In: Avron M (ed) Proc 3rd Int Congr Photosynth I. Elsevier, Amsterdam, pp 13–24
6. Delepelaire P, Chua NH (1979) LDS-PAGE of thylakoid membranes at 4°C: characterization of two additional chlorophyll a-protein complexes. Proc Natl Acad Sci USA 76: 111–115
7. Fleischhacker Ph, Senger H (1978) Adaptation of the photosynthetic apparatus of Scenedesmus obliquus to strong and weak light conditions II. Physiol Plant 43:43–51
8. Friedrich H (1965) Meeresbiologie. Borntraeger, Berlin, p 51
9. Grahl H, Wild A (1973) Lichtinduzierte Veränderungen im Photosyntheseapparat von Sinapis alba. Ber Dtsch Bot Ges 86:341–349
10. Hatch MD, Slack GR (1970) Photosynthetic CO_2-fixation pathways. Annu Rev Plant Physiol 21:141–162

11. Holmes MG, Smith H (1977) The function of phytochrome in the natural environment-II. The influence of vegetation canopies on the spectral energy distribution of natural daylight. Photochem Photobiol 25:539–545
12. Humbeck K, Schumann R, Senger H (1984) The influence of blue light on the formation of chlorophyll-protein complexes in Scenedesmus. This volume
13. Lichtenthaler HK, Buschmann C, Rahmsdorf U (1980) The importance of blue light for the development of suntype chloroplasts. In: Senger H (ed) The blue light syndrome. Springer, Berlin Heidelberg New York, pp 485–494
14. Malkin S, Armond PA, Mooney HA, Fork DC (1981) Photosystem II photosynthetic unit sizes from fluorescence induction in leaves. Plant Physiol 67:570–579
15. Malkin S, Fork DC (1981) Photosynthetic units of sun and shade plants. Plant Physiol 67:580–583
16. Melis A, Harvey GW (1981) Regulation of photosystem stoichiometry, chlorophyll a and chlorophyll b content and relation to chloroplast ultrastructure. BBA 637:138–145
17. Nobel PS, Zaragoza LJ, Smtih WK (1975) Relation between mesophyll surface area, photosynthetic rate and illumination level during development of leaves of Plectranthus parviflorus Henckel. Plant Physiol 55:1067–1070
18. Senger H, Fleischhacker Ph (1978) Adaptation of the photosynthetic apparatus of Scenedesmus obliquus to strong and weak light conditions I. Physiol Plant 43:35–42
19. Wild A (1979) Physiologie der Photosynthese Höherer Pflanzen, Anpassung an die Lichtbedingungen. Ber Dtsch Bot Ges 92:341–364
20. Wild A, Zerbe R, Schulte-Wulwer-Leidig A (1981) A comparison of the activities of nitrate and nitrite reductase under weak and strong light conditions. Ber Dtsch Bot Ges 94: 701–707

Chlorophyll a/Chlorophyll b Ratios of *Chlorella vulgaris* in Blue or Red Light

W. KOWALLIK and R. SCHÜRMANN [1]

Introduction

From the experiments of Voskresenskaya in the early 1950's it is well known that in higher plants growth in blue light results in a greater formation of protein, while that in red light leads to an accumulation of carbohydrates. In the meantime these results have been confirmed repeatedly also for several lower plants, including the "model autotroph" *Chlorella*. Therefore, by now this effect of different wavelengths of the visible spectrum appears to be a universal phenomenon in the plant kingdom (for references see Kowallik [17]. Efforts to elucidate these specific influences of light quality have preferably concentrated on examinations of changes in the activities of photosynthetic as well as respiratory enzymes, which in some cases indeed have been found increased on blue irradiation (Ruyters this Vol.). Additionally, attention has also been focused on possible regulations in the formation of the photosynthetic apparatus. Chlorophyll biosynthesis and the formation of the different chlorophyll protein species have been in the center of these investigations, which sometimes revealed clear differences between blue- or red-treated plants [1, 3, 19, 21, 24].

We here report on blue-dependent changes in the ratio of chlorophyll a/chlorophyll b which might indicate alterations in the ratio of the two photosystems. If so, short wavelength visible radiation might be necessary to produce an optimally equipped photosynthetic apparatus in *Chlorella*.

Material and Methods

For all experiments the unicellular green alga *Chlorella vulgaris*, No. 211-11 h of the Culture Collection at the University of Göttingen, has been used. The organism was grown autotrophically in an inorganic nutrient solution [16] in glass tubes of 100 ml capacity. The suspensions were continuously aerated from the bottom with an air + 2% CO_2 gas mixture. Growth temperature was 30°C. Stock cultures were grown at an optical density at λ 800 nm of 1.2–1.3 in white light of 4500 lx. For the experiments with monochromatic light such suspensions were diluted to an OD_{800} of 0.350 at which they were kept by intermittent dilution in long-term experiments.

1 Fakultät für Biologie, Universität Bielefeld, 4800 Bielefeld 1, FRG

Blue Light Effects in Biological Systems
ed. by H. Senger

Blue light was produced by passing the beam of a Xenon high pressure arc (Osram 450 W/XBO or 150 W/XBO) through a blue Plexiglas filter, cutting off all wavelengths longer than 550 nm or, in case of the determination of a wavelength dependence, through interference filters with halfband widths of 10-15 nm.

Red light resulted from filtering the radiation of a tungsten lamp (Osram Concentra, 235 V, 150 W) through a Plexiglas filter transmitting all wavelengths longer than 580 nm.

For determination of the respective quantum fluence rates the QSM 11-2500 spectrometer (Techtum instruments) was used.

Total nitrogen was determined by the Kjeldahl, total carbohydrates by the anthrone method [22]. The chlorophylls were measured in methanol extracts according to McKinney [20] and Holden [12], the carotenoids by using TLC after Evangelatos and Akoyunoglou [9].

Results

When exposed to blue or to red light of fluence rates of 260 μ Einstein m^{-2} s^{-1} each, i.e., of 6200 or 4200 μW cm^{-2}, respectively, autotrophically growing cells of the unicellular green alga *Chlorella vulgaris* double their dry matter equally within 8 h. Their composition, however, is different: corresponding with earlier reports [16], cells from blue light cultures contain more protein per unit dry weight than those from red light cultures, while the latter exhibit an increased level of carbohydrates. Another difference exists in the amounts of photosynthetic pigments: Total chlorophylls as well as total carotenoids are both enhanced by 25 to 35% in blue-grown cells (Table 1).

Table 1. Protein, carbohydrate, total chlorophylls, and total carotenoids mg^{-1} dry weight of *Chlorella vulgaris* grown autotrophically in red (4200 μW cm^{-2}) or in blue (6200 μW cm^{-2}) light of equal dry matter production

	protein	carbohydrate	chlorophylls	carotenoids
	µg/mg			
red	380	350	5,73	3,41
blue	450	250	7,77	4,24

The increase in total chlorophylls derives from increases in chlorophyll a and in chlorophyll b as well, but the enhancement in chlorophyll b is much more pronounced. Therefore, the ratio of both these pigments differs markedly: While in blue-grown cells chlorophyll a/chlorophyll b is about 2.9/1, it is greater than 5.0/1 in red-grown algae. Since the ratio of around 3.0/1 of the blue-grown algae is hardly different from that in cells cultivated in white light leading to equal dry matter production, transfer of white-grown cells to blue or red light results in an almost unchanged ratio in short wavelength light, but in a rather fast increase in chlorophyll a/chlorophyll b in red

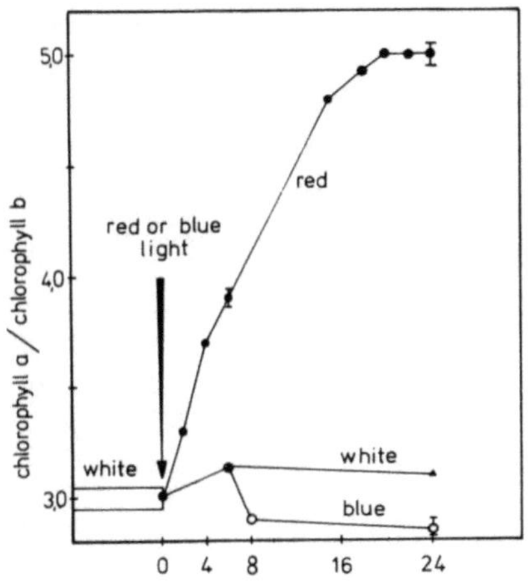

Fig. 1. Changes in chlorophyll a/b ratios in autotrophically growing *Chlorella vulgaris* on transfer from white to red (4200 μW cm^{-2}) or blue (6200 μW cm^{-2}) light of equal dry matter production

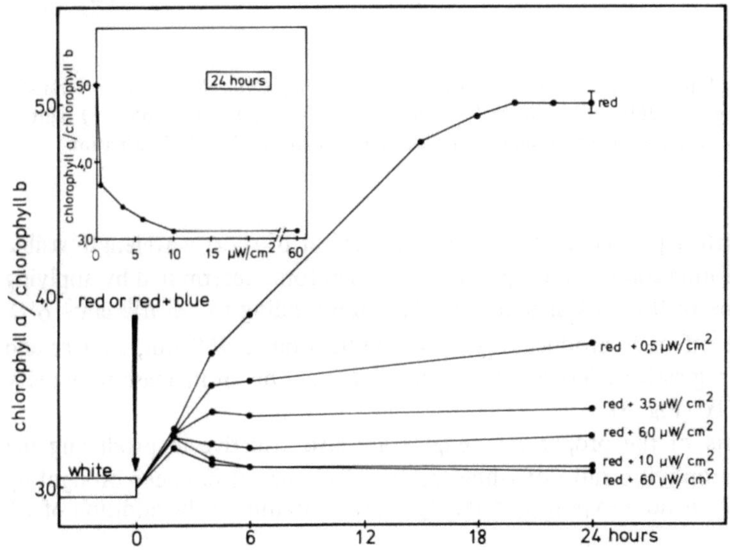

Fig. 2. Changes in chlorophyll a/b ratios in autotrophically growing *Chlorella vulgaris* on transfer from white to strong red (4200 μW cm^{-2}) or strong red plus weak blue light. *Inset* Energy dependence of the reducing effect of blue light on the chlorophyll a/b ratio in red light

light. It takes 15-20 h until the above value of about 5.0/1 is finally reached and stabilized (Fig. 1). During this time about 80% of the chlorophyll a, but only about 50% of the chlorophyll b of blue illuminated cells is formed.

In trying to elucidate the effect on chlorophyll b of those wavelengths absent in the red light field used we first added varying amounts of blue light of λ 453 nm. Figure 2 illustrates that already traces of that radiation are very effective in keeping down the initial ratio of about 3.0/1 thus maintaining the situation characteristic for white-grown cells. Less than 10 μW cm^{-2} added to 4200 μW cm^{-2} red light are sufficient for saturation.

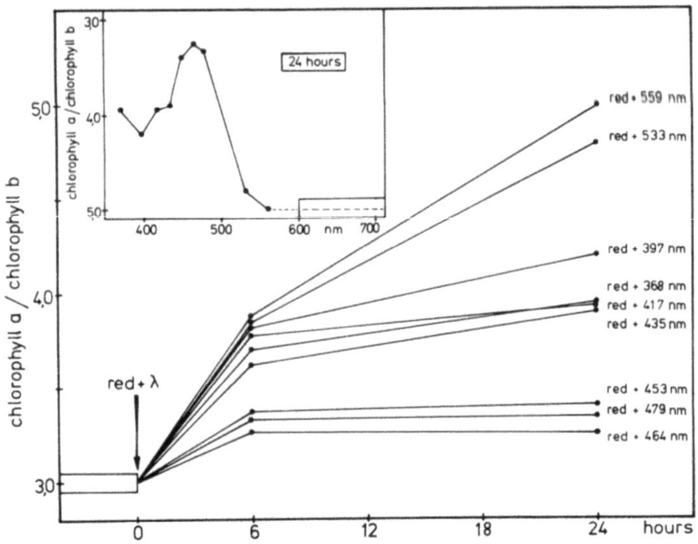

Fig. 3. Changes in chlorophyll a/b ratios in autotrophically growing *Chlorella vulgaris* on transfer from white to strong red (260 μ Einstein m^{-2} s^{-1}) plus weak (0.15 μ Einstein m^{-2} s^{-1}) light of different wavelengths. *Inset* wavelength dependence of that ratio after 24 h of illumination

The wavelength dependence of the enhancing effect of short wavelength visible radiation on the formation of chlorophyll b was, therefore, determined by applying equal fluence rates of 0.15 μ Einstein m^{-2} s^{-1}, corresponding to ~3 μW cm^{-2} of λ 453 nm. Greatest effect is produced by wavelengths around 460 nm, shorter and longer ones are progressively less effective, but there may be an increase in activity again in the near UV (Fig. 3).

Small amounts of the proper wavelengths are also effective in producing the chlorophyll a/chlorophyll b ratio of white-grown cells, when it has been brought up to about 5.0/1 by previous exposure to red light. After 30 min of the addition of 1%

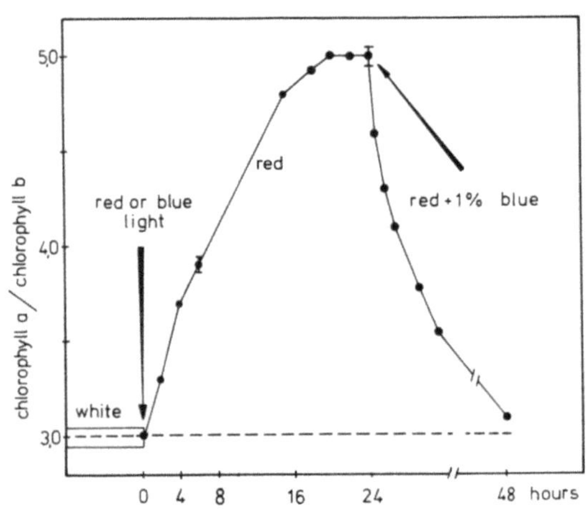

Fig. 4. Influence on the chlorophyll a/b ratio of red-grown *Chlorella vulgaris* of the addition of little blue light (1%)

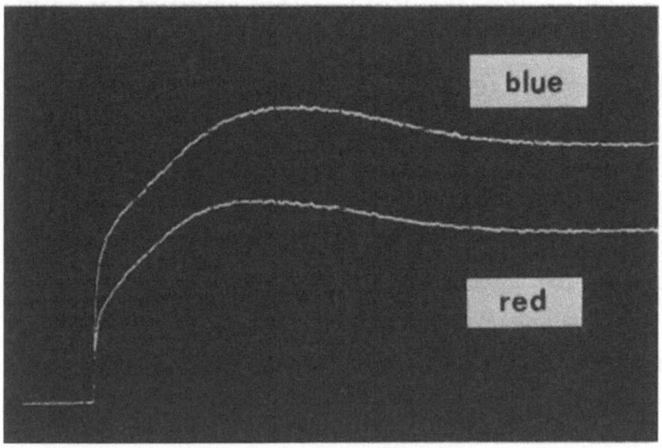

Fig. 5. Fluorescence (λ 691 nm) of red- or blue-grown *Chlorella vulgaris* on an equal chlorphyll a basis

of λ 453 to red light there is a drop from 5.3/1 to about 4.5/1 already. Thereafter the effect slows down, and it takes finally nearly 20 h to establish the low ratio of 3.0/1 (Fig. 4).

To obtain preliminary information on possible alterations of the two photosynthetic photosystems, we measured the fluorescence of blue- and of red-grown cells. As seen in Fig. 5 the stationary fluorescence at λ 691 nm is greater by about 30% in blue- than in red-grown algae.

Discussion

The differences in the amounts of major components, especially the photosynthetic pigments in blue- or red-grown *Chlorella* cells raise the question of any significance of these changes on the photosynthetic capacity. Preliminary experiments indeed show that the O_2-production per total chlorophyll of blue-grown algae is 15-25% greater than that of red-grown cells in white light of different intensities; the difference is even more pronounced (30-40%) on a chlorophyll a basis.

This might be interpreted to show that some factor containing an appropriate amount of chlorophyll b is necessary to secure optimal photosynthesis. Chlorophyll b is assumed to be mostly associated with the light-harvesting chlorophyll protein (LHCP) [2, 25], for which there is much evidence to be functionally related mainly to photosystem II [4, 15, 25]. It thus appears possible that due to a reduced capacity of photosystem II both photosystems are misbalanced in red light and that blue light is somehow necessary to install their proper proportion. A greater capacity of photosystem II of blue-grown cells is certainly indicated by their enhanced fluorescence observed. Investigations on isolated reactions of the photosystems I and II and on the chlorophyll proteins are under way to clarify these ideas further, which at this meeting received additional support by comparable results with a pigment mutant of *Scenedesmus obliquus* (Humbeck and Senger this Vol.). Besides looking for specific light effects on chlorophyll formation proper also Tobin's [26] results of an increase by white light of poly(A)mRNA for the LHCP in *Lemna gibba* should be kept in mind, all the more since Bartling and Kowallik (this Vol.) present evidence for an increase by blue light in total poly(A) mRNA of a chlorophyll-free *Chlorella* mutant. It is certainly fascinating to speculate that it is only the short end of the visible spectrum which is active in bringing about Tobin's increase in the LHCP specific mRNA species, an idea supported by Hundrieser and Richter's [14] observation of an enhancement by blue light in mRNA for ribulose-bisphosphate carboxylase in cell cultures of tobacco.

The increase by blue radiation in the amount of carotenoids has not been followed up closer in this paper. It is therefore only necessary to recall the well-known enhancing effect of short wavelength visible radiation on carotenoids in several fungi [11], but also in unicellular algae [6-8, 10, 27], and – with restrictions – occasionally also in higher plants (e.g., [19]).

Summary

In *Chlorella vulgaris* autotrophically grown in red light the amount of chlorophyll a is only about 80%, that of chlorophyll b even 50% of that in cells cultivated in blue light of equal dry matter production. This results in chlorophyll a/chlorophyll b ratios of about 3/1 in blue-, but about 5/1 in red-grown cells. Addition of only 0.2% of blue to red light leads to the low ratio of pure blue light cultures. Wavelengths around 460 nm are most effective in increasing the amount of chlorophyll b. Influences of short wavelength visible radiation on an optimal constitution of the photosynthetic apparatus are discussed.

Acknowledgments. We thank Prof. G. Schmid for performing the fluorescence measurements and Miss Petra Gayk for typing the manuscript.

References

1. Akoyunoglou G, Anni H, Kalosakas K (1980) The effect of light quality and the mode of illumination on chloroplast development in etiolated bean leaves. In: Senger H (ed) The blue light syndrome. Springer, Berlin Heidelberg New York, pp 473–484
2. Armond PA, Staehelin LA, Arntzen CJ (1977) Spatial relationship of photosystem I and photosystem II, and the light-harvesting complex in chloroplast membranes. J Cell Biol 73:400–418
3. Brinkmann G, Senger H (1980) Blue light regulation of chloroplast development in Scenedesmus mutant C-2A'. In: Senger H (ed) The blue light syndrome. Springer, Berlin Heidelberg New York, pp 526–540
4. Butler WL, Kitajima M (1974) A tripartite model for chloroplast fluorescence. In: Avron M (ed) Proc 3rd Int Congr Photosynth. Elsevier, Amsterdam, pp 13–24
5. Butler WL, Kitajima M (1975) Energy transfer between photosystem II and photosystem I in chloroplasts. Biochim Biophys Acta 396:72–85
6. Dolphin WD (1970) Photoinduced carotenogenesis in chlorotic Euglena gracilis. Plant Physiol 46:685–691

7. Dresbach Ch (1973) Lichtwirkungen auf die Bildung von Photosynthesepigmenten bei Mutanten von Ankistrodesmus und Chlorella. Thesis, Univ Köln
8. Dresbach Ch, Kowallik W (1974) Eine fördernde Wirkung von Blaulicht auf die Carotinoidbildung einer gelben Chlorella-Mutante. Planta 120:291–297
9. Evangelatos GP, Akoyunoglou G (1980) A rapid separation method of leaf pigments by thin layer chromatography on silica gel. 5th Int Congr Photosynth, Halkidiki, Greece, p 150
10. Gross JA, Stroz R (1969) Photostimulation of carotenoid biosynthesis in a nonphotosynthetic Euglena mutant. Plant Physiol Suppl 44:41
11. Harding RW, Shropshire W (1980) Photocontrol of carotenoid biosynthesis. Annu Rev Plant Physiol 31:217–238
12. Holden M (1976) Chlorophylls. In: Goodwin TW (ed) Chemistry and biochemistry of plant pigments, Vol II. Academic, London New York, pp 1–37
13. Humbeck K, Senger H (1984) Influence of light qualities on the formation of chlorophyll-protein complexes. This volume
14. Hundrieser J, Richter G (1982) Blue light induced synthesis of ribulosebisphosphate carboxylase in cultured plant cells. Plant Cell Rep 1:115–118
15. Ke B, Vernon LP, Chaney TH (1972) Photoreduction of cytochrome b_{559} in a photosystem II subchloroplast particle. Biochim Biophys Acta 256:345–357
16. Kowallik W (1962) Über die Wirkung des blauen und roten Spektralbereichs auf die Zusammensetzung und Zellteilung synchronisierter Chlorellen. Planta 58:337–365
17. Kowallik W (1982) Blue light effects on respiration. Annu Rev Plant Physiol 33:51–72
18. Kowallik W, Bartling D (1984) Increase in RNA by blue light in a chlorophyllfree Chlorella mutant. This volume
19. Lichtenthaler HK, Buschmann C, Rahmsdorf U (1980) The importance of blue light for the development of sun-type chloroplasts. In: Senger H (ed) The blue light syndrome. Springer, Berlin Heidelberg New York, pp 485–494
20. McKinney G (1941) Absorption of light by chlorophyll solutions. J Biol Chem 140:315–322
21. Richter G, Reihl W, Wietoska B, Beckmann J (1980) Blue light-induced development of thylakoid membranes in isolated seedling roots and cultured plant cells. In: Senger H (ed) The blue light syndrome. Springer, Berlin Heidelberg New York, pp 465–472
22. Roe JH (1955) The determination of sugar in blood and spinal fluid with anthrone reagent. J Biol Chem 212:335–343
23. Ruyters G (1984) Effects of blue light on enzymes. This volume
24. Schiff JA (1980) Blue light and the photocontrol of chloroplast development in Euglena. In: Senger H (ed) The blue light syndrome. Springer, Berlin Heidelberg New York, pp 495–511
25. Thornber JP (1975) Chlorophyll-proteins: Light-harvesting and reaction center components of plants. Annu Rev Plant Pyhsiol 26:127–158
26. Tobin EM (1981) White light effects on the m-RNA for the lightharvesting chlorophyll a/b protein in Lemna gibba L.G-3. Plant Physiol 67:1078–1083
27. Wallen DG, Geen GH (1971) Light quality and concentration of proteins, RNA, DNA and photosynthetic pigments in two species of marine plankton algae. Mar Biol 10:44–51

The Influence of Blue Light on the Formation of Chlorophyll-Protein Complexes in *Scenedesmus*

K. HUMBECK, R. SCHUMANN, and H. SENGER [1]

Introduction

Chloroplast development in green algae is most significantly regulated by light. The intensity of white light [10, 19] and intermittent illumination [11] influence the chlorophyll a/b ratio, the biosynthesis of pigment-protein complexes and thus the organization of structure and function of the photosynthetic apparatus.

The influence of light of various spectral regions on chloroplast development has been studied under the aspect of respiration enhancement [3, 12], carbohydrate degradation [3, 13], protein and enzyme formation [4, 16, 17] biosynthesis of tetrapyrrol precursors [8, 21] and pigment accumulation [4], as well as the development of photosynthetic activities [10].

Little attention has been paid to the formation of single chlorophyll-protein complexes under various light qualities. In this contribution we will report about the specific biosynthesis of chlorophyll b and light-harvesting chlorophyll-protein complexes in the unicellular green alga *Scenedesmus obliquus*. For these studies we employed either wild-type cells adapted for long periods to blue or red light, or a pigment mutant greening only under light.

Material and Methods

The green alga *Scenedesmus obliquus* strain D_3 was grown autotrophically under homocontinuous conditions [19]. Cultures were illuminated at least for 7 days with red light (5.2×10^{-9} Einstein cm^{-2} s^{-1}, red Plexiglas filter: Röhm & Haas 502) or blue light (5.2×10^{-9} Einstein cm^{-2} s^{-1} filter combination: Schott DAL 444, DAL 424, AL 423). Cells were aerated by bubbling 3% CO_2 in air through the cultures.

The culture conditions for heterotrophic growth of mutant C-2A' of *Scenedesmus obliquus* were the same as described by Senger and Bishop [18]. After 84 h of dark growth, glucose (final concentration: 0.5%) was added and the cells were irradiated with red or blue light for 15 h (filter: Schott NAL 644 or NAL 452, both 4.0×10^{-9} Einstein cm^{-2} s^{-1}).

1 Fachbereich Biologie/Botanik, Universität Marburg, Lahnberge, 3550 Marburg, FRG

Blue Light Effects in Biological Systems
ed. by H. Senger

The chlorophyll content was determined after extraction with hot methanol using the formula of Holden [9] and MacKinney [15].

Analysis of chlorophyll-protein complexes by LDS-PAGE was carried out using the system of Delepelaire and Chua [6]. Their distribution was quantified by scanning at 670 nm.

The preparation of chloroplast particles and the separation on a sucrose density gradient after digitonin treatment were carried out according to Brinkmann and Senger [5].

Results and Discussion

Cells of the wild-type strain of *Scenedesmus obliquus* form chlorophyll in the light or darkness if carbon source and energy are provided. If energy is supplied for autotrophic cultures in the form of blue or red light, the cells will adapt to these conditions after a few days. Since growth is different under the light conditions applied, cells were grown as homocontinuous cultures in a turbidostat. These conditions ensured a homogenous material and provide each cell continuously with the same average light intensity.

Cells adapted under these conditions to either blue or red light show significant differences in the amount and distribution of chlorophylls (Table 1).

Table 1. Content of total Chl, Chl a, Chl b and the ratio of Chl a/Chl b in wild-type cells of the green alga *Scendesmus obliquus* grown 7 days in blue or red light (5.2×10^{-9} Einstein cm^{-2} s^{-1}) under homocontinuous conditions or in mutant C-2A' cells of *Scenedesmus obliquus* greened under red or blue light (4.0×10^{-9} Einstein cm^{-2} s^{-1}) in the presence of glucose (0.5%) within 15 h. For experimental details see Material and Methods

	Homocontinuous cultures of wild-type cells grown under		Mutant C-2A' greened under	
	Red light	Blue light	Red light	Blue light
Total Chl (μg/μl PCV)	9.3	13.8	2.3	2.7
Chl a (μg/μl PCV)	7.9	10.4	1.8	2.0
Chl b (μg/μl PCV)	1.4	3.4	0.5	0.7
Ratio Chl a/Chl b	5.6	3.1	3.6	2.9

Whereas the total chlorophyll is increased in the blue light by about 50% as compared to the red light the chlorophyll b increases by 150%. This clearly indicates the favorization of chlorophyll b formation in blue light. Accordingly the chloropyll a/b ratio drops from 5.6 in the red to 3.1 in the blue.

Analysis of the chlorophyll-protein complex distribution by LDS-PAGE according to Delepelaire and Chua [6] from cells grown under blue or red light yields four

Fig. 1. Scanning diagram of chlorophyll-protein complexes prepared by LDS-gel electrophoresis according to Delepelaire and Chua [6]. Chlorophyll content was measured at 670 nm. The wild-type cells of *Scenedesmus obliquus* grew under red or blue light (5.2 x 10^{-9} Einstein cm^{-2} s^{-1}) for 7 days in a turbidostat. For experimental details see Material and Methods

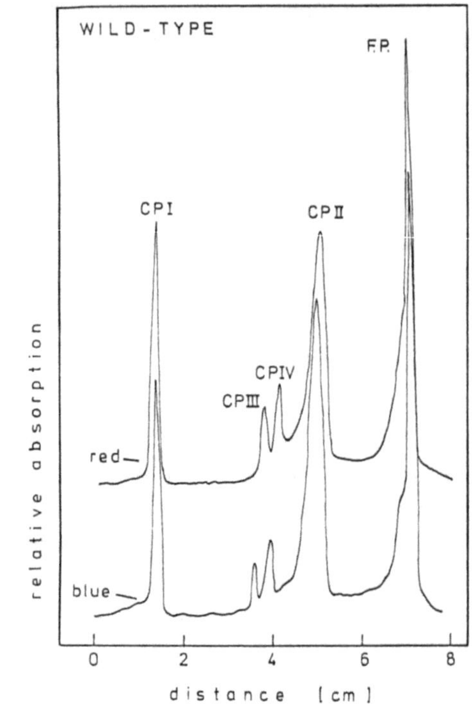

Table 2. Distribution of chlorophyll (%) among the chlorophyll-protein complexes and the free pigment (F.P.) prepared from thylakoids of wild-type cells of *Scenedesmus obliquus* grown under red or blue light (5.2 x 10^{-9} Einstein cm^{-2} s^{-1}). For experimental details see Material and Methods

	CP I	CP II	CP III + IV	F.P.
Red	15.0	37.1	10.1	37.8
Blue	12.9	44.7	6.6	35.8

different complexes plus the free pigment. Scanning profiles of the gels (Fig. 1) demonstrate a considerable increase in CP II complexes attributed to the light-harvesting complex deriving from cells grown under blue light.

Quantitative analysis of the area of the scanning profiles (Table 2) demonstrates an increase of 20% in CP II whereas CP III and CP IV complexes decrease considerably.

The same effect can be seen when chloroplast particle preparations of blue or red light-adapted cells were separated on a sucrose density gradient after digitonin treatment (Fig. 2). Under these conditions fraction I contains free pigments. Fraction II includes the reaction center of PS II and the LHC, fraction III the PS I and fraction IV the complete undissociated photosynthetic apparatus [5]. The distribution pattern clearly indicates that fraction II increases in the blue whereas fraction III demon-

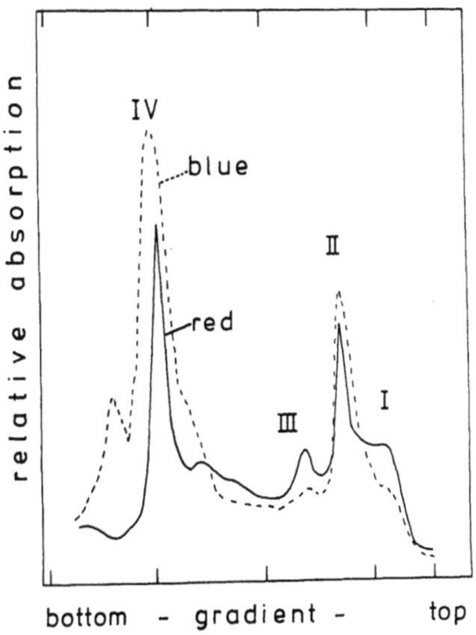

Fig. 2. Distribution of particle fractions of thylakoid preparations from wild type cells of *Scenedesmus obliquus* irradiated with blue or red light and separated on a sucrose density gradient after treatment with digitonin. Transmission was scanned from a photograph. Growth and irradiation conditions as in Fig. 1. For experimental details see Material and Methods

strates a relative decrease as compared to the preparation of cells grown under red light. These results support the findings made on the distribution patterns of chlorophyll-protein complexes.

The result that blue light favors the formation of LHC's is in agreement with the photosynthetic behavior expressed in the light intensity curves [10].

The blue light-adapted cells have a higher response per incident quanta at low quantum flux rates, but reach the saturation level much more quickly than those grown in red light.

Other evidence for the preferential formation of chlorophyll b and LHC under blue light derives from experiments with the greening pigment mutant C-2A' of *Scenedesmus obliquus.*

This mutant develops only traces of chlorophyll when grown heterotrophically in the dark and greens completely in the light [18]. When cells were exposed to either red or blue light during greening the cooperation of various light qualities has been demonstrated previously [20].

Whereas the previous investigations have neglected the formation of specific chlorophyll species and chlorophyll-protein complex pattern we have considered this effect in the current contribution.

As can be taken from Table 1 the chlorophyll formation under red or blue light differ only by 15%. Whereas chlorophyll a increases only by 10% chlorophyll b increases by 40%. Thus the ratio of chlorophyll a/b decreases in cells grown in the blue compared to those grown in red light by about 25%.

The scanning diagram of the chlorophyll-protein complexes derived from cells greened under the different light conditions (Fig. 3) shows according differences. Only CP II complexes increase in the blue whereas all other complexes decrease in comparison (Table 3). The enhanced formation of chlorophyll b and LHCP under blue light in the greening pigment mutant is not quite as pronounced as in the wild type. This might derive from the fact that the wild-type cells were adapted to the

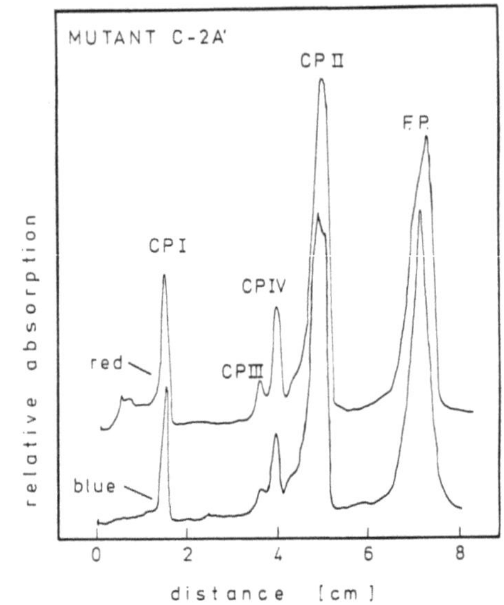

Fig. 3. Scanning diagram of chlorophyll-protein complexes prepared by LDS-gel electrophoresis according to Delepelaire and Chua [6]. Chlorophyll content was measured at 670 nm. The mutant C-2A' cells of *Scenedemus obliquus* greened under red or blue light (4.0 x 10^{-9} Einstein cm^{-2} s^{-1}). For experimental details see Material and Methods

Table 3. Distribution of chlorophyll (%) among the chlorophyll-protein complexes and the free pigment (F.P.) prepared from thylakoids of mutant C-2A' cells of *Scenedesmus obliquus* greened under red or blue light (4.0 x 10^{-9} Einstein cm^{-2} s^{-1}) for 15 h. For experimental details see Material and Methods

	CP I	CP II	CP III + IV	F.P.
Red	13.7	36.5	9.7	40.1
Blue	8.3	40.6	7.5	43.6

light conditions for 7 days whereas the greening pigment mutant was exposed to different light qualities for only 15 h.

Conclusions

The formation of thylakoid membranes is dependent upon light in angiosperms [1, 2]. Even in green algae, which can form chlorophyll in complete darkness, light has a significant regulatory function when cells are illuminated [20]. Action spectra for chlorophyll formation in pigment mutants of *Scenedesmus obliquus* [4, 18] indicate that blue light is most favorable for the biosynthesis of chlorophyll.

In this paper we report a considerable increase in chlorophyll b and the LHC of *Scenedemus* thylakoids under blue light. This is in good agreement with the report on enhanced chlorophyll b formation under blue light in *Chlorella* [14].

Since "white" light contains various amounts of blue light, in these studies red light was always applied as the control. From the two spectral regions blue light is more similar to white light than to red light. This is true for the chlorophyll a/b ratio (white 3.5; blue 3.1; red 5.6) and the amount of chlorophyll/cell mass under similar intensities [10]. From these facts one must conclude that the blue part of white light is most effective for the regulation of chlorophyll biosynthesis in green algae.

The regulation of the size of photosynthetic units (chlorophyll per reaction centers) is linked to the irradiation with blue light. The decreasing ratio of chlorophyll a/b (Table 1), the percent increase of light-harvesting protein complexes (Tables 2, 3) and the course of the photosynthetic light intensity curve [10] indicate an increase of the photosynthetic unit size under blue light compared to red light.

The adaptation to shade conditions either by low white or blue light is directed toward better quantum absorption. Whereas this seems to be realized under low intensities of white light by the increase in the number of photosystems of the same size [7, 10], the cells react to blue light by the enlargement of the light-harvesting system.

Abbreviations

CP = chlorophyll-protein complex; F.P. = free pigment; LDS-PAGE = lithium dodecyl sulfate polyacrylamide gel electrophoresis; LHC = light-harvesting chlorophyll

Acknowledgments. We thank the Deutsche Forschungsgemeinschaft (SFB 103) for generous financial support, Mrs. I. Koss for technical assistance and Mrs. I. Krieger, Mrs. P. Schreiber and Mr. H. Becker for their help during the preparation of the manuscript.

References

1. Akoyunoglou G (1981) Assembly of functional components in chloroplast photosynthetic membranes. In: Akoyunoglou G (ed) Photosynthesis, vol V. Balaban Int Sci Serv, Philadelphia, pp 353–366
2. Boardman NK (1981) Thylakoid membrane formation in the higher plant chloroplast. In: Akoyunoglou G (ed) Photosynthesis, vol V. Balaban Int Sci Serv, Philadelphia, pp 325–339
3. Brinkmann G, Senger H (1978) The development of structure and function in chloroplasts of greening mutants of *Scenedesmus* IV. Plant Cell Physiol 19:1427–1437
4. Brinkmann G, Senger H (1980) Blue light regulation of chloroplast development in *Scenedemus* mutant C-2A'. In: Senger H (ed) The blue light syndrome. Springer, Berlin Heidelberg New York, pp 526–540
5. Brinkmann G, Senger H (1981) Isolation and characterization of a photochemically intact photosynthetic apparatus in *Scenedesmus obliquus.* In: Akoyunoglou G (ed) Photosynthesis, vol III. Balaban Int Sci Serv, Philadelphia, pp 337–346
6. Delepelaire P, Chua NH (1979) Lithium dodecyl sulfate/polyacrylamide gel electrophoresis of thylakoid membranes at 4°C: Characterizations of two additional chlorophyll a– protein complexes. Proc Natl Acad Sci 76:111–115
7. Fleischhacker Ph, Senger H (1978) Adaptation of the photosynthetic apparatus of *Scenedesmus obliquus* to strong and weak light conditions II. Physiol Plant 43:43–51
8. Hase E (1980) Effects of blue light on greening in microalgae. In: Senger H (ed) The blue light syndrome. Springer, Berlin Heidelberg New York, pp 512–525

9. Holden M (1965) Chlorophylls. In: Goodwin TW (ed) Chemistry and biochemistry of plant pigments. Academic, London New York, pp 461–488
10. Humbeck K, Senger H (1984) The blue light factor in sun and shade plant adaptation. This volume
11. Humbeck K, Akoyunoglou G, Senger H (1984) The effect of intermittent light on chloroplast development in a greening pigment mutant of the green alga *Scendesmus obliquus*. Isr J Bot in press
12. Kowallik W (1982) Blue light effects on respiration. Annu Rev Plant Physiol 33:51–72
13. Kowallik W, Schätzle S (1980) Enhancement of carbohydrate degradation by blue light. In: Senger H (ed) The blue light syndrome. Springer, Berlin Heidelberg New York, pp 344–360
14. Kowallik W, Schürmann R (1984) Chlorophyll a/chlorophyll b ratios of *Chlorella vulgaris* in blue or red light. This volume
15. MacKinney G (1941) Absorption of light by chlorophyll solutions. J Biol Chem 140:315–322
16. Ruyters G (1982) Effects of blue-light on pyruvate kinase activity during chloroplast development of unicellular green algae. Photochem Photobiol 35:229–231
17. Schiff JA (1980) Blue light and the photocontrol of chloroplast development in *Euglena*. In: Senger H (ed) Springer, Berlin Heidelberg New York, pp 395–511
18. Senger H, Bishop NI (1972) The development of structure and function in chloroplasts of greening mutants of *Scenedesmus* I. Plant Cell Physiol 13:633–649
19. Senger H, Fleischhacker Ph (1978) Adaptation of the photosynthetic apparatus of *Scenedesmus obliquus* to strong and weak light conditions I. Physiol Plant 43:35–42
20. Senger H, Rühl D (1984) Mode of coaction of photosensing pigments. In: De Greef J (ed) Proc 6th Int Congr Photosynth. Chloroplast Development
21. Senger H, Klein O, Dörnemann D, Porra RJ (1980) The action of blue light on 5-aminolevulinic acid formation. In: Senger H (ed) The blue light syndrome. Springer, Berlin Heidelberg New York, pp 541–551

Carotenogenesis

E. L. Schrott [1]

Introduction

The aim of this review is not to be an updated version of the survey given at the 1st Symposium by Rau and published in the Proceedings [36]. Since in this article our basic knowledge on blue light effects on carotenogenesis in microorganisms is reflected comprehensively, the present review attempts only to supplement our knowledge by covering the most recent literature and by not being formally restricted to microorganisms. For purpose of continuity, some references will date further back. Intensive reviews, which also deal with carotenogenesis, have appeared of late [11, 14, 37, 46, 47] to which the reader is also referred. There will be some unavoidable overlap with other topics in this volume, as carotenogenesis, of course, shares basic principles with other blue light effects.

The aim of the present paper is also to extend the range of subjects concerning carotenogenesis to some closely related aspects.

Photoreceptor Systems

Cryptochrome

For three species of microorganism the action spectra suggest porphyrin(s) to be the photoreceptor(s) [11]. But in most organisms investigated so far, except angiosperms, the "blue (UV) light photoreceptor" is responsible for photocarotenogenesis. Although this evidence is derived only from the action spectra [36] and no statement can be made on the singularity of this pigment system, it was named cryptochrome for convenience. The pertinent action spectra are similar to in vitro absorption spectra of defined flavins and carotenoids. However, absorption spectra are highly dependable upon the solvent used and upon complexation [12, 28, 45] (also see Song this Vol.) and therefore do not congruently reflect the in vivo state. Action spectra might also be perturbed by screening cell components (in particular in the UV region) depending on the organism or the physiological state of the organism. Nevertheless,

1 Botanisches Institut der Universität München, Menzinger Str. 67, 8000 München 19, FRG

Blue Light Effects in Biological Systems
ed. by H. Senger

flavin(s)/flavoprotein(s) or carotenoid(s)/carotenoprotein(s) were the candidates of choice to be the elusive cryptochrome [50, 51].

The controversy between flavins and carotenoids is still pending, despite continuing research in this field. Support for carotenoids as the photoreceptor pigment came from recent experiments on *Phycomyces* [49]. On the other hand, in the same organism, riboflavin could be replaced by roseoflavin (an analog) and achieve 0.1% recovery of the natural phototropism [33]. Other recent results on phototropism favor carotenoids over flavins in corn seedlings [56]. As from earlier data, a single ubiquitous cryptochrome therefore seems to be doubtful.

Carotenogenesis in *Fusarium aquaeductuum* is not affected by several flavin inhibitors or quenchers of flavin triplet state (Schrott unpublished), but these results do not exclude a flavin as the photoreceptor pigment: the triplet state might not be the active state or, simply, uptake of the drugs was insufficient.

Over the years research on the chemical nature of the cryptochrome(s) was stimulated by the discovery that blue light is able to induce absorbance changes (LIAC's) due to flavin-mediated photoreduction of a b-type cytochrome (cf. [8]). LIAC's are still the subject of continuing discussion concerning their physiological relevance (cf. [47]). From the results obtained using diphenylester herbicides which affect both phototropism in vivo and LIAC in vitro to a similar extent, the conclusion is drawn that phototropism and LIAC may act through the same photoreceptor system (Leong and Briggs personal communication). Concerning carotenogenesis, investigations are under way (Schrott unpublished) to prove or disprove a link to physiological responses of those LIAC's orginally found in *Neurospora* (cf. [3]), which are apparently different from those attributed to the nitrate reductase system (cf.[32]).

Valuable help to solve the LIAC dilemma comes from a mathematical approach [27] and from a method introducing phase partition instead of differential centrifugation in order to obtain a fraction enriched in plasmalemma [5, 59].

Action of the Cryptochrome(s). Although there are serious arguments to be acknowledged concerning carotenoids being capable of explaining blue light responses [48], evidence in favor of a flavoprotein is preponderant [35]. The question to answer is how the flavin(s) may act photochemically in a photoreceptor system. Data available so far are consistent in general with a reaction scheme already proposed by Lewis et al. [26]. Reactions which a flavin system may undergo were summarized recently by Strasser and Butler [53].

For carotenogenesis, Rau [36] proposed a reaction scheme (Fig. 1) based mainly on the findings from *Fusarium* and *Neurospora* but also considering LIAC's and articial flavin systems [39]. This hypothetical scheme seems to be conceivable, taking into account that during evolution redoxreactions were used successfully for harvesting sunlight (i.e., photosynthesis) and for the use of the energy potential of organic molecules to form ATP (i.e., respiration). The use of a light-driven flavin-mediated electron transport chain to produce a trigger substance evoking induction is therefore plausible. Only much later in evolution were the higher plants able to use, in addition, a more versatile photoreceptor system, manely phytochrome. Support for redoxreactions being involved in the cryptochromic photoact comes from experiments on *Neurospora* carotenogenesis [2]. These experiments also show that nitrate reductase can hardly be regarded as a key element of the photoreceptor system.

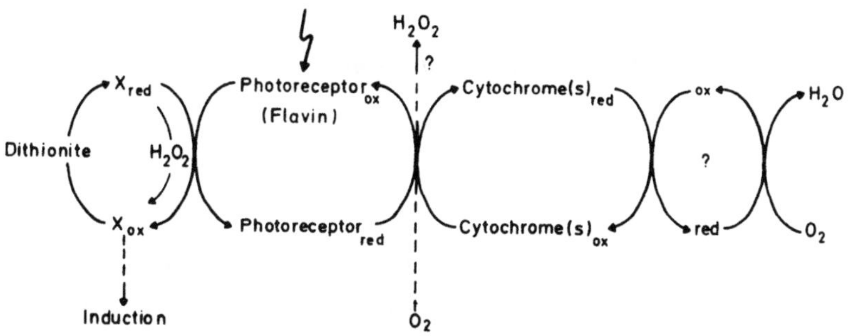

Fig. 1. Scheme illustrating proposed events during the "primary reaction" of chryptochrome [36]

The proposed scheme (Fig. 1) also illustrates that we know even less of the subsequent reactions leading to induction (the "transduction chain"). Results presented by Kritsky et al. [24] suggest that in *Neurospora* a negative relationship exists between the cellular level of cyclic AMP and the rate of accumulation of carotenoids.

Biphasic fluence-response curves for photoinduced carotenogenesis were reported for *Phycomyces* [19], several *Fusarium* species [40] and for *Neurospora* [41]. For *Phycomyces* it was concluded that two different photoreceptor pigments are involved in the response [18, 19]. In *Neurospora* the two phases are separated from each other by an interposing plateau indicating a temporary "blindness" of the photoreceptor system and that competence for further photoinduction is restored by a temperature-dependent reaction [42]. Investigations on the nature of this reaction, using a large variety of inhibitors, failed to bring any elicidation (Schrott unpublished). Afanasieva et al. [1] reported on findings claiming that the temperature at which *Neurospora* was cultivated, controlled formation of functionally active photoregulatory systems. Conclusively, temperature optima for photoreceptor formation, its function and for the subsequent carotenoid formation would be different in *Neurospora* (see also [13, 41]).

UV[2]

The fluence-response curve for *Neurospora* can be altered not only by temperature but also high fluence "black light" (λ_{max} 350 nm, contains some UV-B; see [41]). The action spectrum for *Neurospora* carotenogenesis shows, besides the action in the blue, a peak in the UV-A region (and this photoresponse therefore was classified as cryptochromal) but it also shows action in the UV-B [9]. Similar characteristics were reported on carotenogenesis in *Verticillium* [16]. The UV-B peak might be due to fluorescence (as, of course, the UV-A peak) but might also be the manifestation of a different photoreceptor pigment or system. A UV-B photoreceptor was reported for some developmental processes (cf. [58], see also [10]). The question arising,

2 UV is defined as < 380 nm with subsections denoted as UV-A (320-400 nm) UV-B (280-320 nm) and UV-C (<280 nm).

whether two different photoreceptors may be involved in carotenogenesis in *Neurospora,* remains to be solved. Dual control is a widespread phenomenon in the case of blue/UV light and phytochrome [31, 58].

Mycochrome

When UV-inducible morphogenetic processes are reversed by blue light, or blue light-suppressible developments are reversed by a subsequent "black light" (predominantly UV-A) irradiation, the pigment system responsible for the control was termed "mycochrome" (cf. [6, 25]).

No case was reported so far where mycochrome controlled photocarotenogenesis. However, there is a possibility that a blue light induction attributed to cryptochrome is in fact UV-reversible, and therefore would be another type of mycochrome action. Also, a low threshold for UV light coupled with contamination of the blue light used with some UV would mimic crytochrome action, although a response is mycochromal. For photocarotenogenesis in *Fusarium* and *Neurospora* interactions of UV-A and blue light could be excluded (Huber-Willer et al. in preparation).

Phytochrome

Carotenogenesis in angiosperms is under phytochrome control. For fungi, the presence of phytochrome was a matter of intensive debate during the last few years. The reader is referred to the comprehensive review of Gressel and Rau [11] and a most recent paper of Valadon et al. [55] for more detailed information.

Since carotenogenesis in *Verticillium* (cf. [55]) is controlled by an interaction of phytochrome with a UV/blue light accepting system (most likely not cryptochromal) and since presence of phytochrome was reported for *Neurospora* [21], it had to be clarified whether photocarotenogenesis in *Neurospora* and *Fusarium*, so far ascribed to cryptochrome, is indeed purely cryptochromal. The thorough investigating disclosed no interaction of the response with red light [44].

Chemoinduction

Besides light, in a few fungi and one bacterium, certain chemicals have been found to stimulate or even induce cartenogenesis (cf. [36]). Most recently the discovery was made that veratrole (1,2-dimethoxybenzene) and an impressive number of derivatives are able to give rise to massive carotenoid synthesis in *Phycomyces* (Cerdá-Olmedo personal communication). On the other hand, these chemicals had no such effect in *Fusarium moniliforme.*

Localization of the Photoreceptor(s)

Speculations, more than facts, are still dominating our attempt in the understanding of the subcellular organisation of signal perception and transduction. Both facts and speculations were compiled of late by Gressel and Rau ([11], see also [15]).

Level of Photocontrol

Lag-phases as part of the carotenoid accumulation kinetics in all fungal species investigated, indicate the involvement of a de novo synthesis of the carotenogenic enzymes, rather than activation of preformed enzymes by light [in that case the enzyme(s) involved could themselves be the photoreceptor(s), either the carotenogenic enzymes directly or other enzymes processing proteins to active carotenogenic enzymes].

Some support for a de novo synthesis comes from inhibitor studies (cf. [11]) and more direct evidence from results on the relative amount of polyadenylated RNA being synthesized [43]. Detailed information was yet lacking. In order to verify a de novo synthesis of specific proteins, Mitzka-Schnabel et al. (this Vol.) used polyadenylated RNA of induced and non-induced mycelia of *Fusarium* and *Neurospora* as a template in an in vitro translation system. The products were separated by two-dimensional gel electrophoresis. The gel pattern turned out to be of excellent overall reproducibility, manifesting differences in specific m-RNA's depending on the light pretreatment. Comparison of these in vitro labeled proteins with those labeled in vivo and separated by the same electrophoretical procedure, indicated that at least some of them correlate in their molecular weights. Although involvement of these specific proteins, appearing after photoinduction, in the biosynthesis of the carotenoids remains unclear, these results can be considered as one step further to reach direct evidence for the de novo synthesis of carotenogenic enzymes.

Carotenogenic Enzymes

Carotenogenesis in fungi is most amenable to investigations on the causality of its photocontrol. Drawbacks, however, are that the carotenogenic enzymes are elusive so far and the only access to assay its activities is by means of a cell-free system. Among organisms exhibiting blue light control of carotenoid formation, cell-free systems were only reported for *Mycobacterium* [20]. For *Neurospora,* cell-free systems were developed recently [30, 52] which were capable to incorporate mevalonic acid into phytoene and to a much lesser extent into more unsaturated carotenoids.

Whereas carotenogenic activities were only minimal in extracts from dark-grown cultures, there was a severalfold increase following a short in vivo illumination; this increase was inhibited by cycloheximide treatment prior to illumination [30]. In the meantime the performance of this cell-free system was improved up to a 50-fold efficiency of incorporation into unsaturated carotenoids. Substantial progress also was made in solubilization of the membrane-bound enzymes revealing comparable synthesizing activities. Purification of the enzymes is attempted.

Results obtained using cell-free systems also show that the first step(s) in the carotenoid biosynthesis pathway which undergo photocontrol are among those leading to the formation of the first C_{40} compound, i.e. phytoene [30, 52]; for discussion on steps in carotenogenesis regulated by light also see [14].

Site of Carotenoid Synthesis and Storage

Using *Neurospora* cell fractions, obtained by differential centrifugation, in a cell-free system it was shown that maximum carotenogenic activity (ca. 80%) was localized in two membrane fractions previously shown to contain plasma membranes and, in particular, membranes of the endoplasmic reticulum [30]. These two fractions also contained most of the particulate bound carotenoids whereas the lipid layer on top of the supernatant contained the bulk of carotenoid pigments [29].

These findings are consistent with results on the site of storage of carotenoids in *Phycomyces* [38] and on the site of synthesis in chromoplasts of *Narcissus* [22], in *Capsicum* fruit plastids [4] and in spinach chloroplasts [7].

Whether carotenoids are stored in lipid globules or whether they appear in the lipid layer as the result of the extraction procedure is not yet known.

Function of the Carotenoids

Based on experimental observations made in mutants lacking colored carotenoid pigments, their role in protecting cells or cell organelles against photosensitized oxidation is well established for photosynthetic organisms [23]. Over the last 20 years these observations were extended also to nonphotosynthetic bacteria. Surprisingly, it was found that not only UV-C could be cause of photodynamic damage but also UV-A and even visible light [57].

This could mean that light active in photocarotenogenesis potentially leads to cell damage as long as sufficient amounts of protective carotenoids are absent. However, all of the photocarotenogenesis-exhibiting organisms investigated so far contain at least small amounts of carotenoids even when grown in the dark. Hence, depending on the organism, photocarotenogenesis starts off from different levels of carotenoid content, the level probably being in correlation with the sophistication of the organisms' control system (as an example see [42]). Therefore one can hypothesize that (preformed) carotenoids may protect photocarotenogenesis; when the hazardous illumination continues, the induced carotenoids enhance this protection potential.

The results on the protective function of carotenoids obtained from single-cell fungi and from conidiospores were equivocal and somewhat contradictory [17, 54]. The protective function in mycelia of *Fusarium* was suggested by Huber and Schrott

[17]. The results obtained also indicated a repair mechanism to overcome the damage caused by UV-A. However, it remains unclear whether the repair mechanism is light-inducible also, and therefore solely responsible for the protection and not the concomitantly synthesized carotenoids. The authors also demonstrated that even white light can cause damage. For *Phycomyces* it was reported that white light inhibits mycelial growth [34]; inhibition was by far the greatest in mutant strains lacking colored carotenoids.

Genetics of Carotenogenesis

The genetical aspects of carotenogenesis will be covered by Cerdá-Olmedo in this Volume.

References

1. Afanasieva TP, Belozerskaya TA, Filipovich SY, Chernyshova EK, Kritsky MS (1980) Influence of cultivation temperature on the development of functionally active photoregulatory systems in cells of *Neurospora crassa* and *Physarum polycephalum*. Prikl Biokhim Mikrobiol (Appl Biochem Microbiol) 16:156–161 (in Russian)
2. Belozerskaya TA, Burikhanov SS, Chernyshova EK, Kritsky MS, Lvov NP (1982) Does nitrate reductase play a key role in photoinduction of carotenoid synthesis in *Neurospora crassa*? Neurospora Newslett 29:14–15
3. Briggs WR (1980) A blue light photoreceptor system in higher plants and fungi. In: De Greef JA (ed) Photoreceptors and plant development. Univ Press, Antwerpen, pp 17–28
4. Camara B, Brangeon J (1981) Carotenoid metabolism during chloroplast to chromoplast transformation in *Capsicum annuum* fruit. Planta 151:359–364
5. Caubergs R, Widell S, Larsson C, De Greef JA (1983) Comparison of two methods for the preparation of a membrane fraction of cauliflower inflorescences containing a blue light reducible *b*-type cytochrome. Physiol Plant 57:291–295
6. Cohen Y (1979) Modified mycochrome theory to explain the photoinhibition of spore formation in UV noninducible fungal plant pathogens. Phytoparasitica 7:44–45
7. Costes C, Burghoffer C, Joyard J, Block M, Douce R (1979) Occurence and biosynthesis of violaxanthin in isolated spinach chloroplast envelope. FEBS Lett 103:17–21
8. De Fabo E (1980) On the nature of the blue light photoreceptor: Still an open question. In: Senger H (ed) The blue light syndrome. Springer, Berlin Heidelberg New York, pp 187–197
9. De Fabo EC, Harding RW, Shropshire W Jr (1976) Action spectrum between 260 and 800 nanometers for the photoinduction of carotenoid biosynthesis in *Neurospora crassa*. Plant Physiol 57:440–445
10. Drumm-Herrel H, Mohr H (1981) A novel effect of UV-B in a higher plant (*Sorghum vulgare*). Photochem Photobiol 33:391–398
11. Gressel J, Rau W (1983) Photocontrol of fungal development. In: Shropshire W Jr, Mohr H (eds) Photomorphogenesis. Encyclopedia of plant physiology. New Ser, vol 16. Springer, Berlin Heidelberg New York, pp 603–638
12. Hager A (1970) Ausbildung von Maxima im Absorptionsspektrum von Carotinoiden im Bereich um 370 nm: Folgen für die Interpretation bestimmer Wirkungsspektren. Planta 91: 38–53

13. Harding RW (1974) The effect of temperature on photo-induced carotenoid biosynthesis in *Neurospora crassa*. Plant Physiol 54:142–147
14. Harding RW, Shropshire W Jr (1980) Photocontrol of carotenoid biosynthesis. Annu Rev Plant Physiol 31:217–238
15. Haupt W (1980) Localisation and orientation of photoreceptor pigments. In: Lenci F, Colombetti G (eds) Photoreception and sensory transduction in aneural organisms. Plenum Press, New York, pp 155–172
16. Hsiao KC, Björn LO (1982) Aspects of photoinduction of carotenogenesis in the fungus *Verticillium agaricinum*. Physiol Plant 54:235–238
17. Huber A, Schrott EL (1980) Photokilling and protective mechanisms in *Fusarium aquaeductuum*. In: Senger H (ed) The blue light syndrome. Springer, Berlin Heidelberg New York, pp 299–308
18. Jayaram M, Leutwiler L, Delbrück M (1980) Light induced carotene synthesis in mutants of *Phycomyces* with abnormal phototropism. Photochem Photobiol 32:241–245
19. Jayaram M, Presti D, Delbrück M (1979) Light-induced carotene synthesis in *Phycomyces*. Exp Mycol 3:42–52
20. Johnson JH, Reed BC, Rilling HC (1974) Early photoinduced enzymes of photoinduced carotenogenesis in a *Mycobacterium* species. J Biol Chem 249:402–406
21. Klein RM, Klein DT (1962) Interaction of ionizing and visible radiation in mutation induction in *Neurospora crassa*. Am J Bot 49:870–874
22. Kreuz K, Beyer P, Kleinig H (1982) The site of carotenogenic enzymes in chromoplasts from *Narcissus pseudonarcissus* L. Planta 154:66–69
23. Krinsky NI (1979) Carotenoid protection against oxidation. Pure Appl Chem 51:649–660
24. Kritsky MS, Sokolovsky VY, Belozerskaya TA, Chernysheva EK (1982) Relationship between cyclic AMP level and accumulation of carotenoid pigments in *Neurospora crassa*. Arch Microbiol 133:206–208
25. Kumagai T (1980) Blue and near ultraviolet reversible photoreaction in conidial development of certain fungi. In: Senger H (ed) The blue light syndrome. Springer, Berlin Heidelberg New York, pp 251–260
26. Lewis SC, Schiff JA, Epstein HT (1961) Photooxidation of cytochromes by a flavoprotein from *Euglena*. Biochem Biophys Res Commun 5:221–225
27. Lipson ED, Presti D (1980) Graphical estimation of cross sections from fluence-response data. Photochem Photobiol 32:383–391
28. Massey V (1980) Possible photoregulation by flavoproteins. In: Lenci F, Colombetti G (eds) Photoreception and sensory transduction in aneural organisms. Plenum Press, New York, pp 253–269
29. Mitzka-Schnabel U, Rau W (1980) The subcellular distribution of carotenoids in *Neurospora crassa*. Phytochemistry 19:1409–1413
30. Mitzka-Schnabel U, Rau W (1981) Subscellular site of carotenoid biosynthesis in *Neurospora crassa*. Phytochemistry 20:63–69
31. Mohr H (1980) Interaction between blue light and phytochrome in photomorphogenesis. In: Senger H (ed) The blue light syndrome. Springer, Berlin Heidelberg New York, pp 97–109
32. Ninnemann H (1982) Photoredcution of cytochrome b_{557} of partially purified *Neurospora* nitrate reductase via its internal flavin. Photochem Photobiol 35:391–398
33. Otto MK, Jayaram M, Hamilton RM, Delbrück M (1981) Replacement of riboflavin by an analogue in the blue-light photoreceptor of *Phycomyces*. Proc Natl Acad Sci USA 78: 266–269
34. Presti DE (1982) Photoprotection by carotenoids in *Phycomyces*. Phycomyces 5:44–46
35. Presti D, Delbrück M (1978) Photoreceptors for biosynthesis, energy storage and vision. Plant Cell Environ 1:81–100
36. Rau W (1980) Blue light-induced carotenoid biosynthesis in microorganisms. In: Senger H (ed) The blue light syndrome. Springer, Berlin Heidelberg New York, pp 283–298
37. Rau W (1983) Photoregulation of carotenoid biosynthesis. In: Porter JW, Spurgeon SL (eds) Biosynthesis of isoprenoid compounds. Wiley, New York, pp 123–157

38. Riley GJP, Bramley PM (1976) The subcellular distribution of carotenoids in *Phycomyces blakesleeanus* C115 car-42 mad-107(-). Biochim Biophys Acta 450:429–440
39. Schmidt W (1982) Light-induced redoxcycles of flavins in various alcohol/acetic acid mixtures. Photochem Photobiol 36:699–703
40. Schrott EL (1980) Dose response and related aspects of caroteneogenesis in *Neurospora crassa*. In: Senger H (ed) The blue light syndrome. Springer, Berlin Heidelberg New York, pp 309–318
41. Schrott EL (1980) Fluence response relationship of carotenogenesis in *Neurospora crassa*. Planta 150:174–179
42. Schrott EL (1981) The biphasic fluence response of carotenogenesis in *Neurospora crassa*: Temporary insensitivity of the photoreceptor system. Planta 151:371–374
43. Schrott EL, Rau W (1977) Evidence for a photoinduced synthesis of poly(A) containing mRNA in *Fusarium aquaeductuum*. Planta 136:45–48
44. Schrott EL, Huber-Willer A, Rau W (1982) Is phytochrome involved in the light-mediated carotenogenesis in *Fusarium aquaeductuum* and *Neurospora crassa*? Photochem Photobiol 35:213–216
45. Schwenker U, Gingras G (1973) A carotenoprotein from chromatophores of *Rhodospirillum rubrum*. Biochem Biophys Res Commun 51:94–99
46. Senger H (1982) The effect of blue light on plants and microorganisms. Photochem Photobiol 35:911–920
47. Senger H, Briggs WR (1981) The blue light receptor(s): Primary reactions and subsequent metabolic changes. In: Smith KC (ed) Photochemical and photobiological reviews, vol VI. Plenum Press, New York, pp 1–38
48. Shropshire W Jr (1980) Carotenoids as primary photoreceptors in blue light responses. In: Senger H (ed) The blue light syndrome. Springer, Berlin Heidelberg New York, pp 172–186
49. Shropshire W Jr, Whitaker BD (1981) The spectral sensitivity of photocarotenogenesis in *Phycomyces* mycelia. Am Soc Photobiol Ann Meet MAM-C10
50. Song P-S (1980) Spectroscopic and photochemical characterization of flavoproteins and carotenoproteins as blue light photoreceptors. In: Senger H (ed) The blue light syndrome. Springer, Berlin Heidelberg New York, pp 157–171
51. Song P-S (1980) Primary photophysical and photochemical reactions: Theoretical background and general introduction. In: Lenci F, Colombetti G (eds) Photoreception and sensory transduction in aneural organisms. Plenum Press, New York, pp 189–210
52. Spurgeon SL, Turner RV, Harding RW (1979) Biosynthesis of phytoene from isopentenyl pyrophosphate by a *Neurospora* enzyme system. Arch Biochem Biophys 195:23–29
53. Strasser RJ, Butler WL (1980) Interactions of flavins with cytochrome c and oxygen in excited artificial systems. In: Senger H (ed) The blue light syndrome. Springer, Berlin Heidelberg New York, pp 205–211
54. Thomas SA, Sargent ML, Tuveson RW (1981) Inactivation of normal and mutant *Neurospora crassa* conidia by visible light and near-UV: Role of $^1O_2$, carotenoid composition and sensitizer location. Photochem Photobiol 33:349–354
55. Valadon LRG, Osman M, Mummery RS, Jerebzoff-Quintin S, Jerebzoff S (1982) The effect of monochromatic light in the range 350 to 750 nm on the caroteneogenesis in *Verticillium agaricinum*. Physiol Plant 56:199–203
56. Vierstra RD, Poff KL (1981) Role of carotenoids in the phototropic response of corn seedlings. Plant Physiol 68:798–801
57. Webb RB (1977) Lethal and mutagenic effects of near-ultraviolet radiation. In: Smith KC (ed) Photochemical and photobiological reviews, vol II. Plenum Press, New York, pp 169–261
58. Wellmann E (1983) UV radiation in photomorphogenesis. In: Shropshire W Jr, Mohr H (eds) Photomorphogenesis. Encyclopedia of plant physiology. New Ser, vol 16. Springer, Berlin Heidelberg New York
59. Widell S, Larsson C (1981) Separation of presumptive plasma membranes from mitochondria by partition in an aqueous polymer two phase system. Physiol Plant 51:368–374

Blue/UV Light Effects on Anthocyanin Synthesis

H. DRUMM-HERREL[1]

Introduction

Anthocyanins belong to a class of flavonoids which is nearly ubiquitous in plants except in algae, fungi, lichens and bryophytes. Higher plants differ of course genetically with regard to their potential to synthesize anthocyanin. Moreover, the extent of pigment synthesis and accumulation is influenced by many environmental variables, such as nutrients, temperature, availability of water, infections, and – in particular – light.

The present report emphasizes the specific effect of short wavelength light (i.e., BL and UV) on the synthesis of juvenile anthocyanin in seedlings of angiosperms.

Significance of Juvenile Anthocyanin

Why do plants synthesize juvenile anthocyanin? Let us consider the absorbance of an extract from 3-day-old mustard seedlings (Fig. 1). Light is essential for synthesis of anthocyanin while it only slightly increases the contents of flavonols. Obviously the mustard seedling's flavonol synthesis is almost constitutive while its anthocyanin synthesis is totally light-dependent.

We consider mass production of green/blue/UV light-absorbing pigments, anthocyanin and/or other flavonoid compounds in outer tissue layers, an adaptive mechanism to protect a plant against damaging levels of sunlight (optical adaptive response, [1]).

Flavonols in the epidermis protect a plant against damaging levels of UV, in particular UV-B, while the formation of anthocyanin in peripheral cell layers of seedlings is presumed to be a photoprotective means to attenuate the light absorbed by porphyrins, in particular by protochlorophyll. In fact, the time course of light-mediated anthocyanin synthesis is correlated with presumably damaging exposure to high sunlight prior to the formation of chloroplasts. The epidermal cells of the mustard cotyledon acquire competence toward light with respect to anthocyanin synthesis approximately 27 h after sowing [10]. Competence disappears approximately 36 h later [11].

1 Biological Institute II, University of Freiburg, Schänzlestr. 1, 7800 Freiburg, FRG

Blue Light Effects in Biological Systems
ed. by H. Senger

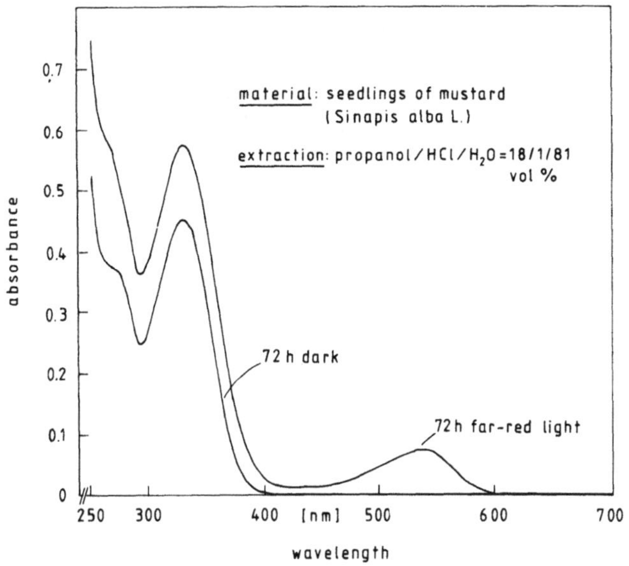

Fig. 1. Absorbance of 3-day-old mustard seedling cotyledons (*Sinapis alba*). Seedlings were either kept in complete darkness or in far-red light (3.5 W m^{-2}). Twenty pairs of cotyledons were extracted with 10 ml of the extraction medium: propanol/ HCl/H_2O = 18/1/81vol %

Data in Favor of the Presumed Photoprotective Function of Juvenile Anthocyanin (and Flavonols)

Oelze et al. [9] have shown that a photodelay in Chl-a and -b accumulation is observed in mustard (*Sinapis alba*) cotyledons during the first 3 h after the onset of white light even at medium fluxes (3 5 00lx) and in red light at a fluence rate of 6.7 W m^{-2}. They were able to demonstrate that a pretreatment of two red light pulses or a 12 h far-red light pretreatment, both operating through phytochrome, can completely prevent the photodelay. The operation of phytochrome in building up some protection device can be separated from the phytochrome effects on the rates during the pre-steady state or steady state phase of Chl accumulation. The following data are compatible with the notion that the formation of anthocyanin is the protective device.

Figure 2 shows the kinetics of Chl-a accumulation in the mustard seedling in three different light qualitaties (UV-B, yellow-green GG14-light, RL) in fully etiolated and in far-red pretreated seedlings. During the first 2 h of irradiation the dark-grown seedlings show a particularly strong photodelay in green light, whereas in UV-B no photodelay is detectable (protection by flavonols, see Fig. 1). In RL a minor photodelay is detectable which is overcome after 4 h. If the seedlings are pretreated for 2 days with far-red light, any photodelay is no longer detectable in all three light qualities. On the other hand, after 4 h of irradiation with UV-B the photodestructive action of this spectral range comes into play in the etiolated as well as in the far-red light pretreated seedlings.

Table 1 confirms and extends these findings. Sixty-h-old seedlings were pretreated in three different ways before they received three different light treatments. Seedlings were either completely etiolated, or had received a pretreatment with two

Fig. 2. Accumulation of chlorophyll (ide)-a in the cotyledons of the mustard seedling (*Sinapis alba*). Seedlings were either etiolated (*closed symbols*) or pre-irradiated (*open symbols*) for 48 h with far-red light (3.5 W m^{-2}). Chlorophyll (ide)-a accumulation was measured in three different light qualities: in yellow-green light (GG14-color glass with a cut off at 480 nm, 300 W m^{-2}), in red light (RL_{10}, 6.7 W m^{-2}), or in UV-B (Philips TL 40 W/12 lamp, peak emission at 315 nm, fluence rate 3 W m^{-2})

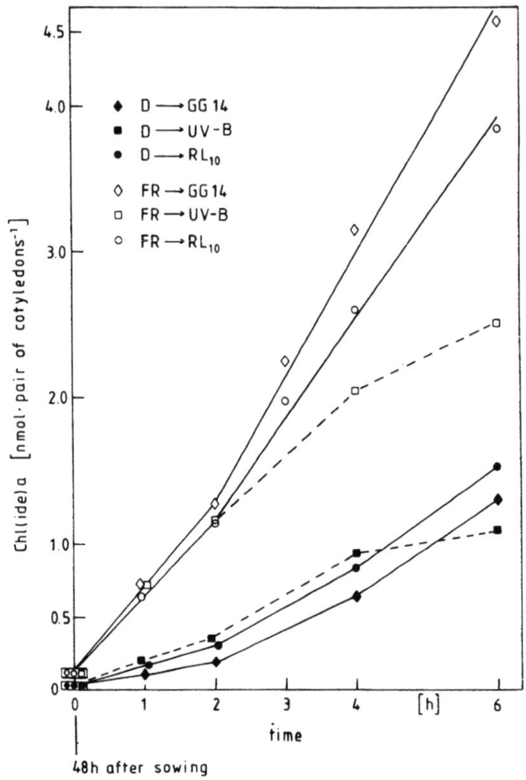

red light pulses separated by 12 h of darkness or a pretreatment for 12 h with far-red light between 36 and 48 h. Anthocyanin contents at the beginning of the different light treatments are included in the Table. In the etiolated system there is a dramatic difference in the rate of Chl-a accumulation during the first 2 h depending on the light quality. In the pulse-pretreated seedlings there is still a photodelay visible in the yellow-green region. In seedlings with the 12 h far-red light pretreatment, which induces a high amount of anthocyanin, the damaging effect of high light fluxes is obviously completely overcome in all three light qualities.

Clearly, there is only a correlation between the amount of anthocyanin and the extent of photostability of the Chl-producing system. However, there is no reason why this correlation should not be interpreted as a causal relationship.

Do Plastids Play a Role in Anthocyanin Synthesis?

The subject with which we tried to give an answer to this question was the tomato seedling (*Lycopersicon esculentum* Mill., cv. Hilds Matina). Anthocyanin synthesis in the tomato hypocotyl is characterized by a strong responsivity to BL/UV [4]. The herbicides Norflurazon (SAN 9789) and Difunon (EMD-IT-5914) were used to inhibit synthesis of colored carotenoids almost completely. Growth, development and

Table 1. Amount of Chl-a accumulated in the cotyledons of the mustard seedling between 60 and 62 h after sowing in different light qualities after different pretreatments (dark, 2 RL pulses, 12 h FR). Anthocyanin content at the beginning of the different light treatments is indicated

Treatment	Chlorophyll accumulated (nmol pair of cotyledons^{-1})			Amount of anthocyanin (A_{535})
	Content	Rate	% of rate in UV-B	
60 h D	0.102	–	–	0.029
60 h D + 2 h RL_1	0.280	0.178	89	
60 h D + 2 h RL_{10}	0.255	0.153	76	
60 h D + 2 h GG14-light	0.168	0.066	33	
60 h D + 2 h UV-B	0.303	0.201	100	
36 h D + 2 x [5 min RL_{10}] + 12 h D	0.235	–	–	0.367
36 h D + 2 x [5 min RL_{10}] + 12 h D + 2 h RL_{10}	0.933	0.698	91	
36 h D + 2 x [5 min RL_{10}] + 12 h D + 2 h GG14	0.710	0.475	62	
36 h D + 2 x [5 min RL_{10}] + 12 h D + 2 h UV-B	1.005	0.770	100	
36 h D + 12 h FR + 12 h D	0.235	–	–	0.712
36 h D + 12 h FR + 12 h D + 2 h RL_{10}	1.415	1.180	100	
36 h D + 12 h FR + 12 h D + 2 h GG14	1.415	1.180	100	
36 h D + 12 h FR + 12 h D + 2 h UV-B	1.415	1.180	100	

anthocyanin synthesis are not affected (see Fig. 9 in Mohr et al. this Vol.). Moreover, even though the carotenoid content was reduced to less than 2% of normal, Norflurazon and Difunon treatments did not reduce seedling sensitivity toward BL/UV. The fluence rate response function for BL/UV-mediated anthocyanin synthesis is still linear and even slightly steeper rather than lower with herbicides [5].

What happens to the plastids in those cells that synthesize anthocyanin in the presence of herbicides, i.e., in the virtual absence of carotenoids?

Anthocyanin in the hypocotyl of tomato is synthesized only in a single cell layer, namely the subepidermis. This is a common feature in many species of dicotyledonous seedlings [8].

Plastids in the anthocyanin producing cell layer exhibit the characteristics of amyloplasts (Fig. 3). The amyloplasts of nontreated plants grown either in BL or UV contain numerous thylakoids and grana. In the herbicide-treated virtually carotenoid-free plants the amyloplasts become badly damaged in BL, presumably in connection with the photodecomposition of chlorophyll. It can be seen that thylakoids and plastidal ribosomes are almost completely lacking. The plastidal envelope seems to remain intact. Despite this severe damage, there is no impairment of anthocyanin synthesis as shown by fluence rate-response curves with and without herbicides. It can

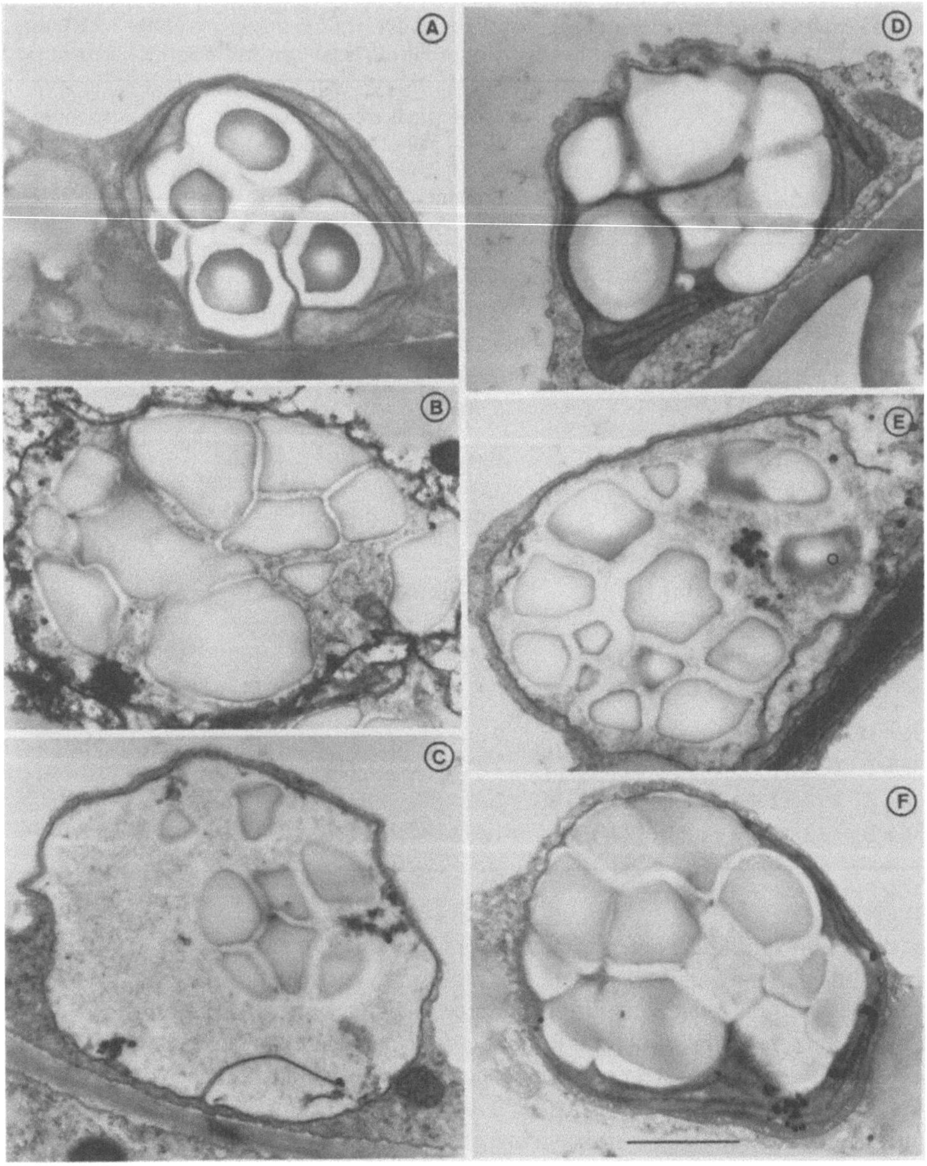

Fig. 3A-F. Electron micrographs of amyloplasts of the subepidermal layer of tomato hypocotyls (*Lycopersicon esculentum*), cv. Hilds Matina). **A-C** 3 days dark + 24 h BL. **A** water control; **B** kept on Difunon (2 x 10^{-5} M); **C** kept on Norflurazon (10^{-5} M). **D-F** 3 days dark + 24 h UV. **D** water control; **E** kept on Difunon (2 x 10^{-5} M); **F** kept on Norflurazon (10^{-5} M). Scale: 1 μm. Micrographs from Dr. R. Bergfeld

be concluded that the BL/UV-mediated formation of juvenile anthocyanin depends neither on intact plastids nor on the presence of detectable amounts of carotenoids. The carotenoid-free amyloplasts in UV are less damaged than in BL. They still produce some chlorophyll. This is probably due to the strong attenuation of the UV in the flavonol-containing epidermis.

The Problem Proper: BL/UV Light Effects on Anthocyanin Synthesis

The experimental approach has been based on the argument that mass pigment synthesis in a seedling requires an investment of matter and energy. Thus, for reasons of survival, a seedling may be expected to form considerable amounts of protective pigment only when necessary and only to the extent required for protection ("economy principle"). Regulation by light of chlorophyll synthesis in higher plants clearly shows that during seedling development a particularly strong selection pressure exists in favor of economic usage of limited resources. In order to respond properly, the seedling has to sense the light conditions in its environment *precisely.* What is the photosensory system through which a plant can detect those light conditions that require (and justify in view of the economy principle) the mass synthesis of anthocyanin in a seedling?

In principle, the answer to this question has already been given by Mohr et al. (this Vol.). Some thoroughly investigated case studies on light-mediated anthocyanin synthesis will be presented which illustrate the general principle, i.e., the coaction of BL/UV photoreceptors and phytochrome.

The classical example: The milo seedling [*Sorghum vulgare* Pers. cv. Weider (hybrid)].

The mesocotyl of the milo seedling does not produce anthocyanin in complete darkness. As described originally by Downs and Siegelman [2] even long-term red or far-red light does not lead to any anthocyanin synthesis, while white light causes rapid pigmentation. This was reinvestigated by Drumm and Mohr [3]. The action of white light can be replaced by irradiating with BL/UV. The expression of the BL/UV effect can be blocked reversibly by a short far-red light pulse. This means that the expression of the BL/UV effect is controlled by phytochrome. In experiments with dichromatic irradiation with two kinds of light to strongly modulate the level of P_{fr} on a constant background of BL/UV it was found that the BL/UV photoreaction as such is not affected by the presence or virtual absence of P_{fr} (Table 2). With and almost without P_{fr} during the 3 h irradiation period, the same amount of anthocyanin is synthesized as long as a RL pulse saturating the phytochrome photoequilibrium is given after the irradiation. Our present interpretation is that phytochrome (P_{fr}) is the effector molecule which causes anthocyanin synthesis through activation of competent genes. The BL/UV effect has to be considered as a transient "responsivity amplification" with a half-life of approximately 6 h (Fig. 4). In the case of the milo mesocotyl there is no detectable responsivity toward P_{fr} without the operation of a BL/UV photoreceptor.

2nd example: cell cultures of parsley (*Petroselinum hortense*). Wellmann [12, 13] and Wellmann and Schopfer [14] have described the phenomenon that the expression of an inductive effect of UV light (λ_{max} < 300 nm) on flavonoid and concomitant en-

Table 2. The effect of simultaneous irradiation with two different light qualities (UV and RG 9-light) on anthocyanin formation in the mesocotyl of *Sorghum* seedlings. The intense RG 9-light was applied to decrease the level of P_{fr} as far as possible. (After [3])

Treatment after sowing	Amount of anthocyanin (A at 510 nm)
60 h D + 3 h UV (1.5 W m^{-2}) + 5 min RL + 24 h D	0.189
60 h D + 3 h [UV + RG 9-light (200 W m^{-2})] + 24 h D	0.055
60 h D + 3 h [UV + RG 9-light (200 W m^{-2})] + 5 min RL + 24 h D	0.183

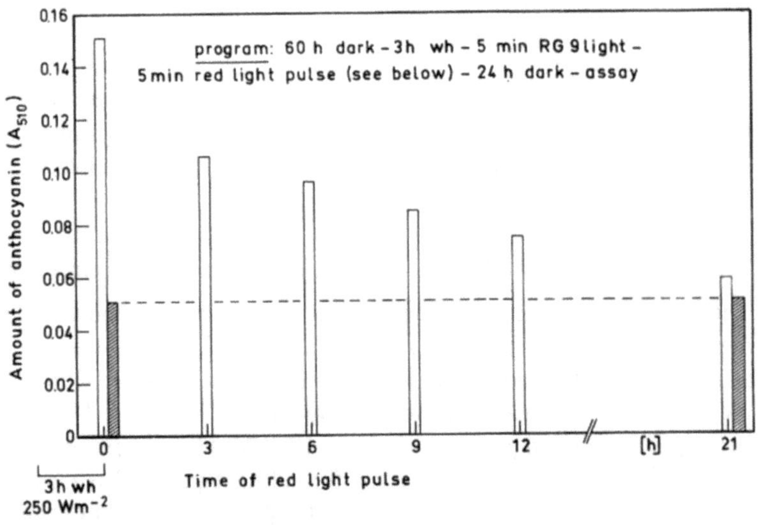

Fig. 4. A study on the half-life of the white light effect on anthocyanin synthesis in the mesocotyl of the milo seedling (*Sorghum vulgare* Pers. cv. Weider). Seedlings were irradiated for 3 h with white light (wh) at 60 h after sowing. Immediately following the wh period they received a 5 min RG 9-light pulse (10 W m^{-2}). A RL pulse (6.7 W m^{-2}) was given either immediately after the RG 9-light pulse or after dark periods of various durations. Anthocyanin was determined 24 h after the standard RL pulse. *Shaded bars* represent controls which received a 5 min RG 9-light pulse only. (After [3])

zyme biosynthesis in cell suspension cultures of parsley as well as in intact seedlings is controlled by phytochrome. But, in contrast to the milo system, Wellmann [12] found that the visible part of the spectrum is without any inductive effect. The peak of the action spectrum is below 300 nm. From a formal point of view Wellmann's system is very similar to the milo system except that the photoreceptors must be different.

3rd example: The maize seedling (*Zea mays* L. cv. Pioneer 3369A). Duke and Naylor [6] have studied the photoinduction of anthocyanin synthesis in root and mesocotyl of maize. They found that continuous white or blue light induced large amounts of anthocyanin in both organs, whereas continuous far-red light induced almost no pigment, at least in the case of the mesocotyl (Fig. 1 in Duke and Naylor [6]). After a 12 h white light treatment, the response is under phytochrome control (Fig. 3 in Duke and Naylor [6]). If the irradiation is terminated by a 5 min far-red pulse, the final amount of anthocyanin is greatly decreased and this effect can be reverted by giving a red light pulse after the far-red light pulse. It seems that the maize system acts very similarly to the milo system, except that it responds weakly to P_{fr} without a blue or white light pretreatment. But the white or blue light pretreatment causes a strong "responsivity amplification" for P_{fr}.

4th example: The tomato seedling (*Lycopersicon esculentum* Mill., cv. Hilds Matina). As stated above, in white light the hypocotyl of the tomato seedling produces large amounts of anthocyanin in the subepidermal layer. A single light pulse has no detectable effect on the production of anthocyanin even though a 12 h exposure to white light, BL or UV leads to a strong anthocyanin synthesis. Twelve h of high fluence rate red or far-red light (operating through phytochrome) exert only a small effect compared to white light. The effect of RL and FR – including long-term light treatments – was previously shown by Mancinelli and Rabino [7] to be exclusively mediated through phytochrome.

Anthocyanin synthesis after a 12 h BL or UV exposure continues in darkness at a considerable rate only if P_{fr} is available. When most of the P_{fr} is returned to P_r by a saturating RG 9-light pulse further anthocyanin synthesis is almost negligible (see Fig. 3 in [4]). Even though phytochrome is involved in the response, a high effectiveness of phytochrome depends on a preceding light absorption by a blue/UV photoreceptor. These results are similar to the data obtained with the milo mesocotyl, except that in milo, even long-term exposure to red and far-red light is totally ineffective. It seems that the corn seedling and tomato belong to the same category.

Summarizing this latter part: In all cases studied so far with regard to anthocyanin synthesis we always find a coaction of BL/UV and light absorbed by phytochrome. But phytochrome (P_{fr}) is hardly or not at all capable of operating without a preceding application of BL/UV. On the other hand, it seems that the action of BL/UV is restricted to establishing or amplifying responsivity to P_{fr}.

We conclude that a P_{fr}-mediated modulation of activity of competent genes, leading to anthocyanin synthesis, is the core of the light response in all cases. However, the different species differ greatly with regard to the red, blue and UV light-dependent processes they must perform in order to establish sensitivity toward phytochrome (P_{fr}) or to amplify sensitivity toward P_{fr}.

Abbreviations

P_{fr}, far-red absorbing form of the phytochrome system. BL, blue light. UV, UV light. RL_1, red light (0.67 W m^{-2}). RL_{10}, red light (6.7 W m^{-2}). FR, far-red light (3.5 W m^{-2}). GG14-light, yellow-green light produced with a GG 14 color glass with a cut off at 480 nm (300 W m^{-2}). UV-B light emitted by a Philips TL 40 W/12 lamp, peak emission at 315 nm (3 W m^{-2}). RG 9-light, long wavelength far-red light produced with KG 3 and RG 9 color glass (10 W m^{-2}). Chl, chlorophyll.

References

1. Caldwell MM (1981) Plant response to solar ultraviolet radiation. In: Lange OL, Nobel PS, Osmond CB, Ziegler H (eds) Encyclopedia of plant physiology. New Ser, vol 12A. Springer, Berlin Heidelberg New York, pp 170–197
2. Downs RJ, Siegelman HW (1963) Photocontrol of anthocyanin synthesis in milo seedlings. Plant Physiol 38:25–30
3. Drumm H, Mohr H (1978) The mode of interaction between blue (UV) light photoreceptors and phytochrome in anthocyanin formation of the *Sorghum* seedling. Photochem Photobiol 27: 241–248
4. Drumm-Herrel H, Mohr H (1982) The effect of prolonged light exposure on the effectiveness of phytochrome in anthocyanin synthesis in tomato seedlings. Photochem Photobiol 35:233–236
5. Drumm-Herrel H, Mohr H (1982) Effect of blue/UV light on anthocyanin synthesis in the absence of bulk carotenoids. Photochem Photobiol 36:229–233
6. Duke SO, Naylor AW (1976) Light control of anthocyanin biosynthesis in *Zea* seedlings. Physiol Plant 37:62–68
7. Mancinelli AL, Rabino I (1978) The high irradiance responses of plant photomorphogenesis. Bot Rev 44:129–180
8. Nozzolillo C (1972) The site and chemical nature of red pigmentation in seedlings. Can J Bot 50:29–34
9. Oelze-Karow H, Rösch H, Mohr H (1983) Prevention by phytochrome of photodelay in chlorophyll accumulation. Photochem Photobiol 37:565–569
10. Steinitz B, Drumm H, Mohr H (1976) The appearance of competence for phytochrome-mediated anthocyanin synthesis in the cotyledons of *Sinapis alba* L. Planta 130:23–31
11. Wagner E, Mohr H (1966) "Primäre" und "sekundäre" Differenzierung in Zusammenhang mit der Photomorphogenese von Keimpflanzen (*Sinapis alba* L.). Planta 71:204–221
12. Wellmann E (1971) Phytochrome-mediated flavone synthesis in cell suspension cultures of *Petroselinum hortense* after preirradiation with ultraviolet light. Planta 101:283–286
13. Wellmann E (1976) Specific ultraviolet effects in plant morphogenesis. Yearly review. Photochem Photobiol 24:659–660
14. Wellmann E, Schopfer P (1975) Phytochrome-mediated de novo synthesis of phenylalanine ammonia-lyase in cell suspension cultures of parsley. Plant Physiol 55:822–827

Development

Blue Light Effects in Cell Cultures

G. RICHTER, J. HUNDRIESER, M. GROSS, S. SCHULTZ, K. BOTTLÄNDER,
and CH. SCHNEIDER[1]

1 Introduction

Greening is a complex process which is controlled by light on various levels. Generally, the photoreceptor systems involved respond to both blue and red light. In cell cultures of tobacco (*Nicotiana tabacum* var. Samsun), however, the development of functionally active chloroplasts is induced and maintained solely by blue light [4, 12]. This system offers a good chance to study the sequence of events taking place in a major cellular differentiation as well as to elucidate the mode of photocontrol which is effective. The present treatise will deal with the formation of chlorophylls and the synthesis of plastid proteins which exert their functions in the thylakoid membranes and in the stroma of chloroplasts. An attempt has also been made to relate the influence of blue light on transcription and translation.

Blue Light-Induced Changes in Plastid Membranes

Chlorophyll Synthesis: Role of Light Quality and Quantity

The blue light-dependent transition of leucoplasts to chloroplasts in tobacco cells cultured in a modified MS-medium [6] can be roughly estimated by following the synthesis of chlorophylls. It starts on day 2 if dark-grown cells are exposed to blue irradiation, and increases in an approximate linear fashion over about 15 days (Fig. 1). The same holds true for white light. In red light, however, only small amounts of chlorophyll are formed over the same period of time. From day 15 on, i.e., in the stationary phase of the cell cycle, a slight increase is observed in some cell culture lines, in others no chlorophyll is detectable at all in red light. The cause could be a certain physiological variability in respect to red light-promoted chlorophyll synthesis in the tobacco cells. Indications are that this difference is reflected in the formation of those polypeptides to which the chlorophylls are complexed in the developing plastid membranes (see below).

After 14 days the accumulation of chlorophylls in blue light is significantly higher than in white light of about the same energy fluence rate (EFR). This superiority of blue light is observed, too, when the increase of chlorophylls per day was plotted

1 Institut für Botanik, Universität Hannover, Herrenhäuser Straße 2, 3000 Hannover 1, FRG

Blue Light Effects in Biological Systems
ed. by H. Senger

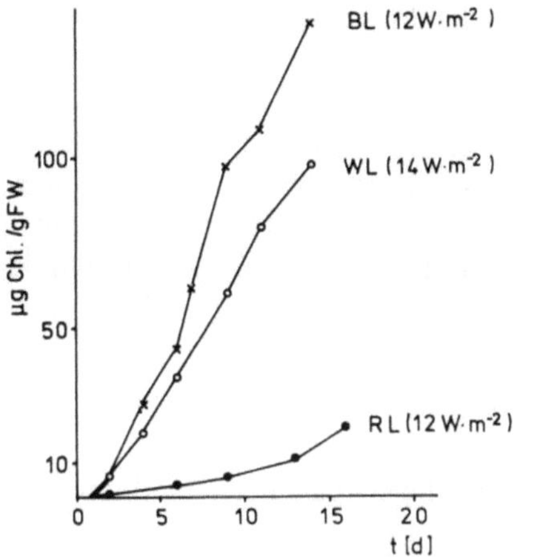

Fig. 1. Time course of chlorophyll synthesis in tobacco suspension cultured cells irradiated with blue light (*BL*), white light (*WL*) or red light (*RL*) of about equal energy fluence rate (*EFR*) expressed as μg total chlorophyll g^{-1} fresh weight (*FW*)

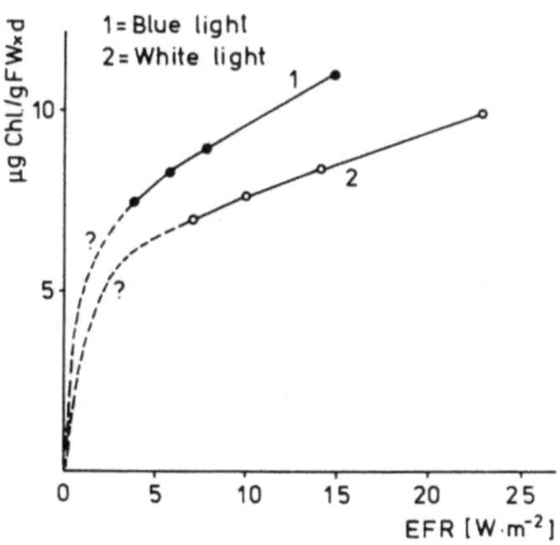

Fig. 2. Chlorophyll synthesis per day of suspension cultured cells in blue or white light as a function of energy fluence (*EFR*)

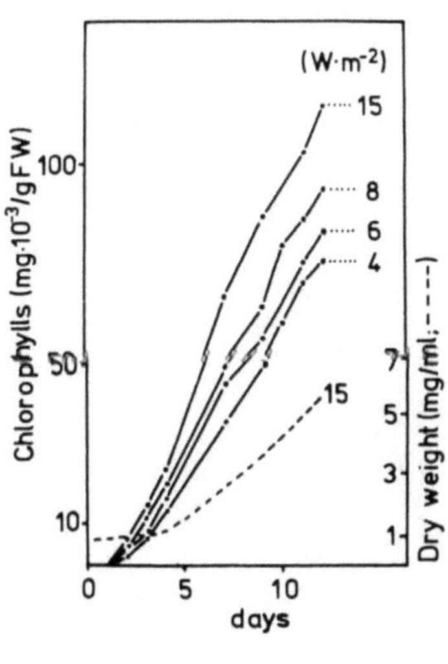

Fig. 3. Chlorophyll accumulation over 15 days in cultured tobacco cells irradiated with blue light of different EFR ($W\ m^{-2}$) expressed as mg $\times$ 10^{-3} total chlorophyll g^{-1} fresh weight (*FW*). For comparison, the increase of cell mass (as dry weight) has been added for EFR of 15 $W\ m^{-2}$

against EFR (5–23 W m^{-2}) for both blue and white light-irradiated cells. For all values of EFR applied in white light only about 78% of the daily chlorophyll production in blue light are registered (Fig. 2). In blue light a close relationship between the rate of chlorophyll synthesis and the EFR applied was found (Fig. 3). For various values of EFR the yield per g fresh weight (FW) increases about linearly over 12 days, i.e., during the logarithmic phase of the culture cycle. This does not mean that cell division is a prerequisite of chlorophyll synthesis – on the contrary, it may well continue during the stationary phase of the cycle (see below).

Influence of the Carbon Source on Chlorophyll Formation

Besides blue light the carbon source is another significant parameter in chloroplast development and growth of the cultured tobacco cells since these processes are manifoldly accelerated if the liquid nutrient medium is supplemented with either CO_2 or sucrose [3]. In a few instances glucose could successfully replace sucrose in maintaining greening and growth of cell cultures [5]. Therefore, it seemed worthwhile to elucidate the role of the two sugars in this respect. Independently of the initial supply the amount of sucrose drops with a rate of 0.2–0.5 g per day and liter (Fig. 4). The time point when the test becomes negative depends on the amount given at the start.

The daily rate of chlorophyll synthesis, however, is not affected by these differences in the sucrose level. From these data follows that a certain quantity of sucrose as low as 3 g l^{-1} is necessary to induce and maintain chlorophyll synthesis in blue light-irradiated tobacco cells. Chlorophyll synthesis proceeds with a constant rate during and beyond the stationary phase up to day 43 even though the medium is depleted of sucrose from day 32 onwards. The process can be prolonged for more than 50 days by adding sucrose periodically to the otherwise unaltered medium (Fig. 5) amounting to a chlorophyll accumulation of about 330 μg g^{-1} FW, which corresponds to one third of the content of young tobacco leaves on FW basis.

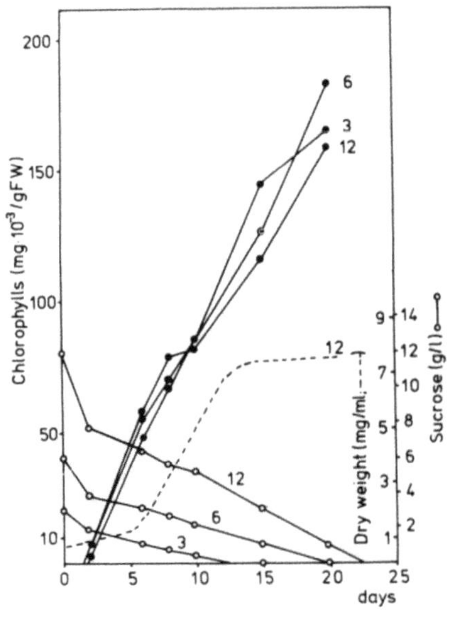

Fig. 4. Chlorophyll formation and sugar uptake by tobacco cells growing in blue light (12 W m^{-2}) and initially supplied with three different concentrations of sucrose: 3,6 and 12 g l^{-1} medium. Cell proliferation (as dry weight) was measured with 12 g l^{-1} sucrose

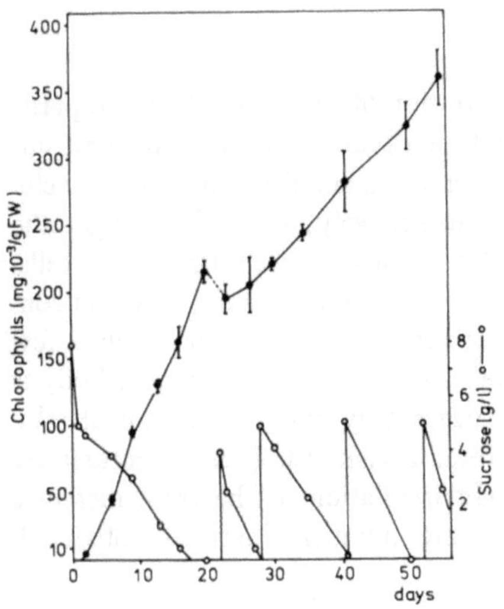

Fig. 5. Chlorophyll formation and sugar uptake by tobacco cells growing in blue light ($12\ W\ m^{-2}$) in a medium periodically supplemented with equal amounts of sucrose to reach a level of about $5\ g\ l^{-1}$ medium

Starting with glucose in the medium the various amounts added are rapidly taken up by the tobacco cells within 1–5 days ($1.5\ g\ d^{-1}\ l^{-1}$) when irradiated with blue light (Fig. 6). The initial supply turns out to be the crucial parameter for yield and maintenance of chlorophyll synthesis. Only with a concentration as high as $10\ g\ l^{-1}$ does an approximate correlation between chlorophyll formation and cell growth (expressed as dry weight increase) exist for about 10 days; then chlorophyll disappears rather rapidly. Moreover, the cells produce insoluble fibrous material which forms massive aggregates and interferes with the growth of cells; these stop dividing, become brown and eventually die.

Apparently glucose is readily taken up by the cells and metabolized, thus establishing conditions which prevent chlorophyll formation. In contrast, sucrose affects the metabolism of the tobacco cells in favor of greening in blue light; a retarded uptake, together with a reduced availability as primary carbon source, could explain the discrepancy in the efficiency of both sugars. From these findings it is obvious that sucrose cannot replace glucose in blue light-induced chlorophyll synthesis.

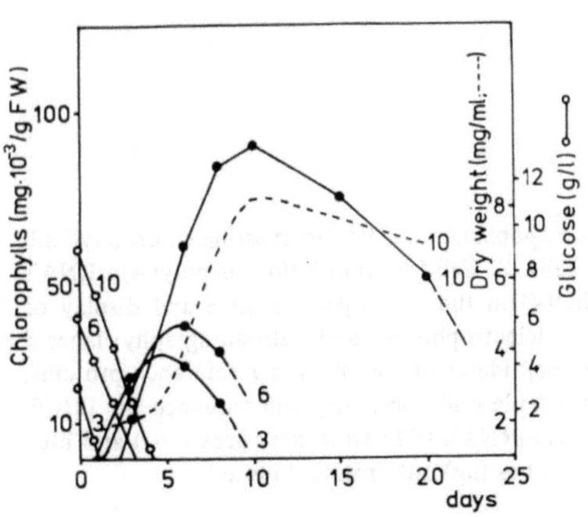

Fig. 6. As described in Fig. 4, except: Cells were initially supplied with three different concentrations of glucose: 3, 6 and $10\ g\ l^{-1}$ medium. Cell proliferation was measured with $10\ g\ l^{-1}$ glucose

Changes in Plastid Morphology

Information on structural changes during the conversion of leucoplasts to chloroplasts during blue light-induced greening of cultured tobacco cells are limited. In a first survey Bergmann and Berger [4] observed that stroma and grana thylakoids were exclusively formed in blue light and in white light, respectively, while in red light the plastids lacked these specific substructures thus resembling those in dark-grown cells. By using suspension-cultured cells derived from the same callus culture, we could confirm and extent these findings. Plastids from dark-grown cells contain merely a few isolated internal membranes as well as few tubuli and vesiculi; a prolamellar body is not detectable. The transition of these plastids triggered by blue light is indicated by the formation of grana stacks and numerous stroma thylakoids. These structures are not observed in plastids after prolonged red light irradiation of the cells; merely a parallel array of a few membranes surrounded by a large number of vesiculi is observed.

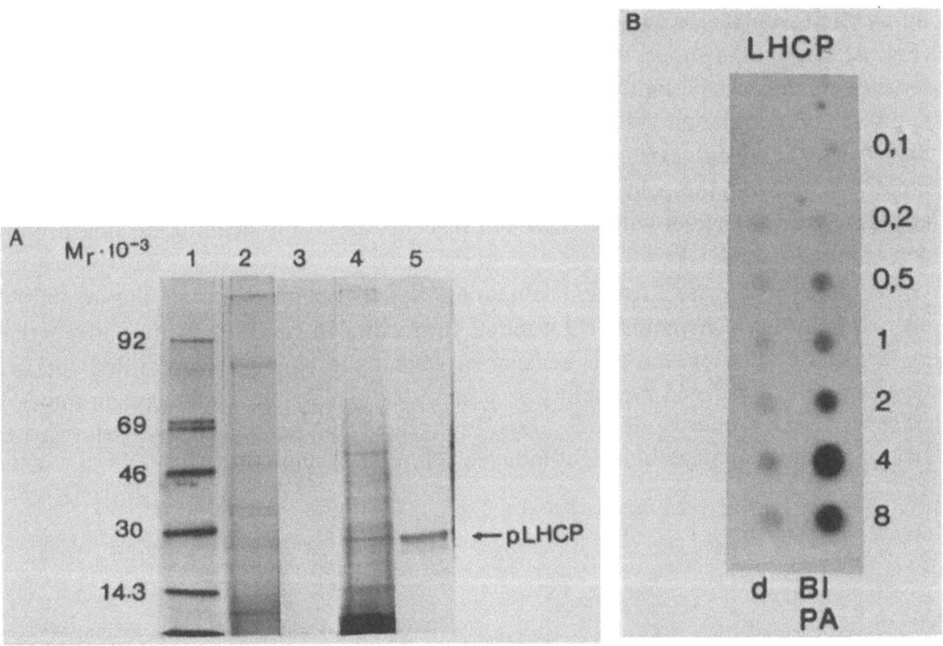

Fig. 7A,B. Accumulation of mRNA coding for the apoproteins of light-harvesting chlorophyll a,b protein (*LHCP*) in blue light-irradiated tobacco cells. **A** Cell-free translation of poly(A)mRNA's from dark-grown (*2*) and blue light-treated cells (*4*) in the wheat-germ system and display of ^{35}S-labeled polypeptides by SDS-polyacrylamide gelelectrophoresis and autoradiography; lanes *3* and *5* immunoprecipitation of the precursor polypeptide(s) of LHCP; lane *1* reference proteins; **B** Hybridization on nitrocellulose to the ^{32}P-labeled plasmid containing the sequence for LHCP; hybrids were detected by autoradiography; poly(A) mRNA's (*PA*) from dark-grown (*d*) and blue light-irradiated cells (*Bl*) were dotted in various quantities (μg) onto nitrocellulose

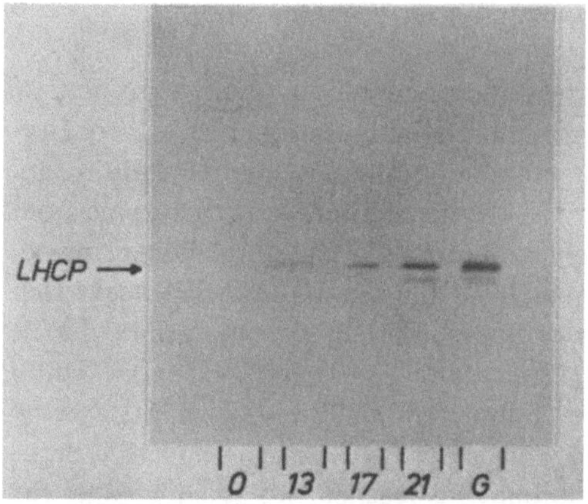

Fig. 8. Synthesis of light-harvesting chlorophyll a,b protein (*LHCP*) in tobacco cells capable of limited chlorophyll formation after prolongated red light irradiation. Immunoreaction of plastid membrane polypeptides with the antibody to LHCP. After SDS-polyacrylamide gel electrophoresis the polypeptides were transferred to nitrocellulose, and the immunoreactive band detected with peroxidase-conjugated anti-immunoglobulins according to Towbin et al. [15]. Plastid membrane proteins were isolated from cells after irradiation for 0, 12, 17 and 21 days. A similar preparation from cells exposed for 21 days to white light (*G*) served as control

Appearance of Chlorophyll-Protein Complexes in Plastid Membranes

Light-Harvesting Chlorophyll a,b Protein. Irradiation with blue light of cultured tobacco cells does not only induce the synthesis of polypeptides and pigments but also their insertion into the developing thylakoid membranes as chlorophyll-protein complexes. Several of these have been isolated and characterized [13]. As to their composition they compare well with those of higher plant species [1]. The accumulation of their protein moieties was tested indirectly by incubating tobacco cells either irradiated for 0–12 days with blue and red light, respectively, or kept in darkness with ^{35}S-methionine for 24 h prior to the extraction of plastid membrane proteins; these were separated electrophoretically. In the blue light-treated cells from day 3 onwords the specific labeling increases drastically up to day 12; this effect is not observed in red light-treated or dark-grown cells. Prominent among those membrane components is a polypeptide of M_r = 25,000, the main protein of the light-harvesting chlorophyll a,b protein (LHCP). Its accumulation is accompanied by the formation of the mRNA sequence coding for the specific precursor polypeptide ($M_r \approx 29{,}000$) which was identified by its in vitro translation product as well as by hybridization with the corresponding cDNA as probe (Fig. 7). In dark-grown or in red light-irradiated cells the low level of this mRNA sequence is not enhanced.

In tobacco cells capable of a limited chlorophyll synthesis during prolongated red light irradiation the formation of LHCP is indicated by the appearance of its main polypeptide concomitantly with the pigments (Fig. 8).

Expression of Other Plastid Membrane Proteins

NADPH-Protochlorophyllide Oxidoreductase. During blue light-induced greening of cultured tobacco cells a decrease or disappearance of polypeptides occurs, too. Outstanding among these is the enzyme protein of NADPH-protochlorophyllide oxidoreductase (EC 1.6.99.-; PCR) with $M_r \approx 36{,}000$. Preparations of this enzyme from etioplast membranes of higher plant catalyze in vitro the transformation of protochlorophyllide to chlorophyllide which represents the crucial light-dependent step during chlorophyll synthesis. In seedlings of barley and oats a rapid decline of PCR activity upon illumination of the etiolated plants has been observed [9]; concomitantly the enzyme protein disappears from the plastid membranes [14], and the specific mRNA species is no longer transcribed [2]. Paradoxically, these light-induced changes take place before the optimal level of chlorophyll is reached. We asked whether the photocontrol of PCR by blue light in tobacco cells is of the same type as described for monocotyledonean species.

The analysis of plastid membrane proteins from dark-grown and blue light-irradiated tobacco cells reveals that only in preparations from the former is the band of PCR present in the electropherogram (Fig. 9A). It was identified by the method of immune blotting ([11], Fig. 9B). A positive result was also obtained with plastid membranes of red light-treated cells while in greening cells raised in intermittent blue or

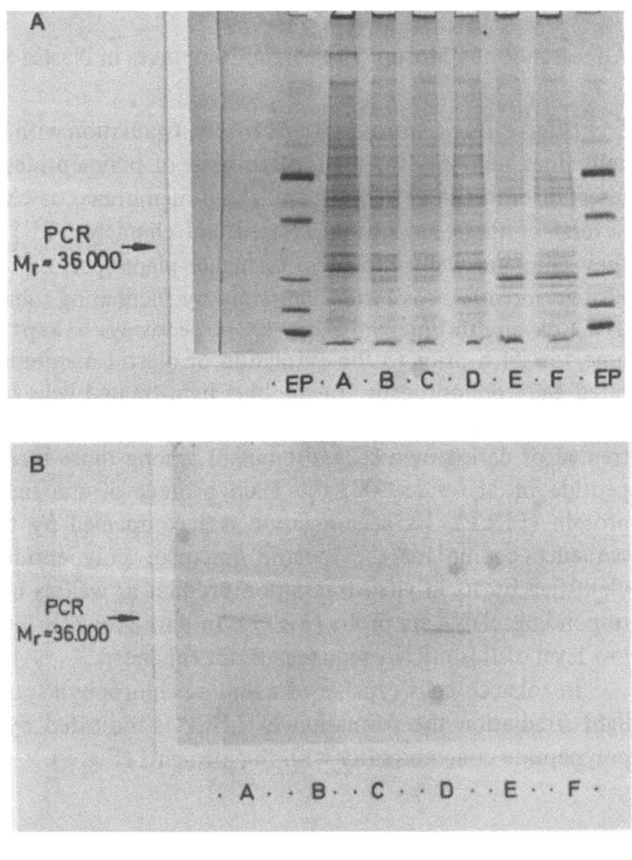

Fig. 9A,B. Detection of NADPH-protochlorophyllide oxidoreductase (*PCR*) in plastid membrane of tobacco cells grown in darkness for 14 days (lane *A*), in blue light (lane *B*), in white light (lane *C*), in red light (lane *D*), in intermittent (12:12 h) blue light (lane *E*) and white light (lane *F*) for the same period of time; *EP* reference proteins. **A** SDS-electrophoresis in polyacrylamide gradient gel (7.5–15%); **B** immune reaction with the antibody to PCR visualized with peroxidase-conjugated anti-immunoglobulins

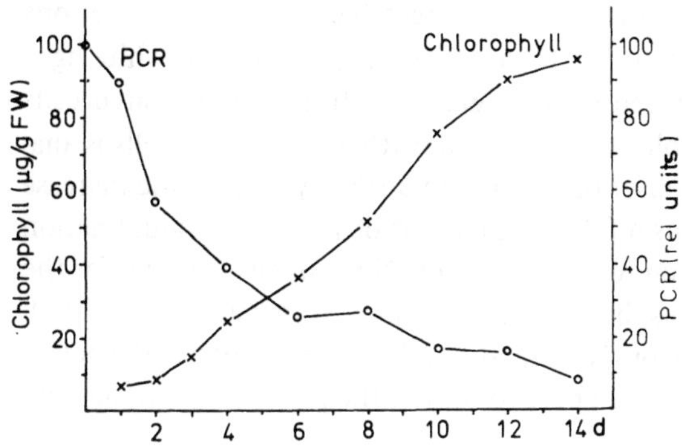

Fig. 10. Effect of blue light irradiation (EFR: 8 W m^{-2}) on the amount of NADPH-protochlorophyllide oxidoreductase (*PCR*) in cultured tobacco cells. Determination by scanning of the band specifically stained by reaction of the peroxidase-conjugated anti-immunoglobulins (see Fig. 9). The accumulation of chlorophylls in these cells over the same period of time has been added

white light the level of immunoreactive PCR is below the limit of detection. When the amounts of PCR were plotted against illumination time (Fig. 10) a rapid decline during the first 4 days in blue light is registered, which is followed by a progressive decrease to about 10% of the initial concentration after 14 days of culture. Thus a residual amount of PCR is apparently present in the fully developed thylakoid membranes of the tobacco cells. This conclusion, however, rests solely on the data obtained with the immune blotting technique being only semiquantitative; moreover, the antibody against PCR of barley was employed; thus a restricted cross-reactivity with the PCR protein of tobacco cannot be excluded. Despite these reservations, the mode of regulation found for PCR of tobacco cells in blue light compares well with that described for tomato [11], where the enzyme protein as well as the specific mRNA for the PCR precursor polypeptide are detectable in strongly reduced amounts even after 12 days of illumination. This accord is confirmed by the finding that fully greened tobacco cells, i.e., after 14 days in blue light, still dispose of translatable mRNA for the precursor of PCR ($M_r \approx 41{,}000$).

32,000 M_r Plastid Membrane Protein. Blue light irradiation of dark-grown tobacco cells stimulates the formation of several major polypeptides and their insertion into the plastid membrane [13]. Characteristic among these is a 32,000 M_r polypeptide which has been found in a number of plants and has probably a function in photosystem II [10]. From day 3 onwards the amount of this polypeptide increases drastically in accordance with the process of greening. Red light fails to bring about a similar enhancement; the pattern of plastid membrane proteins remains practically unchanged over a period of 12 days. The apparent blue light-induced de novo synthesis of the 32,000 M_r polypeptide is reflected on the level of the corresponding mRNA which is a major plastid species and encodes a precursor polypeptide of about 35,000 M_r. These findings led us to assume that the stimulatory effect of blue quanta on the level of the mRNA could account for the observed accumulation of the membrane polypeptide. The major question was whether a blue light-dependent activation/inactivation of stored mRNA sequences or a blue light-induced change in the steady-state concentration of the specific mRNA sequence is responsable for the de novo synthesis registered. The answer was sought by employing a cloned chloroplast DNA fragment containing the structural sequence for the 35,000 M_r precursor polypeptide in a "dot hybridization" assay (details are given Richter this Vol.). Hybridizable mRNA sequences for the 35,000 M_r precursor polypeptide are only detectable in RNA preparations [total, plastid and poly(A)-minus fraction] of tobacco cells kept in blue or white light for several days, i.e., being in an advanced state of greening. In red light-treated cells hybrid formation is almost undetectable. When the total RNA of tobacco cells is analyzed by the same technique from the onset of irradiation until day 4, a drastic increase in the steady-state concentration of this mRNA sequence is observed with illumination time. About the same behavior has been found for the mRNA sequence coding for the large subunit (LSU) of ribulose-1,5-bisphosphat carboxylase/oxygenase (see below). These results, although preliminary, strongly suggest that blue light controls the actual sequence level of plastid mRNA's encoding polypeptides of the plastid membranes and the stroma.

Changes in the Pattern of Stroma Proteins

Ribulose-1,5-Bisphosphate Carboxylase/Oxygenase

A massive synthesis of ribulose-1,5-bisphosphate carboxylase/oxygenase (EC 4.1.1.39; RuBPCase) occurs in dark-grown tobacco cells from day 3 onwards when exposed to blue light [7]. It leads to an about tenfold increase within 8–10 days. In contrast, the low level in red light-treated cells remains unchanged throughout the same period of time. Thus they compare well with dark-grown cells where the subunit proteins of RuBPCase can be determined merely by an optimized separation technique. The blue light-induced formation of RuBPCase is reflected on the level of transcription: the mRNA coding for the precursor polypeptide of the small subunit (pSSU; $M_r \approx 20{,}000$) is obviously only present in illuminated tobacco cells from which it can be isolated and translated in a cell-free system. In darkness its amount is insignificant. These findings led us to assume that the stimulatory effect of blue light on the level of the mRNA's for both the subunits of RuBPCase (SSU, LSU) could account for the massive synthesis of the holoenzyme. In order to gain further evidence in favor of such a pretranslational control, the steady-state concentration of the LSU mRNA was assessed in irradiated and in dark-grown tobacco cells by hybridization, employing the corresponding structural sequence; it had been isolated from chloroplast DNA and cloned in *E. coli* ([8]; Richter this Vol.). The results clearly show that the level of LSU mRNA is high in greening cells respective in their plastids, but almost undetectable in chlorophyll-free cells.

These results together with those depicted above for the precursor polypeptide of the 32,000 M_r thylakoid membrane protein suggest that blue light may have a strong effect on transcription of specific genes; however, the possibility also exists that the stability of the transcripts is affected. In order to solve this problem, suitable systems must be employed where only newly synthesized transcripts are assessable: isolated nuclei or transcriptionally active DNA-RNA polymerase complexes from plastids. The latter system has been chosen by us in a first attempt to evaluate the transcription capacity of such a complex from chloroplasts of greening tobacco cells. Indeed, several RNA species with different molecular masses were formed (details in Richter this Vol.).

Acknowledgments. The authors are grateful to Prof. L. Bergmann for supplying callus cultures. We thank Prof. K. Kloppstech for providing us with the antibodies to LHCP and PCR and for introducing us to the immune blotting technique, Dr. G. Link for the gift of the plasmids pSA 204 and pSA 452, and Prof. R.G. Herrmann for giving us the LHCP DNA fragment.

References

1. Anderson B, Anderson JM, Ryrie IJ (1982) Transbilayer organization of the chlorophyll-proteins of spinach thylakoids. Eur J Biochem 123:465–472
2. Apel K (1981) The protochlorophyllide holochrome of barley (Hordeum vulgare L.) Phytochrome induced decrease of translatable mRNA coding for the NADPH-oxidoreductase. Eur J Biochem 120:89–93
3. Bergmann L (1967) Wachstum grüner Suspensionskulturen von Nicotiana tabacum var. „Samsun" mit CO_2 als Kohlenstoffquelle. Planta 74:243–249
4. Bergmann L, Berger Ch (1966) Farblicht und Plastidendifferenzierung von Zellkulturen von Nicotiana tabacum var. "Samsun". Planta 69:58–69
5. De Klerk-Kiebert YM, Kneppers TJA, Bakker PAHM, Schale HH (1982) Comparison of dry matter synthesis and photosynthetic characteristics of chlorophyllous and non-chlorophyllous cell suspension cultures of soybean (Glycine max. L.). Z Pflanzenphysiol 105:445–456
6. Gross M, Richter G (1982) Influence of sugars on blue light-induced synthesis of chlorophyll in cultured plant cells. Plant Cell Rep 1:288–290
7. Hundrieser J, Richter G (1982) Blue light-induced synthesis of ribulosebisphosphate carboxylase in cultured plant cells. Plant Cell Rep 1:115–118
8. Link G (1981) Cloning and mapping of the chloroplast DNA sequences for two messenger RNAs from mustard (Sinapis alba L.). Nucleic Acids Res 9:3681–3684
9. Mapleston RE, Griffiths WT (1980) Light modulation of the activity of protochlorophyllide reductase. Biochem J 189:125–133
10. Mattoo AK, Pick U, Hoffman-Falk H, Edelman M (1981) The rapidly metabolized 32 000-dalton polypeptide of the chloroplast is the "proteinaceous shield" regulating photosystem II electron transport and mediating diuron herbicide sensitivity. Proc Natl Acad Sci USA 78: 1572–1576
11. Meyer G, Bliedung H, Kloppstech K (1983) NADPH-protochlorophyllide oxidoreductase: Reciprocal regulation in mono- and dicotyledonean plants. Plant Cell Rep 2:26–29
12. Richter G, Reihl W, Wietoska B, Beckmann J (1980) Blue light-induced development of thylakoid membranes in isolated seedling roots and cultured plant cells. In: Senger H (ed) The blue light syndrome. Springer, Berlin Heidelberg New York, pp 465–472
13. Richter G, Beckmann J, Gross M, Hundrieser J, Schneider Ch (1982) Blue light-induced synthesis of chloroplast proteins in cultured plant cells. In: Prog Clin Biol Res, vol 102B. Cell function and differentiation, part B: Biogenesis of energy transducing membranes and membrane and protein energetics (FEBS vol 65). Liss, New York, pp 267–276
14. Santel H, Apel K (1981) The protochlorophyllide holochrome of barley (Hordeum vulgare L.). The effect of light on the NADPH-protochlorophyllide oxidoreductase. Eur J Biochem 120:95–103
15. Towbin H, Staehelin T, Gordon J (1979) Electrophoretic transfer of proteins from acrylamide gels to nitrocellulose sheets: Procedure and some applications. Proc Natl Acad Sci USA 76: 4350–4354

Blue Light Effect on Chloroplast Development in Higher Plants

G. AKOYUNOGLOU [1,2] and H. ANNI [3]

Introduction

The etioplast to chloroplast differentiation in angiosperms depends on light. Light is not only necessary for the transformation of PChl(ide) to Chl(ide), but it regulates also the synthesis and assembly of the thylakoid components into structural or functional units, as well as their relative concentration. Light intensity, mode of illumination and light quality affect in different ways the chloroplast development; thus depending on the conditions used, thylakoids with differences in composition, structure and function can be formed. The differences may concern the size and relative concentration of the PSI and PSII units, the relative concentration of the electron transport components, or the relative abundance of stacked and unstacked lamellae.

Two kinds of approach have been used in studying the effect of blue light on chloroplast development. Etiolated leaves or cotyledons were exposed either to continuous illumination with blue or red light, or to intermittent illumination with red or blue light followed by continuous illumination; the effect of light exposure on development, or on the final developmental stage reached by chloroplasts was then studied [4, 7, 10, 13, 14, 16, 18, 19]. In our studies we exposed etiolated bean leaves (with one cotyledon removed) first to intermittent blue or red light, and then we transferred them to continuous blue or red light; during this treatment we followed the changes in composition, structure and function of the developing thylakoids. Etiolated bean leaves greened under the same mode of illumination, but with white light, served as control. As is known [3, 5, 8, 9], the rate of Chl a formation in intermittent light is very slow; thus Chl a regulates thylakoid development: during intermittent illumination, all thylakoid polypeptides are synthesized, but since Chl a is present in limiting amount, only a few of these polypeptides can be stabilized in the thylakoid by forming Chl a-pigment-protein complexes; the rest are digested. This is especially true for the apoprotein of the light-harvesting complex of PSI and PSII units; thus, the PSI and PSII units formed, contain only the core of the units [1, 3]. Not only the size of the PSI and PSII units formed in intermittent light is small, but also their relative concentration, i.e., the PSII/PSI ratio, which is about 2.5 times larger under these conditions than in the green control [2].

---

1 Fachbereich Biologie, Universität Marburg/Lahn, Lahnberge, D–3550 Marburg, FRG
2 On leave from the Nuclear Research Center "Demokritos", Athens, Greece
3 Biology Department, Nuclear Research Center, "Demokritos", Athens, Greece

Blue Light Effects in Biological Systems
ed. by H. Senger

Previous experiments have shown that the rate of Chl a formation in blue or red intermittent light is almost the same with that in white intermittent light [4]. Since the rate of Chl a synthesis is not affected in a different way by blue, red or white intermittent light, it cannot be responsible for any further changes in thylakoid composition, due to the quality of light used. Thus, the effect of blue or red light on the synthesis and/or assembly of other thylakoid components can be distinguished and studied by exposing etiolated plants to intermittent illumination. Our results showed that blue or red light regulate the synthesis of some of the thylakoid components, affecting thylakoid development, as well as the final stage of development attained. In addition, it seems that blue or red light act in a cooperative way, so that for normal chloroplast development, both light qualities are required.

Materials and Methods

Six-day etiolated plants (*Phaseolus vulgaris*, Red kidney var.) grown at 22° ± 1°C and 90% humidity were used. The handling of the seedlings was as described previously [8]. The leaves were exposed first to intermittent light-dark cycles (2 min blue or red light – 98 min dark), and then they were transferred to blue or red continuous light [4].

The blue light source consisted of Sylvania lamps (F 20T12B) with a 3 mm blue Plexiglas filter (Röhm Darmstadt Germany, no. 627); band width 380–550 nm; emission maximum at 450 nm. The red light source consisted of Philips lamps (TL 40W/15) with a 3 mm red Plexiglas (Röhm Darmstadt Germany, no. 555); band width 600–700 nm; emission maximum at 660 nm. Light intensity for both wavelengths was 22.5 μmol m^{-2} s^{-1}. The intensity used was the highest possible, but it was adequate to phototransform most of the regenerated PChl(ide) to Chl(ide) during the 2 min light period of the cycles.

In another set of experiments the leaves were exposed to continuous monochromatic light (interference filters: Schott AL with peaks at 443, 483, 652 and 694 nm; light intensity: 20 μmol m^{-2} s^{-1}).

The pigments were extracted with 80% acetone as previously described [8] and the Chl concentration was estimated according to Mackinney [15].

For photochemical activity measurements the plastids were isolated as described before [1, 6]; the PSII activity was determined as the rate of the $H_2O \rightarrow DCIP$ or $DPC \rightarrow DCIP$ reduction [17]; the PSI activity was monitored by the rate of the MV-mediated O_2 uptake in a DCMU poisoned reaction mixture containing DCIP and Na ascorbate [12]. Fluorescence measurements were done as previously described [1]. The Chl concentration in all assays was 5 μg ml^{-1}.

Results and Discussion

Effect of Blue Light on the Final Stage of Development

Many workers have studied the effect of blue light on chloroplast development, and especially on the final stage of its development [4,7,10,13,14,18,19]. In these studies

Table 1. Differences in composition, structure and photosynthetic activity of plastids from plants grown under blue or red light [4,7, 10,13,14,18,19]

Blue light / Red light	
Structure:	Less grana stacks per chloroplast Fewer thylakoids per granum
Chl (a+b):	Less chlorophyll g^{-1}, fresh or dry wt.
Chl a/Chl b:	Higher ratio
Chl (a+b)/car:	Higher ratio
xanth/b-car:	Smaller ratio
Phylloquinone K_1:	Higher concentration
Plastoquinone-9:	Higher concentration
Plastoglobuli:	Higher number (∿ 3/1)
Cytf/Chl:	Higher ratio
PSII activity:	Higher rates
Rate of development depends on the tissue and the tissue age	

various plant species and various light intensities were used, and the effect of blue light was compared to that of the red light. A number of such results are summarized in Table 1. As is evident, there is a clear difference between the chloroplasts formed in blue light and those formed in red light; this indicates that blue light has indeed a distinct effect on chloroplast development.

Effect of Blue Light on Chloroplast Development

Pigment Formation

The rate of Chl formation in blue or red continuous light is initially the same (Fig. 1). Later, however, the rate decreases; this occurs earlier in blue than in red light, and so the final Chl level reached is much lower in blue light (see Table 2). In red or white continuous light, however, the rate of Chl formation, as well as the final level reached, are similar.

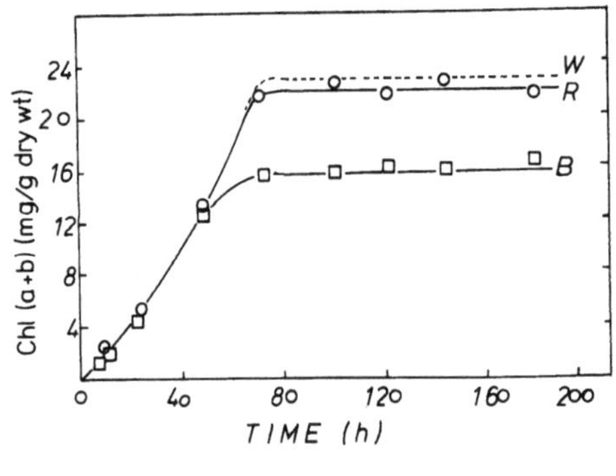

Fig. 1. Time course of Chl a+Chl b accumulation in 6-day etiolated bean leaves upon illumination with continuous blue, red or white light

Table 2. Chlorophyll content and dry weight of 6-day etiolated bean leaves exposed to red or blue light-dark cycles (LDC) or to continuous light (CL)

	Dry wt (mg/2 leaves)	Chl (a+b) (mg g^{-1} dry wt)	b-car	lut	vx
				(% Molar ratio)	
R CL	47.0	22.2	29	–	–
B CL	40.5	16.0	39	–	–
86 RLDC	57.0	4.4	18	48	22
86 BLDC	40.5	3.4	21	43	23
28 B or RLDC+BCL	40.5	13.0	30.5	52	8.5
28 B or RLDC+RCL	47.0	17.0	32.5	46.5	5.5

b-car: b-carotene; lut: lutein; vs: violoxanthin

The rate of Chl formation in blue or red intermittent light is also similar (Fig. 2). Transfer of blue intermittent light plants or red intermittent light plants to continuous blue or red light induces an increase in Chl accumulation. In this case also the higher level is reached in red than in blue continuous light.

Similar to our previous observations with white light [5], the amount of Chl formed in blue or red continuous light depends on preexposure to intermittent light: less Chl is formed in plants transferred to continuous light after a long preexposure to intermittent light.

The effect of blue or red light on carotenoid formation is similar to that on Chl, i.e., higher amount of carotenoids are formed in red than in blue light (Table 2). The relative concentration of the individual carotenoids is also affected by light quality. More b-carotene is formed in blue than in red light. It is interesting to note also that the effect of red intermittent light on leaf growth is more pronounced than that of red

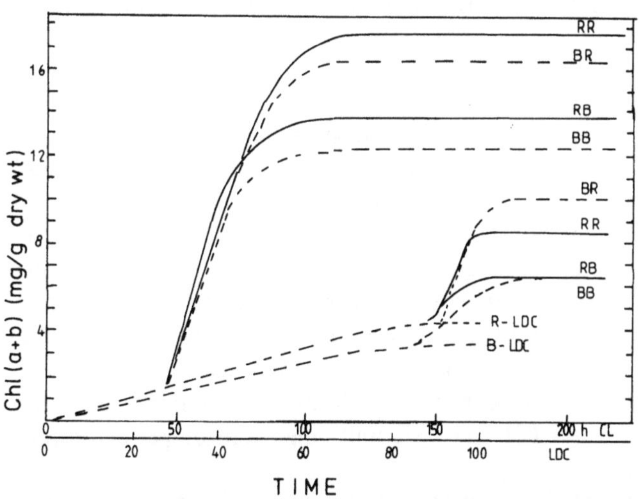

Fig. 2. Time course of Chl a+Chl b accumulation in 6-day etiolated bean leaves exposed first to intermittent blue or red light (B-LDC, R-LDC) and then transferred to continuous illumination with blue or red light (B-CL, R-CL) BB; B-LDC+B-CL; RR: R-LDC+R-CL; BR: B-LDC+R-CL; RB: R-LDC+B-CL

Table 3. PSI and PSII activity of plastids obtained from 6-day etiolated bean leaves exposed to blue, red, or white light-dark cycles (LDC)

	PSI activity mg^{-1} Chl h^{-1}			PSII activity mg^{-1} Chl h^{-1}		
	Blue	Red	White	Blue	Red	White
28 LDC	455	395	485	676	852	950
57 LDC	304	310	480	550	806	950
86 LDC	220	223	475	545	703	890

PSII activity: μmol DCIP reduced; PSI activity: μmol O_2 consumed

continuous light. Since leaf growth is controlled by phytochrome, this indicates that red intermittent light is more effective in activating or increasing the concentration of the active form of phytochrome (P_{fr}) than red continuous light.

Formation of the Photosystem I and II Units

Table 3 shows the changes in the PSI and PSII activity of plastids obtained from etiolated leaves exposed to intermittent illumination with blue, red or white light. The PSI activity mg^{-1} Chl is initially the same in all samples; later, however, as exposure to intermittent light is prolonged, the activity of the plastids in blue or red intermittent light plants decreases, while that of plastids in white intermittent light plants remains almost constant. On the contrary, the PSII activity mg^{-1} Chl is different in the three samples from the start of exposure to intermittent light (28 LDC): it is higher in the white light sample and lower in the blue light sample. As exposure to intermittent light is prolonged, the activity decreases in all cases, but nevertheless the difference observed remains.

Since PS activity mg^{-1} Chl is a measure of the relative concentration of PSI and PSII units mg^{-1} Chl, our results suggest that less PSI and PSII units are formed mg^{-1} Chl in blue and in red intermittent light plastids than in the white light ones. The relative concentration of PSI and PSII units per chloroplast in each sample can be estimated by expressing the PSI and PSII activity on a g dry weight basis. Such estimations are shown in Table 4. As shown, the concentration of PSI and PSII units in all samples increases as exposure to intermittent light is prolonged; however, their concentration is always lower in the blue than in the red intermittent light, and much lower in both blue and red light plastids than in the white light ones. So, during intermittent light, the plastids formed have less PSI and PSII units mg^{-1} Chl or per chloroplast when blue or red light is used.

Taking into account (a) that the amount of Chl accumulating during the intermittent light in all three samples is almost the same; (b) that the rate of Chl a synthesis in white intermittent light is very low; and (c) that the Chl a formed in intermittent light is used for the formation of new and small in size PSI and PSII units, containing

Table 4. PSI and PSII activity of plastids obtained from 6-day etiolated bean leaves exposed to blue, red, or white light-dark cycles (LDC)

	PSI activity g^{-1} dry wt			PSII activity g^{-1} dry wt			PSII/PSI		
	Blue (% of Control)	Red	White	Blue (% of Control)	Red	White	Blue	Red	White
28 LDC	22	22	18	23	34	43	1.04	1.52	2.5
57 LDC	27	36	33	37	66	80	1.29	1.83	2.4
86 LDC	29	36	51	46	81	117	1.80	2.24	2.3

PSII activity: μmol DCIP reduced; PSI activity: μmol O_2 consumed

the core of the units, we can explain our results as follows: in blue or red intermittent light the Chl a formed is not only used for the formation of new units, but also for the increase of the unit size, i.e., for the formation of the light-harvesting components of PSI and PSII. This means that in blue or red intermittent light the rate of PSI and PSII unit formation is much lower than that in white intermittent light, so that Chl a becomes available to bind to the apoprotein of the LHC of PSI and to the LHC of PSII. Since even at this low Chl concentration not all of it is used for the formation of new PSI and PSII units, small in size, but some is also used for the formation of the light-harvesting antenna components, one can conclude that some of the PS unit core polypeptides are formed at a lower rate in blue or red than in white intermittent light. The rate of formation of these polypeptides, therefore, regulates, in addition to Chl a, the development of the thylakoid.

It seems, therefore, that a simultaneous exposure to both blue and red light is required to stimulate the synthesis of all thylakoid polypeptides, and that both light qualities act cooperatively during development of the thylakoid. Blue or red light affect not only the size and concentration of the PSI and PSII units, but also their PSII/PSI ratio, as shown in Table 4. The ratio is higher in white-LDC and much lower in blue-LDC. Moreover, while in white-LDC the ratio remains constant throughout the light exposure, in blue- and red-LDC this ratio changes with time of exposure, indicating that the rate of PSI and PSII formation is not constant under these conditions.

The absence of blue or red light during development does not affect only the process of thylakoid development, but also the final stage reached. This is evident from experiments where the etiolated bean leaves were first exposed to intermittent blue, red or white light for various periods of time, and then were transferred to continuous light, and the final concentration of the PSI and PSII units per chloroplast were estimated. As shown in Tables 5 and 6, the concentration of PSI and PSII units per chloroplast and the PSII/PSI ratio are affected by the quality of light used during intermittent illumination, as well as after transfer to continuous illumination. The relative concentration of the PSI and PSII units is much smaller when blue light is used.

Table 5. PSI and PSII activity of plastids obtained from 6-day etiolated bean leaves exposed to blue, red, or white light-dark cycles (LDC) and then transferred to continuous light (CL)

	Chl (a + b) mg g^{-1} dry wt	PSI activity g^{-1} dry wt (% of Control)	PSII activity g^{-1} dry wt (% of Control)	PSII/PSI
28 BLDC+BCL	13	56	54	0.96
28 RLDC+BCL	13	64	86	1.34
28 BLDC+RCL	17	74	100	1.35
28 RLDC+RCL	17	78	75	0.96
28 WLDC+WCL	22	100	100	1.00

PSII activity: μmol DCIP reduced; PSI activity: μmol O_2 consumed

Table 6. PSI and PSII activity of plastids obtained from 6-day etiolated bean leaves exposed to blue, red, or white light-dark cycles (LDC) and then transferred to continuous light (CL)

	chl (a+b) mg g^{-1} dry wt	PSI activity g^{-1} dry wt (% of Control)	PSII activity g^{-1} dry wt (% of Control)	PSII/PSI
86 BLDC+BCL	6.2	27	70	2.56
86 RLDC+BCL	6.4	28	60	2.10
86 BLDC+RCL	9.9	44	63	1.43
86 RLDC+RCL	8.5	40	59	1.46
86 WLDC+WCL	8.0	50	100	2.00

PSII activity: μmol DCIP reduced; PSI activity: μmol O_2 consumed

Fluorescence Yield

Changes in the fluorescence yield (F_{max}/Chl) when intermittent light plants are transferred to continuous light seem to monitor the degree of organization of the pigment-protein complexes in the thylakoid PSI and PSII units. The changes in F_{max}/Chl of

Table 7. Fluorescence characteristics of plastids obtained from 6-day etiolated bean leaves exposed to blue, red, or white light-dark cycles (LDC) and then transferred to continuous light (CL)

	F_{max}/Chl BLDC+BCL	RLDC+BCL	BLDC+RCL	RLDC+RCL	WLDC+WCL
28 LDC	89	65	89	65	76
28 LDC + 24 h CL	277	280	309	201	243
28 LDC + 90 h CL	185	176	180	160	147
86 LDC	89	65	89	65	76
86 LDC + 46 h CL	390	240	342	225	380
86 LDC + 90 h CL	203	122	203	173	204

Table 8. Greening of 6-day etiolated bean leaves in continuous monochromatic light for 48 h

Illum.	Chl (a+b) $\mu m\, g^{-1}$ fr wt	Chl a / Chl b	F_{max} / Fo	t 1/2 (ms)	F_{max} / Chl
443	1470	3.4	2.95	95	7.4
483	1380	4.5	2.51	105	6.3
652	1950	3.25	2.55	110	6.4
694	810	4.9	2.90	145	7.3

plants exposed first to intermittent blue, red or white light and then transferred to continuous light are shown in Table 7. In all cases the F_{max}/Chl follows the same pattern, i.e., it increases to a maximum, then it decreases and finally it levels off. In blue light, however, the maximum value, as well as the final value, reached are much higher than in the red light, indicating a different degree of organization of the pigment-protein complexes, and, in addition, a different rate for this organization process.

Effect of Monochromatic Light on Chloroplast Development

In preliminary experiments we studied the effect of monochromatic light of various wavelengths in the blue or red region of the spectrum on Chl formation and on the fluorescence characteristics of the chloroplasts. The results are shown in Table 8. It is interesting to note that each monochromatic light has a different effect on the Chl a/Chl b ratio and on the half-rise time of the fluorescence kinetics in the presence of DCMU. The Chl a/Chl b ratio is lower in the 443 nm and 652 nm light, and higher in the 483 nm and 694 nm light. Since the Chl a/Chl b ratio is affected by the relative concentration and size of the PSII unit, the results indicate a difference in the effect of the 443 nm light compared to the 483 nm one on PSII unit size, and also a difference between the effect of 652 nm light and 694 nm light. More experiments are required, however, before any conclusion can be reached.

Conclusion

The results of the present study clearly indicate that blue as well as red light are necessary for the synthesis of certain thylakoid polypeptides. In the absence of one of these light qualities, the rate of synthesis of these polypeptides is small; thus, they become the regulating factor for further thylakoid development, assembly and stabilization of the various components. This has as a consequence the formation of thylakoids with different composition, structure and function. In addition, for normal chloroplast development both light qualities are necessary, suggesting that the blue and red photoreceptors act in a cooperative way.

Abbreviations

PSI: Photosystem I, PSII: Photosystem II; Chl: Chlorophyll; LDC: Light-Dark Cycles of 2 min light-98 min dark; B: blue light; R: red light; W: white light; MV: methyl viologen; DCIP: dichlorophenol indophenol; DPC: diphenyl carbazide.

References

1. Akoyunoglou G (1977) Development of the Photosystem II unit in plastids of bean leaves greened in periodic light. Arch Biochem Biophys 183:571–580
2. Akoyunoglou G (1981) Assembly of functional components in chloroplast photosynthetic membranes. In: Akoyunoglou G (ed) Photosynthesis, vol V. Chloroplast development. Balaban Int Sci Serv, Philadelphia, pp 353–366
3. Akoyunoglou G (1982) Reorganization of the thylakoid components during chloroplast development in higher plants. In: Akoyunoglou G et al (eds) Cell function and differentiation, part B. Liss, New York, pp 171–188
4. Akoyunoglou G, Anni H, Kalosakas K (1980) The effect of light quality and the mode of illumination on chloroplast development in etiolated bean leaves. In: Senger H (ed) The blue light syndrome. Springer, Berlin Heidelberg New York, pp 473–484
5. Akoyunoglou G, Argyroudi-Akoyunoglou JH (1978) Control of thylakoid growth in *Phaseolus vulgaris*. Plant Physiol 61:834–837
6. Akoyunoglou G, Michelinaki-Maneta M (1974) Development of photosynthetic activity in flashed bean leaves. In: Avron M et al (eds) Proc 3rd Int Congr Photosynth. Elsevier, Amsterdam, pp 1885–1896
7. Anni H, Akoyunoglou G (1981) The effect of blue and red light on the development of the photosynthetic units during greening of etiolated bean leaves. In: Akoyunoglou G (ed) Photosynthesis, vol V. Chloroplast development. Balaban Int Sci Serv, Philadelphia, pp 885–894
8. Argyroudi-Akoyunoglou JH, Akoyunoglou G (1970) Photoinduced changes in the Chl *a* to Chl *b* ratio in young bean leaves. Plant Physiol 46:247–249
9. Argyroudi-Akoyunoglou JH, Akoyunoglou A, Kalosakas K, Akoyunoglou G (1982) Reorganization of the Photosystem II unit in developing thylakoids of higher plants after transfer to darkness. Plant Physiol 70:1242–1248
10. Buschmann C, Meier D, Kleudgen HK, Lichtenthaler HK (1978) Regulation of chloroplast development by red and blue light. Photochem Photobiol 27:195–198
11. Castorinis A, Akoyunoglou G, Argyroudi-Akoyunoglou JH (1982) Correlation between the organization of the pigment-protein complexes and the Chl *a* fluorescence yield of chloroplasts, during development in *Phaseolus vulgaris*. Photobiochem Photobiophys 4:283–291
12. Hall DO, Reeves SG, Baltscheffsky H (1971) Photosynthetic control in isolated spinach chloroplasts with endogenous and artificial electron acceptors. Biochem Biophys Res Commun 43:359–366
13. Lichtenthaler HK, Buschmann C (1978) Control of chloroplast development by red light, blue light and phytohormones. In: Akoyunoglou G et al (eds) Chloroplast development. Elsevier, Amsterdam, pp 801–816
14. Lichtenthaler HK, Buschmann C, Rahmsdorf U (1980) The importance of blue light for the development of sun-type chloroplasts. In: Senger H (ed) The blue light syndrome. Springer, Berlin Heidelberg New York, pp 485–494
15. Mackinney G (1941) Absorption of light by chlorophall solutions. J Biol Chem 140:315–322
16. Senger H (1982) The effect of blue light on plants and microorganisms. Photochem Photobiol 35:911–920

17. Vernon LP, Shaw ER (1969) Photoreduction of 2,6-dichlorophenol indophenol by diphenylcarbazide: a photosystem 2 reaction catalyzed by Tris-washed chloroplasts and subchloroplast fragments. Plant Physiol 44:1645–1699
18. Wild A (1979) Physiology of Photosynthesis in higher plants. The adaptation to light intensity and light quality. Ber Dtsch Bot Res 92:341–364
19. Wild A, Holzapfel A (1980) The effect of blue and red light on the content of Chl, cyt-f, soluble reducing sugars, soluble proteins and the nitrate reductase activity during growth of the primary leaves of *Sinapis alba.* In: Senger H (ed) The blue light syndrome. Springer, Berlin Heidelberg New York, pp 444–451

Control of the Activity of Photosynthetic Apparatus in Higher Plants

N. P. VOSKRESENSKAYA[1]

Introduction

It is established that the intensity of solar radiation, its spectral composition and periodicity of illumination control a wide range of vital activities of green plants.

Attention is now focused on the role of light of different quality in the regulation of chloroplast organization and its photosynthetic activity.

Some evidence of a specific role of BL, as compared with RL, in the regulation of photosynthetic apparatus organization was reported at the First International Conference on the Blue Light Effects on Plants and Microorganisms [1, 3, 13, 29] and in the review of Senger [20].

This paper gives a brief outline of the major effects of RL and BL both on photosynthetic CO_2 gas exchange, and on some parameters of the individual stages of photosynthesis. Some experiments will be described that may, to a certain extent, explain the reason for the difference in action of BL and RL on the activity of photosynthetic apparatus.

Materials and Methods

Pea and barley seedlings were grown in phytothrone at 22°C and at a 16 h photoperiod. Fluorescent lamps of RL (600–700 nm, λ_{max} 660 nm) or BL (400–580 nm, λ_{max} 480 nm) were used for the illumination of plants. Light fluxes were equalized by the number of quanta. In one series of experiments, the plants were grown at a light flux equal to 17 nE cm^{-2} s^{-1}; in the other, under varied fluxes of BL or RL. The active mature leaves were used in all the experiments. Isolation of chloroplasts, NADP and SiMo reduction assays and spectral measurements in the cytochrome absorption region were done as described by Voskresenskaya et al. [23] and Drozdova and Voskresenskaya [6]. The activity of ferredoxin: $NADP^+$-reductase was assayed by the diaphorase reaction [2]. Ribulose bisphosphate carboxylase was prepared as described by Hall and Tolbert [7]. The content of soluble ferredoxin was estimated by the method of Haslett et al. [9]. Chloroplasts and chromatin, which were later used to RNA poly-

1 K.A. Timiryazev Institute of Plant Physiology of USSR Academy of Sciences, Moscow, USSR

Blue Light Effects in Biological Systems
ed. by H. Senger

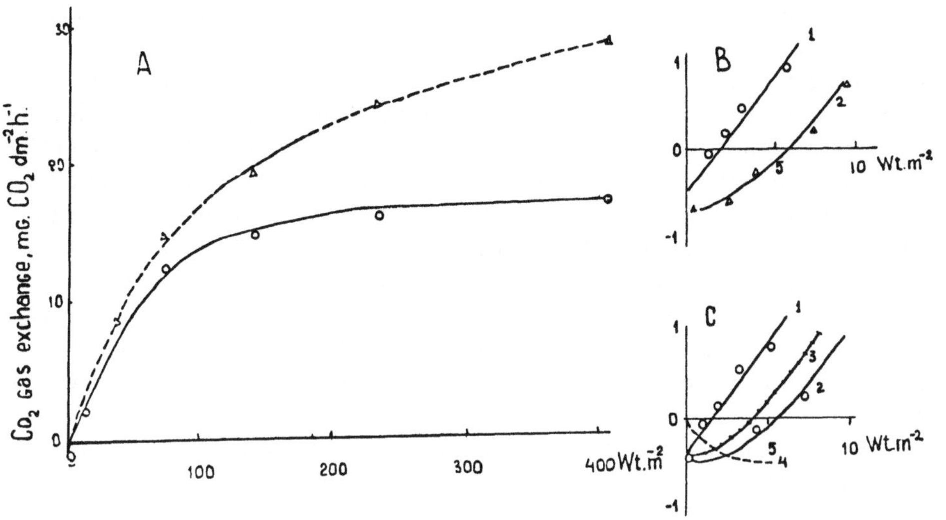

Fig. 1A–C. Light curves of photosynthesis of barley leaves. Plants grown under blue (△- - -△), or red (○——○) light. **A** Light curves measured under white light; **B, C** Photosynthesis nearly light compensation point: **B** Measured under red light for RL plants (*1*) or under blue light for BL plants (*2*); **C** Measured for RL plants: under red light (*1*) or under blue light equal intensity (*2*) or equal quanta (*3*) with red light. Blue light induced CO_2 evolution (*4*)

merase assay, were isolated as described by Romanko et al. [18]; RNA synthesis was estimated by 8-^{14}C-adenine incorporation [12]. An "Infralit-2" infrared gas analyzer was employed for determinations of CO_2 gas-exchange rates (CO_2 concentration in the open system was 300 ppm, humidity in the experimental chamber, 63%; temperature, 25°C). Chlorophyll a fluorescence was excited by blue light (4×10^4 erg cm^{-2} s^{-1}). The reaction center content of photosystem II (PSII) was estimated as described by Malkin et al. [14]. The reaction center content of PSI and that ofplastosemiquinone radicals (PQ $^{\overline{\cdot}}$) were estimated by EPR spectroscopy as described by Voskresenskaya et al. [24].

Results

CO_2 Gas Exchange and the Activities of Carbon Metabolism Enzymes

The light curves of barley leaf photosynthesis were measured under white light. CO_2 gas-exchange in plants grown under BL (BL plants) was always higher than in plants grown under RL (RL plants) (Fig. 1). The difference between the BL and RL plants in favor of the former is particularly obvious on the ascending part of light curve and on the plateau. Light saturation in the RL plants always occurs as early as at 200 W m^{-2} and in BL plants is not even observed at 400 W m^{-2}. By the light curve patterns the BL plants may be referred to as sun-type plants and RL plants to shade-type plants. This differentiation of BL and RL plants agrees with that suggested previously on the basis of other features of the photosynthetic apparatus [13].

During the measurement of the light curve of CO_2 gas exchange under the same light as was used for plant growth, the difference in the plateau levels remained; however some peculiarities were noted as to the location of light compensation points (LCP). The value is significantly higher in BL plants than in RL plants (Fig. 1B). On determining the LCP value under BL, one can also observe an increase in LCP in RL plant leaves (Fig. 1C). The increase of LCP is obviously associated with fast activation of oxidative reactions in photosynthesizing cells under the action of BL. The activation has a low light saturation (Fig. 1C,4). Rapid enhancement of oxidative processes within the cell, which is due to BL effect, has been comprehensively studied by Kovallik [11] in the experiments with green and nongreen mutant of *Chlorella* strains; the enhancement was also shown to have a low light saturation. The phenomenon may have a direct bearing on the enhancement of LCP for photosynthesis under blue and white light.

Hence, at equal energy conditions for photosynthesis under RL, chloroplasts are formed that have lower photosynthetic activities. It is possible that one of the reasons for the enhancement of photosynthetic activities in BL plants may be an elevation in the level of ribulose bisphosphate carboxylase (RBPC) and other Calvin cycle enzymes [15, 22]. It is now known that BL also stimulates the biosynthesis of some other enzymes; e.g., phosphoenol pyruvate carboxylase [10], pyruvate kinase [16] and NADP-linked glyceraldehyde-3-phosphate dehydrogenase [5]. It is likely that BL stimulates the biosynthesis of the above enzymes, thus affecting the functioning of the metabolic pathways.

Electron Transfer Reactions in BL and RL Plants

Isolated chloroplasts of BL plants have been established to possess a significantly higher potential activity (measured under saturating white light) of photoreduction of DCPIP, ferricyanide, and NADP during the electron transfer from water. Similar difference was obtained for all these activities in BL- and RL plants grown both under low [8, 13] and relatively high light intensity [27]. Therefore, there are grounds for believing that changes in the organization of electron transport chain (ETC) developing under BL that lead to enhanced electron transfer have low light saturation characteristic of regulatory reactions.

It was shown also that the rate of DCMU-independent evolution of oxygen, which is only coupled with photoreduction of SiMo within PSII was higher in chloroplasts of BL plants [6]. More active evolution of oxygen in this case correlates well with the fact that the content of prenylquinones [13] and that of $PQ^{\overline{\cdot}}$ localized at the donor side of PSII is higher in BL plants [25]. The difference between chloroplasts of both variants increase still greater when NADP was used instead of SiMo as photoelectron acceptor from water [6]. So, there must be an additional event that limits the electron transfer to PSI, between PSI and PSII in RL plants. The event may reflect changes of the rates of redox conversions of cytochrome f induced by illumination. It has been found that in BL plant chloroplasts, photooxidation of the cytochrome prevails over photoreduction. Both during cyclic and noncyclic electron transfer, electron release within PSI proceeds at a greater rate in BL plants [23]. One should also not overlook higher values of the cytochrome f/chlorophyll ratio in BL plants [29].

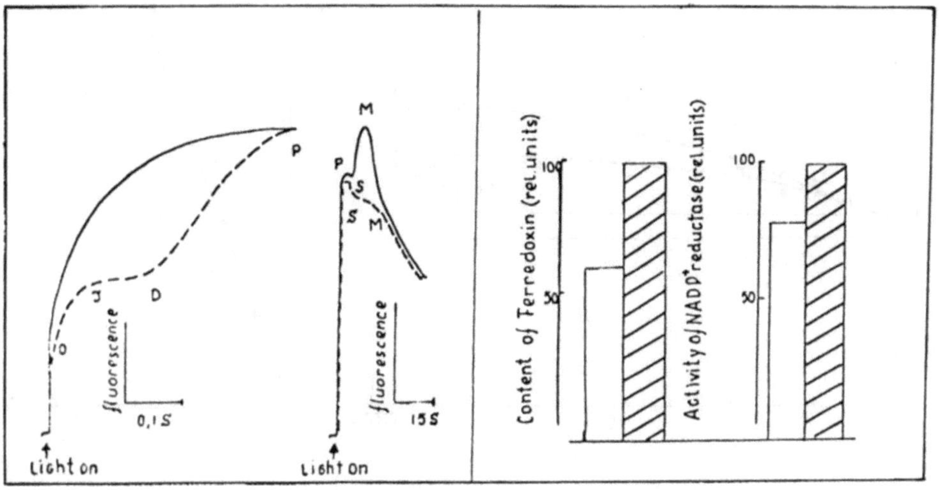

Fig. 2. Time courses of fast and slow chlorophyll a fluorescence transients (*left*) and the content of ferredoxin and Fd: $NADP^+$ reductase (*right*) in BL and RL plants. BL plants, left – – –, right ▨; RL plants, left –––, right ▭

Thus, the main alterations responsible for enhanced electron transfer from water in BL plants are localized at the level of PSII reactions and at the coupling site with PSI reactions. Effects induced by light of different quality may, however, also affect electron transfer within the PSI. With the isolated chloroplasts, no changes can be observed only because some components of electron transfer chain are lost during the isolation of the organelles. Thus, the activity of ferredoxin: NADP-reductase was found to be higher in BL than in RL plants leaf gomogenate. This was also true of the ferredoxin content in a leaf tissue. The more distinct character of the I-D and P-S phases on the kinetic curve of fluorescence induced in barley leaves (Fig. 2) also favors a better organization of PSI in BL-plants. These phases are believed to reflect the reoxidation of Q at the expense of electron transfer within the acceptor part of PSI. Therefore, it is BL rather than RL which in the intact leaf affects the net rate of electron transfer from water by enhancing the biosynthesis of certain components (including those of PSI), similar to the carbon metabolism enzymes.

The Action of Light Intensity on the Development of BL and RL Effects

As mentioned above, the changes in the ETC activity induced by BL are observed both in plants grown at a very low and at a relatively high intensities of light. In all these works the concentrations of chlorophyll and reaction centers PSI [24] in chloroplast suspensions of BL or RL plants were pre-equalized.

Fig. 3a–c. The content of chlorophyll (**a**), reaction centers of photosystems I (**b**) and II (**c**) in leaves of plants grown under various intensity of blue or red light (*left*) and dependence of ΔN blue on ΔN red (*right*). ○- - -○ blue light plants ○––○ red light plants

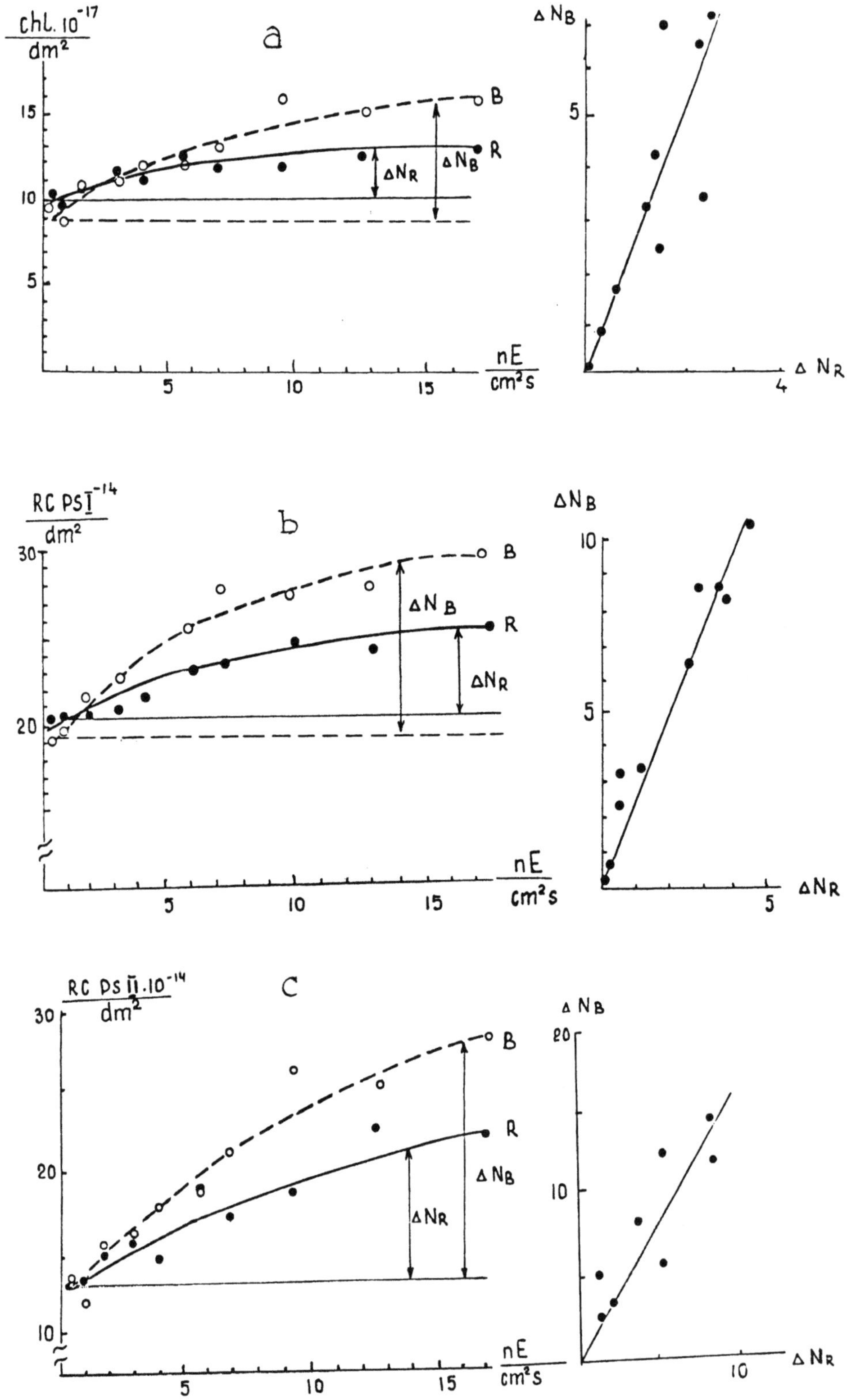
a
chl. 10^-17 / dm²
15
10
5
B
R
ΔN_R
ΔN_B
5
10
15
nE / cm²s
ΔN_B
5
ΔN_R
4
b
RC PS I^-14 / dm²
30
20
B
R
ΔN_B
ΔN_R
5
10
15
nE / cm²s
ΔN_B
10
5
ΔN_R
5
c
RC PS II.10^-14 / dm²
30
20
10
B
R
ΔN_B
ΔN_R
5
10
15
nE / cm²s
ΔN_B
20
10
ΔN_R
10

Fig. 3a–c

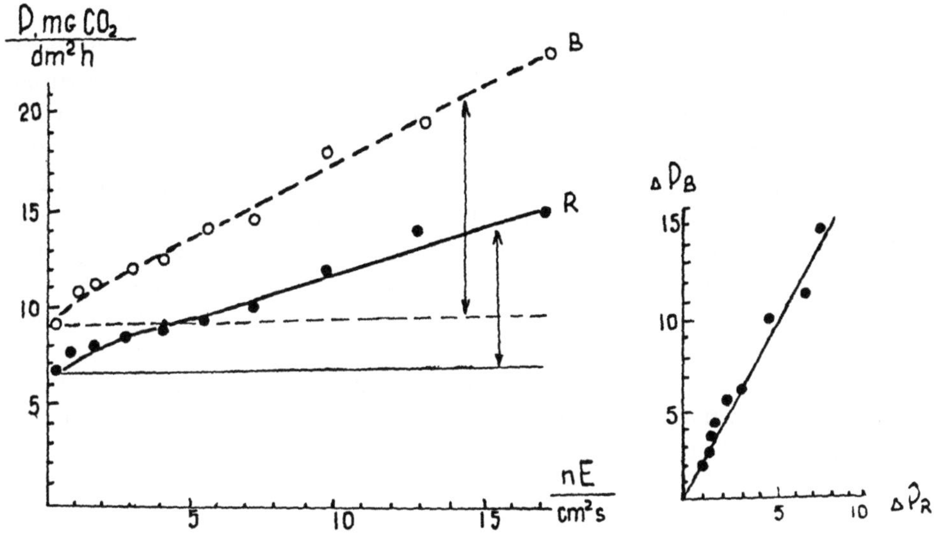

Fig. 4. Photosynthesis in leaves of plants grown under various intensity of blue or red light (*left*) and dependence of ΔP_{blue} on ΔP_{red} (*right*), blue light plants o- - -o red light plant •—•

In whole leaves of plants grown under 17 nE cm^{-2} s^{-1}, the content of chlorophyll and reaction centers of PSI and PSII are higher in BL plants, regardless of whether the data are calculated per unit of fresh weight or per unit of leaf surface. With decrease of BL or RL fluxes under growth of plants, the progressive diminution of the content of chlorophyll and reaction centers of PSI and PSII is observed. Simultaneously, the difference between BL and RL plants tends to gradually decrease. At the lowest intensities of light, the difference disappears (Fig. 3, left). It is noteworthy that the lower value for chlorophyll content in BL plants as compared with RL plants has already been obtained at low light intensity [13, 29]. So the positive effect of BL on the content of chlorophyll and that of reaction centers of PSI and PSII depends on light intensity.

In the case of photosynthetic CO_2 gas-exchange (measured at saturating intensity of white light), the decrease of BL and RL fluxes during the growth of plants leads to decrease of the plateau level of photosynthesis. But at any flux, the level remains significantly higher in BL plants (Fig. 4, left). Thus, there is no direct relationship between regulatory action of BL on CO_2 gas-exchange and some indices such as chlorophyll concentration in the leaf and the number of reaction centers of PSI and PSII. One can see, however, the relationship between the low energy BL effects on CO_2 gas-exchange and the organization of photosynthetic electron transfer chain. Both processes remain significantly higher in BL plants whatever the intensity of light [4].

It should be noted that at the lowest light flux used for illumination of plants during growth, all the parameters investigated still had quite a pronounced value, and later on they changed only insignificantly with further increase of light flux (Figs. 3, 4 left). This agrees with the idea that a very small quantity of light is necessary for the formation of well-functioning chloroplasts and that BL optimizes the conditions of chloroplast formation at these light intensities, this shapes the level of the photosynthetic plateau. Further increase in light intensity facilitates the quantitative increment of the parameters. If one consider increments of various parameters taken at an intensity of light as a function of the value at 0.5 nE cm^{-2} s^{-1}, a direct correlation is observed between the increment values of the investigated parameters under BL and RL within the range of light intensities tested (Figs. 3, 4 right). The linear relationship of the parameters in BL and RL plants indicates that, over a wide range of light intensities, a precise ratio of BL effects to those of RL is maintained. This fact corroborates the excitation of the same photoreceptor(s) (or photoprocess) by blue and red rays. BL is more effective in this respect. This can be explained by a greater absorption quotient of the photoregulatory pigment over the blue region of the spectrum and, as a consequence, by accumulation of the active form of the pigment or, alternatively, by the accumulation of constituents of the photosynthetic apparatus at the background of an interaction between the photosynthetic process and the BL-activated processes of protein biosynthesis and other biosyntheses.

Effect of BL and RL on Nucleic Acid Metabolism and RBPC Biosynthesis

How can the mechanism of BL effects on the activities of photosynthetic apparatus be explained now? First of all, it should be noted that the main reason for the light-quality-dependent adaptive changes indicated above must be sought in the different action of BL and RL on the genome of the green cell. It is well known that the chloroplast biogenesis involves two genomes, that of the nucleus and of the chloroplast itself. The photoreceptor excited by the BL is suggested to exist at least in the cytoplasm of *Euglena* [19] and *Chlorella* [21]. The greater quantity of information delivered from the nucleus, which is the effect of the photoreceptor, enhances the possibilities of biosynthesis of the components of chloroplast.

We have made an attempt to check these possibilities using a higher plant (barley leaves) [25]. A labeled preparation of total RNA was isolated from BL and RL leaves, fractionated in sucrose density gradient, and the radioactivities of the individual fractions were determined. The radioactivity incorporated in the heavy RNA fractions where precursors of rRNA and mRNA may be localized and that observed in the region of a gradient just between the positions of the small ribosomal RNA and tRNA, was shown to be higher in BL plant leaves (Fig. 5). The difference was more pronounced in 5-day-old leaves and it is less expressive in 9-day-old aging leaves. Thus, the data obtained by us with higher plants corroborate the different effects of BL and RL on the distribution of the label in RNA precursors.

Further, chromatin and chloroplasts were isolated from leaves of BL and RL plants and the activities of RNA polymerases (RNAP) were measured at the rates of ^{14}C-ATP incorporation in RNA. Simultaneously the activity and the content of RBPC were determined.

Figure 6 shows that the activity of RNAP in chloroplasts is almost similar both in 5- and 7-day-old barley leaves of BL and RL plants. This is not the case with chromatin-bound RNAP, where the activity was always significantly higher in the BL plants. The RBPC activity was also higher in the plants. Essentially the same data were obtained for 8-day-old leaves (Table 1). A high activity of chromatin-bound RNAP in

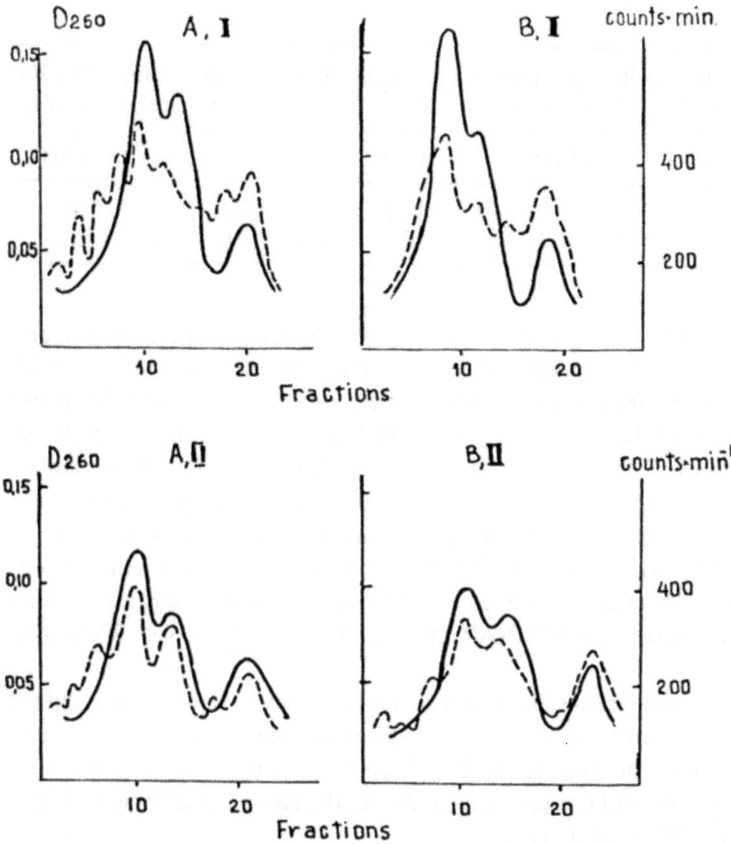

Fig. 5A,B. Sedimentation profiles in sucrose density gradient of total RNA from leaves of BL and RL plants. Plants grown under blue light (A) and red light (B). *Dotted line* absorption in UV; *solid line* radioactivity, *I* age of leaves – 5 days, *II* age of leaves – 9 days

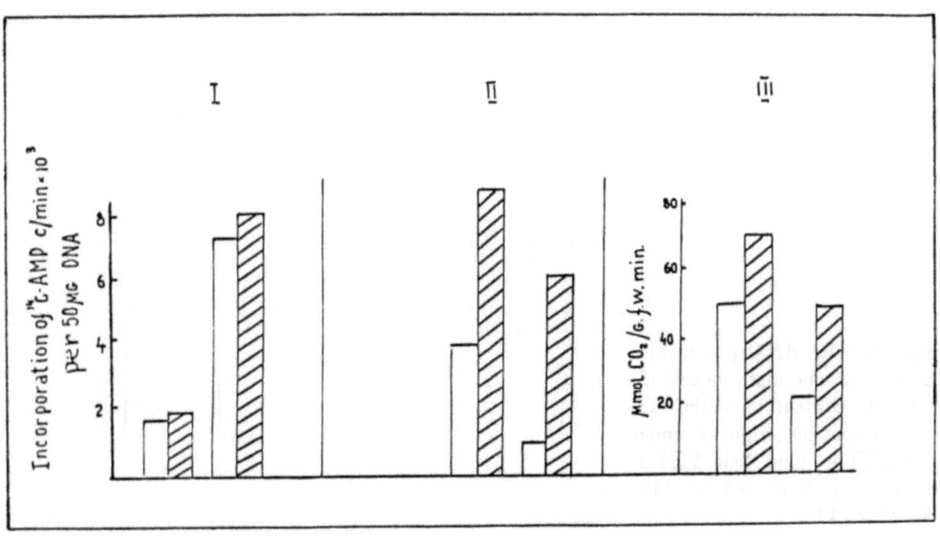

Fig. 6. The activities of RNA polymerase of chloroplasts (*I*) chromatin-bound (*II*) and of RBPC. Red light-grown plants ▭. Blue light grown plants ▨

Table 1. The blue and red light effect on the activity of RNA polymerase and on the amounts of RBPC

Plant illumination	RNA polymerase of chromatine counts min^{-1} per 50 μg DNA	RNA polymerase of chloroplasts counts min^{-1} per 50 μg DNA	Soluble protein mg g^{-1} fr wt	Blue/red, %	RBPC mg g^{-1} fr wt	Blue/red,	Specific activity of RBPC
Red light	950	2660	12.2		6.7		1.2
Blue light	3350	2860	20.0	164	9.6	143	1.3

BL plants correlates well with a higher content of soluble protein and that of RBPC in the BL plants. The identical specific activities of RBPC polypeptides in BL and RL plants indicate that the increase in the enzyme activity, which is observed in BL plants is due to accumulation or de novo synthesis of RBPC rather than to altered properties of the enzyme. Besides, the less active synthesis of RPBC in the RL plants may be conditioned by an insufficient, as compared with BL plants, supply of some cytoplasmic components to the chloroplast. A small subunit of RBPC, that is coded for by nuclear DNA and that is synthesized on 30S ribosomes in the cytoplasm, may be one of such a compounds. This suggestion seems to be true at least of young leaves, since it is the activity of chromatin-bound RNAP rather than that of chloroplast RNAP that is enhanced by the BL.

In the next series of the experiments, the leaves of RL and BL plants were cut off and transferred for 24 h either to the same conditions of illumination as those used during the plants growth, or from RL to BL. Then the activity of chromatin-bound RNAP and the content of RBPC were assayed. Again, the two parameters were higher in leaves replaced from RL to BL (Fig. 7).

A mechanism of BL action on the activity of chromatin-bound RNAP is still obscure. Some evidence has been obtained by us in the experiments where the effect of heterologous supernatants on the RNAP activity in BL and RL plants was studied.

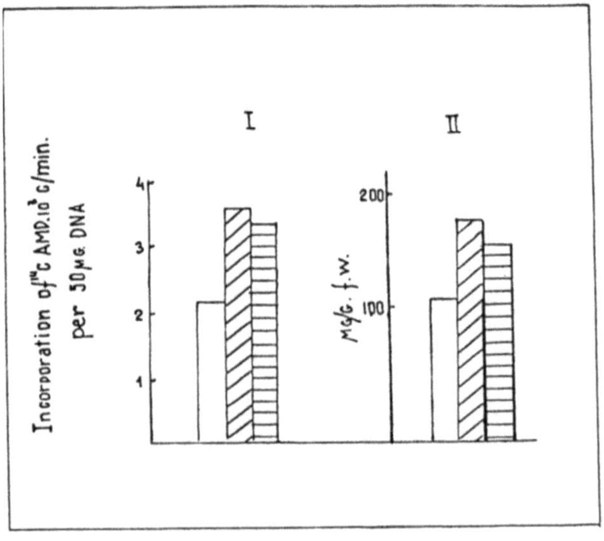

Fig. 7. The RNA polymerase activity of chromatin-bound (*I*) and the content of RBPC (*II*) in leaves of plants grown under red ▭ or blue light ▨ or transferred from red to blue light for 24 h

Table 2. RNA polymerase activity of barley leaves grown under red and blue light

	Light conditions	C^{14} incorporation into AMP, counts/on 50 μg DNA/ 20 min — Leaves homogenized in buffer	Leaves homogenized in heterological supernatant
RNA polymerase of chromatin	Red light	942	2250
	Blue light	2842	1446
RNA polymerase of chloroplasts	Red light	10853	9702
	Blue light	11687	5625

The activity of chromatin-bound or chloroplast RNAP obtained by homogenization of the leaves in a buffer (Table 2) served as standard.

In this case, the chromatin-bound RNAP activity was usually higher in the BL plant leaves, as compared with RL plants and no spectral dependence was observed for chloroplast RNAP. After homogenization of RL plants leaves, in the supernatant of BL plant leaves the chromatin-bound RNAP activity is enhanced more than twofold approaches the value obtained with BL plant leaves in the control. There is no change, however, in the activity of chloroplast RNAP. These data indicate that a soluble product extracted in the supernatant is formed in the plant leaves during the BL action. The product specifically affects the activity of chromatin-bound RNAP. The supernatant of RL plant leaves inhibits the activities of both RNAP. One is to assume then that a soluble nonspecific inhibitor of nuclear and chloroplast RNAP is present in the supernatant of the RL plant leaves. The inhibitor may also be responsible for faster aging of chloroplast in the RL plants [22, 27].

All this evidence has prompted us to the conclusion that BL plays an important role in the regulation of nucleic acid metabolism and protein biosynthesis both in intact cells and chloroplasts of higher flowering plants. During the growth of plants BL affects total RNA composition by enhancing the synthesis of heavy species that may contain in addition to other species, mRNA. The chromatin-bound RNAP activity was found to be highly sensitive to BL. This effect may be one of the reasons for the elevated synthesis of a small subunit of RBPC and other chloroplast constituents in BL plants. If so, our data favor involvement of BL in the regulation of protein biosynthesis at the level of transcription.

Conclusion

The evidence obtained by us indicates in toto that both phytochrome and a pigment excited by BL are involved in the control of the chloroplast structure and function. The pigment is likely to be localized in the cytoplasm. The main function of the pigment appears to be a direct (or mediated) interaction with the components of the protein synthesis system. The latter leads to induction of certain proteins [1, 17] or to activation of protein biosynthesis both in the cytoplasm and in chloroplast [10, 11,

16, 22]. Contrary to BL plants, the formation of active chloroplasts is restricted in RL plants grown under RL of varying intensity. As a result, RL plant chloroplasts age faster and lose the capacity to effectively utilize the high intensity light [22]. The property of chloroplasts to assemble which involves a system of pigments comprising both phytochrome and a pigment excited only by BL, seems to have been acquired by higher plants in the course of the evolution under the natural conditions of solar radiation.

References

1. Akoyunoglou G, Anni H, Kalosakas K (1980) The effect of light quality and the mode of illumination on chloroplast development in ethiolated bean leaves. In: Senger H (ed) The blue light syndrome. Springer, Berlin Heidelberg New York, pp 473–484
2. Avron M, Jagendorf AR (1956) A TPNH diaphorase from chloroplasts. Arch Biochem Biophys 65:475–490
3. Björn LO (1980) Blue light effects on plastid development in higher plants. In: Senger H (ed) The blue light syndrome. Springer, Berlin Heidelberg New York, pp 455–464
4. Bukhov NG, Rojkovsky AD, Chetverikov AG, Voskresenskaya NP (1983) The influence of prolonged action of light intensity and quality on the content of chlorophyll, reaction centers of PS I and PS II and on the activity of CO_2 gas exchange in barley seedlings. Thesis of the 21 coordination conference of CMEA, Problem 1.18.1, Hungaria, in press
5. Conradt W, Ryters G (1980) Blue-light effects on enzymes of the carbohydrate metabolism in chlorella. 2. Glyceraldehyde-3-phosphate dehydrogenase (NADP dependent). In: Senger H (ed) The blue light syndrome. Springer, Berlin Heidelberg New York, pp 368–371
6. Drozdova JS, Voskresenskaya NP (1981) Videlenie O_2 chloroplastami gorocha v prisutstvii kremnievomolibdenovii kisloti posle dlitelnoi adaptazii rastenii k krasnomu i sinemu svetu. Dokl Akad Nauk SSSR 256:244–247
7. Hall NP, Tolbert NE (1978) A rapid procedure for the isolation of ribulose bisphosphate carboxylase/oxygenase from spinach leaves. FEBS Lett 96:167–169
8. Harnischfeger G, Treharne K, Feierabend J (1974) Studies on the primary photosynthetic processes of plastids from wheat grown under light of different spectral quality. Plant Sci Lett 3: 61–66
9. Haslett BG, Cammack R, Whatley FR (1973) Quantitative studies on Ferredoxin in greening bean leaves. Biochem J 136:697–703
10. Kamiya Y, Miyachi (1975) Blue-light induced formation of phosphoenolpyruvate carboxylase in colorless Chlorella mutant cells. Plant Cell Physiol 16:729–736
11. Kowallik W, Schätzle S (1980) Enhancement of carbohydrate degradation of blue light. In: Senger H (ed) The blue light syndrome. Springer, Berlin Heidelberg New York, pp 344–360
12. Kulaeva ON, Selivankina SYu, Romanko EG (1981) Videlenie summarnogo preparata RNK iz rastenii. Nauka, Moskva, pp 171–185
13. Lichtenthaler HK, Buschmann C, Rahmsdorf U (1980) The importance of blue light for the development of suntype chloroplasts. In: Senger H (ed) The blue light syndrome. Springer, Berlin Heidelberg New York, pp 485–494
14. Malkin Sh, Armond PA, Mooney HA, Fork DC (1981) Photosystem II. Photosynthetic unit sizes from fluorescence induction in leaves. Plant Physiol 67:570–579
15. Poyarkova NM, Drozdova JS, Voskresenskaya NP (1973) Effect of blue light on the activity of carboxylating enzymes and $NADP^+$-dependent glyceraldehyde-3-phosphate dehydrogenase in bean and maize plants. Photosynthetica 7:58–66
16. Ruyters G (1980) Blue light effects on enzymes of the carbohydrate metabolism in Chlorella. I. Pyruvate kinase. In: Senger H (ed) The blue light syndrome. Springer, Berlin Heidelberg New York, pp 361–367

17. Richter G, Beckmann J, Gross M, Hundriesser J, Schneider Chr (1982) Blue-light induced synthesis of chloroplast proteins in cultured plant cells. Cell function and differentiation, part B. Giss, New York, pp 267–276
18. Romanko EG, Selivankina SG, Ovcharov AK (1982) Uchastie zitokinin-svyazyvayuschich belkov is listjev jachmenja v aktivazii zitokininom sinteza RNK v isolirovannich jadrach i chloroplastach. Phiziol Rast 29:524–531
19. Schiff JA (1980) Blue light and the photocontrol of chloroplast development in Euglena. In: Senger H (ed) The blue light syndrome. Springer, Berlin Heidelberg New York, pp 495–525
20. Senger H (1982) The effect of blue light on plants and microorganisms. Photochem Photobiol 35:911–920
21. Steup M (1975) Die Wirkung von blauem und rotem Licht auf die Synthese ribosomal RNA bei Chlorella. Arch Microbiol 105:143–147
22. Voskresenskaya NP (1979) Photoregulatornii aspecti metabolisma rastenii. Nauka, Moskva
23. Voskresenskaya NP, Drozdova JS, Krendeleva TE (1977) Effect of light quality on the organization of photosynthetic electron transport chain of pea seedlings. Plant Physiol 59:151–154
24. Voskresenskaya NP, Drozdova IS, Moskạlenko AA, Chetverikov AG, Zelniker YL (1982) Perestroiki fotosinteticheskogo apparata pod vlijaniem dlitelnogo dejstvija krasnogo i sinego sveta. Fiziol Rast 29:447–456
25. Voskresenskaya NP, Drozdova IS, Romanko EG, Selivankina SYu, Kuroedov VA, Scharenkova ChA, Nichiporovich AA (1983) Sinii svet kak vactor regulazii aktivnosti RNK polimeraz prorostkov jachmenja. Fiziol Rast in press
26. Voskresenskaya NP, Nechaeva EB, Vlasova MP, Nichiporovich AA (1968) Vosstanovlenie photosyntheticheskogo apparata v starejuschich listyach jachmeniya sinim svetom i kinetinom. Physiol Rast 15:890–897
27. Voskresenskaya NP, Oshmarova JS (1969) Photosynthesis and some photosynthetic electron transport chain reactions in pea seedlings grown in red or blue light. In: Metzner H (ed) Progress of photosynthesis research, vol III. Metzner, Tübingen, pp 1669–1674
28. Weissbach A, Horecker BL, Hurwitz J (1956) The enzymatic formation of phosphoglyceric acid from ribulose diphosphate and carbon dioxide. J Biol Chem 218:N2, 795–810
29. Wild A, Holzapfel A (1980) The effect of blue and red light on the content of chlorophyll, cytochrome f, soluble reducing sugars, soluble proteins and the nitrate reductase activity during growth of the primary leaves of sinapis alba. In: Senger H (ed) The blue light syndrome. Springer, Berlin Heidelberg New York, pp 444–454

Blue Light Effects on Morphogenesis and Metabolism in *Acetabularia*

R. SCHMID[1]

Introduction

Most of the investigations on photomorphogenesis deal with higher plants or multicellular organisms. Because of the high degree of complexity of these organisms, it was of special interest to extend photomorphogenetic studies upon more primitive organisms and attention has been driven to lower fungi and unicellular algae. Among the unicellular algae *Acetabularia* proves to be particularly suitable. First the morphogenetic events during the vegetative phase of its life cycle are well defined, and second the strict localization of the single nucleus inside the rhizoid allows, by a simple procedure, the production of enucleated cell fragments for investigations on regulatory mechanisms.

The morphogenetic events during the growth of the stalk of the cell are the production of lateral hair whorls in regular intervals and finally the formation of the species-specific cap, which marks the end of the vegetative growth phase.

The formation of whorls and caps requires blue light in all *Acetabularia* species so far investigated. Furthermore blue light is necessary to maintain high metabolic activity at least in *Acetabularia mediterranea.* If not mentioned in particular, the statements here are concerned with this species.

Effects of Blue Light

Effects on Growth and Morphogenesis

If *A. mediterranea* is cultivated under blue light conditions, growth rate and morphogenesis continue as normal, i.e., as in white light. Hair whorls are formed in regular intervals and the cap develops without delay. In red light the stalk growth rate decreases and is zero after two weeks of irradiation [4, 5]. Whorls and caps are not formed under these conditions. The slow stalk elongation in red light cannot be increased by higher light intensities [5].

---

1 Institut für Pflanzenphysiologie und Zellbiologie, Freie Universität Berlin, Königin-Luise-Str. 12–16, 1000 Berlin 33, FRG

Dedicated to the late Prof. Dr. H. Clauss

Blue Light Effects in Biological Systems
ed. by H. Senger

When cells in red light are supplied additionally with blue light of a few seconds up to 1 h each day, they start to grow continuously again and form whorls and caps. Under these conditions, the growth rates depend on the logarithm of the amount of the daily blue light [5, 9]. Furthermore, with increasing supply of blue light the begin of the development of the cap is more and more advanced [9].

Similar behavior is described for *A. crenulata* [35], but *A. calyculus* and *A. cliftonii* continue to grow also in red light [9, 29]. Whorls and caps, however, develop only when blue light is present. The number of whorls per length of the stalk depends on the logarithm of the amount of blue light, again supplied by daily pulses in addition to red light [9]. Thus the lack of morphogenetic events under red light conditions is not only due to the inhibition of growth, as might be expected from the results with *A. mediterranea.* There is definitely a specific influence of blue light on the realization of morphogenesis.

It has to be noted that the literature data concerning growth of *A. mediterranea* and *A. crenulata* diverge. As reported from Richter's laboratory [23, 24] the cells of both species also grew continuously in red light. Such differences seem to be resolved, since during the last years we were able to isolate strains of *A. mediterranea* which do not show growth enhancement by blue light.

Effects on the Metabolism

In addition to the decrease of the growth rate in red light there are changes in metabolic activities. One of the most obvious is the change in the photosynthetic activity. Both, O_2-production and CO_2-fixation were measured [6, 8, 24, 26, 31, 36]. Under red light conditions photosynthetic rates decrease with a half-life of about 10 days. Transfer of the cells to blue light or addition of small intensities of blue light to continuous red light rapidly reverse these changes [10, 26, 31] (Fig. 1). The Hill reaction with chloroindophenols as oxidants, which measures the activity of the electron transport system, including photosystem I and II [3], shows a similar behavior [7]. A more

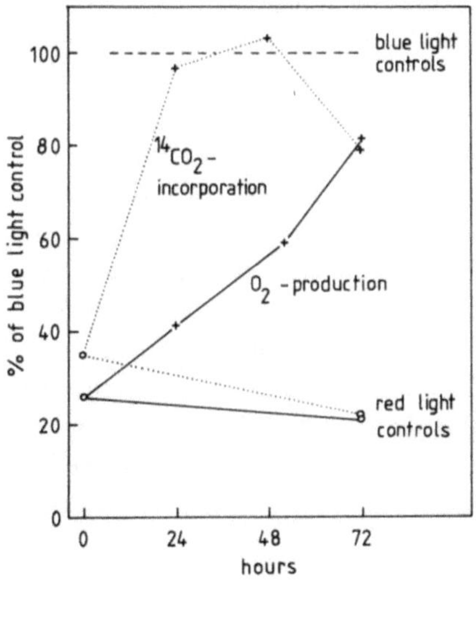

Fig. 1. Induction of photosynthesis by blue light after 3 weeks of irradiation with red light. The data are from [27] and [32]. Controls from continuous blue light are set to 100%

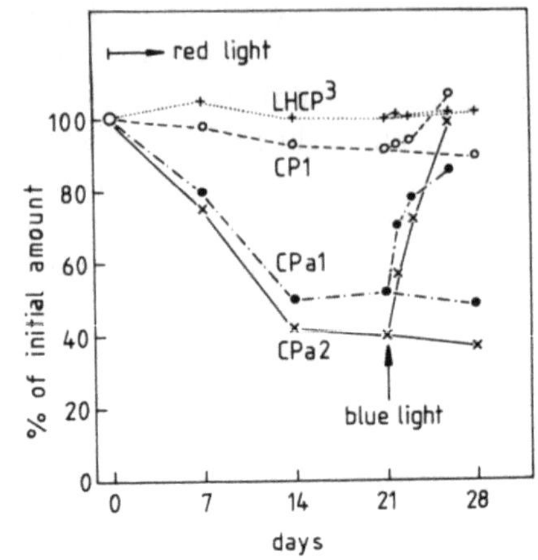

Fig. 2. Relative changes in the amount of chlorophyll-protein complexes in red light and after induction by blue light. Values at the beginning of the experiment are set to 100%. SDS-disc PAGE was used to separate the proteins. Peaks were integrated for quantitative evaluation

detailed analysis of the components of the photosynthetic apparatus [33] led us to the conclusion that it is a light-dependent step within photosystem II which is mainly affected by blue light.

In SDS-PAGE of thylakoid membranes we found that two minor chlorophyll-protein complexes underwent blue light-dependent quantitative changes which correlate well with the behavior of total photosynthesis [33] (Fig. 2). These complexes (named CPa1 and CPa2 according to the nomenclature of [1]) are discussed as components of the reaction center of photosystem II [13, 25]. Our conclusion is that the amount of photosystem II reaction center is the only rate-limiting factor common to both red and blue light. The quantitative changes determine the change of the rates of photosynthesis in either light condition.

The dark respiration in *A. mediterranea* is affected by blue light in a way similar to photosynthesis, i.e., in red light it falls to low levels and can be activated rapidly by blue light [33]. Therefore the light compensation point of the apparent photosynthesis is lowered in red light. This could explain the enormous accumulation of starch. Transfer to blue light reverses the situation and leads to rapid starch degradation [6, 8].

The changes in dark respiration could partly be due to the changes in the specific activity of pyruvate kinase [19]. However, they may also be attributed to the reduction of β-carboxylation which supplies the TCA cycle with intermediates, as is discussed for *Chlorella* [16]. For *Acetabularia* an indication for such an interpretation is derived from the clear and reproducible difference in the kinetics of induction of photosynthesis when measured either as O_2-production or CO_2-fixation (Fig. 1). To my understanding this difference can only be the consequence of a comparatively rapid activation of a β-carboxylating enzyme, although, so far, we were not able to measure any activity of such enzymes (PEP carboxylase, PEP carboxy kinase.

We have also been interested in other metabolic activities and it has been found that sucrose and fructosan contents also depend on blue light [6, 8, 31]. UDPG pyrophosphorylase, an enzyme involved in sucrose synthesis, shows clear changes in specific activity due to blue light irradiation [19]. A more detailed description is presented below.

Cells cultivated in red or blue light differ in their content of RNA and protein, but both components show a strict linear correlation independent of the light condition [27]. The rate of protein synthesis is discussed below.

Other Blue Light Effects

Number of Chloroplasts. Cells of *A. mediterranea* and *A. cliftonii* contain different numbers of chloroplasts when grown in red or blue light [29, 30]. Under red light conditions the multiplication of the chloroplasts is inhibited. After transfer of *A. mediterranea* from red to blue light the number of chloroplasts per cell doubles once between the 8th and the 24th hour, possibly due to one synchronous division [30]. Nevertheless, the ratios of total cell protein to the plastid and thylakoid protein remained constant under all light conditions [30].

Intracellular Movements. In *Acetabularia,* as in *Vaucheria* [2] and *Bryopsis* [18], chloroplasts accumulate in stalk regions which are locally illuminated with blue light [20, 21]. The chloroplasts are said to move actively toward the blue light spot (Paque personal communication) and this phenomenon was attributed to chloroplast phototaxis. In *Vaucheria,* on the other hand, the accumulation of the chloroplasts is the consequence of their immobilization by blue light [2] and in *Bryopsis* retardation of the chloroplasts by blue light was observed [18]. Thus, in these organisms the chloroplasts are rather trapped inside the blue light sector and the mechanism in *Acetabularia* seems to be different.

Literature data on the velocity of protoplasmic streaming are contradictory. In *A. crenulata* streaming rates decreased in red light and blue light was activating [37]. On the other hand it was described for *A. mediterranea* that after a period of 3 weeks in darkness, no protoplasmic streaming was observed. A switch to red light activated streaming, while blue light did not [11, 22]. It is difficult to decide whether the two species behave differently with respect to the activation of protoplasmic streaming by light or whether the discrepancies are due to the different experimental conditions. Furthermore it is well established that the mobile system in *Acetabularia* shows a circadian rhythm [17]. Since blue light sets the circadian clock in *Acetabularia* (see below), the beginning of blue light means "beginning of the day" after a preceding dark period. Red light does not affect the circadian rhythm. The interpretation of these experiments is thus rendered difficult, and it is obvious that contradictory results may derive from experiments done at different times of the day.

As reported recently, the transport of rRNA from the nucleus into the stalk is inhibited in red light [11, 22]. The inhibition can only be overcome by blue light.

Effects on the Circadian Rhythm. After prolonged red light irradiation blue light could "induce" the circadian rhythm of photosynthesis [10] in a cell population. The circadian chloroplast migration is a rhythm which is characterized by streaming of a large portion of the chloroplasts up and down the stalk once a day [17] and it can easily be measured in single cells. It has been shown that this rhythm continues under red light conditions and that blue light is setting the phase [28]. The circadian rhythms of photosynthesis and chloroplast migration are tightly coupled [34] and the "induction" of the photosynthetic rhythm therefore should be due to the synchronization of individual cells being out of phase. Only recently it was reported that blue light induced the circadian rhythm of the apico-basal electrical potential after long dark treatments. Red light induces the potential but the rhythmic properties are lacking and begin only with the onset of blue light [22].

A second effect of blue light on the circadian clock is the extension of the period of the rhythm, when a background of red light is present [28]. The length of the period increases with increasing blue light intensities.

Light Dependence and Photoreception

The Role of Red Light

Before going into details of blue light photoreception in *Acetabularia* it is necessary to give some notes on the role of red light in order to eliminate misunderstandings.

By absorption of the chlorophyll, photosynthesis works in blue light as well as in red light. The photoreceptor for blue, however, is not active in red light conditions. In our experiments red light is therefore used mainly as a source of energy for carbon assimilation and serves as an alternative to switching off blue light. The regulatory role of blue light can thus be studied without interacting effects of photosynthesis. It is important to note that red light itself exerts no positive photomorphogenetic effect in *Acetabularia,* and there is at present no indication for the action of phytochrome. All the effects which occur in the presence of pure red light are therefore due to the absence of blue light.

Anyhow, the application of pure red light is a highly artificial condition which does not occur in the natural environment and one should not try to see any adaptive value in the reactions of the cells to red light.

Action Spectroscopy

Action spectroscopy was for a long time difficult in *Acetabularia* because of the slowness of most of the reactions. Thus, only crude action spectra have been published for the enhancement of growth [6] and for the reinforcement of photosynthetic oxygen production after a preceding red light illumination [36]. It was obvious from these action spectra that only light of wavelengths below 520 nm is effective and that maximum response is at about 450 nm.

Only recently we were able to establish a suitable and sensitive experimental system to study the light dependence of growth and morphogenesis in *Acetabularia* [32]. Very young cells, when illuminated for 1 week with continuous red light, start to grow and form a hair whorl after induction by a low fluence blue light pulse. The number of cells which finally form a whorl in the course of a further day in red light is dependent on the fluence of blue light and the law of reciprocity holds for this effect. Fluence response curves for various wave lengths are parallel within the limits of error

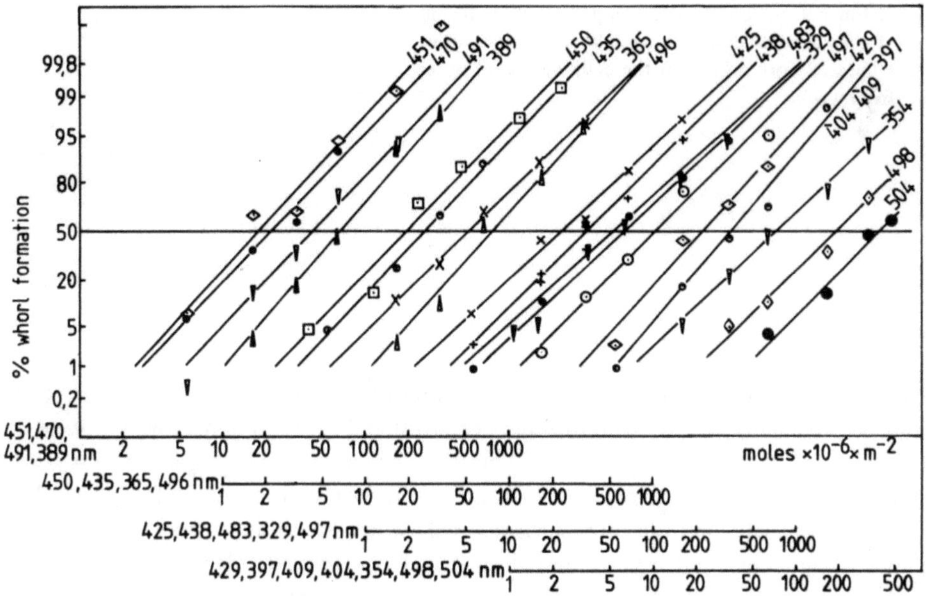

Fig. 3. Fluence-response curves (probit plots) at different wave lengths for the induction of whorl formation

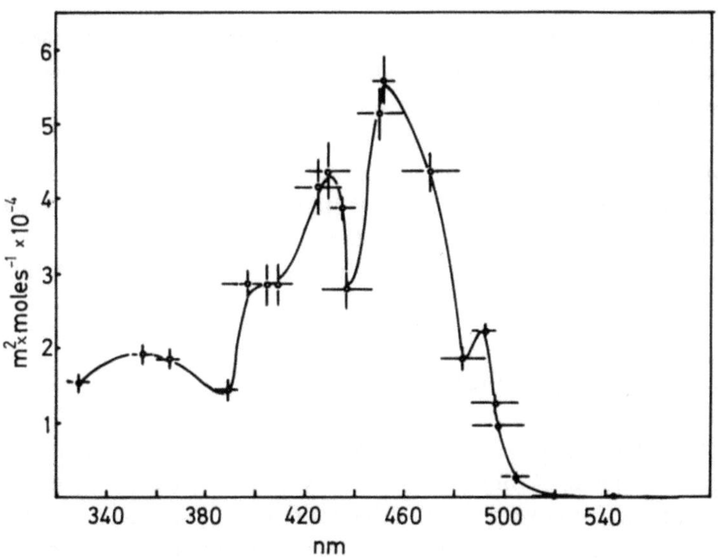

Fig. 4. Action spectrum for hair whorl formation. The fluence necessary for 50% whorl formation is used for this spectrum. *Horizontal bars* half band widths of the filters. *Vertical bars* deviations of the linear regressions of the fluence response curves

Fig. 5. Fluence-response relationships for whorl formation of cells irradiated with blue light either at the base or the apex. Before induction the cells were kept in red light for 3 weeks

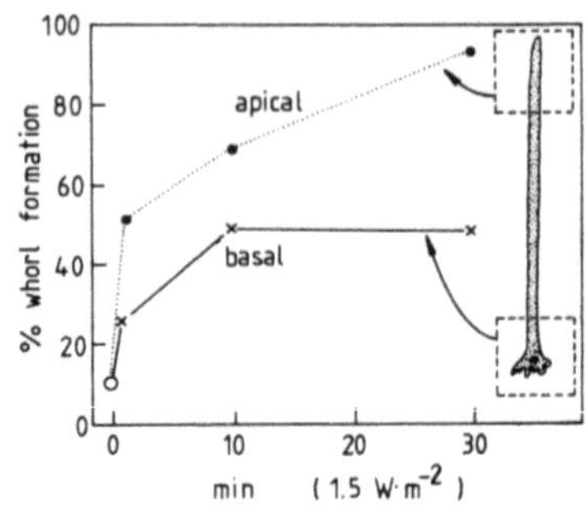

(Fig. 3). Thus, only one photoreceptor is responsible for this effect. The action spectrum shows a detailed fine structure between 400 and 500 nm and a low action peak at 360 nm (Fig. 4). It is typical for a blue-UV photoreceptor but again cannot solve the question about its chemical nature, as comprehensively discussed by DeFabo [12].

Although only one photoreceptor is involved in growth and whorl formation, it cannot be excluded that other blue light effects in *Acetabularia* depend on a different photoreceptor. Interactions may also occur with a second photoreceptor, for instance, with a possible green light receptor which inhibits the electrogenic chloride pump in *Acetabularia* [14].

Uneven Distribution of the Photoreceptor Along the Cell?

When morphogenesis is induced after prolonged red light by irradiating either base or tip of the cells, higher responsiveness is observed upon illumination of the tip (Fig. 5). Both irradiations show different saturation levels for blue light. Furthermore, chloroplasts accumulate in a blue light spot independent of its localization on the cell [20]. Photoreception of blue light can therefore occur over the entire length of the cell and is not restricted to the actively growing tip region. Different levels of saturation after irradiation of either the apex or the base should rather be due to uneven distribution of components of the transduction chain or to hindrance of the flow of information from the base toward the growing cell tip.

Changes in Responsiveness and the Requirement for Blue Light

We have measured the induction of hair whorl formation in young cells which were pre-treated for 1 week with red light. Blue light, necessary to induce about 60% of the cells to form whorls, was split into two pulses of different fluence. The response to the second pulse begins to decrease when it is given more than 10 min after the first pulse. It reaches a minimum after 1h and is followed by a refractory period of several hours in which a gradual increase of the response to the second pulse is observed [32] (Fig. 6). Therefore, for the induction of whorls the law of reciprocity is valid only within the first 10 min after the beginning of blue light. Fluence-response curves measured 1 h after a first short blue light pulse compared to fluence-response curves without blue pre-irradiation indicate that the responsiveness to the second pulse is not fully lost.

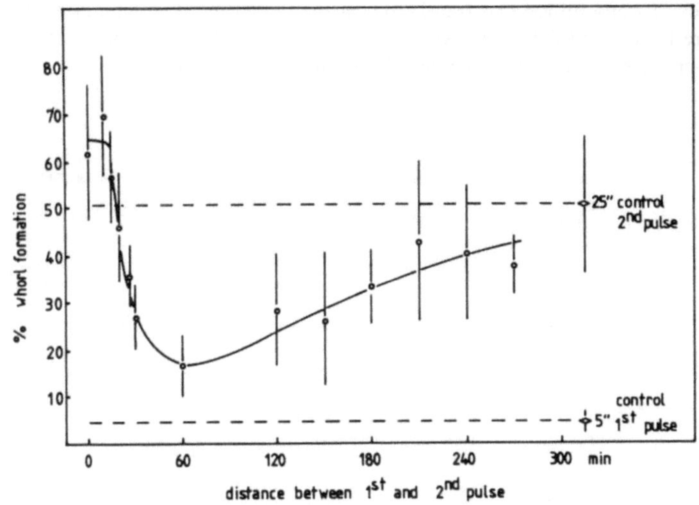

Fig. 6. Time-dependent changes in the responsiveness to a second pulse of blue light after a first short blue light pulse. Cells were 1 week in red light prior to blue light induction. *Bars* are the standard deviation

Higher fluences of blue light can induce all of the cells to form hair whorls [32] (Fig. 7).

Repeated pulses of blue light in 30-min intervals could not prevent the refractory period. We believe, therefore, that the decrease in responsiveness does not represent an adaptation to low light conditions. On the other hand, it is evident that blue light induces a sequence of reactions, which temporarily lower the responsiveness to further blue light.

Although this has not yet been studied in detail, there is some indication that the various responses to blue light in *Acetabularia* also need different amounts of blue light for their induction and maintenance. Very low fluences are required, for example, for the induction of growth and whorl formation. On the other hand, photosynthesis

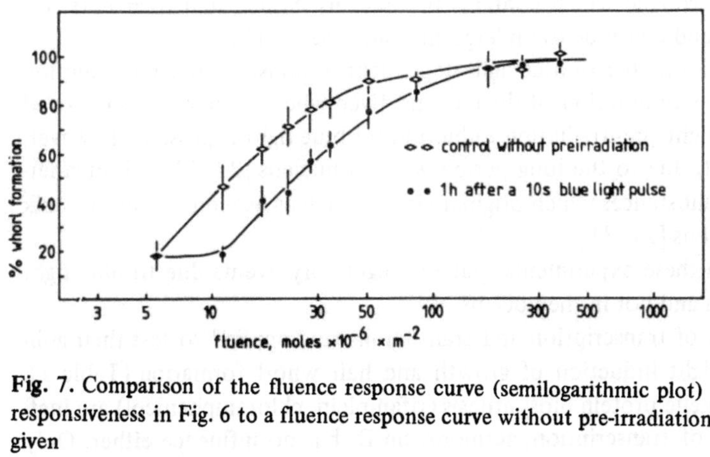

Fig. 7. Comparison of the fluence response curve (semilogarithmic plot) at the point of minimum responsiveness in Fig. 6 to a fluence response curve without pre-irradiation. Standard deviations are given

Table 1. Effect of inhibitors of transcription and translation on the induction of growth. Prior to the induction the cells were kept in red light for 3 weeks. They were incubated in the presence of the inhibitors for one further day either in red or in blue light. (After Clauss unpublished)

Tip region of the cell	Control red light	Blue light					
		—	Actinomycin D 10 µg / ml	Rifampicin 10 µg / ml	Cycloheximide 0.25µg / ml	Puromycin 30 µg / ml	Chloramphenicol 100 µg / ml
	98 %	—	2 %	—	100 %	98 %	5 %
	—	5 %	5 %	10 %	—	2 %	5 %
 whorl	2 %	95 %	93 %	90 %	—	—	90 %

can hardly be induced by several hours of blue light after previous red light irradiation. Continuous low irradiation of blue light in addition to red light is most effective (preliminary results). It is evident, at least with respect to the circadian clock, that the organism senses blue light in two ways: first by quanta counting in the case when the rhythm is synchronized by the begin of blue light, and second by measuring the light intensity of continous irradiations when considering the changes of the period of the rhythm [28].

Regulation

Enucleation and Use of Inhibitors of Protein Biosynthesis

Acetabularia is particularly suitable for studies of the posttranscriptional regulation of protein biosynthesis because it is very simple to obtain large anucleate cell fragments by amputation of the rhizoid which contains the nucleus. Enucleated fragments survive over long periods and continue to undergo morphogenesis [15].

It was shown for a number of blue light-dependent responses that their induction is not prevented by the amputation of the nucleus. Even when the nucleus is removed prior to the long red light pre-irradiation induction by blue light is possible, however, to a diminished extent, due to the long period without nucleus [24, 27, 31]. Furthermore, morphogenetic substances which originate in the nucleus accumulate in the stalk under red light conditions [27, 31].

It is evident from these experiments that the regulatory events due to blue light occur in the cytoplasm and not in the nucleus.

Various inhibitors of transcription and translation were applied to test their ability to suppress blue light induction of growth and hair whorl formation (Table 1). Inhibitors of the organelle protein biosynthesis (rifampicin, chloramphenicol) are ineffective. The inhibitor of transcription, actinomycin D, has no influence either. Only

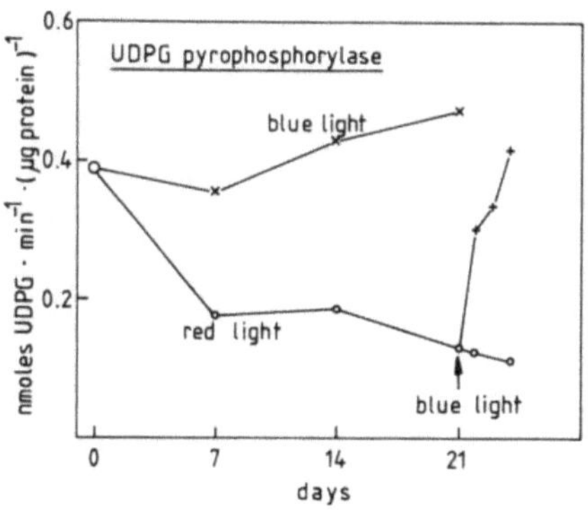

Fig. 8. Changes in the specific activity of UDPG pyrophosphorylase in red and blue light and after induction by blue light. At each point of time cells were homogenized and the activity of the enzyme was determined in the crude extract

when the cytosolic translation is blocked by cycloheximide and puromycin is induction of morphogenesis and growth fully prevented. Thus, translational events in the cytosol play a role in the transduction of the blue light stimulus.

Regulation of UDPG Pyrophosphorylase

Further insight into regulatory mechanisms can be obtained by studies of the regulation of specific enzymes. UDPG pyrophosphorylase proved to be very suitable for this purpose.

The specific activity (on a total cell protein basis) of UDPG pyrophosphorylase in crude extracts from the irradiated cells does not change under blue light conditions but it continuously decreases in red light. A switch from red to blue light causes a rapid rise to blue light levels [19] (Fig. 8). This inductive effect of blue light is the same in nucleate cells as well as in anucleate fragments (Fig. 9). However, cycloheximide (0.5 μg ml^{-1} of culture medium) supresses blue light induction and the activity remained on the level of the red light controls (Fig. 10).

These results do not prove de novo synthesis and provide no information about mechanisms of regulation.

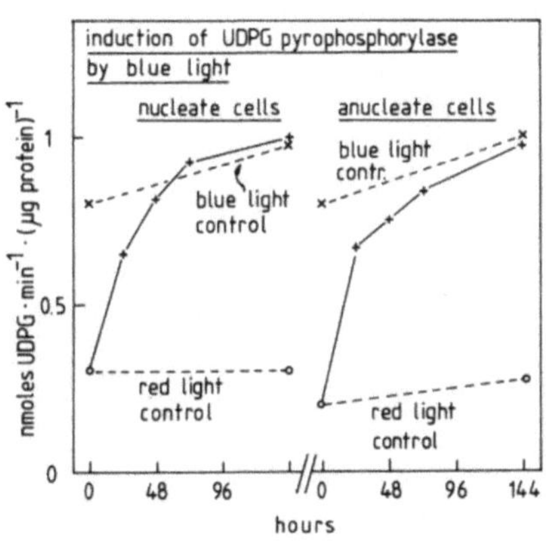

Fig. 9. Induction of UDPG pyrophosphorylase activity in nucleate and anucleate cells. A pre-irradiation of 3 weeks of red light was used. Cells were enucleated 2 days before the beginning of blue light

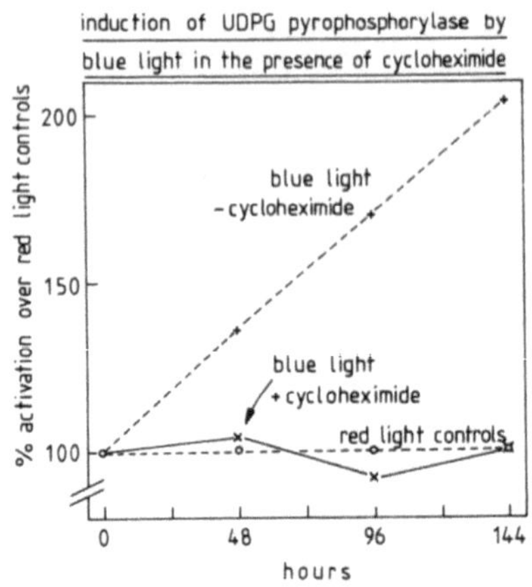

Fig. 10. Effect of cycloheximide (0.5 $\mu g\ ml^{-1}$) on the induction of UDPG pyrophosphorylase activity after 3 weeks of pre-irradiation with red light. All red light controls with and without cycloheximide were set to 100%

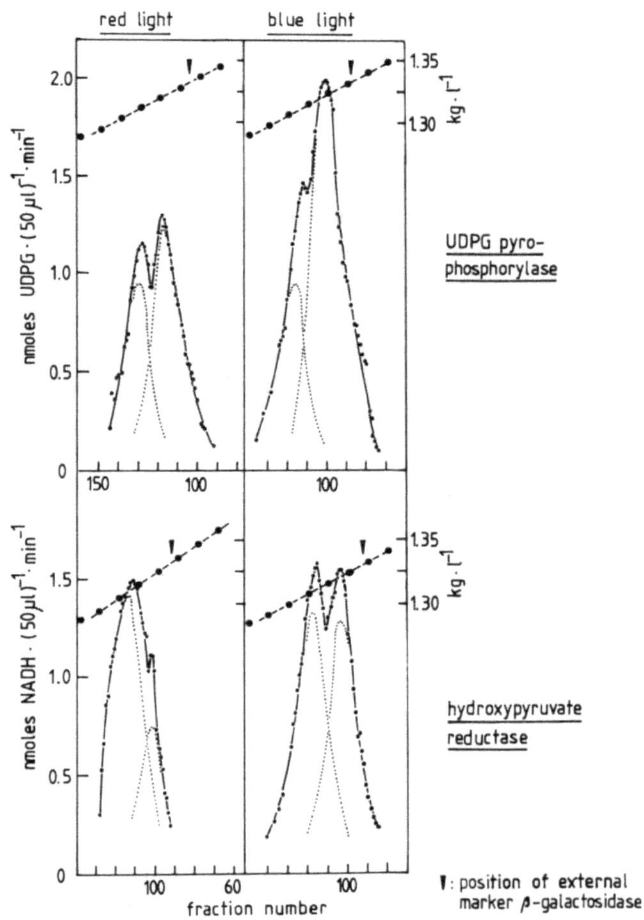

Fig. 11. Density labeling of UDPG pyrophosphorylase and hydroxypyruvate reductase during the induction by blue light. Before the beginning of the experiment the cells were pre-illuminated with red light for 3 weeks. Then they were incubated in a deuterated amino acid mixture for 3 days in red or in blue light. Crude extracts of the cells were centrifuged in a buffered $RbCl/D_2O$ gradient at 300,000 *g* in a fixed angle rotor

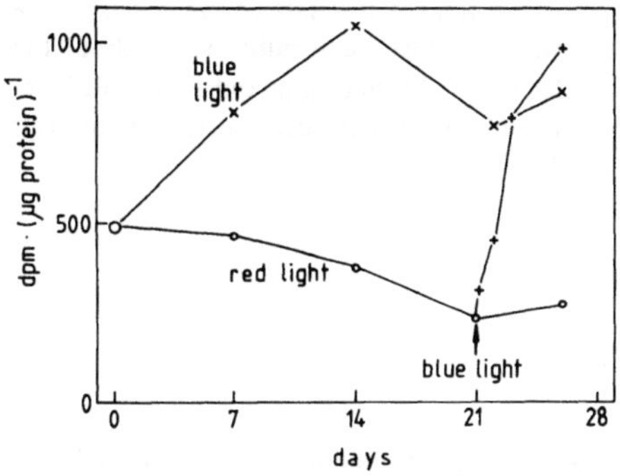

Fig. 12. Activity of the cytosolic translation in red and blue light. Proteins were labeled in a ^{3}H-amino acid mixture for 1 h in the presence of chloramphenicol. Then they were precipitated from a homogenate by trichoroacetic acid. The radioactivity was determined from the supernatant after hydrolysis of the precipitate by pronase P

The question about de novo synthesis was investigated by density labeling experiments (Fig. 11). The cells were kept for 3 weeks in red light. Then they were incubated in a deuterated amino acid mixture for 3 days either in red or in blue light. Crude extracts were centrifuged in a $RbCl/D_2O$ gradient.

During the inductive phase in blue light an about fourfold increase in the rate of synthesis of UDPG pyrophosphorylase is obvious when compared to the red light controls. This might suggest a specific stimulation of UDPG pyrophosphorylase synthesis. As a control we measured the rates of synthesis of hydroxypyruvate reductase (Fig. 11) the specific activity of which is unaffected by blue light (data not shown). This enzyme also shows a similar increase of its rate of synthesis by blue light. In fact the increase in the synthesis of both enzymes appears to be due to a fourfold stimulation of the cytosolic translation by blue light (Fig. 12).

Considering only the rates of synthesis of the enzymes, it is impossible to explain the specific induction of UDPG pyrophosphorylase. Both enzymes differ, however, with respect to their destruction. The half-life of UDPG pyrophosphorylase is about one fourth of that of hydroxypyruvate reductase.

Our present working hypothesis is that the blue light-dependent changes in UDPG pyrophosphorylase activity are mainly due to the overall changes in the rate of cytosolic translation. The latter decreases continuously in red light and is activated by blue light. Under these conditions those few enzymes which have a short half life are more affected than enzymes with an extremely low but average destruction. They can follow changes in the rate of translation much more rapidly and their relative level (specific activity) is dependent on these changes. A mechanism like that may be the basis for most of the blue light effects on the metabolism in *Acetabularia*. However, we cannot exclude further specific regulation on the translational level.

If this working hypothesis for enzyme regulation reflects the true situation in either light condition, the effect of blue light on the translational apparatus is rather unspecific. It could be due to a simple regulation step near the site of photoreception. In any case this way of the analysis of the blue light signal transduction seems to be promising.

Furthermore, various blue light responses have been shown for *Acetabularia* and at least some of them differ in their dose-response relationship. Further analysis of these relationships may prove that the effects may be divided into different groups of blue light dependency. Such an analysis would help to clarify different regulatory mechanisms.

Acknowledgments. Thanks are due to the Deutsche Forschungsgemeinschaft for generously supporting our investigations. Much of the recent results could not have been published without the diligent work of my co-workers E. Idziak, I. Maass, B. Nickl, M. Tünnermann and H. and R. Wennicke. I thank K. Giesecke for many helpful discussions and for proofreading of the manuscript and Mrs. R. Reisberg for typewriting.

References

1. Anderson JM, Waldron JC, Thorne WW (1978) Chlorophyll-protein complexes of spinach and barley thylakoids. Spectral characterization of six complexes resolved by an improved electrophoretic procedure. FEBS Lett 92:227–233
2. Blatt MR, Briggs WR (1980) Blue light induced cortical fiber reticulation concomitant with chloroplast aggregation in the alga *Vaucheria sessilis.* Planta 147:355–362
3. Böhme H, Reimer S, Trebst A (1971) On the role of plastoquinone in photosynthesis: The effect of dibromothymoquinone, an antagonist of plastoquinone, on non-cyclic and cyclic electron flow systems in isolated chloroplasts. Z Naturforsch 26b:341–352
4. Clauss H (1963) Über den Einfluß von Rot- und Blaulicht auf das Wachstum kernhaltiger Teile von *Acetabularia mediterranea.* Naturwissenschaften 50:719
5. Clauss H (1968) Beeinflussung der Morphogenese, Substanzproduktion und Proteinzunahme von *Acetabularia mediterranea* durch sichtbare Strahlung. Protoplasma 65:49–80
6. Clauss H (1970) Effect of red and blue light on morphogenesis and metabolism of *Acetabularia mediterranea.* In: Brachet J, Bonotto S (eds) Biology of *Acetabularia.* Academic Press, London New York, pp 177–191
7. Clauss H (1970) Der Einfluß von Rot- und Blaulicht auf die Hillaktivität von *Acetabularia* Chloroplasten. Planta 91:32–37
8. Clauss H (1972) Der Einfluß von Rot- und Blaulicht auf die Photosynthese von *Acetabularia mediterranea* und auf die Verteilung des assimilierten Kohlenstoffs. Protoplasma 74:357–379
9. Clauss H (1973) Regulation der Morphogenese von *Acetabularia* durch Licht. In: Vom Gen zum Genus – Beiträge zur Zellforschung. Freie Univ Berlin, Pressedienst Wiss, pp 45–59
10. Clauss H (1979) Auslösung der circadianen Photosynthese-Rhythmik bei *Acetabularia* durch Blaulicht. Protoplasma 99:341–346
11. Dazy AC, Puiseux-Dao S, Durand M, Santa-Maria A (1981) The effects of blue and red light on streaming recovery and RNA transport after dark treatment in *Acetabularia mediterranea.* Protoplasma 105:354
12. DeFabo E (1980) On the nature of the blue light photoreceptor: Still an open question. In: Senger H (ed) The blue light syndrome. Springer, Berlin Heidelberg New York, pp 187–197
13. Delepelaire P, Chua NH (1979) Lithium dodecyl sulfate/polyacrylamide gel electrophoresis of thylakoid membranes at 4°C: Characterizations of two additional chlorophyll a-protein complexes. Proc Natl Acad Sci USA 76:111–115
14. Gradmann D (1978) Green light (550 nm) inhibits electrogenic Cl^- pump in the *Acetabularia* membrane by permeability increase for the carrier ion. J Membr Biol 44:1–24
15. Hämmerling J (1934) Über formbildende Substanzen bei *Acetabularia mediterranea,* ihre räumliche und zeitliche Verteilung und ihre Herkunft. Roux' Arch Entwicklungsmech Org 131:1–81
16. Kamiya Y, Miyachi S (1974) Effects of blue light on respiration and carbon dioxide fixation in colourless *Chlorella* mutant cells. Plant Cell Physiol 15:927–937
17. Koop H-U, Schmid R, Heunert H-H, Milthaler B (1978) Chloroplast migration: A new circadian rhythm in *Acetabularia.* Protoplasma 97:301–310
18. Mizukami M, Wada S (1981) Action spectrum for light-induced chloroplast accumulation in a marine coenocytic green alga, *Bryopsis plumosa.* Plant Cell Physiol 22:1245–1255
19. Nickl B, Maass I, Schmid R (1982) Posttranscriptional control of enzyme activity in *Acetabularia.* Induction by blue light. In: Eur Symp Photomorphogen Plants. Reading UK Book Abstr, p 56

20. Paques M, Sironval C, Bonotto S (1979) On chloroplast movement in the stalk of *Acetabularia mediterranea*. In: Bonotto S, Kefeli V, Puiseux-Dao S (eds) Developmental biology of *Acetabularia*. Elsevier, Amsterdam, pp 155–162
21. Paques M, Brouers M (1981) Chloroplast phototaxis in *Acetabularia mediterranea*. Protoplasma 105:360–361
22. Puiseux-Dao S, Borghi H, Dazy AC, Durand M (1983) Effects of white, red, and blue light on the electrical activity, the cellular streaming, the rRNA transport and the morphogenesis in *Acetabularia*. In: Abstr XIth Int Seaweed Symp, Quindao, China
23. Richter G (1962) Die Wirkung von blauer und roter Strahlung auf die Morphogenese von *Acetabularia*. Naturwissenschaften 49:238
24. Richter G, Kirschstein MJ (1966) Regeneration und Photosynthese-Leistung kernhaltiger Zellteilstücke von *Acetabularia* in blauer und roter Strahlung. Z Pflanzenphysiol 54:106–117
25. Satoh K (1982) Further characterization of the photosystem II chlorophyll a-protein complex purified from digitonin extracts of spinach chloroplasts. Polypeptide composition. In: Akoyounoglou G (ed) Photosynthesis, vol III. Balaban Int Sci Serv, Philadelphia, pp 607–616
26. Schael U, Clauss H (1968) Die Wirkung von Rotlicht und Blaulicht auf die Photosynthese von *Acetabularia mediterranea*. Planta 78:98–114
27. Schmid R (1978) Zur Frage der Beteiligung des Zellkerns an den blaulichtabhängigen Photomorphosen bei *Acetabularia mediterranea*. Dissertation, Freie Univ Berlin
28. Schmid R (1981) Effects of blue light on the circadian rhythm in *Acetabularia mediterranea*. Protoplasma 105:364
29. Schmid R, Clauss H (1974) Die Vermehrung der Chloroplasten von *Acetabularia* im Rot- und Blaulicht. Protoplasma 82:283–287
30. Schmid R, Clauss H (1975) Multiplication and protein content of chloroplasts of *Acetabularia mediterranea* in blue light after prolonged irradiation with red light. Protoplasma 85:315–325
31. Schmid R, Clauss H (1977) Is the nucleus involved in the blue light mediated photomorphoses in *Acetabularia mediterranea?* In: Woodcock CLF (ed) Progress in *Acetabularia* research. Academic Press, London New York, pp 255–269
32. Schmid R, Idziak E, Tünnermann M (1982) Changes in the blue light sensitivity of morphogenesis of *Acetabularia* after a short blue light pulse. In: Eur Symp Photomorphogen Plants. Reading UK Book Abstr, pp 66
33. Schmid R et al (1984) in preparation
34. Schweiger H-G, Broda H, Wolff D (1981) Simultaneous recording of two circadian rhythms in an individual cell of *Acetabularia*. Protoplasma 105:365
35. Terborgh JW (1965) Effects of red and blue light on the growth and morphogenesis of *Acetabularia crenulata*. Nature (London) 207:1360–1363
36. Terborgh JW (1966) Potentiation of photosynthetic oxygen evolution in red light by small quantities of monochromatic blue light. Plant Physiol 41:1401–1410
37. Williams NS, Terborgh JW (1970) Protoplasmic streaming in red light potentiated by small quantities of blue light in *Acetabularia crenulata*. Plant Physiol Suppl 46:1

Movement and Growth

Photomovement

D. P. HÄDER[1]

Introduction

Many microorganisms orient themselves in their photoenvironment using three basic photoresponses:

1. *Phototaxis* is an oriented movement with respect to the light direction. Many flagellates show a positive phototactic orientation (movement toward the light source) in low light intensities and a negative phototaxis (away from the light source) in higher light intensities [31, 50].

2. *Photokinesis* describes the dependence of the speed of movement on the steady state light intensity, independent of the light direction. If the velocity at a given intensity is higher than in the dark control, photokinesis is called positive and if lower, negative. Some organisms do not move in darkness or in very bright light [13].

3. A sudden change in light intensity causes a *photophobic response.* Either a step-up or a step-down in light intensity or both can induce a response, the nature of which is a characteristic of the organism, e.g., blue-green algae reverse the direction of their movement while many flagellates turn sidewise or rotate on the spot [48]. Both a temporal change (e.g., when the light is suddenly switched on or off) as well as a spatial change (e.g., when an organism enters a dark or bright area) can induce a response as in the case of the green flagellate *Euglena gracilis* [54].

In the following, I will discuss new methods, recent results, and novel models for the three photoresponses starting with phototaxis.

Phototaxis

Oriented movement with respect to the light direction can be assayed using one of two techniques:

a) *Population techniques* measure the effect of mass movement (Fig. 1). When a homogeneous suspension of the ciliate *Stentor* is unilaterally irradiated in a cuvette the

1 Philipps Universität Marburg, Fachbereich Biologie — Botanik, Lahnberge, 3550 Marburg, FRG

Blue Light Effects in Biological Systems
ed. by H. Senger

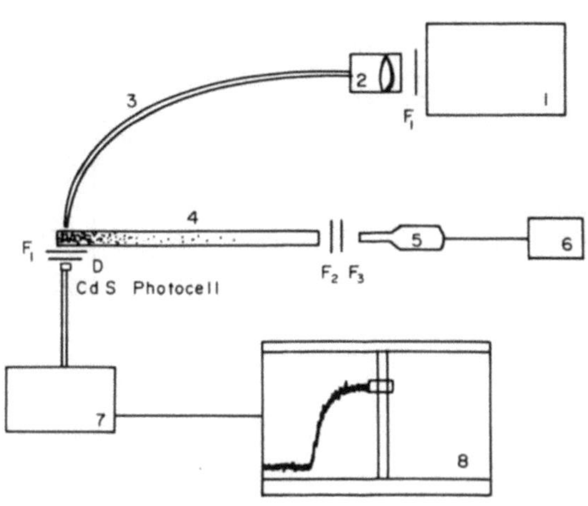

Fig. 1. Apparatus to measure negative phototaxis of *Stentor coeruleus.* The organisms, originally homogenously distributed in a cuvette (*4*) swim away from the light source (*5* with transformer *6*). When they accumulate at the far end they attenuate the infrared measuring beam (*1–3*) which is detected by a photocell. The signal is amplified (*7*) and recorded (*8*) with respect to time [4]

organisms will move away from the light source due to negative phototaxis and accumulate at the far end of the cuvette [4]. The attenuation of the measurement beam perpendicular to the actinic beam due to the accumulating organisms can be recorded with respect to time. Population techniques have the advantage that they integrate over the responses of a large number of organisms. A disadvantage is that the interpretation of the results can be ambiguous: when, e.g., a drug is added which potentially inhibits phototactic orientation, the failure of the population to accumulate could result from an inhibited motility instead of from an impaired ability to detect the light direction. These problems can be overcome by using (b). *Single cell analysis.* The movement vectors of individual cells during a given time interval are plotted often with the help of time-lapse cinematography [58] or with a video system [16, 32, 59]. The angles of deviation from the incident light direction can be grouped in sectors and plotted as histograms. The data analysis is based on directional statistics for which computer programs have been developed [15, 32].

Phototactic Orientation in Flagellates

Using a videomicrographic system, Watanabe and Furuya [59] analyzed positive phototaxis in *Cryptomonas* in a single cell analysis. The cells gradually oriented toward the light source within 4 s after the onset of the actinic light without showing a phobic response. A wavelength of 570 nm produced an optimal effect. *Euglena gracilis,* in contrast, has a "typical" blue light photoreceptor. It shows both positive and negative phototaxis (Fig. 2; [5, 22, 32]) depending on the light intensity. Using a similar technique, Morel-Laurens and Feinleib [39] compared the negative phototactic orientation of a wild type strain of *Chlamydomonas* with an "eyeless" mutant (ey 627), which shows an impaired ultrastructure in the stigma region. While the two strains do not differ in their step-up photophobic behavior, the mutant is strongly impaired in its phototactic orientation. This result can be seen as a support for the hypothesis that the stigma area is involved in photoperception in this organism.

Fig. 2. Phototactic orientation of *Euglena gracilis* in parallel broad band blue light. At 30 mW m^{-2} (*left blocks*) more organisms move toward the light source (*above line*, positive phototaxis) deviating from the light direction by no more than ± 30° (*front block*), ± 60° (*center block*) or ± 90° (*rear block*). In the dark control (*center blocks*), there is an equal distribution and in 30 W m^{-2} (*left blocks*) most organisms move away from the light source [32]

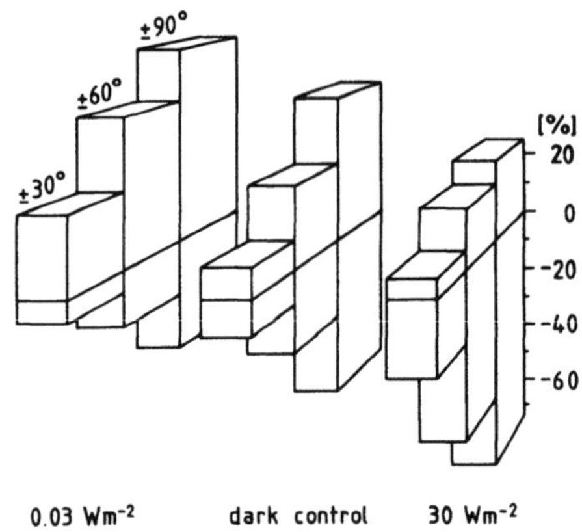

Freeze-fracture EM studies of a number of flagellated algae have demonstrated a structural connection between the flagellar root system and the outer chloroplast membrane in the stigma region [34, 35, 37, 52]. This could act as a connecting link between the photoreceptor and the motor apparatus of the cell. In many flagellates, the stigma lies inside the chloroplast envelope and consists of one or more layers of densely packed lipophilic globules brightly colored by carotenoids [36, 37].

Based upon these findings, Foster and Smyth [12] have proposed a novel mechanism to detect light direction in some flagellated algae. The stigma lipid droplet layers have a different refractive index than the aqueous layers in between and are stacked at a distance of a quarter of the wavelength of the actinic light (Fig. 3). A light wave entering the stack perpendicular to the outer membrane is partially reflected at each layer. The reflected waves constructively interfere with the incoming waves to form a signal of maximal strength at the plasma membrane. This suggests that the photoreceptor could be localized in one of the membranes overlaying the stigma. In fact, both the plasmalemma and the outer chloroplast membrane carry an array of intramembraneous particles with a different size and distribution over the stigma than in any other area of the cell [37, 52].

Most flagellates rotate during their forward movement. In lateral light, the photoreceptor senses a modulated signal which eventually triggers the flagella (or one of them) to effect a course correction. Smyth and Berg [55] have found a decrease in the flagellar beat frequency in *Chlamydomonas* upon changes in the light intensity. The sensory transduction chain of phototaxis which eventually leads to a modulation of the flagellar activity seems to involve cation fluxes in many microorganisms including flagellates and ciliates [7, 42, 44, 47, 53, 57]. In *Euglena gracilis,* the photoreceptor is thought to be localized in the paraflagellar body, a swelling near the base of the longer flagellum [1], rather than in or near the stigma [27]. Doughty and Diehn [8, 9] have proposed a model to explain the sensory transduction chain in phobic responses which are thought to be the basis for phototactic orientation in *Euglena.* Light is supposed to activate a Na^+/K^+ pump in the flagellar membrane. The sodium concentration, in turn, controls a Ca^{2+} channel which allows the influx of Ca^{2+}. The Ca^{2+} binds to the axoneme to effect a change in the flagellar beating pattern [6, 10]. An alternative model assumes that the light signal causes a contraction of circular and longitudinal

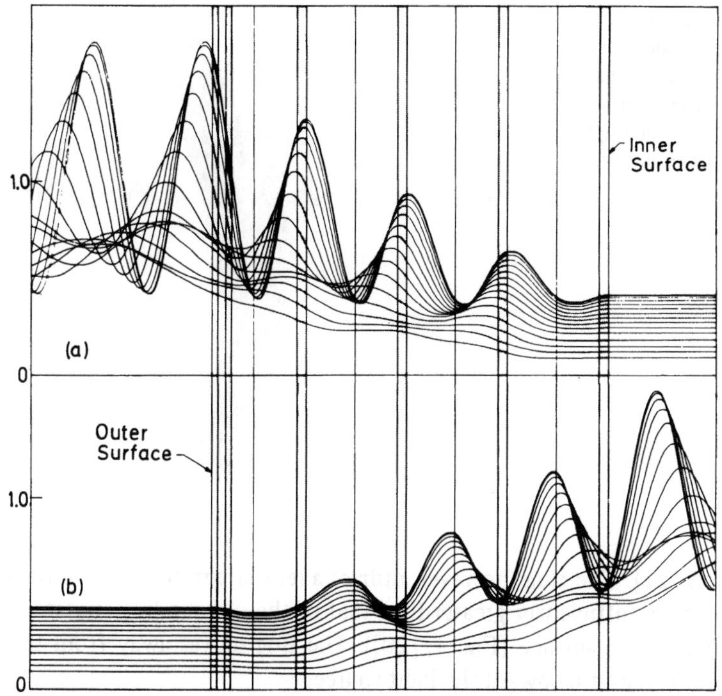

Fig. 3a,b. Light intensity distribution in the stigma and overlapping membranes in light impinging from the left (**a**) and right (**b**, through the cell). *Bars from left to right* plasmalemma, chloroplast double membrane, four double membranes of stigma. *Curves* are drawn for different angles of incidence at 5° intervals from normal incidence (*upper curves*) to 70° (*lowest curves*) [12]

microtubules and microfilaments around the reservoir which mechanically readjusts the position of the flagellum [38].

The closely related *Euglena mutabilis* lacks an emergent flagellum; therefore, motility is restricted to gliding movement. In addition, there are a number of other striking differences: while *E. gracilis* shows both positive and negative phototaxis, *E. mutabilis* shows solely positive orientation. Furthermore, the action spectrum extends throughout the visible spectrum [23].

Phototaxis in Slime Molds

Phototactic orientation in the acellular slime mold *Physarum polycephalum* is covered in a separate article (Schreckenbach this Vol.). The cellular slime mold, *Dictyostelium discoideum,* shows positive phototaxis in its multicellular pseudoplasmodium (slug) stage [2] which under certain conditions can deviate from the light direction by a certain angle [11]. In its amoebal stage, it shows positive phototaxis in low light intensities [24] and negative in higher [25]. It is remarkable that the action spectra of both stages in the life cycle are different [28, 49]. Furthermore, the detection of light direction is based on different mechanisms. When a bright microbeam is focused onto the moving front of an amoeba, it retracts the pseudopodium and establishes a new front

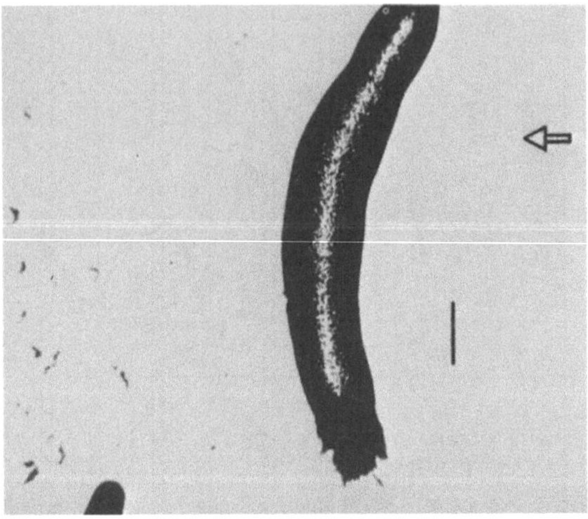

Fig. 4. Slug of *Dictyostelium discoideum* (strain NC4). The parallel observation beam from below is focused into a focal band at the upper surface. The slug turns toward the actinic light (*arrow*). *Bar* represents 0.1 mm. (Original from Häder)

in a different direction [21]. The slug, however, utilizes a lens effect. Since the refractive index of the plasma is sufficiently greater than that of the surrounding air, lateral light is focused onto the rear surface (Fig. 4; [20]). The slug moves away from this bright band which results in a turn toward the light source.

Photokinesis

In many microorganisms, the speed of movement depends on the light intensity. Recently photokinesis has been studied in the ciliate, *Blepharisma japonicum* [30, 33]. It saturates at about 300 lx and seems to involve a different photoreceptor than the photophobic response also present in this organism. Photokinetic effects may also be involved in the light avoidance response described in the colorless ciliates *Paramecium caudatum* and *P. tetraurelia* [29]. The cells accumulate in the shaded half of a glass tube. When associated with symbiontic *Chlorella*, however, *Paramecium bursaria* accumulates in light fields [45, 46, 51]. In this case, the algal partner seems to cooperate in the photoperception, since under DCMU inhibition of the algal photosynthetic electron transport chain the ciliates fail to photoaccumulate.

Photophobic Responses

Photophobic responses can be elicited by both temporal or spatial changes in light intensity [54, 56]. In the blue-green alga, *Phormidium uncinatum*, a sudden change in light intensity causes a reversal of movement. The filaments glide with a mechanism not yet fully understood. The two ends of a filament are morphologically identical, but a temporary physiological polarity dictates the direction of movement. This polarity

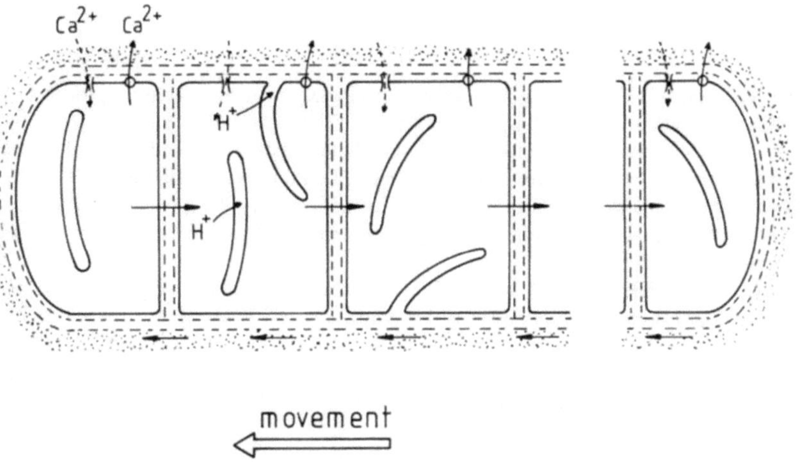

Fig. 5. Hypothetical model for the sensory transduction in photophobic responses of the cyanobacterium, *Phormidium uncinatum.* In light, protons are transported into the thylakoid vesicles which may be in contact with the plasmalemma. Ca^{2+} is pumped out of the cells into the medium or slime sheath (*stippled*) or cell wall (*dashed lines*). When the front cells (*leftmost*) enter a dark area the proton gradient breaks down which causes voltage-dependent calcium specific channels to open. The massive cation influx causes a depolarization of the cytoplasm. Longitudinal currents from cell to cell and back in the slime sheath are indicated by *small arrows.* (After Häder)

involves an electrical gradient between the front and rear ends [14, 18, 19]. These potentials can be measured using extracellular electrodes [3, 40, 41].

The current model of the sensory transduction of photophobic responses involves an amplification based on gated cation currents (Fig. 5). In light, the photosynthetic electron transport causes a vectorial transport of protons out of the cytoplasm into the thylakoid vesicles (which in this prokaryotic organism are not enclosed in a chloroplast membrane). When the front cells enter a shaded area, the proton gradient breaks down suddenly, which causes a slight depolarization of the cytoplasmic compartment. The sudden potential change is supposed to open voltage-dependent, calcium-specific gates which allow a massive influx of calcium along a gradient previously established by calcium pumps [17, 26, 40]. The gated cation influx causes a considerable amplification of the initial depolarization. Both calcium conductance blockers (e.g., lanthanum ions and ruthenium red) and specific calcium ionophores (e.g. A23187) inhibit photophobic responses as well as removal of calcium ions from the medium. The calcium uptake can directly be demonstrated using $^{45}Ca^{2+}$ [26]. The photophobic response is a simple, but rather effective mechanism which allows the organisms to stay in areas with appropriate light conditions [43].

References

1. Banchetti R, Rosati G, Verni F (1980) Cytochemical analysis of the photoreceptor in *Euglena gracilis* Klebs (Flagellata Euglenoidina). Monit Zool Ital (NS) 14:165–171
2. Bonner JT (1982) Comparative biology of cellular slime molds. In: Loomis W (ed) The development of *Dictyostelium discoideum.* Academic Press, London New York, pp 1–33
3. Chailakhyan LM, Glagolev AN, Glagoleva TN, Murvanidze GV, Potapova TV, Skulachev VP (1982) Intercellular power transmission along trichomes of cyanobacteria. Biochim Biophys Acta 679:60–67
4. Chang T-H, Walker EB, Song P-S (1981) Effects of ionophores and ruthenium red on the phototaxis of *Stentor coeruleus* as measured by simple devices. Photochem Photobiol 33:933–936
5. Colombetti G, Häder D-P, Lenci F, Quaglia M (1982) Phototaxis in *Euglena gracilis:* Effect of sodium azide and triphenylmethyl phosphonium ion on the photosensory transduction chain. Curr Microbiol 7:281–284
6. Colombetti G, Lenci F, Diehn B (1982) Responses to photic, chemical, and mechanical stimuli. In: Buetow DE (ed) The biology of *Euglena,* vol III. Physiology. Academic Press, London New York, pp 169–195
7. Colombetti G, Lenci F, Song P-S (1982) Effects of K^+ and Ca^{2+} ions on motility and photosensory responses of *Stentor coeruleus.* Photochem Photobiol 36:609–611
8. Doughty MJ, Diehn B (1979) Photosensory transduction in the flagellated alga, *Euglena gracilis.* I. Action of divalent cations, Ca^{2+} antagonists and Ca^{2+} ionophore on motility and photobehavior. Biochim Biophys Acta 588:148–168
9. Doughty MJ, Diehn B (1982) Photosensory transduction in the flagellated alga, *Euglena gracilis.* III. Induction of Ca^{2+}-dependent responses by monovalent cation ionophores. Biochim Biophys Acta 682:32–43
10. Doughty MJ, Grieser R, Diehn B (1980) Photosensory transduction in the flagellated alga, *Euglena gracilis.* II. Evidence that blue light effects alteration in Na^+/K^+ permeability of the photoreceptor membrane. Biochim Biophys Acta 602:10–23
11. Fisher PR, Williams KL (1981) Bidirectional phototaxis by *Dictyostelium discoideum* slugs. FEMS Microbiol Lett 12:87–89
12. Foster KW, Smyth RD (1980) Light antennas in phototactic algae. Microbiol Rev 44:572–630
13. Häder D-P (1979) Control of locomotion. Photomovement. In: Haupt W, Feinleib ME (eds) Encyclopedia of plant physiology, vol VII. Physiology of movements. Springer, Berlin Heidelberg New York, pp 268–309
14. Häder D-P (1981) Electrical and proton gradients in the sensory transduction of photophobic responses in the blue-green alga, *Phormidium uncinatum.* Arch Microbiol 130:83–86
15. Häder D-P (1981) Computer-based evaluation of phototactic orientation in microorganisms. EDV Med Biol 12:27–30
16. Häder D-P (1982) Coupling of photomovement and photosynthesis in desmids. Cell Motil 2: 73–82
17. Häder D-P (1982) Gated ion fluxes involved in photophobic responses of the blue-green alga, *Phormidium uncinatum.* Arch Microbiol 131:77–80
18. Häder D-P, Burkart U (1982) Mathematical simulation of photophobic responses in blue-green algae. Math Biosci 58:1–17
19. Häder D-P, Burkart U (1982) Enhanced model for photophobic responses of the blue-green alga, *Phormidium uncinatum.* Plant Cell Physiol 23:1391–1400
20. Häder D-P, Burkart U (1983) Optical properties of *Dictyostelium discoideum* pseudoplasmodia responsible for phototactic orientation. Exp Mycol 7:1–8
21. Häder D-P, Claviez M, Merkl R, Gerisch G (1983) Responses of *Dictyostelium discoideum* amoebae to local stimulation by light. Cell Biol Int Rev 7:611–616
22. Häder D-P, Colombetti G, Lenci F, Quaglia M (1981) Phototaxis in the flagellates, *Euglena gracilis* and *Ochromonas danica.* Arch Microbiol 130:78–82

23. Häder D-P, Melkonian M (1983) Phototaxis in the gliding flagellate, *Euglena mutabilis.* Arch Microbiol 135:25–29
24. Häder D-P, Poff KL (1979) Light-induced accumulations of *Dictyostelium discoideum* amoebae. Photochem Photobiol 29:1157–1162
25. Häder D-P, Poff KL (1979) Photodispersal from light traps by amoebas of *Dictyostelium discoideum.* Exp Mycol 3:121–131
26. Häder D-P, Poff KL (1982) Dependence of the photophobic response of the blue-green alga, *Phormidium uncinatum,* on cations. Arch Microbiol 132:345–348
27. Heelis DV, Heelis PF, Bradshaw F, Phillips GO (1981) Does the stigma of *Euglena gracilis* play an active role in the photoreception processes of this organism? A photochemical investigation of isolated stigmata. Photobiochem Photobiophys 3:77–81
28. Hong CB, Häder MA, Häder D-P, Poff KL (1981) Phototaxis in *Dictyostelium discoideum* amoebae. Photochem Photobiol 33:373–377
29. Iwatsuki K, Naitoh Y (1982) Photoresponses in colorless *Paramecium.* Experientia 38:1453–1454
30. Kraml M, Marwan W (1983) Photomovement responses of the heterotrichous ciliate *Blepharisma japonicum.* Photochem Photobiol 37:313–319
31. Lenci F, Häder D-P, Colombetti G (1984) Photosensory responses in freely motile microorganisms. In: Lenci F, Colombetti G (eds) Membranes and sensory transduction. Plenum Press, New York London, in press
32. Lipson ED, Häder D-P (1984) Video data acquistion for movement responses in individual organisms. Photochem Photobiol 39:437–441
33. Matsuoka T (1983) Distribution of photoreceptors inducing ciliary reversal and swimming acceleration in *Blepharisma japonicum.* J Exp Zool 225:337–340
34. Melkonian M (1982) Structural and evolutionary aspects of the flagellar apparatus in green algae and land plants. Taxon 3:255–265
35. Malkonian M (1983) Functional and phylogenetic aspects of the basal apparatus in algal cells. J Submicrosc Cytol 15:121–125
36. Melkonian M, Robenek H (1979) The eyespot of the flagellate *Tetraselmis cordiformis* Stein (*Chlorophyceae*): Structural spezialization of the outer chloroplast membrane and its possible significance in phototaxis of green algae. Protoplasma 100:183–197
37. Melkonian M, Robenek H (1980) Eyespot membranes of *Chlamydomonas reinhardii:* A freeze-fracture study. J Ultrastruct Res 72:90–102
38. Mikolajczyk E, Kuznicki L (1980) Body contraction and ultrastructure of *Euglena.* Acta Protozool 20:1–24
39. Morel-Laurens NML, Feinleib ME (1983) Photomovement in an "eyeless" mutant of *Chlamydomonas.* Photochem Photobiol 37:189–194
40. Murvanidze GV, Gabai VL, Glagolev AN (1982) Taxic responses in *Phormidium uncinatum.* J Gen Microbiol 128:1623–1630
41. Murvanidze GV, Glagolev AN (1982) Electrical nature of the taxis signal in cyanobacteria. J Bacteriol 150:239–244
42. Nakamura S (1981) Two different backward-swimming mutants of *Chlamydomonas reinhardii.* Cell Struct Funct 6:385–393
43. Nelson DC, Castenholz RW (1982) Light responses of *Beggiatoa.* Arch Microbiol 131:146–155
44. Nichols KM, Rikmenspoel R (1980) Flagellar waveform reversal in *Euglena.* Exp Cell Res 129: 377–381
45. Niess D, Reisser W, Wiessner W (1981) The role of endosymbiotic algae in photoaccumulation of green *Paramecium bursaria.* Planta 152:268–271
46. Niess D, Reisser W, Wiessner W (1982) Photobehaviour of *Paramecium bursaria* infected with different symbiotic and aposymbiotic species of *Chlorella.* Planta 156:475–480
47. Nultsch W (1979) Effect of external factors on phototaxis of *Chlamydomonas reinhardii.* III. Cations. Arch Microbiol 123:93–99
48. Nultsch W, Häder D-P (1979) Photomovement of motile microorganisms. Photochem Photobiol 29:423–437

49. Poff KL, Häder D-P (1984) An action spectrum for phototaxis by pseudoplasmodia of *Dictyostelium discoideum.* Photochem Photobiol 39:433–436
50. Poff KL, Hong CB (1982) Photomovement and photosensory transduction in microorganisms. Photochem Photobiol 36:749–752
51. Reisser W (1981) Host-symbiont interaction in *Paramecium bursaria:* Physiological and morphological features and their evolutionary significance. Ber Dtsch Bot Ges 94:557–563
52. Robenek H, Melkonian M (1981) Comparative ultrastructure of eyespot membranes in gametes and zoospores of the green alga *Ulva lactuca* (Ulvales). J Cell Sci 50:149–164
53. Satir BH, Garofalo RS, Gilligan DM, Maihle NJ (1980) Possible functions of calmodulin in protozoa. Ann NY Acad Sci 356:83–91
54. Shimmen T (1981) Quantitative studies on step-down photophobic response of *Euglena* in an individual cell. Protoplasma 106:37–48
55. Smyth RD, Berg HC (1982) Change in flagellar beat frequency of *Chlamydomonas* in response to light. Cell Motil Suppl 1:211–215
56. Song P-S (1981) Photosensory transduction in *Stentor coeruleus* and related organisms. Biochim Biophys Acta 639:1–29
57. Uematsu-Kaneda H, Furuya M (1982) Effects of calcium and potassium ions on phototaxis in *Cryptomonas.* Plant Cell Physiol 23:1377–1382
58. Vesely P, Maly M, Cumpelik J, Pluta M, Tuma V (1982) Improved spatial and temporal resolution in an apparatus for time-lapse phase contrast cine light micrography of cells in vitro. J Microsc 125:67–76
59. Watanabe M, Furuya M (1982) Phototactic behavior of individual cells of *Cryptomonas* sp. in response to continuous and intermittent light stimuli. Photochem Photobiol 35:559–563

Chloroplast Displacement Response to Blue Light Pulses

T. WALCZAK[1], J. ZURZYCKI[2] and H. GABRYS[2]

Introduction

The great majority of plant cells containing chloroplasts are able to translocate the chloroplasts within the cell in dependence on illumination conditions. In most plants this system of translocation is controlled by blue and near-UV absorbing photoreceptor system (review: [14]). Our information on the process of light-dependent chloroplast rearrangements is related predominantly to the photoreceptor problem, its nature, localization and orientation in a cell, and to the smaller extent also to the end

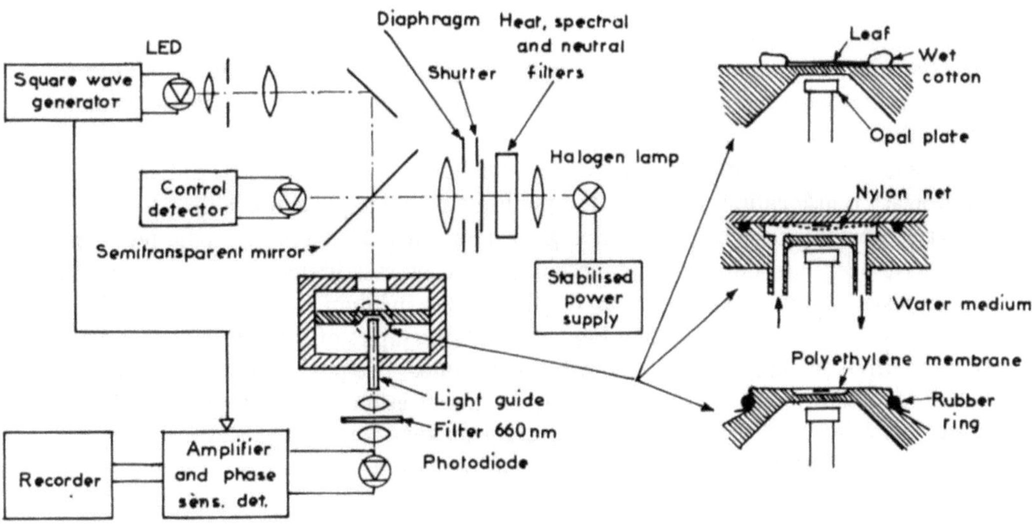

Fig. 1. Diagram of the recording photometer (*right*). *Left* modifications used for measurements of the leaves of land plants (*top*), water plants with continuous flow of the medium (*center*) and water plants without exchange of the medium (*bottom*)

1 Department of Nuclear Spectroscopy, Institute of Nuclear Physics, Radzikowskiego 152, 31-142 Kraków, Poland

2 Department of Plant Physiology, Institute of Molecular Biology, Jagellonian University, Al. Mickiewicza 3, 31-120 Kraków, Poland

Blue Light Effects in Biological Systems
ed. by H. Senger

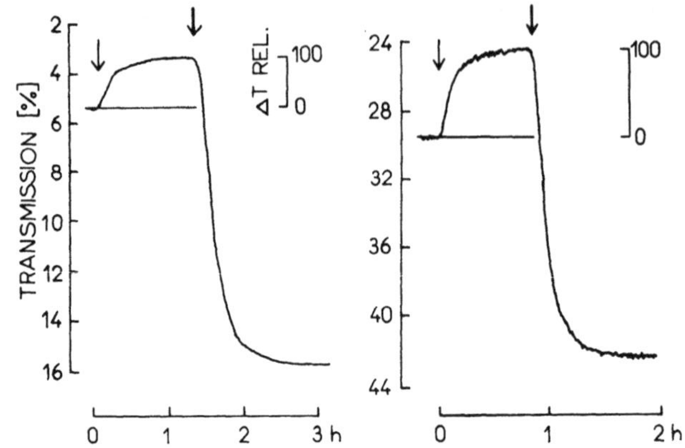

Fig. 2. Examples of the recordings of chloroplast translocations as measured photometrically. *Left: Tradescantia viridis, right: Lemna trisulca.* Starting position is the dark arrangement. At low fluence rate the light transmission decreases whereas, at high fluence rate the light transmission increases following the low light and high light rearrangement. *Abscissa* time (min). *Ordinate* transmission in percent of incident light and/or relative transmission changes taking as the reference the change from dark to low light arrangement

effector where actomyosine system is supposed to be active [2, 10]. The chain of reactions leading from the excitation of photoreceptor molecules to the translocation is very little understood.

Some time ago we developed a new method for continuous recording of the chloroplast translocation, based on the photometric principle and enabling the study of rearrangements also in the physiological darkness [11]. Recently some modifications have been introduced to the method permitting measurements on land leaves in air, on water plants with or without the flow of the medium etc. The principle of the method is illustrated in Fig. 1, and Fig. 2 presents examples of the recordings for dark-adapted leaves exposed to low and high fluence rates. The possibility of recording the behavior of chloroplasts in the dark made an investigation of the after-effects of light pulses possible. This, in turn, led to postulating some intermediate stages linking the primary photoreception with the end effect of irradiation.

Single Light Pulses

When a leaf, pretreated in the darkness, is exposed to a short pulse of blue light (max 450 nm, half-width 80 nm) of high intensity, certain rearrangements of chloroplasts can be observed in the following dark period. The character of these transient effects depends on the duration of the light pulse. By using the fluence rate of blue light of 30 W m^{-2}, short pulses (10 ms–1 s) induce in the following darkness a temporary decrease in the light transmission indicating a partial rearrangement to the low light position of chloroplasts. Its maximum is obtained in 10–20 min after the light pulse and in 60–300 min the chloroplasts come back to the uniform distribution character-

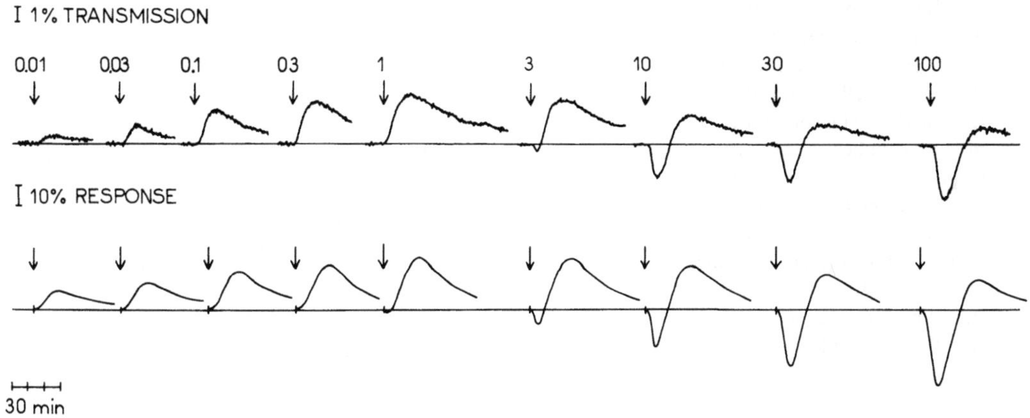

Fig. 3. Examples of the recorded transmission changes in *Lemna* induced by single pulses of blue light (30 W m^{-2}). The pulse durations (in s) are given. *Lower row* computer simulation of the pulse experiment [15]

istic of the darkness. By using the longer pulses of 30–100 s which are, however, still shorter than a lag period of chloroplast translocation, the response of chloroplasts is biphasic consisting of the first wave of a transmission increase followed by the wave of a transmission decrease. It corresponds to a partial translocation of chloroplasts to the high light position (called negative wave) and than partial translocation to the low light position (positive wave) before returning to the dark distribution (Fig. 3a). Very similar responses were recorded both in leaves of *Tradescantia* [5] and in fronds of *Lemna* [15] in spite of their different tissue structures (composed structure of the land plant leaf in the first case and one-layer mesophyll in the other), different medium (air and water respectively) and different general transmission.

The time course of the rearrangement in the darkness points to the existence of a long living intermediate state, generated during the irradiation and slowly decaying in the darkness.

Addition of the Light Effects

The validity of the reciprocity law was confirmed for *Tradescantia* leaves. The positive wave was induced in the range of the half saturation of the response by using constant energy (3 J m^{-2}) applied in the form of pulses from 0.1 to 100 s. The proper fluence rates were obtained with the use of the neutral filters. Figure 4 shows that the magnitude, as well as the kinetics, of the response depends solely on the energy applied and not on a pulse duration or a fluence rate used, even if both parameters are varied by four orders of magnitude [5].

When two light pulses are applied one after the other in the lag period of the rearrangement or even during the increasing phase of the first wave, the response is the same as after one pulse lasting as long as two previous pulses [8]. The effects are fully

Fig. 4. Chloroplast rearrangements in *Tradescantia* leaf induced by single light pulses in the range of half saturation of the reaction. The energy of 3 J m^{-2}, was applied in the form of pulses 0.1, 1, 10 and 100 s duration. The proper fluence rates were obtained by using neutral filters. *Upper curve* the maximal positive response induced by 1 s pulse of 30 W m^{-2} fluence rate [5]

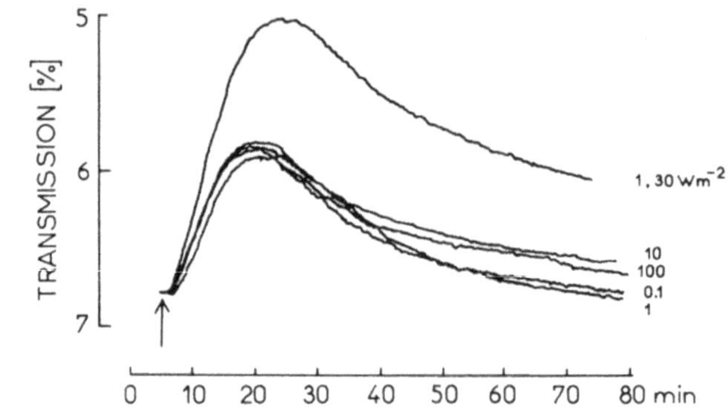

additive when the dark interval between two pulses does not exceed 2.5 min. Separation of the second pulse by a longer dark period results in the induction of the new biphasic response. The longer is the dark interval, the greater is the newly generated negative wave (Fig. 5a). The ability to react to the second pulse as to the first one is restored when the chloroplasts return to the initial dark position.

Such a behavior seems to indicate that the decay of the long-living intermediate state is not significant during the first 2 min of darkness following the first pulse and that it is almost complete only in 60–300 min after the first pulse.

Intermittent Light

As was found earlier in a cell submitted to intermittent irradiation, the chloroplasts react to the mean fluence rate value. The steady-state position is identical to that observed for continuous irradiation with the same average energy [12]. Recent investi-

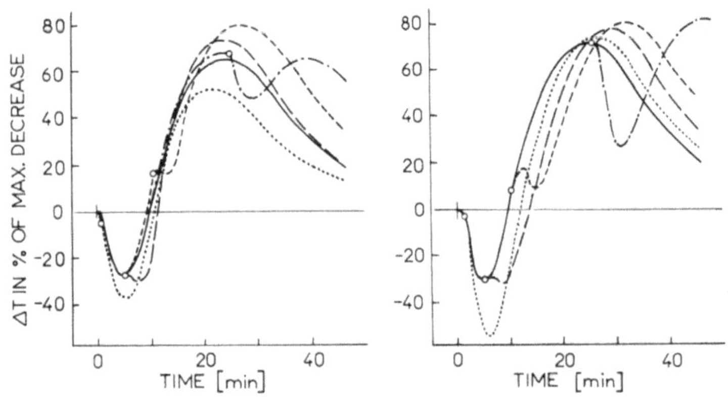

Fig. 5. *Right* light transmission changes in *Lemna* induced by two successive light pulses (30 W m^{-2}, 10 s) using different dark interval (2.5, 6, 10 and 25 min). *Left* computer simulation of the same irradiation programme [8]

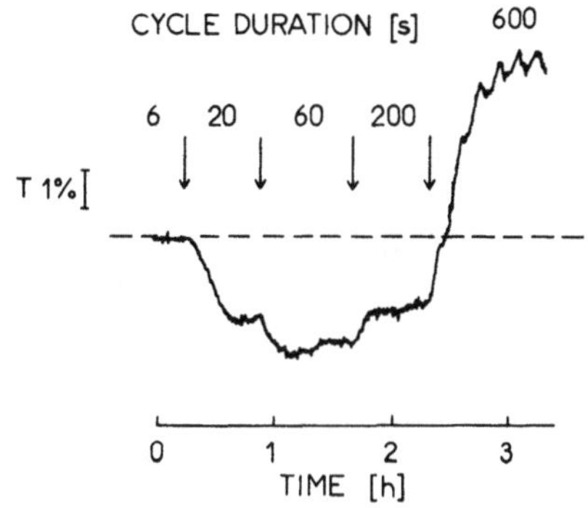

Fig. 6. Effect of intermittent light on chloroplast arrangement. Fluence rate of 30 W m^{-2} applied during 1/10 of the cycle. Cycle durations 6, 20, 60, 200, 600 s. ***Broken line*** average response to continuous irradiation with 3 W m^{-2} [6]

gation showed that this is the case for a frequency of light pulses higher than 0.5 Hz. At a medium frequency and at a higher average fluence rate the response is different from what could be expected from the integration of the applied energy (Gabrys et al. in preparation) (Fig. 6).

This kind of information opens a new possibility of exactly calculating the generation and decay kinetics of the intermediate states.

Model Considerations

Displacement of chloroplasts from one cell area to the other can be regarded as a mechanical response to the local attraction and repulsion forces. There are many reasons to believe that every part of the cell border, along which the chloroplasts can be moved, is equivalent, i.e., generates the attractive and repulsive forces identically under the same fluence rate. The fluence rate, however, is not homogeneous in various cell parts. As the incident light is parallel to the side walls, their irradiation is much weaker than that of the perpendicular walls. Local generation of the attraction and repulsion forces was observed in many experiments where only a part of the cell was illuminated [2, 3, 9, 13]. The calculations of the fluence rates at the side walls situated parallel to the incident light and at perpendicular walls showed that the ratio of both values varies between 1:10 and 1:100, depending on the shape of the cell, refractive index of the medium, wave-length, arrangement of chloroplasts etc. [4]. Assuming, for the sake of simplicity, that the cell is a rectangular box, one arrives at the ratio of fluences between upper and lower walls, and side walls of about 1:50, i.e., about 0.02 of incident energy reaches the side walls. This difference is big enough to give quite different response in generation of attraction and repulsion forces at both types of the cell walls.

The response of chloroplasts to a single light pulse may be explained in the following way (Fig. 7): Let us assume that in a cell kept for a longer time in the darkness the attractive (Y_1) and repulsive (Y_2) forces are equal to zero at every cell wall. The lack of any differences between the parts of the cell results in the equal distribution of chloroplasts along all walls (dark position). The light pulse of an enough energy generates the attraction and repulsion forces which are different at the walls $\perp$ and $\parallel$ to

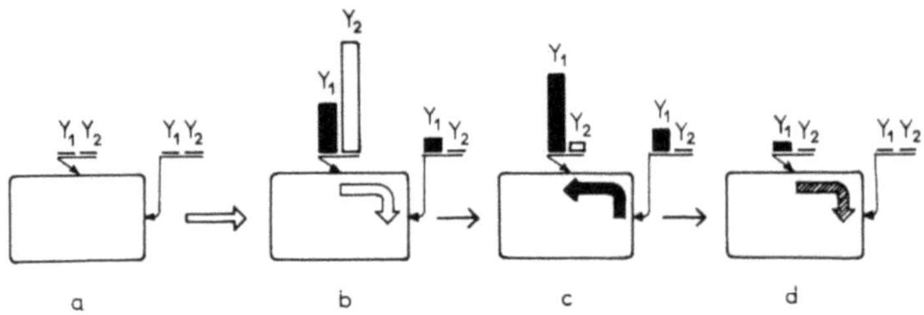

Fig. 7a–d. Schematic presentation of the changes in Y_1 and Y_2 levels after a light pulse. **a** dark-adapted cell; Y_1 and Y_2 at every cell wall close to zero; **b** shortly after the application of a strong light pulse; repulsion agent (Y_2) at the cell walls perpendicular to light dominates resulting in an outflow of chloroplasts from these walls; **c** later on the attractive forces on these walls dominate, chloroplasts tend to take low light arrangement; **d** after a still longer time Y_1 and Y_2 decrease again, the initial arrangement (case **a**) is restored

illumination. These forces decrease in the following dark period with the characteristic kinetics reaching after some time zero level again. If we assume that in a given moment (shortly after a pulse) on the $\perp$ walls the repulsive forces (Y_2) are greater than the attractive ones and on the $\parallel$ walls the situation is reversed (Fig. 7), the chloroplasts will escape from the former and will move to the latter. This corresponds to a negative wave. The repulsive forces decrease faster than attractive ones which leads to a situation depicted in Fig. 7c and induces a positive wave. After further prolongation of the dark time the initial situation is restored and chloroplasts come again to the equal distribution in the cell. Certain relative values and kinetic parameters were ascribed to the functions X and Y and their optimalization was carried out on the basis of the results of the light pulse experiments.

As was found in many studies on the action spectra of negative and positive chloroplast reaction to irradiation, both types of response are controlled by the same photoreceptor [3, 7, 9, 14]. It seems likely that the excited photoreceptor controls in some way the generation of both Y_1 and Y_2 levels. The life-time of the excited photoreceptor is too short to explain the occurence of long-lasting after-effects. Therefore we assume the existence of the first intermediate state X, which is directly related to the degree of photoreceptor excitement but has a much longer life time. Thus the sequence of events taking place in every part of the cell can be described as

$$F \xrightarrow{\downarrow I} F^* \longrightarrow X(I) \begin{array}{l} \nearrow Y_1 \\ \searrow Y_2 \end{array} \qquad (1)$$

where I is the flux energy of the light pulse, F^* number of excited photoreceptor molecules and X (I) the first long-living intermediate state generated as a function of I. The displacement of chloroplasts is a response which integrates the differences in Y_1 and Y_2 in the whole cell, therefore to describe the rearrangement the following scheme can be suggested:

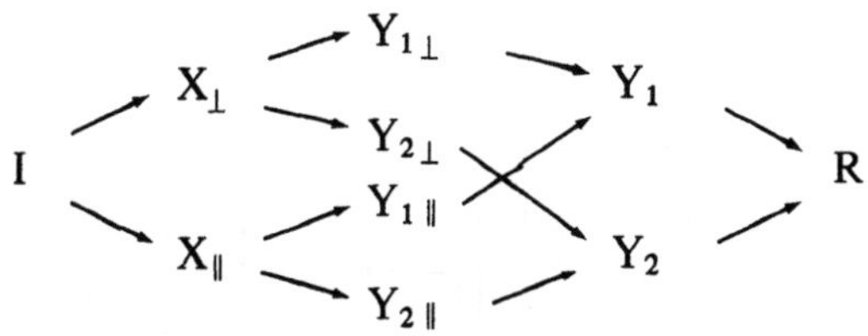

(2)

There are many ways to describe the relationships between I, X, Y_1, Y_2 and R, as well as the kinetic parameters of these factors. The proposed solution [15] has the advantage of being relatively simple and giving a very good fit with the experimental data. We assume in it that X is directly proportional to the fluence rate with the coefficient k_1:

$$X_{\perp} = k_1 I \quad (3)$$
$$X_{\parallel} = k_1 I r \quad (4)$$

where r is the ratio of fluences at ∥ and ⊥ walls respectively. X increases exponentially during irradiation, reaching a constant level after a sufficiently long time and decreases also exponentially after switching off the light. The time constant of formation and decay is identical and equal to 1/a (Fig. 8). The calculations give the value of about 3 min for 1/a. One can expect thus that during short light pulses the increase of X is nearly linear in time. This would explain the validity of reciprocity law (Fig. 4) as well as the additivity of the effects when two light pulses are applied in short time one after the other (Fig. 8).

The dependence of Y_1 and Y_2 on X is defined as:

$$Y_1 = f \log(1 + X) \quad (5)$$
$$Y_2 = f \log(1 + X)(1 - e^{k_2 X}) \quad (6)$$

where f is a normalization factor, k_2 an adjustable parameter. The grounds for assuming a logarithmic relationship were discussed in an earlier paper [15]. The parameter k_2 has a low value, therefore the generation of the repulsive forces Y_2 is significant under high fluence rate or dose (Fig. 9).

The recorded time course of the chloroplast response to light pulses indicates the delay in the reaction. This delay is shorter for the negative response and longer for the positive one. To describe thus the chloroplast movement, one must assume that the effector system responds to the attractive and repulsive forces Y_1 and Y_2 with delays

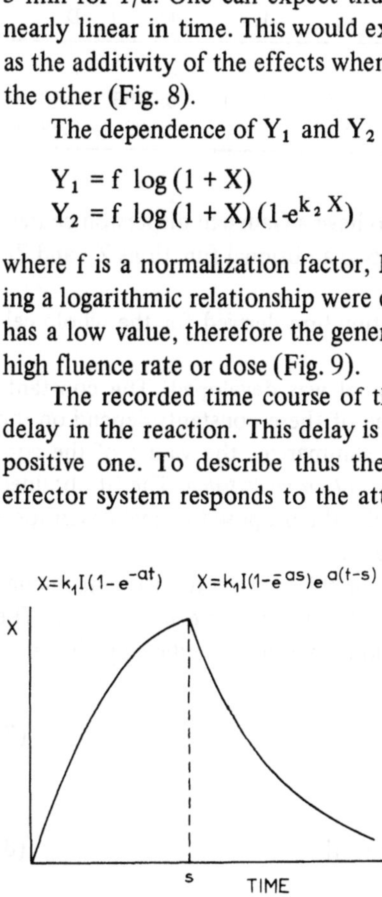

Fig. 8. Increase and decrease of X level in time; *s* pulse duration

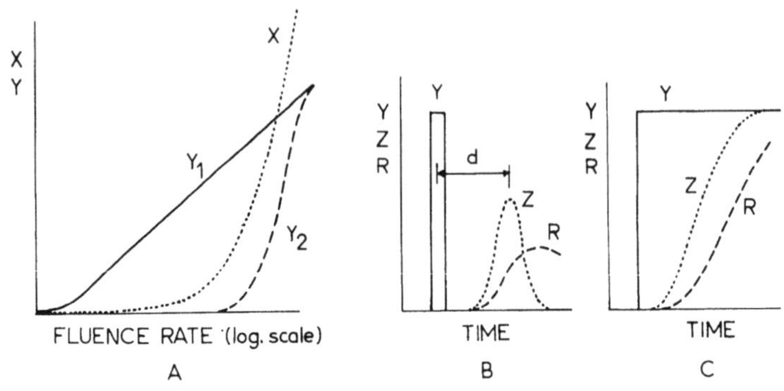

Fig. 9A–C. Scheme of the dependence of the model parameters on the fluence rate (**A**) and on time (**B, C**). **B**-a simplified picture of a pulse experiment where Y (Y_1 or Y_2) is assumed to change instantly to high level and again to zero level; *d* delay (d_1 for Y_1 and d_2 for Y_2); *R* response of chloroplasts. **C** illustration of the beginning of constant irradiation assuming that Y reaches instantly its steady state level

Table 1. Values of constants calculated for experiments on *Lemna trisulca*

	k_1	r	f	k_2	1/a	d_1	d_2	1/c	D
[15]	1.73	0.019	0.318	0.00021	–	–	–	–	–
[15]	12.22	0.019	0.216	0.00021	0.333	5,87	2.651	0.0393	0.8945
[8]	3.3	0.02	0.275	0.00021	0.254	5.87	2.65	0.103	0.8945

d_1 and d_2, respectively. The delays are assumed to have a Gaussian dispersion characterized by a factor of D [15]. To account for that two additional functions Z_1 and Z_2 were introduced into the calculations.

The response of chloroplast followed the Z function derived for the whole cell with the time constant of 1/c.

The computer program for the described model was developed. The constants were calculated by using the least-squares fit. Some of these constants depend on the physiological state of the material used. Table 1 summarizes the values of the constants calculated for various series of experiments on *Lemna trisulca.* The fit obtained is illustrated exemplarily in Figs. 3b and 5b. Generally the proposed formal description gives a satisfactory agreement with the experimental data.

The model describing the steady-state response of chloroplasts to different fluence rates is simpler since the kinetic parameters do not play any role in this case. The fluence-response curve for the steady-state conditions may be described by the function

$$R_{(st)} = (Y_{1\perp} - Y_{1\parallel}) - (Y_{2\perp} - Y_{2\parallel}) \qquad (7)$$

or

$$R_{(st)} = f\,[\log(1 + k_1 I) - \log(1 + k_1 rI)] - f\,[\log(1 + k_1 I)(1\text{-}e^{-k_2 I}) - \log(1 + k_1 rI)(1\text{-}e^{-k_2 rI})] \qquad (8)$$

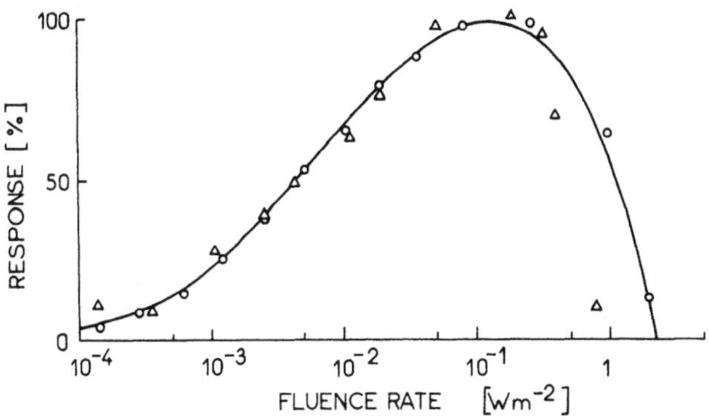

Fig. 10. The response of chloroplasts in *Lemna trisulca* to continuous irradiation with different fluence rates recorded by two methods. *Open circles* photometric values after correction for photometric readings. *Triangles* values of microscopic countings of chloroplasts. *Continuous line* the fluence rate response calculated according to Eq. (8). Discrepancies in the region of the high fluence rates are due to the different geometry of irradiation systems used in both experimental methods respectively [15]

In Fig. 10 the fluence-response curve in *Lemna* cells recorded by two methods is compared to the function calculated with the use of Eq. (8).

Concluding Remarks

The aim of the present study was to postulate some hypothetical intermediate states functioning as an information chain between the excited photoreceptor and the end effect, i.e., displacement of chloroplasts. This simple model involving two states X and Y gave a good fit with the experimental data. The postulated states should correspond to physical phenomena in a cell. One may speculate on the action of light on permeability of plasmalemma inducing the ion fluxes and related electrical phenomena, changes in local concentration or gradients of ions and their significance for the regulation of actomyosine system. At this moment it would be premature to identify the X and Y states with any of these physical processes. We hope that further investigation will make such identification possible.

References

1. Blatt MR, Briggs WR (1980) Blue-light-induced cortical fiber reticulation concomitant with chloroplast aggregation in the alga *Vaucheria sessilis.* Planta 147:355–362
2. Blatt MR, Wessells NK, Briggs WR (1980) Actin and cortical fiber reticulation in Siphonaceous alga *Vaucheria sessilis.* Planta 147:363–375
3. Fischer-Arnold G (1963) Untersuchungen über die Chloroplastenbewegungen bei *Vaucheria sessilis.* Protoplasma 56:495–520

4. Gabryś-Mizera H (1976) Model consideration of the light conditions in noncylindrical plant cells. Photochem Photobiol 25:435–461
5. Gabryś H, Walczak T, Zurzycki J (1981) Chloroplast translocations induced by light pulses. Effects of single light pulses. Planta 152:553–556
6. Gabryś H, Walczak T, Kajfosz J, Zurzycki J (1984) Chloroplast translocations in intermittent light. In preparation
7. Inoue Y, Shibata K (1973) Light-induced chloroplast rearrangements and their action spectra as measured by absorption spectroscopy. Planta 114:341–358
8. Kajfosz J, Walczak T, Zurzycki J (1983) Chloroplast translocations in *Lemna trisulca* induced by two successive blue light pulses. Physiol Veg 21:519–525
9. Mayer F (1964) Lichtorientierte Chloroplasten-Verlagerung bei *Selaginella martensii.* Z Bot 52: 346–381
10. Schönbohm E (1973) Kontraktile Fibrillen als aktive Elemente bei der Mechanik der Chloroplastenverlagerung. Ber Dtsch Bot Ges 86:407–422
11. Walczak T, Gabryś H (1980) New type of photometer for measurements of transmission changes corresponding to chloroplast movements in leaves. Photosynthetica 14:65–72
12. Zurzycka A, Zurzycki J (1954) Studies on the phototactic movements of chloroplasts II. Acta Soc Bot Pol 23:279–288
13. Zurzycki J (1972) Primary reactions in the chloroplast rearrangements. Acta Protozool 11: 189–199
14. Zurzycki J (1980) Blue light-induced intracellular movements. In: Senger H (ed) The blue light syndrome. Springer, Berlin Heidelberg New York, pp 50–68
15. Zurzycki J, Walczak T, Gabryś H, Kajfosz J (1983) Chloroplast translocations in *Lemna trisulca* L induced by continuous irradiation and by light pulses. Kinetic analysis. Planta 157: 502–510

Flavin-Mediated Weak-Light Chloroplast Movement in *Mougeotia*

T. WALCZAK[1], H. GABRYS,[2] and W. HAUPT[3]

Introduction

In *Mougeotia* the two types of chloroplast movement depend in a different way on the wavelength of the orienting light: The high-fluence-rate response (face-to-profile orientation) can be induced exclusively by blue light, and this is an effect of an interaction of phytochrome and a blue-light-absorbing pigment (cf. [10]). In contrast, the low-fluence-rate response (profile-to-face orientation) has its maximum in red light with a typical red/far-red reversibility; thus, phytochrome has been accepted as the main photoreceptor pigment involved ([4], cf. also [5]). There was, however, some controversy, whether or not the lower effect of the blue light region could be explained on the basis of phytochrome alone (e.g. [3], cf. also [5]). Only recently a strong argument in favor of a separate blue-light photoreceptor pigment has been derived from a new experimental approach [2]. These results will be summarized briefly, and additional experiments will be reported.

The rationale of our experiments was the assumption that under strong scattered far-red irradiation virtually all phytochrome in the cell should be converted to the inactive P_r form, and low-fluence-rate red light could no longer establish an effective P_{fr} gradient. A blue-light effect against this very low P_{fr} level, then, should strongly indicate an independent blue-light-sensing pigment.

Materials and Methods

Mougeotia was cultured in petri dishes as described previously [13]. The chloroplasts were preoriented to the profile position with white light irradiated from the sides for 70 min at 17°C.

Two different methods were used to estimate the response: the microscopical counting and the photometric technique. For the first method, the preparations were

1 Institute of Nuclear Physics, Radzikowskiego 152, 31-142 Kraków, Poland

2 Institute of Molecular Biology, Jagiellonian University, Al. Mickiewicza 3, 31–120 Kraków, Poland

3 Institut für Botanik und Pharmazeutische Biologie der Universität Erlangen-Nürnberg, Schloßgarten 4, 8520 Erlangen, FRG

Blue Light Effects in Biological Systems
ed. by H. Senger

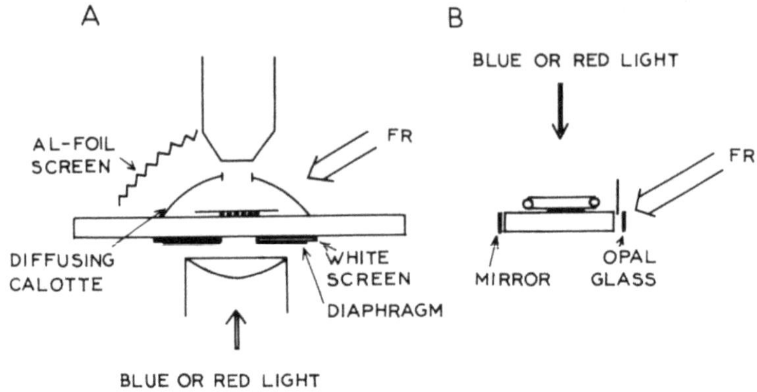

Fig. 1A,B. Experimental set-up (**A**) under the microscope (**B**) in the photometer chamber. For explanation see text. (After [2])

irradiated under the microscope. They consisted of a few trichomes mounted on slides with coverslips and sealed with vaseline at both ends. Simultaneously with the orienting blue or red light from the microscope lamp, part of the preparations obtained strong far-red (727 nm) from an additional halogen lamp, scattered with a celluloid calotte put onto the coverslip (Fig. 1A). The fluence rate of the scattered far-red light, measured at the site of the preparation, was 4–5 W m^{-2}. For more details see our previous paper [2].

For the photometrical investigations, the method of Walczak and Gabryś [12] has been adapted to *Mougeotia*. The preparations consisted of about 10 trichomes mounted parallely on a slide in a drop of culture medium and fastened at both extremes with parafilm stripes. They were covered with a membrane used for the YSI oxygen electrode (Yellow Springs, Ohio) stretched over a metal ring (Fig. 1B). Thus prepared, the algae could survive for up to 2 weeks, and the transmission changes corresponding to the chloroplast responses were reproducible during the first 4 days. For technical reasons the additional strong far-red was applied laterally (Fig. 1B). One side of the 4 mm thick glass plate was irradiated with the scattered far-red reflected on the other side by a mirror. A part of the light underwent the total internal reflection at the glass-water interface, the rest passed through the algae-containing solution and was then reflected at the membrane-air interface. According to our estimations, the spatial divergence of the lateral far-red beam irradiating the *Mougeotia* filaments was about 60°. The fluence rate measured under the surface of the opal glass was about 10 W m^{-2}.

For the photometric method, red light (660 nm, 40 mW m^{-2}) modulated with a frequency of 800 Hz has been used as measuring light during the whole experimental period of 1.5 h. This measuring light, given alone, would produce a strong profile-to-face response; but on the background of the far-red irradiation, both scattered and lateral, no movement at all can be observed in response to the measuring light, as indicated by orienting experiments. Thus, for the blue-light experiments the photometric method has been used only when far-red was given as a background.

The orienting blue or red light was obtained by using interference filters: 449 and 451 nm for blue light in the first and second series, respectively, and 681 nm for red light. Far-red light has been produced by a halogen lamp (12 V, 100 W) equipped with an interference filter 727 nm or with glass filters RG 9 (3 mm) and C-805 (3 mm) from Zeiss, Jena, in the first and second series, respectively.

For the action spectrum, as measured in the presence of strong far-red, interference filters were used with maximum transmittance at 405, 427, 451, 480 and 500 nm, respectively. With a set of neutral density filters the photon-fluence rate has been adjusted to 38 nmol m^{-2} s^{-1}, which corresponds to 10 mW m^{-2} for 451 nm. At this wavelength and this fluence rate, a response of about 60% of maximum possible profile-to-face orientation is found after 1.5 h.

In contrast to most of the previous investigations dealing with profile-to-face movement, the light was not applied as a single inducing pulse but continuously during the whole response [cf. 9, 13].

Results and Discussion

In a first series, the chloroplast orientation has been induced by blue light or red light of different fluence rates, irradiated through the microscope, with and without additional far-red, according to Fig. 1A. The response has been counted in individual cells after 30 and 60 min. The fluence-rate-response curve for the blue-light effect (Fig. 2) suggests a logarithmic dependence, as has been observed for the chloroplast movement in *Lemna trisulca* and in other blue-light-sensitive plants from threshold to saturation [1, 14, 16].

Moreover, the results clearly show that the blue-light effect is completely independent of the presence or absence of strong far-red background light. This far-red, on the other hand, completely abolishes the red-light induction of chloroplast movement, even if this red light is given thousandfold stronger than its threshold fluence rate (Fig. 2). As has been pointed out by Gabryś et al., these results cannot be interpreted in terms of phytochrome action and indicate strongly the presence of an additional sensor pigment underlying the blue-light response [2].

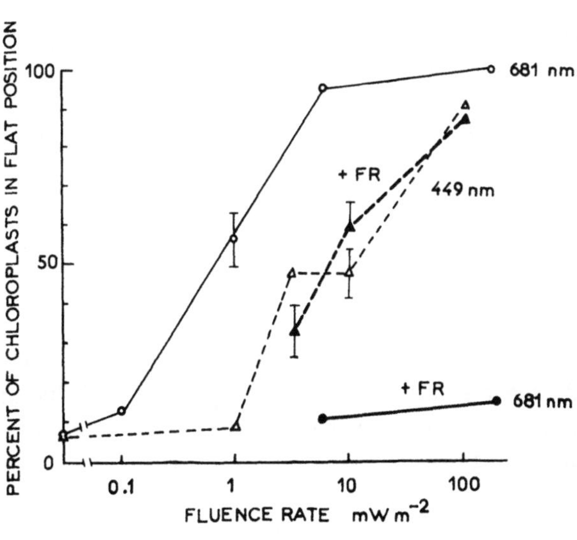

Fig. 2. Fluence-rat-response curves of the chloroplast orientation after 1 h in continuous red or blue light with or without additional far-red. *Open symbols* and *thin lines* inducing blue/red light only; *closed symbols* and *heavy lines* simultaneous irradiation with strong scattered far-red. Standard errors indicated when exceeding 5%. (After [2])

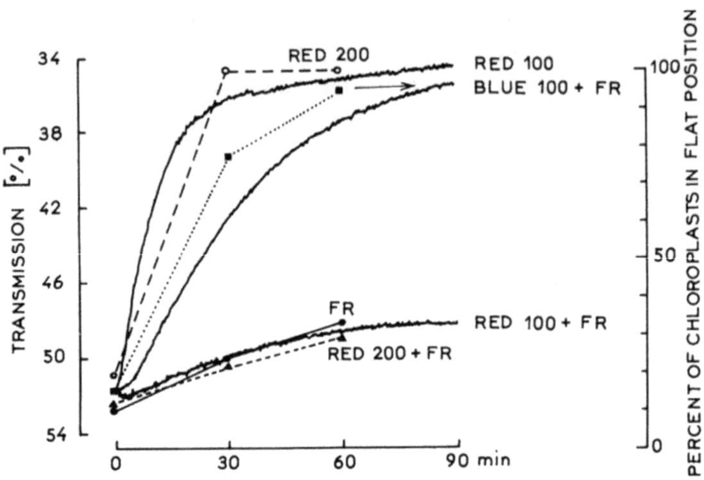

Fig. 3. Time-course of chloroplast responses determined under the microscope (*lines with symbols*) and recorded photometrically (*curves*) for two different preparations. The response was induced by continuous red or blue light with (*closed symbols*) and without (*open symbols*) additional far-red. Blue and red fluence rates in mW m^{-2} are given besides the curves

In a second series, the response to blue light and to red light, respectively, has been followed continuously by the photometric method and compared with the results of the first series (direct microscopic observation). After proper adjustment of the ordinates, the two methods give very similar results (Fig. 3). The general course of the curves shows that small differences in the chloroplast position (its angle of rotation) can be detected much more sensitively than by direct estimation under the microscope. Thus, the new approach by the photometric method turns out to be useful and reliable for more detailed investigations on the blue-light photosensory pigment, which will be reported next.

The kinetics for the response to different wavelengths on the background of far-red light is shown in Fig. 4. All these curves have been recorded for the same preparation. With other preparations, similar differences between the wavelengths have been found. It was therefore possible to derive a preliminary action spectrum for these curves. For this purpose we define the transmission change from t = 0 to t = 60 min or t = 90 min as $\Delta T_\lambda = T_{\lambda,t} - T_{\lambda,o}$ for the given wavelength and express it as a fraction of ΔT_{451}, to obtain the wavelength-dependent response value

$$R_\lambda = \frac{\Delta T \lambda}{\Delta T_{451}} ,$$

which is plotted against the wavelength in Fig. 5. As can be seen, the resulting curves are nearly identical for t = 60 min and t = 90 min, and thus they may be reliable in spite of the large standard errors. Very similar curves result if the ΔT values are corrected for the blank value ΔT_{FR} according to

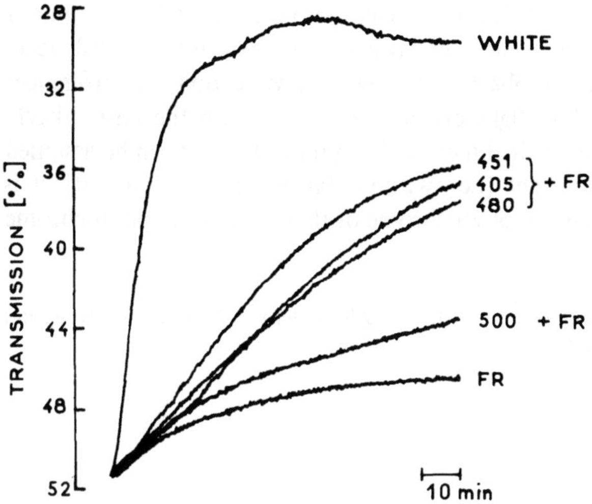

Fig. 4. Photometric recording of chloroplast orientation induced by different wavelengths in one preparation. Simultaneously with the inducing light the preparation was irradiated laterally with strong far-red. The wavelengths are given besides the curves. Additionally the transmission changes for the complete profile-to-face response in white light and for the background far-red are recorded

$$R'_\lambda = \frac{\Delta T_\lambda - \Delta T_{FR}}{\Delta T_{451} - \Delta T_{FR}} \quad ,$$

but in this case the standard errors are still greater (data not shown).

The action spectrum thus obtained shows a maximum between 430 and 450 nm and resembles action spectra for chloroplast movement in several other species, in which only blue light is effective [15]. This points to a flavin as the second photoreceptor pigment for the low-fluence-rate response in *Mougeotia.*

Different from the action spectrum reported in this paper, 480 nm is the most effective wavelength for the high-fluence-rate response of *Mougeotia* [8]. However, a strict comparison cannot be made, because in the latter case we are dealing with an interaction of the blue-light photoreceptor with phytochrome, possibly modified by the photosynthetic system [5, 10]. It is still an open question how those interactions can be explained in vivo [7]. In vitro, an efficient energy transfer has been reported to

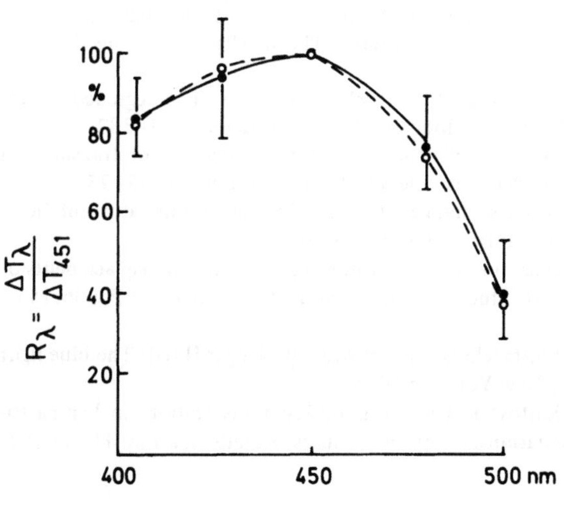

Fig. 5. Action spectrum of the short-wavelength-induced chloroplast orientation in *Mougeotia* in the presence of strong lateral far-red after 60 min (*dashed line*) and 90 min (*solid line*). The ordinate shows relative quantum responsivity. Mean from 12 preparations. For details see text

occur from the singlet excited state of flavin to the chromophore of P_r [6, 11]. In our experiments, however, phytochrome was "inactivated" by the strong far-red background irradiation, and thus we could observe the "pure" flavin action. No differences have been found comparing the blue-light effect with and without the far-red background, either in light sensitivity, or in the kinetics of response. Thus, it can be assumed that also in monochromatic blue light (i.e., without far-red) most of the effect is accounted for by flavin rather than by phytochrome or than by a flavin-phytochrome interaction.

Acknowledgment. This work was supported in part by the Alexander-von-Humboldt-Stiftung and by the Deutsche Forschungsgemeinschaft.

References

1. Gabryś H, Walczak T (1980) Photometric study of chloroplast phototranslocations in leaves of land plants. Acta Physiol Plant 2:281–290
2. Gabryś H, Walczak T, Haupt W (1983) Blue-light induced chloroplast orientation in Mougeotia. Evidence for a separate sensor pigment besides phytochrome. Plante 160:21–24
3. Hartmann KM, Cohnen-Unser I (1973) Carotenoids and flavins versus phytochrome as the controlling pigment for blue-UV-mediated photoresponses. Z Pflanzenphysiol 69:109–124
4. Haupt W (1959) Die Chloroplastendrehung bei Mougeotia I. Über den quantitativen und qualitativen Lichtbedarf der Schwachlichtbewegung. Planta 53:484–501
5. Haupt W (1982) Light-mediated movement of chloroplasts. Annu Rev Plant Physiol 33:205–233
6. Sarkar HK, Song P-S (1982) Blue light induced phototransformation of phytochrome in the presence of flavin. Photochem Photobiol 35:243–246
7. Schäfer E, Haupt W (1983) Blue-light effects in phytochrome-mediated responses. In: Pirson A, Zimmermann M (eds) Encyclopedia of plant physiology, New Ser, vol 16. Springer, Berlin Heidelberg New York, pp 723–744
8. Schönbohm E (1963) Untersuchungen über die Starklichtbewegung des Mougeotia-Chloroplasten. Z Bot 51:233–276
9. Schönbohm E (1969) Untersuchungen über den Einfluß von Photosynthesehemmstoffen und Halogeniden auf die Starklicht- und Schwachlichtbewegung des Chloroplasten von Mougeotia spec. Z Pflanzenphysiol 60:255–269
10. Schönbohm E (1980) Phytochrome and non-phytochrome dependent blue light effects on intracellular movements in fresh-water algae. In: Senger H (ed) The blue light syndrome. Springer, Berlin Heidelberg New York, pp 69–96
11. Song P-S, Sarkar HK, Kim IS, Poff KL (1981) Primary photoprocesses of undegraded phytochrome excited with red and blue light at 77 K. Biochim Biophys Acta 635:369–382
12. Walczak T, Gabryś H (1980) New type of photometer for measurements of transmission changes corresponding to chloroplast movements in leaves. Photosynthetica 14:65–75
13. Weisenseel M (1968) Vergleichende Untersuchungen zum Einfluß der Temperatur auf lichtinduzierte Chloroplastenverlagerungen. Z Pflanzenphysiol 59:56–69
14. Zurzycki J (1967) Properties and localization of the photoreceptor active in displacements of chloroplasts in Funaria hygrometrica. II. Studies in polarized light. Acta Soc Bot Pol 36:143–152
15. Zurzycki J (1980) Blue light-induced intracellular movements. In: Senger H (ed) The blue light syndrome. Springer, Berlin Heidelberg New York, pp 50–68
16. Zurzycki J, Walczak T, Gabryś H, Kajfosz J (1983) Chloroplast translocations in Lemna trisulca L. induced by continuous irradiation and by light pulses. Kinetic analysis. Planta 157: 502–510

Influence of ATPase Inhibitors on Light-Dependent Movement of Chloroplasts in *Vallisneria*

K. SEITZ and B. MAURER[1]

Introduction

Light-dependent intracellular movement responses of chloroplasts (orientation movement and cyclosis) are controlled by the mobility of chloroplasts in cytoplasmic streaming. Light-induced changes in the mobility of chloroplasts can be investigated in centrifugation experiments [6].

Blue light is known to cause changes in the centrifugability of chloroplasts, at low irradiance centrifugability is decreased, at high irradiance centrifugability of chloroplasts is increased. The decrease in centrifugability in the low intensity range is induced by chlorophylls in dependence upon the photosynthetic electron flow, as action spectra and inhibitor experiments have shown. The increase in centrifugability in the high intensity range, on the other hand, is induced by a flavin photoreceptor and can be inhibited by uncouplers of oxidative phosphorylation [7].

The mobility of chloroplasts depends upon the motive mechanism of cytoplasmic streaming, in which an actomyosin like molecular system is operating. Light can thus induce changes in the mobility of chloroplasts by an effect on the actomyosin system of the motive mechanism of cytoplasmic streaming.

Changes in the actomyosin system can be induced by modifications in the availability of ATP or by changes in the intracellular ionic relations [2, 5, 8]. Changes in ions can be induced by the flavin photoreceptor, being located in the cell membrane, by an effect on membrane permeability, or by an effect on transmembrane ion transport in dependence upon ion pumps [1, 11]. Thus an ionogenic ATPase like the plasmamembrane H^+-ATPase could be involved in the action mechanism of light. In this regard the influence of the ATPase inhibitors ABA, DCCD, DES and Vanadate on the light-dependent changes in the centrifugability of chloroplasts has been investigated.

Results and Discussion

The influence of ATPase inhibitors has been investigated in leaves of *Vallisneria* in preparations as described earlier [6]. After a pretreatment of 24 h darkness the prepara-

1 Institut für Botanik und Pharmazeutische Biologie der Universität Erlangen-Nürnberg, Schloßgarten 4, 8520 Erlangen, FRG

Blue Light Effects in Biological Systems
ed. by H. Senger

Table 1. Influence of ATPase inhibitors on light-dependent changes in the centrifugability of chloroplasts

			Dark control % Centifugability	Low intensity- Irradiation	High intensity- Irradiation
				Light-induced change in centrifugability (difference to dark control)	
in 1% ethanol	Ethanol	1%	55 ± 3	– 10	+ 26
	DCCD	10^{-4} M	57 ± 4	– 12	+ 22
	DES	10^{-4} M	59 ± 4	– 11	+ 22
	ABA	10^{-3} M	56 ± 3	– 10	+ 25
in H_2O	H_2O		53 ± 3	– 9	+ 28
	Vanadate	10^{-3} M	61 ± 4	– 12	+ 16
	Arsenate	10^{-3} M	58 ± 2	– 11	+ 19
	DNP	10^{-4} M	64 ± 4	– 14	+ 13

tions were irradiated for 20 min with 454 nm interference filtered light. The irradiance was 15 mJ m^{-2} s^{-1} in the low intensity range and 1800 mJ m^{-2} s^{-1} in the high intensity range. After the irradiation the preparations were centrifuged for 1 min at 1100 *g* (3000 rpm) and then fixed in boiling water. The percentage of centrifugability was evaluated microscopically.

The results of the experiments with the different ATPase inhibitors are summarized in Table 1. The light-induced change in the centrifugability of chloroplasts is given in Table 1 as the difference in % centrifugability between the irradiated sample and an unirradiated dark control.

The results in Table 1 show that ABA, known as an inhibitor of H^+-K^+ exchange at the cell membrane [4], has no effect on the light-dependent changes in the centrifugability of chloroplasts.

DCDD and DES, known as inhibitors of the plasmamembrane H^+-ATPase [3], induce a slight increase in the centrifuagability of chloroplasts in the dark control, but cause only insignificant modifications of the light-induced changes in the centrifugability in the low and high intensity range. This result indicates, if the inhibitors are taken up by cells of *Vallisneria,* that an ionogenic H^+-ATPase is not involved in the action mechanism of the light induced changes in the centrifugability of chloroplasts in *Vallisneria.*

Vanadate is another well-known ATPase inhibitor [10] and has been reported to have only a weak effect on cytoplasmic streaming [9]. Vanadate has a more pronounced influence, it induces increased centrifugability of chloroplasts in the dark control. This causes a reduction of the effect of light in the high intensity range, while the effect of light in the low intensity range is usually promoted.

The influence of Vanadate is similar to that of uncouplers of oxydative phosphorylation like Arsenate and DNP. The increase in the centrifugability of chloroplasts induced by these substances could thus be a consequence of uncoupling of oxydative phosphorylation, if not a rather unspecific membrane damage is the actual cause.

Taken together, the results of these experiments with ATPase inhibitors do not indicate that an ionogenic ATPase is involved in the action mechanism of light-dependent chloroplast movement in *Vallisneria.*

Abbreviations

ABA: abscisic acid
DCCD: dicyclohexyl carbodiimide
DES: diethylstilbestrol
DNP: 2,4 dinitrophenole

References

1. Blatt MR, Weisenseel M, Haupt W (1981) A light-dependent current associated with chloroplasts aggregation in the alga Vaucheria. Planta 152:513–526
2. Britz SJ (1979) Chloroplast and nuclear migration. In: Haupt W, Feinleib ME (eds) Encyclopedia of plant physiology, New Ser, vol VII. Springer, Berlin Heidelberg New York, pp 170–198
3. Cheeseman JM, Lafayette P, Gronewald J, Hanson J (1980) Effect of ATPase inhibitors on cell potential and K^+ influx in corn roots. Plant Physiol 65:1139–1145
4. Dörffling K (1980) Abscisic acid. Prog Bot 42:118–125
5. Haupt W (1982) Light-mediated movement of chloroplasts. Annu Rev Plant Physiol 33:205–233
6. Seitz K (1979) Light induced changes in the centrifugability of chloroplasts: Different action spectra and different influence of inhibitors in the low and high intensity range. Z Pflanzenphysiol 95:1–12
7. Seitz K (1980) Light induced changes in the centrifugability of chloroplasts mediated by an irradiance dependent interaction of respiratory and photosynthetic processes. In: Senger H (ed) The blue light syndrome. Springer, Berlin Heidelberg New York, pp 637–642
8. Seitz K (1982) Chloroplast motion in response to light in aquatic vascular plants. In: Symoens JJ, Hooper S, Compere P (eds) Studies on aquatic vascular plants. R Bot Soc Belgium, Brussels, pp 89–101
9. Shimmen T, Tazawa M (1982) Effect of intracellular Vanadate on electrogenesis, excitability and cytoplasmic streaming in Nitellopsis obtusa. Plant Cell Physiol 23:669–677
10. Simons TJR (1979) Vanadate, a new tool for biologists. Nature (London) 281:337–338
11. Zurzycki J (1980) Blue light-induced intracellular movements. In: Senger H (ed) The blue light syndrome. Springer, Berlin Heidelberg New York, pp 50–68

Phototaxis and Photomorphogenesis in *Physarum polycephalum* Plasmodia

TH. SCHRECKENBACH[1]

Introduction

The true slime mold (myxomycete) *Physarum polycephalum* is well established as a model organism for studies on cell motility, cell cycle regulation and differentiation. The giant multinucleated plasmodia of this species in addition represent an attractive photobiological model system. Visible light has been shown to affect most functions of this organism: it inhibits growth [12] and induces a light avoidance response in the moving plasmodium [5, 26, 33]. Light also controls the differentiation of plasmodia into resting structures; it induces the formation of fruiting bodies (sporulation) [9] and inhibits the alternative pathway of sclerotization [17, 31]. In all of these photoresponses blue light seems to be the dominant active component; in some cases red light has been reported to exert some, although considerably weaker, activity.

All photoresponses observed in *Physarum* plasmodia appear to help the organism to find an optimal environment for growth and reproduction. The natural habitats of myxomycete plasmodia are dark and humid spots where the organism can feed on rotton organic material. Both the photoinhibition of growth and the light avoidance response result in a preferential orientation of the cell toward these nutritionally advantageous places. Starving plasmodia which are not yet competent for the photoinduction of sporulation do not sclerotize in the light, i.e., in an environment unfavorable for the plasmodium which germinates from the dormant structure. Instead, the starving cell gets an additional chance to migrate toward the dark by phototactic orientation. The formation of fruiting bodies, which is induced by light, will preferentially occur at places which are dry and most suitable for spore dispersal.

The *Physarum* system offers various advantages for studies dealing with the mechanism of signal transduction and expression during phototaxis and photomorphogenesis: the mold can be grown axenically in large amounts using either microplasmodial suspension culture or giant macroplasmodia on a solid support [8]. This facilitates all experiments which require the isolation of large amounts of any compound from homogeneous biological material. Several further advantages are offered specifically by the macroplasmodial system: (1) studies on signal transduction during phototaxis can be carried out not only in one individual giant cell but also in systems of lower complexity, namely in isolated plasmodial strands [36] (2) the size of the organism allows

1 Max-Planck-Institut für Biochemie, 8033 Martinsried, FRG

Blue Light Effects in Biological Systems
ed. by H. Senger

the application of microinjection techniques to study the function of signal factors during phototransduction [38] and to investigate regulatory mechanisms of gene expression [1] (3) the development of fruiting bodies proceeds with a high degree of synchrony and thus allows the analysis of a temporal program of gene expression (4) biochemical events during photoinduced fruiting body formation can be correlated with the highly synchronous progression of the mitotic cycle in the nuclei – of which there are some 10^8 – in an indivudual plasmodium.

In the following, a brief summary of the photoresponses in *Physarum* is given. The literature reviewed mainly covers work published since 1979, since earlier publications on blue light effects in *Physarum* have been intensively discussed by Rakoczy [26]. Here, it is not intended to give a detailed review on all light responses in *Physarum* reported during these last few years, but to focus on those topics where we have gained a better understanding of regulatory mechanisms often due to converging results from different laboratories. Any reader who is interested in learning more about specific aspects of the system is referred to one of the detailed review articles that have recently been published [3, 10, 27].

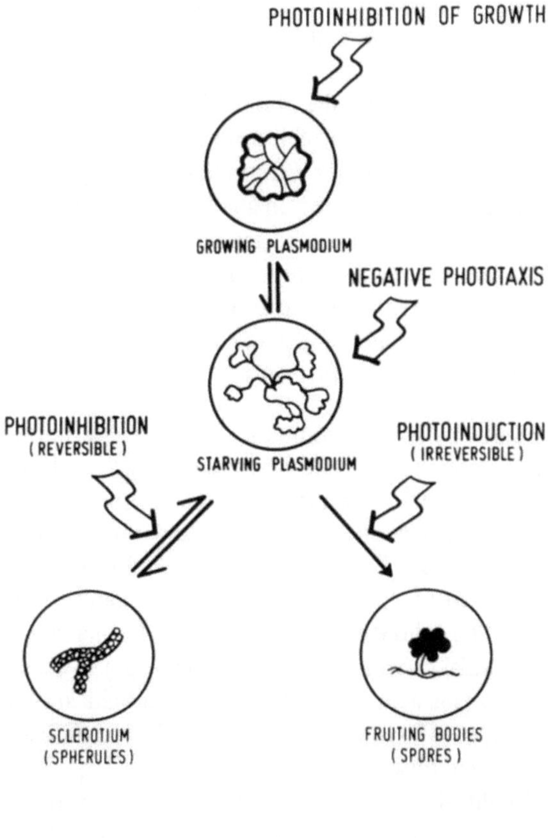

Fig. 1. The influence of light on growth, migration and differentiation of *Physarum* plasmodia

The Phototactic Response

When a plasmodium is transferred from growth medium to nonnutrient agar it converts into a highly motile network of strands (Fig. 1, Fig. 2A,B). Migration of the starving plasmodium is based on the vectorial translocation of the cellular material by the cytoplasmic shuttle streaming which is caused by rhythmic contractions of a complex fibrillar system [11, 20]. Observation of a moving plasmodium under the light microscope reveals the polarization of the cell into highly organized contracting veins and relaxed fronts into which the cytoplasm is protruded with each cycle of contraction (Fig. 2A, B). The involvement of actomyosin in these processes is well established [2, 15] and the specific organization of actomyosin-containing fibrils in specialized areas of a migrating plasmodium has been demonstrated using immunefluorescence techniques [21, 22].

It has been known for a long time that plasmodia of the related species *P. nudum* avoid a field illuminated from above and migrate into the dark [26]. The action spectrum of this response revealed the involvment of a typical blue light receptor absorbing around 375 and 450 nm [5]. The same type of light avoidance response was also observed in plasmodia of a white mutant strain of *P. polycephalum* [33].

In a more recent study the existence of true negative phototactic responses in *P. polycephalum* was demonstrated [14]: migrating plasmodia reacted to low intensity illumination with lateral light with a positive response and to illumination with light of higher intensity ($\geqslant 5$ W cm^{-2}) with a negative one. The negative phototactic response was analyzed in more detail: the authors showed that the plasmodia reacted by either performing a U-turn or by a direct reversal, the mode of reaction depending on the angle of the incident light. The fluence response curves were measured using both a yellow and a white mutant strain and clearly demonstrated an about fivefold increased sensitivity of the white strain. The action spectra calculated from the data exhibited four main peaks between 300 and 500 nm and showed only a very weak response in the red part of the spectrum (Fig. 3). No differences in the action spectra of the pigmented and the white strain were observed. These results are interpreted as being strong evidence against the involvement of the yellow pigments of wild type strains in the photoreception process.

Analysis of the phototactic responses in living plasmodia strongly suggests that blue light induces a repolarisation of the contractile system by converting the relaxed front of the cell into a pressure-generating end. This interpretation is supported by recent studies of Wohlfarth-Bottermann and coworkers who analyzed the effect of blue light on the contraction activity of excised palsmodial strands [4, 35, 36]. The data obtained with this model are illustrated by the scheme in Fig. 4 and can be summarized as follows: (1) illumination of the whole specimen results in an increase in period duration of the longitudinal and radial contraction activity; the response occurs after a lag phase of 2–3 min and is of a transient nature (2) only blue light around 496 nm was found to be active, red light (650 and 720 nm) showed no significant effect. This finding is in contrast to earlier observations of Hato et al. [16]. These authors reported that red and far red light also had an effect on the motive force of isolated strands analyzed by the double chamber method; (3) if only part of the

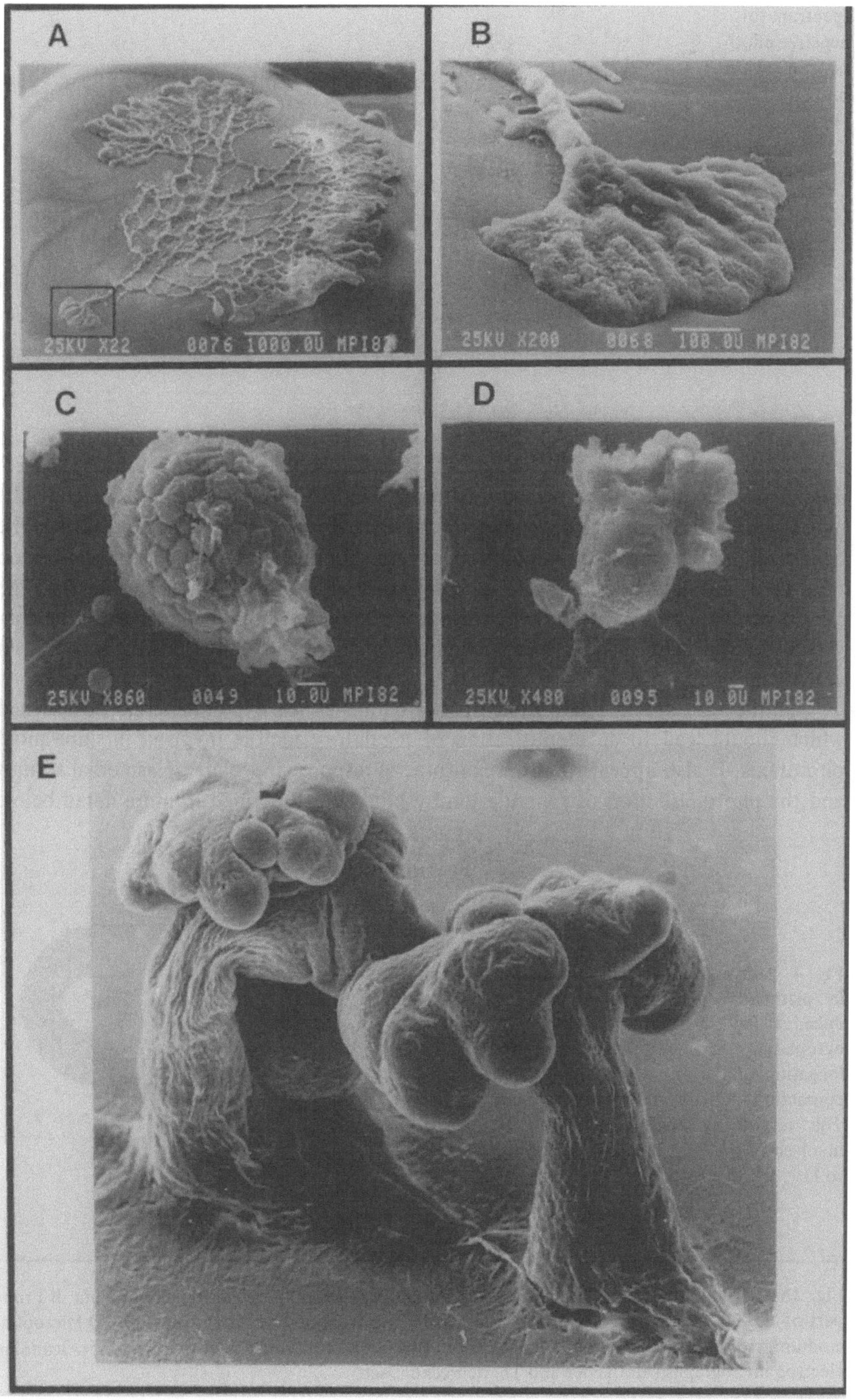
A
25KV X22 0076 1000.0U MPI82
B
25KV X200 0068 100.0U MPI82
C
25KV X860 0049 10.0U MPI82
D
25KV X480 0095 10.0U MPI82
E

Fig. 3. Action spectrum for negative phototaxis in a white mutant strain [14]

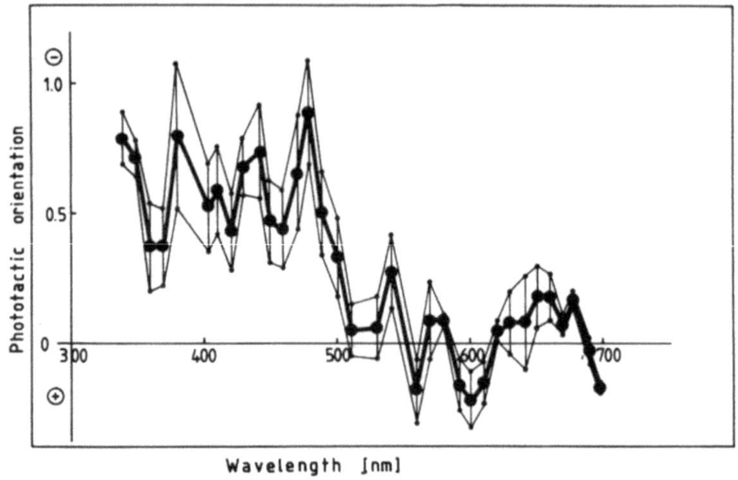

specimen is illuminated the response is transmitted to the nonilluminated area by the cytoplasmic stream, its size depending on the distance from the illuminated site (4) the light response is suppressed by inhibitors of respiration, and stimulated by the addition of α-keto glutarate and pyruvate as respiratory substrates [19] suggesting that the α-keto glutarate and pyruvate dehydrogenase complexes might be involved in the conversion of the light signal into a metabolic signal which finally controls actomyosin activity.

On the basis of these data straightforward biochemical experiments can be designed which should lead to an identification of regulatory factors involved in slime mold phototaxis. It also appears that the phototactic response in isolated plasmodial strands and the photoinhibition of spherulation by blue light described in more detail below

Fig. 4. Pathway of sensory transduction in plasmodial phototaxis. A photoinduced signal X is transported by the cytoplasmic flow and induces a prolongation of the radial and longitudinal contaction activity in isolated strands. The respiratory chain appears to be involved in the light response, according to [19, 36]

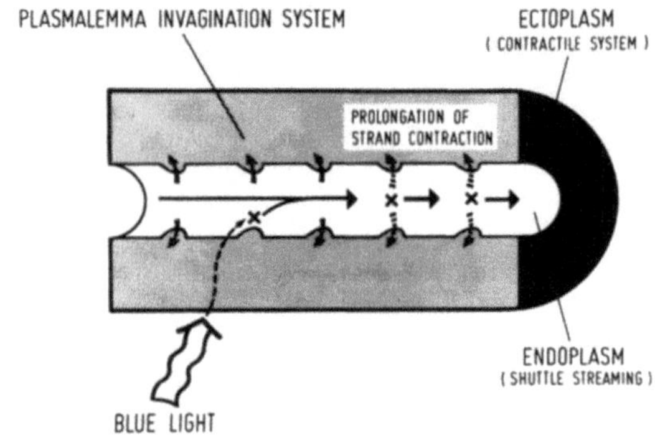

Fig. 2A–E. Stages in the life cycle of *Physarum polycephalum*. **A** Migrating plasmodium; **B** Front part of the plasmodium in (**A**); **C** Spherule cluster representing one microsclerotium; **D** Microplasmodium with photoinduced pseudopodia. **E** Fruiting bodies containing microscopic spores. Scanning electron micrographs (M. Claviez and Th. Schreckenbach)

might share common primary reactions. Biochemical studies on the nature of these processes are certainly facilitated by the availability of systems with different degrees of complexity: for example, intact plasmodia as a source of light activated extracts and excized plasmodial strands, and also isolated complexes containing constituents of the contractile apparatus as assay systems.

Photomorphogenesis

A plasmodium which has been starved for several days becomes competent for the photoinduction of sporulation. Under laboratory conditions this differentiation process is initiated by a 1–3 h pulse of white light and proceeds in an irreversible manner after a certain critical time point of development [28]. Sporulation in *Physarum* is accompanied by a meiosis and completes the life cycle of this organism: the haploid amoebae which eventually germinate from the mature spores will form new plasmodia by sexual fusion. Under conditions which do not allow the formation of fruiting bodies the plasmodium will not die but differentiate into a different type of dormant structure termed sclerotium. This encysted material consists of clusters of multinucleated units called spherules. Upon refeeding, plasmodia will germinate from these spherules and will fuse with each other, thus forming one macroscopic cell. Whereas sporulation does not occur in microplasmodial suspensions, the process of spherulation is more easily studied with this cell type because here the differentiation is completed within about 30 h after transfer to starvation medium, is more easily accessible to biochemical analysis, and is easily monitored under the light microscope. Spherulation in microplasmodia is reversibly inhibited by continuous illumination with blue light [31]. The photoregulation of this process thus appears to be completely different from the photoinduced development of fruiting bodies which proceeds irreversibly after induction by a pulse of light. Observations of Hosoda indicate that the photoinhibition of spherulation in microplasmodia is not an artificial effect but also occurs in macroplasmodia which are illuminated in a state which does not allow fructification [17].

Recent studies described below have presented experimental evidence that during both photomorphogenetic processes in *Physarum* the expression of prominent plasmodial and differentiation-specific proteins is regulated at the mRNA-level and that both systems are suitable for further analysis of the mechanism of gene expression using cloned developmentally regulated genes as specific hybridization probes.

Photoinhibition of Spherulation in Microplasmodia

About 15 h after the transfer of a microplasmodial suspension to starvation medium the formation of morphologically distinct spherule clusters (Fig. 2C) is observed and the transition is completed after about 30 h [31, 33]. Spherulation is inhibited by low concentrations (3 μg ml^{-1}) of the RNA polymerase B-inhibitor α-amanitin; the inhibitory effect is observed only during the first 10 h of starvation, possibly indicating that

Fig. 5. Sclerotization (spherulation) of microplasmodia and its inhibition by blue light [31, 33]

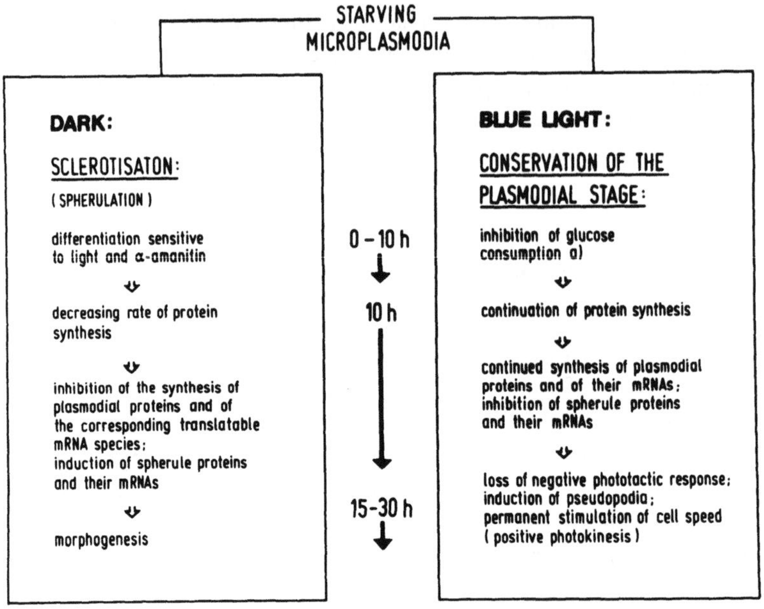

differentiation-specific genes are transcribed during this period ([33], also see summarizing scheme in Fig. 5). Spherulation is characterized by a decreasing rate of protein synthesis and by the appearance of spherule-specific proteins as well as their corresponding mRNA's, as has been shown by in vivo labeling of plasmodial protein and translation in vitro of extracted mRNA [24]. Some of the proteins synthesized in vivo and in vitro were identified as cell wall proteins extractable from the mature spherule (Fig. 6B). A second characteristic feature of the biosynthetic protein profiles during spherulation is the decrease in the synthesis of various plasmodial proteins and their translatable mRNAs, the most prominent of these being plasmodial actin (Fig. 6B).

The effect of blue light on the gene expression of starving microplasmodia is in all respects antagonistic to the starvation stimulus. Under illumination the rate of protein synthesis is maintained at the original level, the formation of spherule proteins and their mRNAs is permanently inhibited and the continued expression of plasmodial proteins, e.g., plasmodial actin, and their mRNA's is allowed (Fig. 6A).

These results clearly demonstrate that continuous illumination of microplasmodial suspensions with blue light conserves the plasmodial state by regulating the formation of distinct stage-specific translatable mRNA-species. The analogy between the effects of light and α-amanitin points to the possibility that blue light might not act at the level of mRNA-processing or transport, but rather might exert a transcriptional control as has been postulated, but not proven, for many blue light-regulated systems [13]. In the *Physarum* system it should be possible to check for such a transcriptional control of blue light by using cloned stage-specific cDNA sequences as probes for molecular hybridization with cellular RNA.

Several lines of evidence indicate that the photoinhibition of spherulation and plasmodial phototaxis might share common reaction pathways: (1) illuminated plasmodia lose the light avoidance response normally observed in starving cells but are

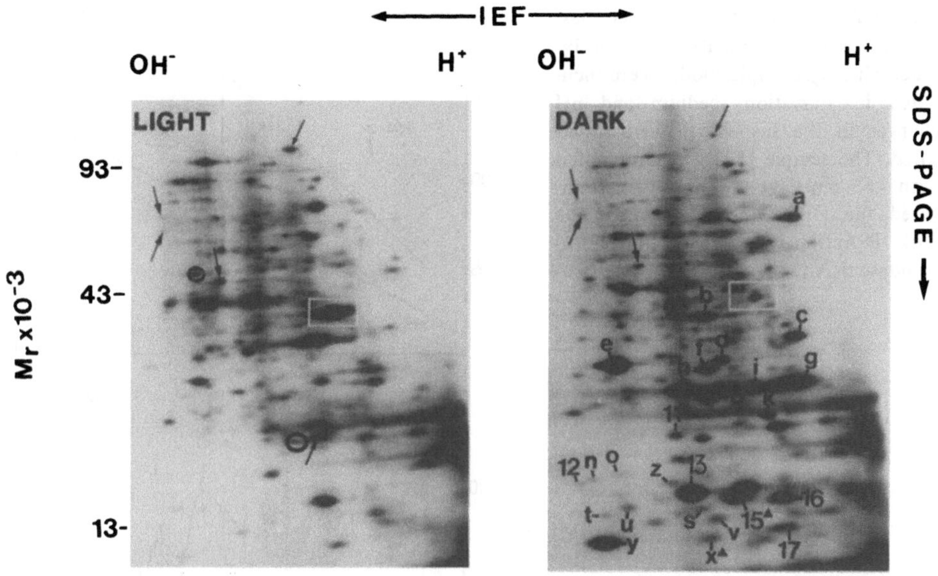

Fig. 6. Blue light inhibits spherulation at the mRNA level. Total RNA was extracted from illuminated microplasmodia and from dark incubated control plasmodia after 25 h of incubation in starvation medium (0% spherulation in the sample, 80% in the dark control). The RNA was translated in vitro and the products were analyzed by two-dimensional electrophoresis. Spherule-specific proteins are marked by letters or numbers, the *numbers* indicating the identity with the corresponding in vivo labeled material. Cell wall proteins are marked by a *triangle,* starvation proteins, present in both samples, by an *arrow* and light-induced proteins by a *circle.* Plasmodial actin is enclosed by a *rectangle* [24]

characterized by a twofold stimulation of their migration rate [33]. Certainly, this positive photokinetic effect is based on a permanent stimulation of the contractile system during the preillumination period and can be demonstrated when the illuminated microplasmodia are converted into macroplasmodia by fusion and subjected to analysis of their motile responses [33]. The stimulation of plasmodial motility is manifested by the induction of extensive pseudopodia in the microplasmodial state ([31], see Fig. 2D). (2) Very probably, the effect of blue light on the contraction activity in isolated plasmodial strands is related to the function of the respiratory chain (see above and [19]). Also in starving microplasmodia a relationship between blue light function and energy metabolism has been observed: the conversion of glucose into trehalose, a high molecular weight polysaccharide and into an acid which is secreted into the medium, is inhibited by illumination ([31], also see Fig. 7a). In contrast to earlier reports of Daniel [7] an influence of light on the respiration of intact plasmodia or isolated mitochondria was not observed [30] but should be reinvestigated using the inhibitors and respiratory substrates as in the study of Korohoda et al. [19]. (3) The photoinhibition of acid production and secretion in starving microplasmodia occurs with a lag phase of 2–3 min, i.e., the lag phase observed in the contractile response of isolated plasmodial strands [19]. Furthermore, the action spectrum of this pH response (Fig. 7B) in a white mutant strain reveals the existence of a blue light receptor similar to the one which mediates the light avoidance response in *P. nudum* [5].

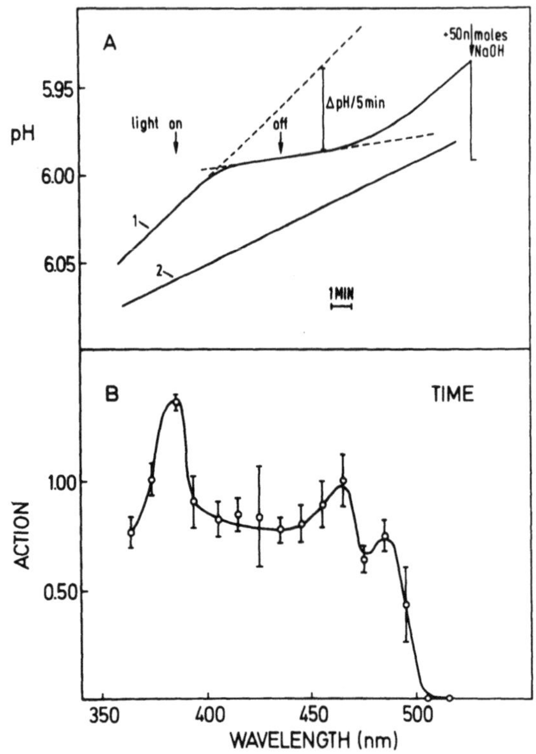

Fig. 7A,B. Acid secretion in a microplasmodial suspension and its inhibition by blue light. **A** Microplasmodia were incubated in starvation medium and pH changes in the suspension were monitored. The sample [12] was illuminated with 465 nm light (24 nE s^{-1}) as indicated, the control [26] was kept in the dark. **B** Action spectrum of the photoresponse in **A** [31]

In summary, further investigation of the blue light effects in microplasmodial suspensions appears worthwhile because this system might lead to a better understanding of primary photometabolic responses and of the regulatory influence of blue light on gene expression.

Photoinduction of Sporulation

Light – according to the present state of knowledge – is the only physiological signal which induces fruiting body formation in *Physarum.* Photoinduction is achieved only when the plasmodium has reached a competent state after a 4-day period of starvation on a balanced salt solution containing niacin [9]. Competence appears to be promoted by yet unknown factors which accumulate in the medium during starvation [34] and possibly depends on a minimal number of mitotic divisions during the starvation phase [6]. The morphological and cytological aspects of fruiting body morphogenesis which is initiated by a 1–4 h pulse of light have been intensively studied by several laboratories (reviews: [3, 10, 27]). The time course of fruiting body formation is schematically outlined in Fig. 8 and is briefly characterized by distinct morphological stages: the plasmodial strand is first cleaved into nodular structures about 10 h after induction; the cytoplasm is then protruded into the vertical direction and pillars arise from the nodules. Morphogenesis is completed by the formation of stalks consisting of dead

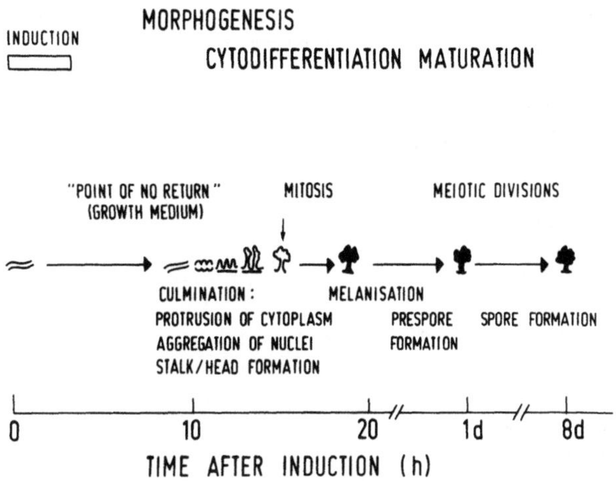

Fig. 8. Time course of fruiting body formation in *Physarum polycephalum.* The "point of no return" marks the time point beyond which growth medium does not prevent sporulation [28]. For details see text and [3, 10, 27]. *Arrow* indicates the morphogenetic stage from which RNA was extracted for the experiment in Fig. 9

fibrillar material and of heads pigmented by melanin and containing the prespore particles. A presporangial mitosis and meiotic divisions accompany maturation of the spore material.

The photoreceptor which mediates this response is not known so far. In the related species *P. nudum,* blue and red light is able to induce sporulation as has been shown in earlier studies of Rakoczy [26]. The shape of the fluence-response curves in this system point toward the existence of two photoreceptors rather than one pigment molecule. Preliminary studies indicate that in *P. polycephalum,* too, both blue and red light are active ([31] Bialczyk and Schreckenbach unpublished).

Wormington and Weaver [38] have reported on the induction of sporulation by the injection of an extracted yellow plasmodial pigment fraction which had been illuminated in vitro. The involvement of this pigment in the physiological induction pathway, however, remains uncertain because (1) white mutant strains exhibit the same, if not a greater photosensitivity as yellow pigmented wild-type strains (Bialczyk and Schreckenbach unpublished) and (2) sporulation is also unspecifically induced by the injection of salts [37]. It therefore appears to be necessary to purify the extracted yellow pigment to homogeneity and analyze its chemical nature in order to clarify its function in slime mold photomorphogenesis.

Photoinduced fruiting body formation in *Physarum* represents an ideal system to study the regulation of gene expression during photomorphogenesis. In previous studies evidence has been presented that sporulation depends on protein and RNA synthesis [28, 29] and that new spore-specific proteins can be identified by pulse-labeling of plasmodial protein during morphogenesis [18]. The complex mixture of all developmentally regulated proteins has, however, been analyzed only by one-dimensional gel electrophoresis which does not allow an unambiguous identification of any of the components. In addition, nothing at all is known so far about the mechanism by which the expression of any of these proteins is regulated.

In a recent study [32] a strategy has been presented which might help to solve some of these problems: pulse-labeling of plasmodial protein during fruiting body formation revealed that the different developmental stages are characterized by distinct biosynthetic protein profiles and by specific mRNA patterns as revealed by translation

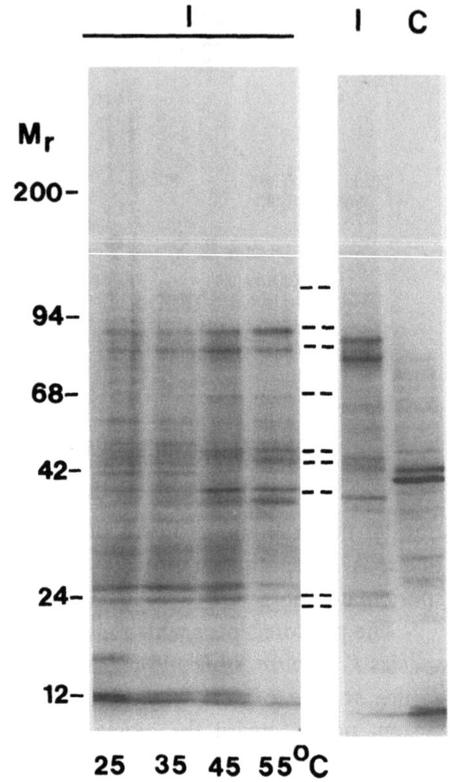

Fig. 9. Thermal elution of developmentally regulated spore-specific mRNA species from poly-U sepharose. Sporulation in plasmodia was induced by a 3 h pulse of white light (*I* induced sample; *C* dark control). Total RNA was extracted at 15 h (the morphogenetic stage is marked by the arrow in Fig. 9), translated in vitro and the products were separated by SDS-polyacrylamide gel electrophoresis (5–20% gel, panels on the *right*). RNA from the sample was bound to poly-U sepharose and eluted using a temperature gradient as described in [23]. Separation of the in vitro translation products of the different fractions is shown on the *left*. The experiment demonstrates that new translatable mRNA species are induced during fruiting body development and that these mRNA molecules have longer poly-A tails than most of the plasmodium-specific sequences. (After H. Putzer and Th. Schreckenbach unpublished)

in vitro of extracted mRNA. An example is given in Fig. 9 which demonstrates that various new translatable mRNA species are induced at 15 h of development and that these new mRNA molecules bind more tightly to poly-U sepharose than plasmodial mRNA indicating that they have longer poly-A tails [23]. Studies in progress deal with the identification of these proteins which apparently are regulated at the mRNA level, and with the construction of a cDNA bank containing photoinduced genes. With the help of these probes it should be possible to check the assumption that pulse illumination of *Physarum* plasmodia by blue or red light activates spore-specific genes and to analyze in more detail the expression in vivo and in vitro of these genes, also using photomorphogenesis mutants.

Conclusions

In spite of considerable efforts, the central questions concerning blue light-mediated responses in microorganisms and plants have been left unanswered so far: (1) What is the chemical nature of the photoreceptor and the mechanism by which this pigment converts the light signal into a chemical signal? (2) Which cytoplasmic factors, if any, participate in the transduction of the light signal to the reactive site? and (3) On which level and by which mechanism is the cellular response regulated? This problem is of special complexity and importance in the case of photomorphogenetic reactions. The nature of the photoreceptor pigment has been elucidated unabiguously only as far as phytochrome, the red light receptor from higher plants and algae is concerned [25]. The latter two questions, however, appear to represent the main unresolved problems of photomorphogenesis in general. Here, the *Physarum* system seems to offer a variety of advantages which make it most suitable for biochemical and molecular biological studies on the mechanism of light signal conversion and expression. In contrast to many plant systems the organism is available as homogeneous biological material in nearly unlimited amounts; it can be handled either in the form of macroscopic giant cells or in suspensions of microscopic cells. Plasmodial motility and actomyosin regulation can be studied using systems of different degrees of complexity and the techniques of mRNA extraction, translation in vitro and transcription analysis in isolated nuclei can be successfully applied. On the basis of the results obtained in different laboratories within the last few years, strategies can be deduced which hopefully lead to an understanding of the photoregulation of cell motility and of the differential expression of distinct developmentally regulated proteins during slime mold photomorphogenesis.

Acknowledgment. The author's work was supported by the Deutsche Forschungsgemeinschaft.

References

1. Adams DS, Noonan D, Jeffery WR (1981) Proc Natl Acad Sci USA 78:83–87
2. Adelman MR, Taylor EW (1969) Biochemistry 12:4964–4975
3. Aldrich HC, Daniel JW (eds) (1982) Cell biology of Physarum and Didymium, vol I, II. Academic Press, London New York
4. Baranowski Z, Shraideh Z, Wohlfarth-Bottermann KE (1982) Cell Biol Int Rep 6:859–865
5. Bialczyk J (1979) Photochem Photobiol 30:301–303
6. Chapman A, Coote JG (1982) J Gen Microbiol 128:1489–1501
7. Daniel JW (1966) In: Cameron IL, Padilla GM (eds) Cell synchrony. Academic Press, London New York, pp 117–152
8. Daniel JW, Baldwin H (1964) In: Prescott DM (ed) Methods in cell physiology, vol I. Academic Press, London New York, pp 9–41
9. Daniel JW, Rusch HP (1962) J Bacteriol 83:234–240
10. Dove WF, Rusch HP (eds) (1980) Growth and differentation in Physarum polycephalum. University Press, Princeton
11. Fleischer M, Wohlfarth-Bottermann KE (1975) Cytobiologie 10:339–365
12. Gray WD (1938) Am J Bot 25:511–522
13. Gressel J (1980) In: Senger H (ed) The blue light syndrome. Springer, Berlin Heidelberg New York, pp 133–153
14. Häder D-P, Schreckenbach Th (1984) Plant Cell Physiol 25:55–61
15. Hatano S, Oosawa F (1966) Biochim Biophys Acta 127:488–498
16. Hato M, Ueda T, Kurihara K, Kobatake Y (1976) Cell Struct Funct 1:269–278
17. Hosoda E (1981) Mycologia 73:689–696
18. Jockusch BM, Sauer HW, Brown DF, Babcock KL, Rusch HP (1970) J Bacteriol 103:356–363
19. Korohoda W, Shraideh Z, Baranowski Z, Wohlfarth-Bottermann KE (1983) Planta 158:54–62
20. Nagai R, Yoshimoto Y, Kamiya N (1978) J Cell Sci 33:205–225

21. Naib-Majani W, Stockem W, Wohlfarth-Bottermann KE, Osborn M, Weber K (1982) Eur J Cell Biol 28:103–114
22. Osborn M, Weber K, Naib-Majani W, Hinssen H, Stockem W, Wohlfarth-Bottermann KE (1983) 29:179–186
23. Palatnik CM, Storti RV, Jacobson A (1979) J Mol Biol 128:371–395
24. Putzer H, Werenskiold K, Verfuerth C, Schreckenbach Th (1983) EMBO J 2:261–267
25. Pratt LH (1979) Photochem Photobiol Rev 4:59–124
26. Rakoczy L (1980) In: Senger H (ed) The blue light syndrome. Springer, Berlin Heidelberg New York, pp 570–583
27. Sauer HW (ed) (1982) Developmental biology of Physarum. Cambridge Univ Press, Cambridge
28. Sauer HW, Babcock KL, Rusch HP (1969) Exp Cell Res 57:319–327
29. Sauer HW, Babcock KL, Rusch HP (1969) Biochim Biophys Acta 195:410–421
30. Schreckenbach Th (1981) In: Rakoczy L (ed) Biology of Physarum. Proc Vth Physarum Meet. Jaciellonian Univ Press, Cracow, pp 24–33
31. Schreckenbach Th, Walckhof B, Verfuerth C (1980) Proc Natl Acad Sci USA 78:1009–1013
32. Putzer H, Werenskiold A-K, Verfürth C , Claviez M, Schreckenbach Th (1984) Submitted
33. Schreckenbach Th, Verfuerth C (1982) Eur J Cell Biol 28:12–19
34. Wilkins AS, Reynolds G (1979) Dev Biol 72:175–181
35. Wohlfarth-Bottermann KE, Block I (1981) Cell Biol Int Rep 5:365–373
36. Wohlfarth-Bottermann KE, Block I (1982) Cold Spring Harbour Symp Quant Biol 46:563–568
37. Wormington WM, Cho CG, Weaver RF (1975) Nature (London) 256:413–414
38. Wormington WM, Weaver RF (1976) Proc Natl Acad Sci USA 73:3895–3899

Red Light-Induced Shift of the Fluence-Response Curve for First Positive Phototropic Curvature of Maize Coleoptiles *

E. SCHÄFER,[2] M. IINO,[1] and W. R. BRIGGS[1]

Introduction

The fluence-response curve for first positive phototropic curvature of dark-grown oat or maize coleoptiles is shifted to higher fluences when the seedlings are exposed to red light (R) prior to phototropic induction with blue light (B) (review: [3]). In the work with maize coleoptiles, it has been shown that R given immediately before the phototropic induction is ineffective, and periods over 1 h are needed for full induction of the shift effect [1]. In the same report, it has also been shown that only a very low fluence of R is needed for the saturation of the effect. The action spectrum for this shift effect is in agreement with the assumption that it is mediated by phytochrome [2]. On the other hand, the R induction was reversible by a subsequent far-red light (FR)pulse although the latter light pulse established more P_{fr} than the inductive light pulse. The result lead to the notion of a paradox. We have reinvestigated the problem of the shift effect with particular respect to relationships between R fluences and the response.

Materials and Methods

Presoaked caryopses of *Zea mays* L. were sown on moist paper in plastic boxes and kept for 2 days under R (0.15 μmol m^{-2} s^{-1}). The resulting seedlings were then selected for uniformity and transplanted into vermiculite-containing boxes. Following an incubation in the dark for a day, experimental preirradiation with R or FR (5 min or 30 s) and phototropic induction with B (30 s) were performed as indicated in each figure legend. The phototropic curvature of the coleoptiles was measured 100 min after the phototropic induction. For description of light sources and other experimental details see [4].

---

* C.I.W.-D.P.B. Publication No 834

1 Carnegie Institution of Washington, Department of Plant Biology, 290 Panama Street, Stanford, CA 94305, USA

2 Present Address: Institut für Biologie II, Universität Freiburg, Schänzlestr. 1, 7800 Freiburg, FRG

Blue Light Effects in Biological Systems
ed. by H. Senger

Results and Discussion

Recently, Iino et al. [4] have shown that R and B elicit positive phototropic curvature of maize mesocotyls, and in addition, negative curvature of the coleoptile. The latter response has been shown to be entirely the gravitropic compensation for the positive mesocotyl curvature. The mesocotyl curvature and the accompanying negative curvature can be eliminated by exposing the seedlings from above to 5 min R (2.7×10^3 μmol m^{-2}) just before phototropic induction. The B-induced positive coleoptile curvature is retained in the R-pretreated plants, and the fluence-response curve for this response (duration of B irradiation is kept at 30 s) is that characteristic of the first positive coleoptile curvature, i.e., a fluence-response curve composed of ascending and descending arms (Fig. 1).

When the 5 min irradiation with R was given 2 h before phototropic induction with B, the fluence-response curve for the curvature response shifted nearly one order of magnitude to higher fluences, thus confirming the earlier findings (Fig. 1). The fluence-response curve for the R-induced shift effect was then obtained with a fixed B fluence of 0.1 μmol m^{-2}, which was chosen as inducing a curvature response in the ascending arm either of the shifted or the nonshifted phototropic fluence-response curve. In addition to the irradiation with R of various fluences given 2 h prior to the phototropic induction, seedlings were also exposed from above to 5 min R (a fixed fluence as above) just before the phototropic induction. This latter R treatment was given since the low R fluences used to test the shift effect may not be sufficient to eliminate the R-sensitive curvature responses which can also be induced by B [4]. As shown in Fig. 2 (upper part), the threshold and the saturation for the shift were found at R fluences of about $10^{-2.5}$ and $10^{0.5}$ μmol m^{-2}, respectively. The fluence response curve for the R-induced shift was also obtained with a fixed B fluence of 6.3 μmol m^{-2}, which induced a curvature response in the descending arm. The threshold and the saturation were found at fluences of 10^{-4} and 10^{-1} μmol m^{-2} (Fig. 2, lower part). This result is in good agreement with that reported by Briggs and Chon [1], who also used the descending part for the fluence-response study.

It therefore became evident that the R fluences needed for the shift are about 1.5 log units lower for the descending part than for the ascending part.

The shift effect for either of the B fluences could be induced by FR (Fig. 3). The result with FR is not inconsistent with the view that the response is mediated by phytochrome, since phytochrome phototransformation ($P_r \rightarrow P_{fr}$) induced by the low R-fluences needed to saturate the shift effects could also be induced by the FR. The difference in spectral responsivities for R and FR is also in agreement with that obtained for phytochrome-mediated growth responses in etiolated oat seedlings [5].

The different R fluence-response relationships in shifting the ascending arm and the descending arm of first positive curvature suggest a complexity in B-sensitive photosystems involved in this curvature response. A possible interpretation of the results is that the decrease in the curvature with increasing B fluences (descending arm) is caused by a different photosystem from that inducing the increase in response with increasing B fluences (ascending arm), and that R modulates these two possible photosystems differently.

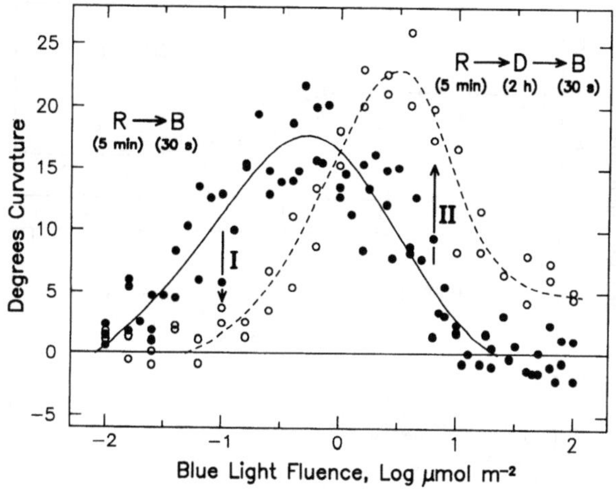

Fig. 1. Fluence-response curves for blue light-induced phototropic curvature of maize coleoptiles, which have been raised for 2 days in red light followed by 1 day in the dark. A 5-min red light pretreatment (2.7×10^3 μmol m^{-2}) was given either immediately before (*solid line*) or 2 h before the phototropic induction with a 30 s blue light pulse (*dashed line*)

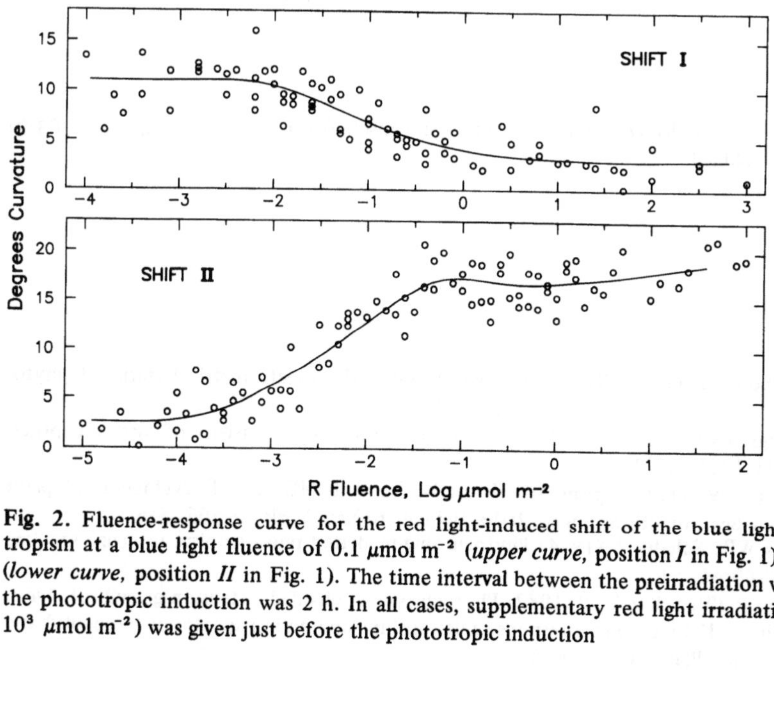

Fig. 2. Fluence-response curve for the red light-induced shift of the blue light-induced phototropism at a blue light fluence of 0.1 μmol m^{-2} (*upper curve*, position *I* in Fig. 1) or 6.3 μmol m^{-2} (*lower curve*, position *II* in Fig. 1). The time interval between the preirradiation with red light and the phototropic induction was 2 h. In all cases, supplementary red light irradiation (5 min, 2.7×10^3 μmol m^{-2}) was given just before the phototropic induction

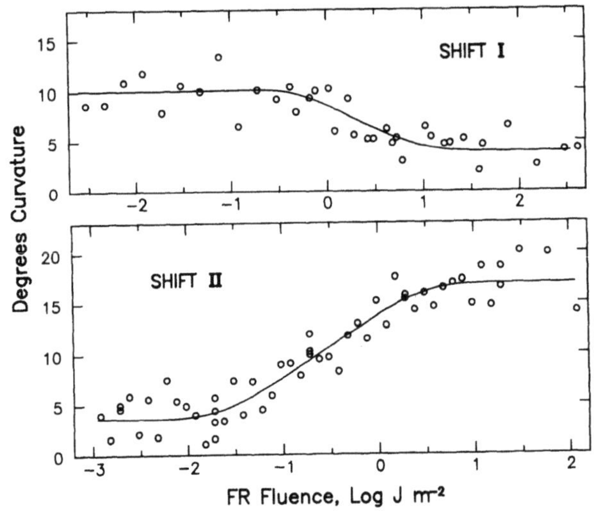

Fig. 3. Fluence-response curve for the far red (RG 9) light-induced shift of the blue light-induced phototropism at a blue light fluence of 0.1 μmol m^{-2} (*upper curve*, position *I* in Fig. 1) or 6.3 μmol m^{-2} (*lower curve*, position *II* in Fig. 1). Other details as in Fig. 2

Acknowledgment. Support by Deutsche Forschungsgemeinschaft (D 2, SFB 206 and Scha 303/2) is gratefully acknowledged.

References

1. Briggs WR, Chon HP (1966) The physiological versus the spectrometrical status of phytochrome in corn coleoptiles. Plant Physiol 41:1159–1166
2. Chon HP, Briggs WR (1966) Effect of red light on the phototropic sensivity of corn coleoptiles. Plant Physiol 41:1715–1724
3. Dennison DS (1979) Phototropism. In: Haupt W, Feinleib ME (eds) Encyclopedia of plant physiology, New Ser, vol VII. Springer, Berlin Heidelberg New York, pp 506–566
4. Iino M, Briggs WR, Schäfer E (1984) Phytochrome-mediated phototropism in maize seedling shoots. Planta 160:41–51
5. Schäfer E, Lassig TV, Schopfer P (1982) Phytochrome-controlled extension growth of *Avena sativa* L. seedlings. II Fluence-rate response relationships and action spectra of mesocotyl and coleoptile responses. Planta 154:231–240

The Possible Role of H^+ and Ca^{2+} in Photobehavior of *Blepharisma japonicum*

V. Passarelli[1], F. Lenci[1], G. Colombetti[1], E. Barone[2], and R. Nobili[2]

Introduction

Photobiological processes in the heterotrich ciliate *Blepharisma japonicum* have been extensively and intensively investigated (e.g., [6, 7] and reference therein); in particular, similarly to several other microorganisms (e.g., [9]), *Blepharisma* has been shown to be able to respond to environmental illumination conditions, clustering in shadowed regions [7].

In a series of preliminary experiments [2, 4] we have investigated light-induced motile reactions in *Blepharisma japonicum* and shown that this cell exhibits step-up photophobic responses (see also [8]; for terminology see [5]). Different from the related photoresponsive ciliate *Stentor coeruleus*, in *Blepharisma* we have not yet been able to detect, in the experimental conditions described below, any oriented photomovement with respect to the direction of the light stimulus (see also [8]).

Recently Kraml and Marwan [8] have reported an action spectrum for the step-up photophobic response which shows a single band centered at about 400 nm. This would exclude the hypericin-type blepharismin as the photoreceptor pigment for the light dependent behavior of *Blepharisma*. However, I R spectra suggest that this blue light pigment too is a hypericin-type molecule [8].

On the basis of what is known about photoreception and photosensory processes in the related protozoan *Stentor coeruleus*, a few working hypotheses can be put forward also about the photosensory transduction chain for *Blepharisma* photobehavior. The photoreceptor pigment for both photophobic responses and phototaxis of *Stentor* is the hypericin-like stentorin [13, 14], which, upon light absorption, would release protons to the cytoplasm [15]. The resulting transient ΔpH would lead to an influx of Ca ions, responsible for ciliary reversal and therefore for the photobehavioral reaction ([12] and reference therein; [3]). We have therefore performed a series of experiments in order to investigate the molecular basis of photosensory responses of *Blepharisma japonicum.*

The Ca-blocking agent Ruthenium Red (RuRed) [3, 12], and the protonophore carbonyl-cyanide-chlorophenyl-hydrazone (CCCP) have been used to study respectively the possible role of calcium ions and of vectorial light-induced proton fluxes in photoperception and phototransduction processes in *Blepharisma.*

1 CNR Istituto di Biofisica, Via S. Lorenzo, 26, 56100 Pisa, Italy
2 Istituto Zoologia Universita, Via Volta, 56100 Pisa, Italy

Blue Light Effects in Biological Systems
ed. by H. Senger

Materials and Methods

Blepharisma japonicum was grown in batch cultures in the dark at 18°C, fed on an *Enterobacter aerogenes* culture in a lettuce extract at pH 6.8 [11]. One-week-old cultures were washed by gentle centrifugation, resuspended in SMB (saline medium for *Blepharisma:* NaCl 1.5 mM; KCl 0.05 mM; $CaCl_2$ 0.4 mM; $MgSO_4$ 0.1 mM; sodium phosphate buffer 2.0 mM; pH 6.8), and kept 24 h in this solution before the experiments were carried on.

Cell motility and photobehavior were analyzed using an IR-sensitive videocamera coupled to a videorecorder and a TV monitor, as described elsewhere [1]. Step-up photophobic responses were elicited exposing the cell to a sudden increase of blue light intensity from 0.0 to 15.0 W m^{-2}. At this light intensity 100% control cells showed phobic reactions.

CCCP and RuRed were purchased from Sigma Chem. Co., and used without further purification. As CCCP was dissolved in ethanol, the control sample was, in this case, treated with the same amount of ethanol as the test sample.

Results

Upon a sudden increase in light intensity ($dI/dt > 0$), *Blepharisma japonicum* exhibits a typical step-up photophobic response: with a delay which usually depends on the intensity of the light stimulus, cells initially stop and then gyrate backward [3, 8].

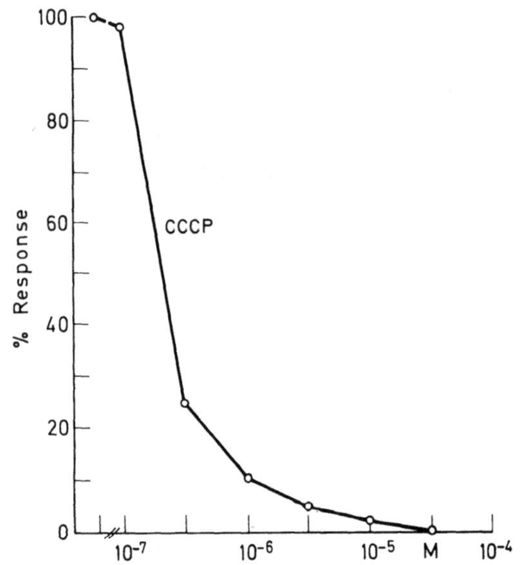

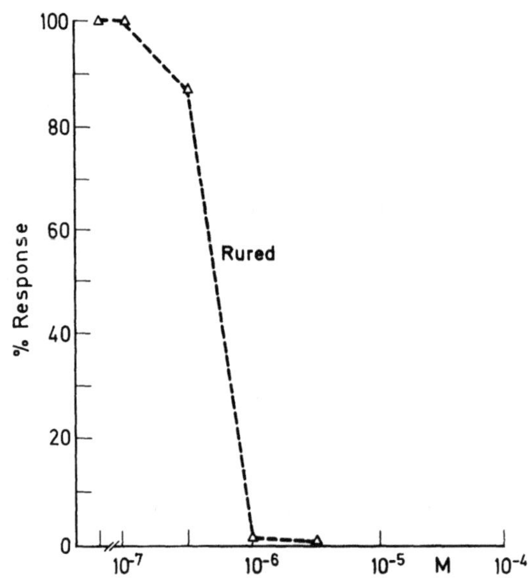

Fig. 1. Effect of the protonophore CCCP on step-up photophobic responses of *Blepharisma japonicum.* No effect of CCCP on cell motility was observed (see text)

Fig. 2. Effect of the Ca-blocking agent RuRed on step-up photophobic responses of *Blepharisma japonicum.* No effect of RuRed on cell motility was observed (see text)

As previously mentioned, at a blue light intensity of 15.0 W m^{-2}, 100% control cells show step-up photophobic responses. CCCP 10^{-7} M only slightly affects the percentage of reacting cells. Further increasing the protonophore concentration, the number of cells showing step-up photophobic response smoothly decreases to zero for CCCP 5×10^{-5} M (Fig. 1). At all CCCP concentrations used, the percentage of swimming cells and their trajectory pattern were the same as in the control sample.

As with CCCP, RuRed in the concentration range used does not significantly affect either motility or swimming patterns of *Blepharisma* with respect to the control. RuRed at 10^{-7} M is definitely ineffective, but at 10^{-6} M it completely inhibits step-up photophobic responses of *Blepharisma* (Fig. 2).

Discussion

The clear-cut effect of CCCP, a well-known protonophore, on the step-up photophobic response of *Blepharisma,* suggests the hypothesis that a light-induced proton gradient might be one of the primary molecular events following light perception in this cell. CCCP, disrupting the light-induced proton gradient, would thus inhibit the cell response to light.

The inhibitory effect of RuRed can be understood, assuming that it blocks Ca influx into the ciliary space responsible for ciliary reversal in control cells.

A general scheme that takes into account the previously mentioned results is reported in Fig. 3. According to this scheme, light would induce vectorial proton translocation across the cell membrane with the formation of a proton gradient. This would in turn cause, probably by membrane depolarization, an increase in membrane conductance to Ca channels, with subsequent Ca influx into the ciliary space and ciliary reversal. If this hypothesis holds true, both metabolic drugs, capable of interfering with proton gradients or Ca channels, should influence the photophobic behavior of *Blepharisma.*

References

1. Barghigiani C, Colombetti G, Franchini B, Lenci F (1979) Photobehavior of Euglena gracilis: Action spectrum for the step-down photophobic response of individual cells. Photochem Photobiol 29:1015–1019
2. Colombetti G, Lenci F, Passarelli V, Barone E, Nobili R, Santangelo G (1983) Photosensory transduction in Blepharisma japonicum. J Protozool in press
3. Colombetti G, Lenci F, Passarelli V, Nobili R, Santangelo G (1982) Risposte fotomotorie nel protozoo ciliato Blepharisma japonicum. Boll Zool 49:47
4. Colombetti G, Lenci F, Song PS (1982) Effects of K and Ca ions on motility and photosensory responses of Stentor coeruleus. Photochem Photobiol 36:609–611
5. Diehn B, Feinleib ME, Haupt W, Hildebrand E, Lenci F, Nultsch W (1977) Terminology of behavioral responses of motile microorganisms. Photochem Photobiol 26:559–560
6. Giese AC (1973) Blepharisma, the biology of a light-sensitive protozoan. Stanford Univ, Stanford

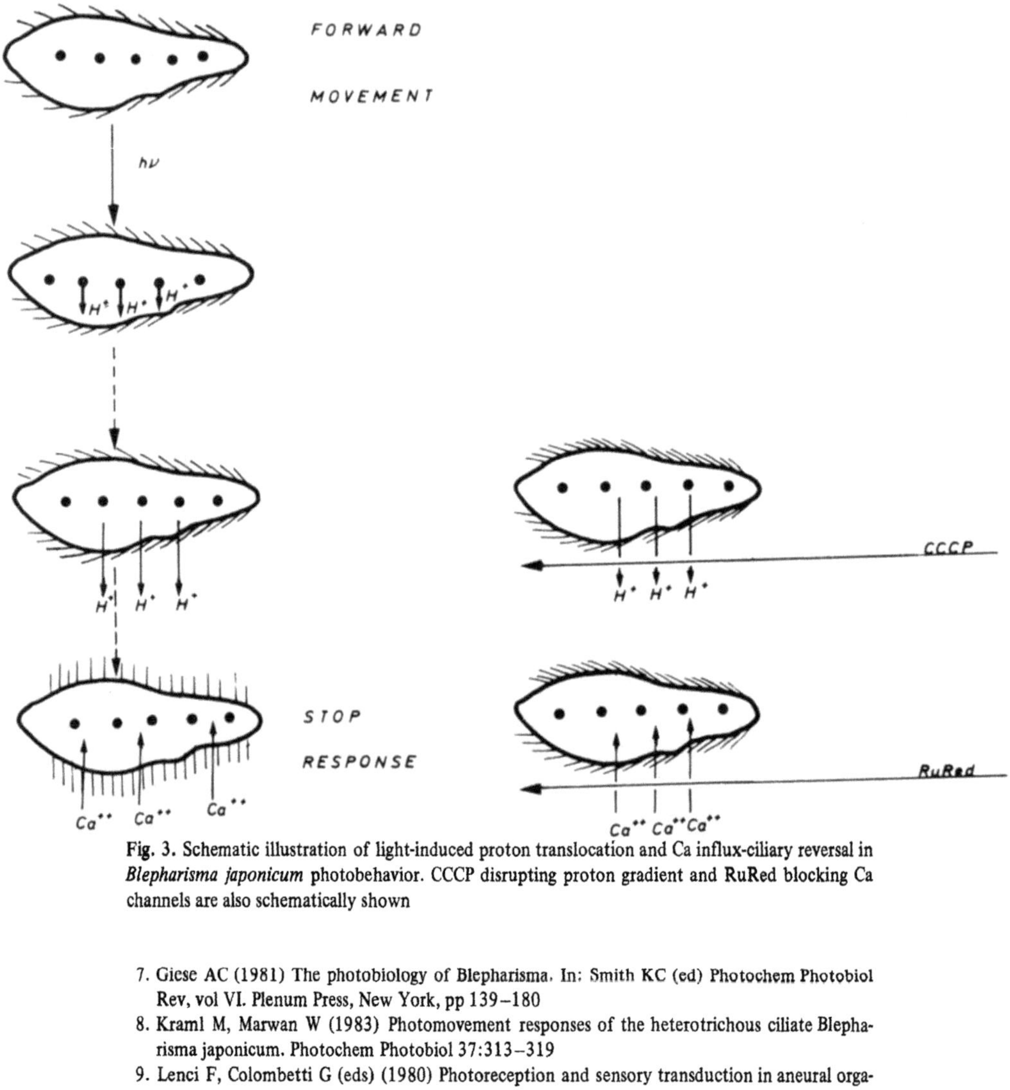

Fig. 3. Schematic illustration of light-induced proton translocation and Ca influx-ciliary reversal in *Blepharisma japonicum* photobehavior. CCCP disrupting proton gradient and RuRed blocking Ca channels are also schematically shown

7. Giese AC (1981) The photobiology of Blepharisma. In: Smith KC (ed) Photochem Photobiol Rev, vol VI. Plenum Press, New York, pp 139–180
8. Kraml M, Marwan W (1983) Photomovement responses of the heterotrichous ciliate Blepharisma japonicum. Photochem Photobiol 37:313–319
9. Lenci F, Colombetti G (eds) (1980) Photoreception and sensory transduction in aneural organisms. Plenum Press, New York
10. Matsuoka T (1983) Distribution of photoreceptors inducing ciliary reversal and swimming acceleration in Blepharisma japonicum. J Exp Zool 225:337–340
11. Miyake A (1975) Control factor of nuclear cycles in ciliate conjugation: Cell-to-cell transfer in multicellular complexes. Science 189:53–55
12. Song PS (1981) Photosensory transduction in Stentor coeruleus and related organisms. Biochim Biophys Acta 639:1–29
13. Song PS, Häder DP, Poff KL (1980) Step-up photophobic response in the ciliate Stentor coeruleus. Arch Microbiol 126:181–186
14. Song PS, Häder DP, Poff KL (1980) Phototactic orientation by the ciliate Stentor coeruleus. Photochem Photobiol 32:781–786
15. Walker EB, Yoon M, Song PS (1981) The pH dependence of photosensory responses in Stentor coeruleus and model system. Biochim Biophys Acta 634:289–308

Blue Light and Stomatal Function

E. ZEIGER[1]

Introduction

Stomata provide plants with a dominant means to adjust to environmental changes [2]. Through their control of pore apertures, stomata maximize plant homeostasis (biological organization) by modulating the extent of physical exchange between the plant and its milieu [32]. Typically, stomata are largely closed at night, when biological activity and exchange with the environment is minimal, and are wide open during most of the day, when the need for gaseous exchange is large. Because of the marked daily changes in light intensity, stomata are remarkably sensitive to photon fluxes and light quality.

In the intact leaf, direct light responses of stomata have been difficult to separate from those of the photosynthesizing mesophyll. Specific light responses of guard cells, however, have been unambiguously demonstrated and have been shown to depend on the activity of guard cell chloroplasts and a separate, blue light-dependent photosystem [23, 32]. This chapter will review available evidence on the phenomenology of the blue light response of stomata and its photobiological properties, and discuss some hypotheses on its mode of action and functional role in stomatal movements.

Differential Responses of Stomata to Blue and Red Light

Comparisons between stomatal responses to broad band, blue and red light have given the earliest indications of a specific blue light effect on stomatal movements (review: [6, 14, 31]). A primary regulation of the photoresponses of guard cells by their chloroplasts would predict quantum efficiencies of blue and red light on stomatal opening matching the absorption spectrum of chlorophyll; instead, blue light has been consistently shown to be more effective than red light. Available information indicates that, under red irradation, only the PAR (photosynthetic active radiation)-dependent system from guard cell chloroplasts is operational whereas, under blue light, both the PAR and blue light-dependent photosystems control the photoresponses of stomata [23, 26, 34].

In addition, fluence rate-response curves of stomatal opening under the two light qualities have shown that the blue light photosystem has no detectable activation

1 Department of Biological Sciences, Stanford University, Stanford, CA 94305, USA

Blue Light Effects in Biological Systems
ed. by H. Senger

threshold and shows a high output at low irradiances, whereas the PAR-dependent system begins to promote stomatal opening only after exposure to minimal light intensities [8, 17, 20, 23, 28, 34]. The implications of this differential sensitivity of the two photoreceptor systems of guard cells for the photoregulation of stomatal opening are discussed later.

Wavelength Dependency and Action Spectra of Stomatal Opening

A detailed wavelength dependence of stomatal opening and rubidium (K^+) uptake in guard cells of *Vicia* showed a maximum at 455 nm both at low and high irradiances [7]. In contrast, a red maximum at around 650 nm was absent at low light intensities but was prominent at high photon fluxes. These photoresponses are consistent with the activation of the blue light photosystem at low light intensities. At higher irradiances, the PAR-dependent system of the guard cell chloroplasts promoted stomatal responses in the red and caused an additive effect in the blue. As a consequence of this interaction, stomatal responses at high irradiances are consistently higher in the blue than in the red.

Wavelength dependences of malate formation in epidermal peels of *Vicia* [17, 19] and of the blue light-dependent transpiration increase in *Avena* [27] have also been reported. Malate biosynthesis is an integral aspect of stomatal opening, with malate functioning as a counterion for K^+ [1]. Malate biosynthesis, like stomatal opening and Rb^+ uptake, also showed a maximum at 450–460 nm; in addition, measurements in the long UV region of the spectrum revealed a second maximum at 380 nm. In agreement with the findings with *Vicia*, the blue light-dependent transpiration increase in *Avena* had a maximum at around 450 nm [27].

The only reported action spectrum on stomatal opening was obtained with detached leaves of *Xanthium strumarium*, in gas exchange experiments [26]. Fluence dose-response curves of stomatal conductances at different wavelengths were clearly nonparallel, providing unambiguous evidence for the activity of more than one photoreceptor in the guard cells. An action spectrum constructed from the obtained fluence response curves showed a peak around 450 nm and a small shoulder at around 660 nm, features which are in notorious agreement with the wavelength dependence of stomatal opening obtained at low irradiances with epidermal peels of *Vicia* [7]. Sharkey and Raschke [26] also investigated in detail the response in the red region of the spectrum and found it identical to the photoresponses of mesophyll photosynthesis, confirming the involvement of chlorophyll in stomatal movements.

The observed ratio between the blue and red peaks of the action spectrum was about 10 [26]. Because, intrinsically, action spectra are essentially independent of quantum fluxes, one could interpret the relative ratio between the blue and red peaks as an expression of the relative contributions of the PAR and blue light-dependent photosystems to the photoresponses of stomata over wide ranges of physiological responses of the leaf. This is, however, not likely to be the case because both photoreceptor systems have presumably contributed to the observed responses under blue light. In addition, the higher threshold and smaller slopes of the red light responses depressed the red shoulder in a dose-dependent fashion [26]. The relative contributions of the two photoreceptor systems, evaluated from calculations of their apparent photon efficiencies, are discussed below.

Dual Beam Experiments

A different approach to the analysis of the photobiological properties of stomata has been the use of dual beam experiments [17, 18, 23, 29, 34]. In experiments with epidermal peels of *Vicia* and *Commelina,* stomatal apertures measured under saturating fluxes of red light were substantially increased upon the additions of small doses of blue light [23]. These observations showed that the activation of the blue light photosystem under conditions in which the PAR-dependent system was light-saturated caused a specific metabolic response leading to a substantial enhancement of stomatal opening, providing another line of evidence for the activity of a blue light photosystem in guard cells. Dual beam experiments have also shown that red irradiation alone is insufficient for the attainment of maximal levels of stomatal opening, which require the activation of the blue light photosystem [18, 23, 34]. These results, and data on the interaction of blue and red light on rates of malate biosynthesis [17] and the light-dependent transpiration increase in grasses [27], show that the normal expression of the photoresponses of stomatal guard cells require the activity of both the blue light photosystem and the guard cell chloroplasts.

An Apparent Quantum Efficiency of Blue and Red Light on Stomatal Opening

Using monochromatic light provided by lasers, an apparent quantum efficiency of blue (457.9 nm) and red (640.0 nm) light was calculated as the point to point slopes in fluence rate response curves under each light quality [34]. At low irradiances, blue light was highly effective in promoting stomatal opening whereas red light showed no response. At moderate to high light intensities, blue and red light showed similar, lower quantum efficiencies. This analysis indicates that, because of its lower activation threshold, the blue light photosystem is the primary mediator of the stomatal photoresponses at low irradiances. As the light intensities increase, the PAR-dependent photosystem is activated and, through its photophosphorylating activity, supplies essential high energy equivalents required for the maintenance of stomatal apertures at moderate to high irradiances [15, 23, 32]. As discussed in the preceding section, the blue light photosystem continues to operate after activation of the PAR-dependent system; further experiments illustrating that point are shown later.

The Photobiological Properties of Stomata from *Paphiopedilum*

Stomata from the genus *Paphiopedilum,* Orchidaceae, are functional despite their lack of chloroplasts [16]. Because of their unique characteristics, *Paphiopedilum* stomata have provided us with a valuable opportunity to test several aspects of the photobiology of guard cells emerging from the notions detailed above. Experiments with epidermal peels of *P. harrisianum* showed that stomata from this genus open in response to blue but not to red light [33], indicating that they possess a seemingly normal blue light photosystem but fail to respond to red light because of the absence of guard cell chloroplasts. Fluence-response curves of stomatal opening under blue light showed kinetics typical of a single photoreceptor system, with light saturation observed between 0.05 and 0.1 mmol m^{-2} s^{-1}. It remains to be established if these observed saturated values of the blue light photoreceptor are typical of all species or are only representative of a system lacking the PAR-dependent photosystem.

In addition, stomatal conductances at saturating levels of white or blue light irradiation in several species of *Paphiopedilum* tested were very low [30]. We reasoned that the evolutionary changes leading to the loss of competent guard cell chloroplasts in this genus were not deleterious because of these low, prevailing levels of stomatal conductance. Light-dependent stomatal opening in *Paphiopedilum* thus appears to be controlled by the sole activity of the blue light photosystem and is devoid of a stringent requirement for the PAR-dependent photosystem, which in species with chlorophyllous stomata, is essential to sustain moderate to high stomatal conductances.

The Blue Light Response of Stomata Under Natural Conditions

Despite the substantial amount of information on the blue light responses of stomatal guard cells accumulated over the last few decades, the study of the function of the blue light photosystem in stomatal movements has only recently been approached. We analyzed that problem under the hypothesis that, because of the observed enrichment of early morning, solar radiation in the blue [4, 9], the activation of the blue light photoreceptor at dawn would result in larger stomatal conductances and a potential increase in the daily carbon gain of the leaf [35]. Measurements of stomatal conductance of attached leaves of *Malva* in the early morning, under blue and red light, characterized a blue light-dependent phase of stomatal opening which enhanced dark levels of opening and was not observed under red light [35]. These results provided the first evidence for a blue light response of stomata under natural conditions. Estimates of the carbon gain ensuing from the enhanced stomatal opening caused by blue light were found to be small. However, the calculations reflected conditions of unlimited water supply and photosynthetic activity over most of the day. The impact of the enhanced carbon gain caused by the early morning, blue light response could be significantly greater in water-stressed leaves, in which midday and afternoon stomatal closing [22] is often observed.

The Functional Role of the Blue Light Photosystem of Guard Cells

In nature, under unstressed conditions, daily courses of stomatal movements show two qualitatively distinct classes of responses. Stomatal conductances increase steadily from predawn, following large increases in light intensity, and reach near-saturation by midmorning [35]. Available evidence indicates that this phase of stomatal opening is dominated by an intrinsic, direct light response of the guard cells [25, 34] which, in the absence of water stress, is largely uncoupled from environmental conditions and from the photosynthetic responses of the mesophyll [32]. During the rest of the day, a different class of stomatal response ensues, in which levels of stomatal conductances are tightly coupled to prevailing photosynthetic rates of the leaf. During this phase, which encompasses most of the daily activity of the plant but only a minor fraction of the whole range of stomatal conductances, stomatal responses to humidity, temperature and intercellular CO_2 concentrations come into play together with the light responses, in a sequence of fine adjustments of stomatal conductances which have been postulated to ensure an optimal relationship between rates of CO_2 uptake and water loss [2, 3].

The known properties of the blue light photosystem of guard cells indicate that it mediates the early photoresponses of stomata. In turn, the PAR-dependent photosystem of guard cell chloroplasts is activated at somewhat higher light intensities and, through its photophosphorylating capacity, supplies high energy equivalents needed to sustain moderate to high levels of stomatal conductances [32]. These combined activities of the two photoreceptor systems of guard cells therefore regulate the light response of stomata controlling the major increases in stomatal opening seen from sunrise to midmorning.

An important characteristic of the PAR-dependent system of guard cell chloroplasts has an additional implication for the coupling between mesophyll photosynthesis and the stomatal responses. Photophosphorylation in guard cell chloroplasts has been shown to be CO_2 sensitive [15], with high intercellular CO_2 concentrations inhibiting photophosphorylation in the guard cells. That feature of the PAR-dependent system makes it optimally suited as a mechanism coupling photosynthesis and conductance under conditions of high conductance levels. With photophosphorylation rates limiting stomatal apertures, changes in intercellular CO_2 concentrations ensuing from fluctuations in photosynthetic activity of the mesophyll would regulate photophosphorylation rates in guard cells and produce fine adjustments in stomatal conductance reflecting the prevailing photosynthetic rates of the leaf [32]. The PAR-dependent system of guard cells thus appears to play a dual role in the regulation of stomatal conductances on the second phase of stomatal movements; on the one hand, it supplies a major source of energy sustaining stomatal apertures at large conductance levels, on the other it provides stomata with a regulatory CO_2 response. What, if any, would be the function of the blue light photosystem in this phase of the stomatal responses?

Recent experiments in our laboratory indicate that, besides its role on the stomatal photoresponses during the early morning, the blue light photosystem also functions in the regulation of steady state levels of stomatal conductance throughout the day.

In these experiments, net photosynthesis and stomatal conductance were measured in intact, attached leaves of *Xanthium strumarium* grown in a green house. In leaves irradiated with broad band, red light, net photosynthesis and stomatal conductance increased with irradiance levels (Fig. 1). Initial rates of stomatal conductance were sluggish (data not shown) but both leaf functions achieved reasonably high rates at high photon fluxes (Fig. 1). After the leaf had reached steady state conditions under red irradiances of about 0.8 mmol m^{-2} s^{-1}, blue light at 0.01 mmol m^{-2} s^{-1} was added. Upon the addition of blue light, stomatal conductance, net photosynthesis and intercellular CO_2 concentrations increased (Fig. 2). Because the time constant of the photosynthetic apparatus is substantially faster than the stomatal responses, a primary response of the mesophyll to the addition of blue light should have resulted in a fast

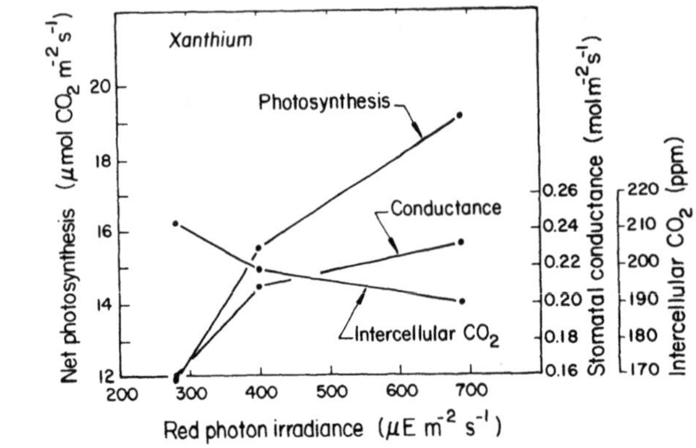

Fig. 1. Net photosynthesis, stomatal conductance and intercellular CO_2 concentrations in an attached leaf of *Xanthium strumarium*, measured in a steady-state, gas exchange system under different red irradiances. The light source was a 500 W incandescent (GE model 3200 k) lamp filtered through a Kodak 1A filter and a water bath. Temperature in the leaf chamber was 20 ± 0.2°C; VPD: 0.9 KPa; ambient CO_2: 325 ppm. Light intensities were measured with a Li-Cor quantum probe

increase in net photosynthesis and a transitory decrease in intercellular CO_2 concentrations, followed by a subsequent adjustment of the levels of stomatal conductance. In contrast, the observed increases in stomatal conductance and intercellular CO_2 concentrations indicate that the photosynthetic mesophyll was largely light-saturated under the red irradiances and that the primary response to the addition of blue light was an increase in stomatal apertures. The higher levels of stomatal conductance resulted

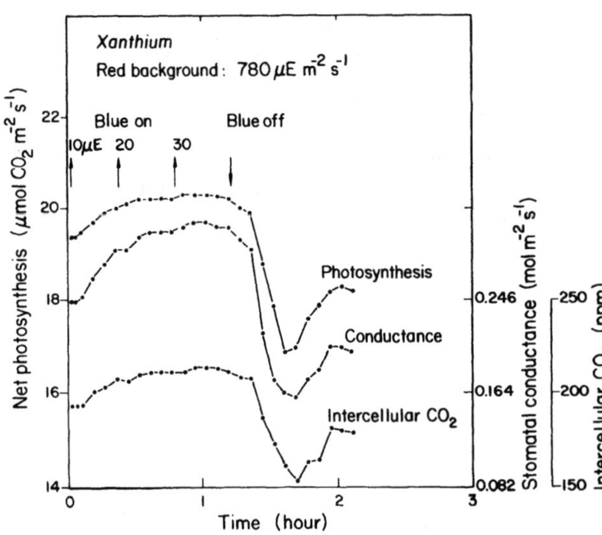

Fig. 2. Responses of net photosynthesis and stomatal conductance, and changes in intercellular CO_2 concentrations, to the addition of blue light to a background of red irradiance. Plant material and experimental conditions as in Fig. 1. The light source for the blue irradiance was a 150 W GE (DCL) projection lamp, filtered through a 2424 blue Plexiglas (Roehm and Haas) and a OCLI, wide band, hot mirror. Addition and removal of blue light indicated by *arrows*

in increasing amounts of intercellular CO_2 concentrations which in turn stimulated net photosynthesis.

Subsequent increases in blue photon fluxes caused further stimulation of stomatal conductance, with an apparent saturation level at around 0.03 mmol m^{-2} s^{-1}. It should be noted, however, that these experimental conditions did not allow an unambiguous measurement of the saturation levels of the blue light response because of the concomitant increases in intercellular CO_2 concentrations that could have elicited a closing response.

We also observed a dramatic response to the removal of the blue irradiation (Fig. 2), which caused drastic decreases in stomatal conductance, net photosynthesis and levels of intercellular CO_2. As in the opening phase, the kinetics of these changes can be unambiguously interpreted as a result of a primary response of stomatal guard cells to the removal of blue light.

These data indicate that prior to the blue light irradiation, stomata had responded to red irradiation with levels of conductance which, although lower than those observed at comparable photon fluxes of white light, yielded intercellular CO_2 concentrations that could sustain significant rates of net photosynthesis. The photoresponses of stomata in this phase of the experiments must have depended on the guard cell chloroplasts.

Addition of blue light caused a sharp increase in stomatal conductance, in a "light on" response not unlike the one observed in the early morning but significantly different from the latter inasmuch as it occured on a background of high photosynthetic rates and a presumed, substantial output of the PAR-dependent system. It is, therefore, apparent that on a background of red light, the activation of the blue light photosystem of the guard cells elicits a "light on" response regardless of the activity of the PAR-dependent system and the levels of stomatal opening. This property of the blue light photosystem is more evident upon the removal of blue light, which causes a "light off" signal of enough magnitude to generate a substantial overshoot (Fig. 2). The overshoot was compensated by the PAR-dependent system of the guard cells only when a significant time interval had elapsed after the removal of blue light. These observations are consistent with the notion of a role of the blue light photosystem as a specific light sensor of stomatal guard cells [32, 34]. Because of its high activation threshold, the PAR-dependent photosystem would be a poor light sensor at low light intensities; its CO_2 sensitivity would further hinder its function as an indicator of light intensities and make the stomata oscillation-prone under conditions in which photosynthetic rates would undergo significant fluctuations, such as in partially cloudy skies [34]. In contrast, the insignificant activation threshold and high output of the blue light photosystem would render it optimally suited as a specific light sensor. With low saturation intensities – an estimate from current information suggests a value of around 0.05 mmol m^{-2} s^{-1} [23, 33] – the blue light photosystem would activate stomatal opening at dawn and remain in the "light on" mode throughout most of the day. Its activity would prevent wide stomatal oscillations in response to short-term fluctuations in light intensity and intercellular CO_2 concentrations and, perhaps more importantly, would gear the guard cell metabolism for reactions associated with daylight activity [23].

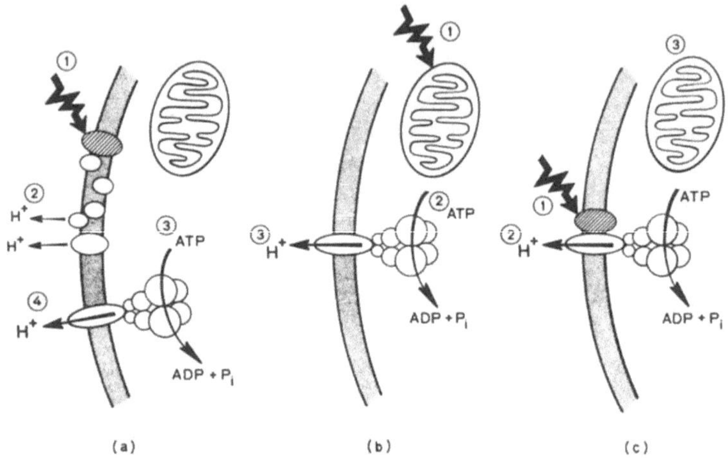

Fig. 3. Hypothetical modes of action of the blue light photoreceptor in guard cells. Absorption of blue light is depicted as step 1. **a** A blue light chromophore within a membrane-bound, electron transport chain that transduces blue quanta into electron fluxes resulting in the extrusion of protons across the membrane. Proton fluxes result in some ion uptake and the activation of a plasmalemma ATPase, which catalyzes the bulk of proton extrusion associated with stomatal opening. **b** Blue light is directly absorbed by the guard cell mitochondria, leading to an enhanced ATP synthesis which drives the plasmalemma ATPase. **c** A blue light photoreceptor is directly associated with the plasmalemma ATPase; its excitation results in a stimulation of proton extrusion

The Metabolic Properties of the Blue Light Photosystem of Guard Cells

The identity of the blue light photoreceptor in the guard cells remains to be elucidated. The spectroscopic data discussed above are all consistent with cryptochrome; a recent report on a normal blue light response of stomata from leaves treated with the herbicide SAN, which inhibits carotene biosynthesis, argues further against the involvement of carotenes in the blue light response of stomata [10].

A primary effect of blue light on respiration has been proposed in several systems [11] including guard cells [12]. On the other hand, the prominence of large ion fluxes during stomatal movements raises the possibility of a flavin chromophore within a membrane-bound electron transport chain, which could transduce photochemical energy into active ion transport and initiate stomatal opening [31, 32, 36]. The characterization of a fluorescing chromophore in guard cell vacuoles [31, 36, 37] with an emission spectrum consistent with a flavin, could be indicative of the presence of a blue light chromophore in the tonoplast. Current knowledge on that aspect of the blue light response of guard cells has been recently discussed elsewhere [32]; further work on that problem should reveal if a flavin or flavoprotein participates in the green vacuolar fluorescence of guard cells.

Three different, hypothetical ways in which a blue light signal could be transduced into ion fluxes mediating stomatal movements are depicted in Fig. 3. Blue light could be absorbed by a specific blue light photoreceptor within a membrane-bound electron transport chain (Fig. 3a). Membrane-bound flavins in an electron transport chain have been postulated in several blue light responses [24]. The scheme in Fig. 3a places the electron transport chain at the plasmalemma but a similar mechanism in the tonoplast, pumping protons into the vacuole, would share most of its putative properties with a plasmalemma-bound chain extruding protons into the cell wall. The known ratios of activity between the blue light and PAR-dependent photosystems of the guard cells is reflected in the hypothetical relationship between the proton pumping by the electron transport chain and a more active proton pumping ATPase, which would catalize the bulk of proton extrusion driving stomatal opening. Proton extrusion by the electron transport chain would serve a dual purpose of generating a small magnitude, electrochemical gradient and regulating cytoplasmic reactions which would activate the ATPase. Potential sources of ATP for the proton pumping ATPase are oxidative phosphorylation by mitochondria and photophosphorylation by guard cell chloroplasts (for simplicity, the latter are not included in the diagram). A second alternative would involve a direct activation of mitochondrial activity by blue light (Fig. 3b). This possibility however is inconsistent with recent findings in other systems, showing that in the light, activation of photophosphorylation suppresses respiration in mitochondria [5, 21]. Finally, blue light could directly activate a plasmalemma ATPase (Fig. 3c), in a fashion similar to the mode of action of the fungal toxin, fusicoccin [13].

The elucidation of the photobiological properties of the blue light photosystem of guard cells is providing us with important insights on stomatal function and the biology of guard cells. In addition, these studies have significant potential for further advances on our understanding of the blue light response in plants. The localization of the blue light photoreceptor of stomata on a single cell type and its presumed connection with specific, reversible ion fluxes could supply important clues on the identity of the blue light photoreceptor and its function in the transduction of light signals into metabolic events associated with the blue light response.

Acknowledgments. This research was supported by the National Science Foundation and the Department of Energy of the United States. I thank Ms. Sally Assmann for valuable discussions, Dr. Cyril Grivet for experimental work, and Ms. Judy Nishimoto for excellent logistic support.

References

1. Allaway WG (1981) Anions in stomatal operation. In: Jarvis PG, Mansfield TA (eds) Stomatal physiology. Soc Exp Biol (SS) 8:71–85. Cambridge Univ Press, Cambridge
2. Cowan IR (1982) Regulation of water use in relation to carbon gain in higher plants. In: Lange OL, Nobel PS, Osmond CB, Ziegler H (eds) Encyclopedia of plant physiology, New Ser, vol 12B. Physiological plant ecology. Springer, Berlin Heidelberg New York, pp 589–613
3. Farquhar GD, Sharkey TD (1982) Stomatal conductance and photosynthesis. Annu Rev Plant Physiol 33:317–345
4. Goldberg B, Klein WH (1977) Variations in the spectral distribution of daylight at various geographical locations on the earth's surface. Sol Energy 19:3–13
5. Hampp R, Goller M, Ziegler H (1982) Adenylate levels, energy charge, and phosphorylation potential during dark-light and light-dark transition in chloroplasts, mitochondria, and cytosol of mesophyll protoplasts from *Avena sativa* L. Plant Physiol 69:448–455
6. Hsiao TC (1976) Stomatal ion transport. In: Lüttge U, Pitman MG (eds) Encyclopedia of plant physiology, New Ser, vol IIB. Transport in plants II. Springer, Berlin Heidelberg New York, pp 195–221

7. Hsiao TC, Allaway WE, Evans LT (1973) Action spectra for guard cell Rb^+ uptake and stomatal opening in *Vicia faba*. Plant Physiol 51:82–88
8. Jarvis PG, Morison JIL (1981) The control of transpiration and photosynthesis by the stomata. In: Jarvis PG, Mansfield TA (eds) Stomatal physiology. Soc Exp Biol (SS) 8:247–279. Cambridge Univ Press, Cambridge
9. Johnson TB, Salisbury FB, Connor GI (1967) Ratio of blue to red light: a brief increase following sunset. Science 155:1663–1665
10. Karlsson PE, Hoglund H-O, Klockare R (1983) Blue light induces stomatal transpiration in wheat seedlings with chlorophyll deficiency caused by SAN 9789. Physiol Plant 57:417–421
11. Kowallik W (1982) Blue light effects on respiration. Annu Rev Plant Physiol 33:51–72
12. Lurie S (1978) The effect of wavelength of light on stomatal opening. Planta 140:245–249
13. Marre E (1979) Fusicoccin: a tool in plant physiology. Annu Rev Plant Physiol 30:273–288
14. Meidner H, Mansfield TA (1968) Physiology of stomata. McGraw-Hill, London
15. Melis A, Zeiger E (1982) Chlorophyll *a* fluorescence transients in mesophyll and guard cells. Modulation of guard cell photophosphorylation by CO_2. Plant Physiol 69:642–647
16. Nelson SD, Mayo JM (1975) The occurrence of functional non-chlorophyllous guard cells in *Paphiopedilum* spp. Can J Bot 53:1–7
17. Ogawa T (1980) Synergistic action of red and blue light on stomatal opening of *Vicia faba* leaves. In: Senger H (ed) The blue light syndrome. Springer, Berlin Heidelberg New York, pp 622–628
18. Ogawa T (1981) Blue light response of stomata with starch-containing (*Vicia faba*) and starch-deficient (*Allium cepa*) guard cells under background illumination with red light. Plant Sci Lett 22:103–108
19. Ogawa T, Ishikawa H, Shimada K, Shibata K (1978) Synergistic action of red and blue light and action spectra for malate formation in guard cells of *Vicia faba* L. Planta 142:61–65
20. Pemadasa MA (1982) Abaxial and adaxial stomatal responses to light of different wavelengths and to phenylacetic acid on isolated epidermes of *Commelina communis* L. J Exp Bot 33:92–99
21. Robinson DC, Wellburn AR (1981) Light quality and hormonal influences upon the rate of ATP formation by mitochondria during greening. Biochem Physiol Pflanz 176:54–59
22. Schulze ED, Lange OL, Evenari M, Kappen L, Buschbom U (1980) Long-term effects of drought on wild and cultivated plants in the Negev Desert. II. Diurnal patterns of net photosynthesis and daily carbon gain. Oecologia 45:19–25
23. Schwartz A, Zeiger E (1984) Metabolic energy for stomatal opening. Roles of photophosphorylation and oxidative phosphorylation. Planta 161:129–136
24. Senger H, Briggs WR (1981) The blue light receptor(s): primary reactions and subsequent metabolic changes. Photochem Photobiol 6:1–38
25. Sharkey TD, Raschke K (1981) Separation and measurement of direct and indirect effects of light on stomata. Plant Physiol 68:33–40
26. Sharkey TD, Raschke K (1981) Effect of light quality on stomatal opening in leaves of *Xanthium strumarium* L. Plant Physiol 68:1170–1174
27. Skaar H, Johnsson A (1978) Rapid, blue light induced transpiration in *Avena*. Physiol Plant 43:390–396
28. Travis AJ, Mansfield TA (1981) Light saturation of stomatal opening on the adaxial and abaxial epidermis of *Commelina communis*. J Exp Bot 32:1169–1179
29. Voskresenskaya NP, Polyakov AM (1976) Regulatory action of blue light on photosynthetic gas exchange: action spectrum of light saturation of CO_2 exchange in Lily of the Valley leaves. Sov Plant Physiol 23:6–11
30. Williams WE, Grivet C, Zeiger E (1983) Gas exchange in *Paphiopedilum:* lack of chloroplasts in guard cells correlates with low stomatal conductance. Plant Physiol 72:906–908
31. Zeiger E (1980) The blue light response of stomata and the green vacuolar fluorescence of guard cells. In: Senger H (ed) The blue light syndrome. Springer, Berlin Heidelberg New York, pp 629–636
32. Zeiger E (1983) The biology of stomatal guard cells. Annu Rev Plant Physiol 34:441–475

33. Zeiger E, Assmann SM, Meidner H (1983) The photobiology of *Paphiopedilum* stomata: opening under blue but not red light. Photochem Photobiol 38:627–630
34. Zeiger E, Field C (1982) Photocontrol of the functional coupling between photosynthesis and stomatal conductance in the intact leaf: blue light and PAR-dependent photosystems in guard cells. Plant Physiol 70:370–375
35. Zeiger E, Field C, Mooney HA (1981) Stomatal opening at dawn: possible roles of the blue light response in nature. In: Smith H (ed) Plants and the daylight spectrum. Academic Press, London New York, pp 391–407
36. Zeiger E, Hepler PK (1977) Light and stomatal function: blue light stimulates swelling of guard cell protoplasts. Science 196:887–889
37. Zeiger E, Hepler PK (1979) Blue light-induced, intrinsic vacuolar fluorescence in onion guard cells. J Cell Sci 37:1–10

Oecological Aspects

Responses of Unicellular Marine Plants to Natural Blue-Green Light Environments

S. W. JEFFREY[1]

Introduction

A blue-green filter covers 70% of the Earth's surface, and the light transmitted through it has been supporting one quarter of the world's plant productivity for millions of years. This filter is the world's oceans, and the plants responsible are the microscopic unicellular marine algae which range in size from 1 to 200 μm, and represent some thousands of species from many algal Divisions. Although there is much current interest in blue light mechanisms of terrestrial plants and micro-organisms, little attention has been paid to this giant blue light experiment. These marine plants are adapted to their specialized environment by possessing a variety of pigment systems capable of efficiently harvesting blue-green light. This paper briefly describes the light environment, the plants and the pigments found in the open ocean. In addition, the question was posed whether blue-green radiation, as well as being used for energy transduction, exerts any specific photo-control of the photosynthetic apparatus. This paper shows that low intensity blue-green light regulates the chloroplast systems of some groups by increasing the total amount of pigments, the number of thylakoid bands, the para-crystalline structure of the pyrenoid and the photosynthetic capacity. Changes in pigment ratios were not significant except in the biliprotein-containing species.

We therefore suggest that the "blue light syndrome" is relevant to this group of marine unicellular plants which account for one quarter of the world's productivity, as well as the terrestrial plants and microorganisms previously studied [28].

Unicellular Marine Plants and Productivity

While global patterns of terrestrial plant production are well known to plant physiologists [23], global patterns of marine plant production and their significance are less familiar. Figure 1 shows the distribution of plant production in the world oceans, and Table 1 distinguishes the quantitative importance of plants in different marine habitats. The production of the three major types of marine plants, the unicellular algae and the attached multicellular seaweeds and seagrasses, together equal about half the terrestrial

1 CSIRO Marine Laboratories, PO Box 1538, Hobart, Tasmania, Australia

Blue Light Effects in Biological Systems
ed. by H. Senger

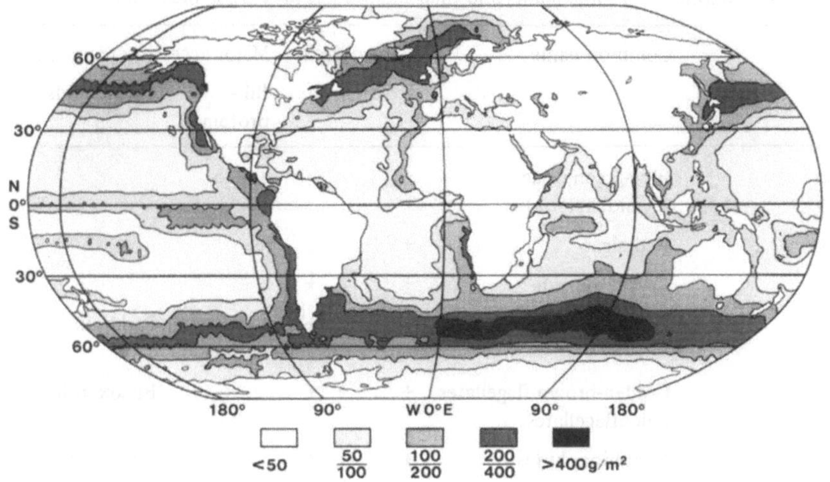

Fig. 1. Primary productivity pattern of the oceans – ranges mapped in g m^{-2} yr^{-1}, dry matter. (After [3])

Table 1. Importance of marine plants in global plant productivity. (After [35])

Plant community	Total net primary production (dry matter)		Chlorophyll	
	Total (10^9 t yr^{-1})	% of Total	Total (10 t)	% of Total
Open ocean	41.5		10.0	
Upwelling zones	0.2		0.1	
Continental shelf	9.6		5.3	
Algal beds and reefs	1.6		1.2	
Estuaries (excluding marsh)	2.1		1.4	
Total aquatic freshwater plants	6.8	3.9	6.5	2.7
Total aquatic marine plants	55.0	31.9	18.0	7.4
Total terrestrial plants	110.7	64.2	219.5	89.9
Global total	172.5	100.0	244.0	100.0

plant production or 31.9% of the global total. However, because of the vast areas involved, the unicellular algae of the open ocean, upwelling zones and continental shelves contribute by far the major proportion (over 90%) of the total marine production.

The unicellular plants that make up these floating populations (phytoplankton) are listed in Table 2, and have recently been reviewed by Jeffrey and Vesk [19]. The most important are the silica-walled diatoms (Bacillariophyta) which account for 5,500–10,000 species [9] and the dinoflagellates (Dinophyta) with about 1,200 species, not all of which, however, are photosynthetic [4, 30]. Green flagellates (Chlorophyta, Prasinophyta and Euglenophyta) were recently found to be significant in the open ocean [12, 15], and various types of golden-brown flagellates (Prymnesiophyta, Rhaphidophyta, Chrysophyta, Cryptophyta) may dominate the phytoplankton at various

Table 2. Unicellular marine planktonic plants and their major light-harvesting pigments

Division	Common name	Chlorophylls				Major accessory pigments	
		a	b	c_1	c_2	Bili-proteins	Carotenoids
Cyanophyta	Blue-green algae (cyanobacteria)	+				+	
Rhodophyta	Red algae	+				+	
Cryptophyta	Cryptomonads	+			+	+	
Dinophyta	Dinoflagellates	+			+[a]		Peridinin
Chrysophyta							
Chrysophyceae	Golden-brown flagellates Silicoflagellates	+		+	+		Fucoxanthin
Rhaphidophyceae	Rhaphidophytes	+		+	+		Fucoxanthin
Prymnesiophyta	Golden-brown flagellates with haptonema; Coccolithophorids	+		+	+		Fucoxanthin
Bacillariophyta	Diatoms	+		+[b]	+		Fucoxanthin
Euglenophyta	Euglenoids	+	+				Siphonaxanthin[c]
Chlorophyta							
Chlorophyceae	Green flagellates	+	+				Lutein
Prasinophyceae	"Scaly" green flagellates	+	+				Siphonaxanthin[c]

[a] Some species with chrysophyte endosymbionts contain fucoxanthin and chl c_1 and c_2 [17]
[b] Some species contain only chl c_2 (Jeffrey and Stauber unpublished)
[c] Some species only
+ = present

times of the year and in particular regions. Unicellular blue-green algae, Cyanophyta [34], are ubiquitous throughout the world's oceans. Seasonal successions of all species, at least in inshore waters, are well documented [27].

Marine Light Environments

Marine light fields are significantly different from those of terrestrial environments. The full visible spectrum of white light is found only at the ocean's surface, since the sun's electro-magnetic energy is selectively absorbed and filtered as it penetrates the upper layers. Infra-red and red light (> 750 nm) is absorbed by the water itself, reducing the solar energy to one half in the first half meter [21]. In clear waters, attenuation of light with depth (Fig. 2) results in the rapid disappearance of red and yellow light, with green, blue and blue-green light (max 490 nm) penetrating the furthest. In clear tropical waters significant blue-green light (e.g., 0.1–1% of the surface intensity) can penetrate to depths of 200 m.

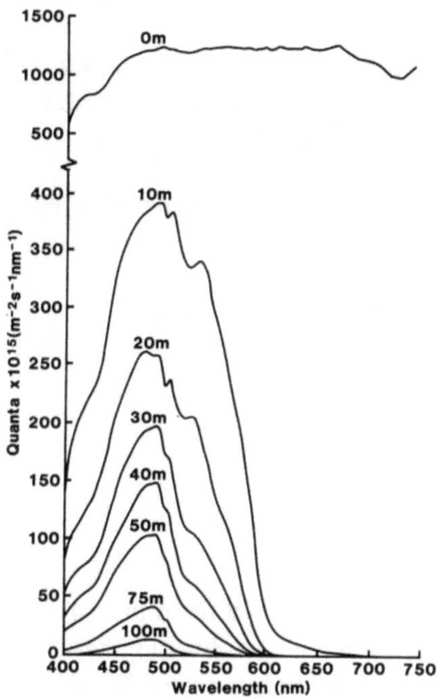

Fig. 2. Distribution of downwelling spectral irradiance. East Australian Current, Station 11, SPRIGHTLY cruise SP11/81

In inshore coastal regions, dissolved yellow humic compounds from land runoff and particulates of organic and inorganic origin further modify light fields by absorbing strongly in the violet and blue-green region. This shifts the maximum penetration of light from blue-green wavelengths (480–490 nm) to predominantly green wavelengths (550 nm) and can reduce the depth of light penetration to a few meters [21]. Most of the hundreds of light profiles taken in tropical and subtropical waters of Australia show characteristic blue-green spectra, with classic "green" waters encountered at sea only during spring blooms of diatoms along the coast (Jeffrey unpublished).

The transmission of individual wavelengths, as a percentage of their values at the surface are shown in Fig. 3. Light quality, as illustrated at the 1% light level, changes progressively with depth, red wavelengths (700, 680 and 650 nm) being eliminated

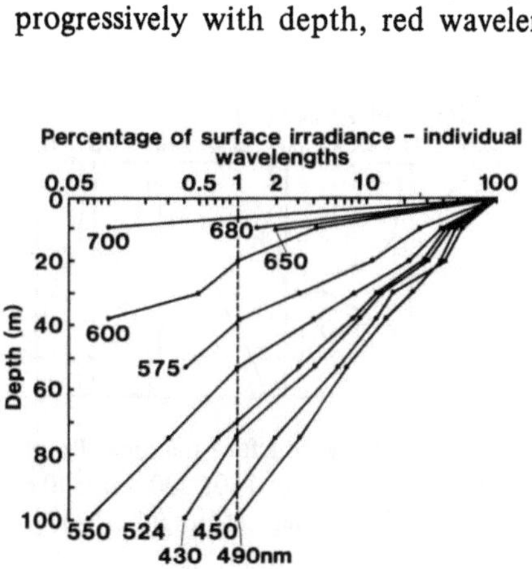

Fig. 3. Depth profiles of specific wavelengths of downwelling irradiance as a percentage of surface irradiance. East Australian Current, Station 11, SPRIGHTLY cruise SP11/81

close to the surface, followed by orange (600 nm), yellow (575 nm), green (550, 524 nm), violet (430 nm), blue (450 nm) and blue-green (490 nm).

The range of underwater light intensities varies markedly with latitude, season, time of day, cloud cover, sea state and phytoplankton biomass. Between latitudes of 32° and 37°S, noon values of downwelling light at the surface (400–750 nm) can vary from 200 to over 1100 × 10^{18} Quanta m^{-2} s^{-1}. Taking an average value of 730 × 10^{18} Quanta m^{-2} s^{-1} at the sea's surface, then a typical depth profile of downwelling light in the East Australian Current in spring months [measured with a (cosine collector) Techtum spectroradiometer] shows integrals of 240 × 10^{18} Quanta m^{-2} s^{-1} at 10 m, 100 (20 m), 18 (30 m), 6 (40 m), 1.8 (50 m), 0.3 (75 m) and 0.06 (100 m). Thus these waters resemble a broad blue-green filter through which light passes in continuously decreasing intensity.

Vertical Distribution of Planktonic Algae

It might be expected that the unicellular algae would congregate close to the sea's surface to obtain the maximum light available. This is rarely the case. Instead, they form marked sub-surface maxima, even in well mixed water columns, as deep as 90–130 m in the open ocean and tropical seas [14], shallowing to 50–80 m in more enriched waters of warm-core eddies [16]. Across the North West Shelf of Australia (Fig. 4) deep chlorophyll maxima are regularly encountered. When surface blooms of diatoms form in coastal waters they are usually short-lived [7, 11] due to sinking and to cell disintegration from photo-oxidation and grazing.

The mechanisms regulating this vertical distribution are not well understood. Phytoplankton can be distributed homogeneously by water mixing, or can be concentrated by hydrodynamic features such as fronts, thermoclines and discontinuity layers. Some motile cells can undergo significant vertical migrations, while in other nonmotile species positive buoyancy can occur by a combination of metabolic (ionic) nutritional and light control. Much more work is still needed to understand this deep layering phenomenon.

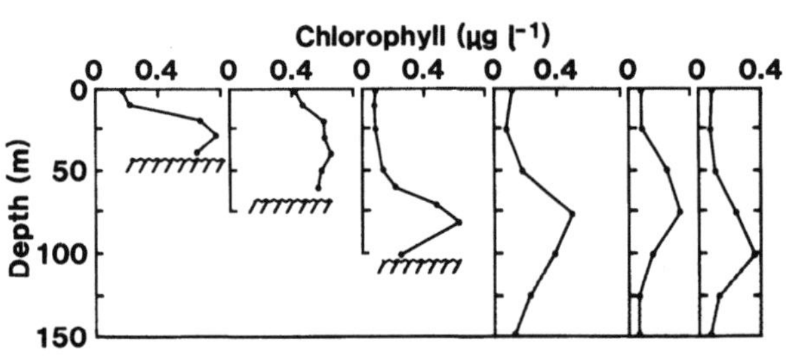

Fig. 4. Depth profiles of phytoplankton chlorophyll a showing deep subsurface maxima. Panels *from left to right* indicate shallow to deep water columns (i.e., 40, 70, 110, 150, 200 and 300 m depth). Transect 1, North West Shelf (Australia) SOELA cruise S06/82 (Hallegraeff unpublished)

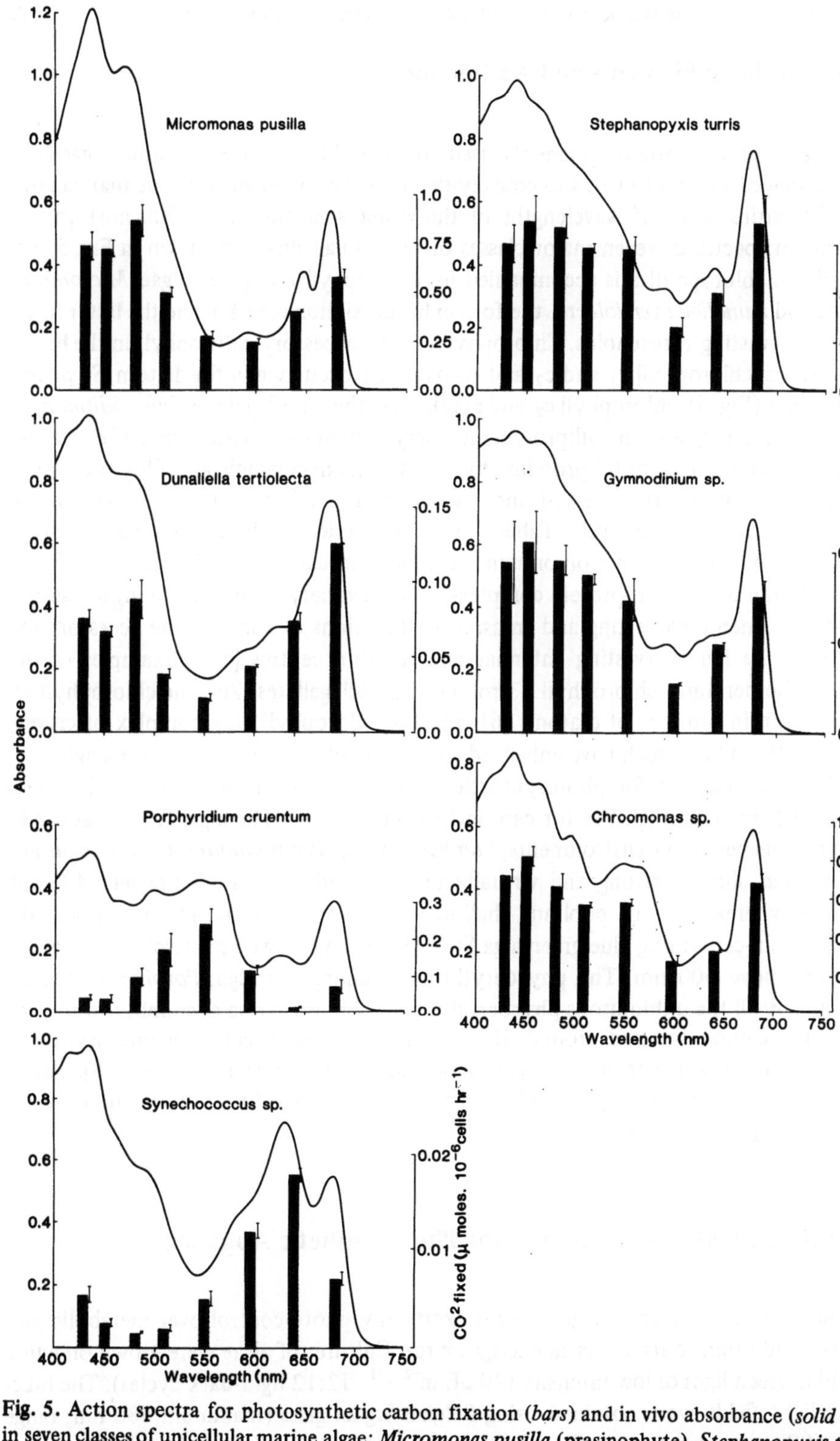

Fig. 5. Action spectra for photosynthetic carbon fixation (*bars*) and in vivo absorbance (*solid line*) in seven classes of unicellular marine algae: *Micromonas pusilla* (prasinophyte), *Stephanopyxis turris* (diatom); *Dunaliella tertiolecta* (chlorophyte); *Gymnodinium* sp. (dinoflagellate); *Porphyridium cruentum* (rhodophyte); *Chroomonas* sp. (cryptophyte); *Synechococcus* sp. (cyanophyte). Exponentially growing algal cells were illuminated for 15 min in $Na_2H^{14}CO_3$ (activity 19.4×10^6 dpm ml^{-1}) under 50 $\mu E\ m^{-2}\ s^{-1}$ of various wavelengths. Narrow band interference filters with half-band widths of 10 nm (Balzers, Liechtenstein) and maximum transmission of 429, 451, 481, 509, 550, 597, 639 and 681 nm were used with quartz halogen lamps (250 W) (Bulleid, Holdsworth and Jeffrey unpublished)

Photosynthetic Pigments and Action Spectra

The major photosynthetic pigments used for light harvesting by marine algae have been reviewed in detail [13]. Collectively the algae have pigment systems that can harvest the entire range of wavelengths of the visible spectrum (350–700 nm). In vivo absorption spectra of seven major classes of unicellular algae are shown in Fig. 5. The ubiquitous chlorophyll a is accompanied by chlorophyll b in green algae, *Micromonas pusilla* and *Dunaliella tertiolecta,* the former having siphonaxanthin and the latter lutein as light-harvesting carotenoids. Chlorophyll c is the accessory chlorophyll in the brown algal classes, chlorophyll c_1 and c_2 and fucoxanthin occurring in the diatom, *Stephanopyxis turris* (Fig. 5), chlorophyll c_2 and peridinin in the dinoflagellate *Gymnodinium* sp., and chlorophyll c_2 and the biliprotein phycoerythrin in the cryptomonad *Chroomonas* sp. Chlorophyll a and the biliproteins phycoerythrin, phycocyanin and allophycocyanin occur in the more primitive red and blue-green algae, *Porphyridium cruentum* and *Synechococcus* sp. respectively. Table 2 gives an overview of these major pigment distributions and further information on minor components can be found in Jeffrey [13].

Isolation of pigment-protein complexes from a variety of unicellular algae has confirmed the energy harvesting and transferring functions of some of the accessory pigments in the light-harvesting antennae and reaction centres [10]. Examples of this include the peridinin chlorophyll a protein of dinoflagellates [26], the chlorophyll a/c – fucoxanthin proteins of diatoms [6], and the chlorophyll a/c_2 complex of cryptomonads [10], all of which have enhanced absorption of blue and green wavelengths.

Further evidence for photosynthetic activity in the blue-green region has been obtained from action spectra for carbon fixation (Fig. 5). The algal types that can be found in deeper water environments, *Stephanopyxis, Gymnodinium, Chroomonas* and *Micromonas* showed strong and virtually uniform carbon fixation between 430 and 550 nm, whereas the tide pool and shallow water green flagellate *Dunaliella,* and the phycocyanin-containing blue-green alga *Synechococcus,* showed relatively more carbon fixation above 600 nm. The phycoerythrin-containing red alga, *Porphyridium* (and presumably all the unbiquitous phycoerythrin-containing marine cyanophytes) showed maximum carbon fixation around 550 nm. These action spectra for photosynthetic carbon fixation which resemble classic action spectra for oxygen evolution [8], clearly show that blue-green and green radiation is effectively used for photosynthesis by the deeper living plant groups.

Blue-Green Light Regulation of the Photosynthetic Apparatus

We questioned whether blue-green light exerts any photo-control over metabolic processes in addition to its use as an energy source. Cultures of algae were grown on white and blue-green light of low intensity (40 $\mu E\ m^{-2}\ s^{-1}$; 12:12 light-dark cycles). The blue-green light fields were obtained using Belgian Signal glass (maximum 480 nm; range 400–600 nm), whose spectral properties resembled marine light fields at about 20 m depth, but with somewhat lower intensities.

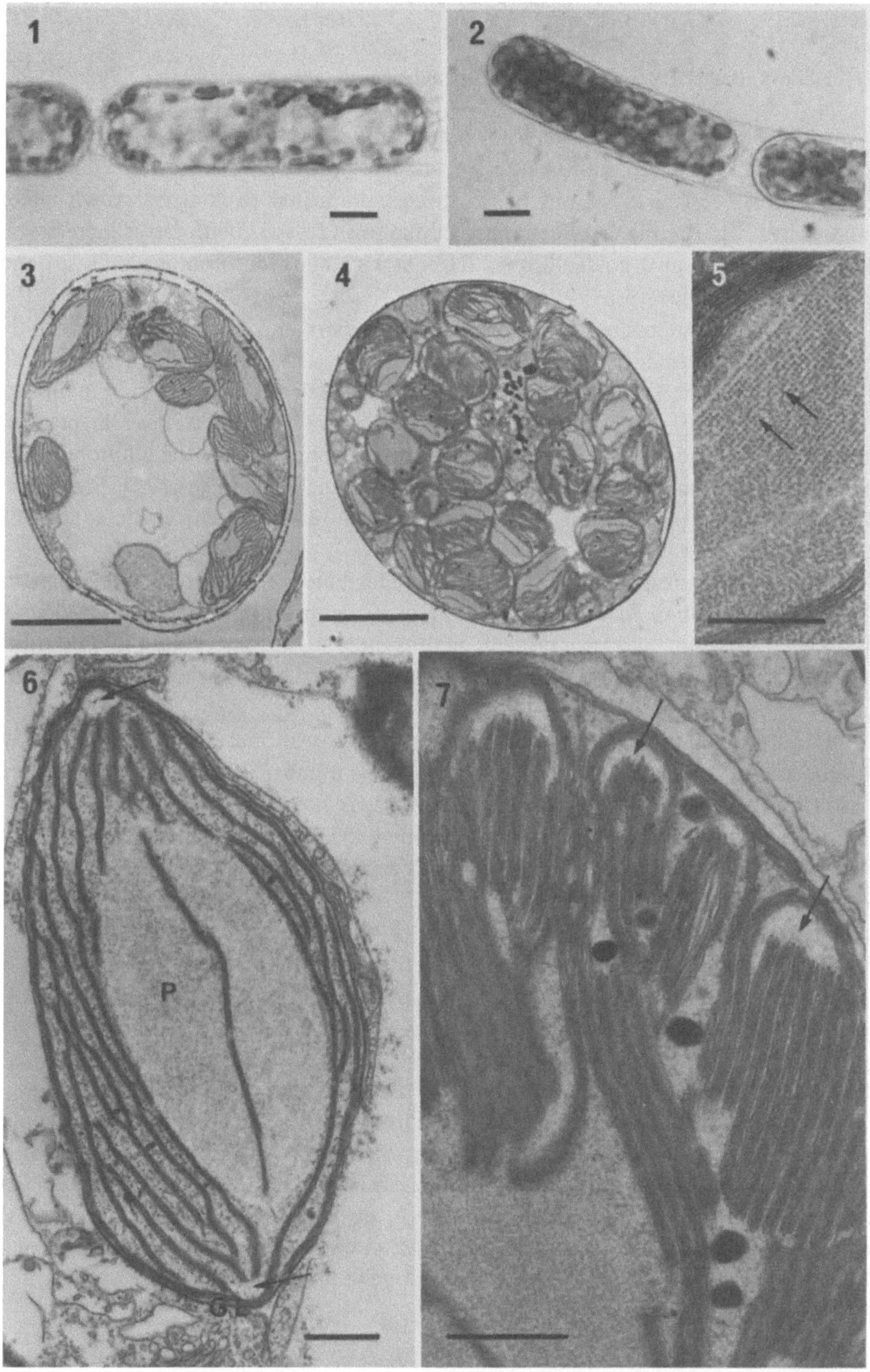

Fig. 6. Cells of the diatom *Stephanopyxis turris*, grown on low intensity white and blue-green light (50 $\mu E\ m^{-2}\ s^{-1}$). Light micrographs show typical white light (1) and blue-green light cell (2) with increased number of chloroplasts in cell on blue-green light (*Bar* = 10 μm). Electron micrographs of transverse section of typical white light cell (3) with chloroplasts arranged around periphery, compared with typical blue-green light cell (4) filled with rounded thylakoid-packed chloroplasts (*Bar* = 10 μm). Detail of dense pyrenoid from blue-green light chloroplast (5) showing linear arrangement of particles (*arrows*) (*Bar* = 0.2 μm). Typical white light chloroplast (6) showing 3-thylakoid band arrangement bounded by girdle lamellae (*GL*); large central pyrenoid (*P*) traversed by a 2–3 thylakoid band; ring genophore (*arrows*) lying beneath girdle lamellae (*Bar* = 0.5 μm). Typical blue-green light chloroplasts (7) with several fibrillar regions (*arrows*) lying beneath girdle lamellae: closely packed 3-thylakoid bands associated with each of possible DNA areas (*Bar* = 0.5 μm) [18]

Growth rates of cultures were not significantly affected by blue-green light, but by the end of log phase, cultures grown under blue-green light showed chlorophyll increases of 17–500% in 21 out of 23 species, compared to controls grown under white light [20]. The major effects were in diatoms (15 spp), with lesser increases in dinoflagellates (3 spp), golden-brown flagellates (3 spp), cryptomonads (1 sp) and green flagellates (1 sp).

The relative proportions of chlorophylls and carotenoids were not significantly altered by blue-green light, except for small increases in chorophyll c:a ratios [31], although Faust et al. [5] recently showed green light stimulation of peridinin biosynthesis in a shallow water dinoflagellate. Stronger evidence for changes in pigment ratios (chromatic adaptation effects) were, however, obtained for the biliprotein-containing cryptomonad under both blue-green [31] and green light [22]. Significant increases in the phycoerythrin : chlorophyll a ratio resembled chromatic adaptation effects in blue-green algae [2].

Other biochemical changes in cultures under blue-green light include increased photosynthetic carbon fixation [18, 29, 32], increased polar lipid [18], increased DNA and RNA [33], and increased protein, photosynthetic oxygen evolution, PEPcarboxykinase activity and Photosystem I activity (Holdsworth, Stauber and Jeffrey unpublished).

The structural basis for the pigment changes was an increase in the number of chloroplasts (Fig. 6.1, 6.2) and an increase in the number of thylakoids per chloroplast (Fig. 6.3–6.7). In the diatom *Stephanopyxis turris* grown on white light the chloroplasts were elongate and arranged around the periphery of the cell (Fig. 6.3). The chloroplasts contained 3-thylakoid bands surrounded by girdle lamellae, beneath which lay a single ring genophore (Fig. 6.6). In cells grown on blue-green light, the numerous chloroplasts were more rounded (Fig. 6.4) and packed with 3-thylakoid bands, which were associated with multiple fibrillar (DNA) regions beneath loops of girdle lamellae (Fig. 6.7). The large central pyrenoid (Fig. 6.6), site of ribulose bisphosphate carboxylase, contained dense paracrystalline areas in cells grown on blue-green light (Fig. 6.5). Similar ultrastructural changes were observed in four other diatoms, a dinoflagellate, and a cryptomonad [31].

In summary, the major effects of low intensity blue-green light on unicellular marine algae are increased pigment synthesis, no significant change in pigment ratios (except in the cryptomonads), more chloroplasts in some centric diatoms, more thylakoids per chloroplast (all species), more DNA regions per chloroplast (four species), increased paracrystalline areas in the pyrenoid, some evidence for increased photosynthetic CO_2 fixation, O_2 evolution, protein and polar lipid, and delayed photobleaching in older cells.

Concluding Remarks

These preliminary results suggest that blue-green light has profound effects in regulating the photosynthetic apparatus of unicellular marine algae. It increases their viability at depth in the sea by enabling enhanced synthesis of thylakoids to occur for more efficient photon capture. Cells possessing such a mechanism for survival at depth would have a distinct competitive advantage.

It was previously thought that these blue-green light responses of unicellular marine algae resembled those of shade species of higher plants. This is not the case. In contrast to marine plants, red light (not blue-green) controls the development of shade-type chloroplasts in terrestrial plants, whereas blue light controls the formation of sun-type chloroplasts [1, 24]. The explanation for the different control mechanisms must lie in the light quality available in the different environments. Terrestrial shade habitats for example, are poor in blue light and rich in red: far-red wavelengths, and therefore could use either red light or the absence of blue as an environmental signal.

The blue-green light responses of the unicellular marine algae may well be a direct adaptation to the ancient light environment where these primitive plants developed. While all plant groups retained some form of blue light photocontrol of metabolic processes [25], the colonization of land and the evolution of phytochrome added the possibility of additional environmental control in the red : far-red region.

Acknowledgments. I wish to thank Dr. Maret Vesk for Fig. 6 and continued collaboration with the electron microscopy, Dr. G.M. Hallegraeff for Fig. 4 and Professor E. Holdsword and Mr. N.C. Bulleid for determining the action spectra measurements.

References

1. Akoyunoglou G, Anni H, Kalosakas K (1980) The effect of light quality and mode of illumination on chloroplast development in etiolated bean leaves. In: Senger H (ed) The blue light syndrome. Springer, Berlin Heidelberg New York, pp 473–484
2. Bogorad L (1975) Phycobiliproteins and complementary chromatic adaptation. Annu Rev Plant Physiol 26:369–401
3. Bunt JS (1975) Primary productivity of marine ecosystems. In: Leith H, Whittaker RH (eds) Primary productivity of the biosphere. Ecol Stud 14. Springer, Berlin Heidelberg New York, pp 169–183
4. Cox ER, Zingmark RG (1971) Bibliography of the Pyrrophyta. In: Roscowski JR, Parker BC (eds) Selected papers in phycology. Dep Bot, Univ Nebraskà, Lincoln, pp 803–808
5. Faust MA, Sager JC, Meeson BW (1982) Response of *Prorocentrum mariaelebouriae* (Dinophyceae) to light of different spectral qualities and irradiances: growth and pigmentation. J Phycol 18:349–356
6. Friedman AL, Alberte RS (1983) Isolation, biogenesis and distribution of a diatom light-harvesting complex. Workshop on molecular structure and function of light-harvesting pigment-protein complexes and photosynthetic reaction centres. Inst Molekularbiol Biophys, Zürich
7. Hallegraeff GM (1981) Seasonal study of phytoplankton pigments and species at a coastal station off Syndney: Importance of diatoms and nanoplankton. Mar Biol 61:107–118
8. Haxo FT (1960) Wavelength dependence of photosynthesis and the role of accessory pigments. In: Allen MB (ed) Comparative biochemistry of photoreactive pigments. Academic Press, London New York, pp 330–360
9. Hostetter HP, Stoermer EF (1971) Bibliography of the Bacillariophyceae. In: Roscowski JR, Parker BC (eds) Selected papers in phycology. Dep Bot, Univ Nebraska, Lincoln, pp 784–796
10. Ingram K, Hiller RG (1983) Isolation and characterization of a major chlorophyll a/c_2 light harvesting protein from a *Chroomonas* species (Cryptophyceae). Biochim Biophys Acta 722: 310:319

11. Jeffrey SW (1974) Profiles of photosynthetic pigments in the ocean using thin-layer chromatography. Mar Biol 26:101–110
12. Jeffrey SW (1976) A report on green algal pigments in the Central North Pacific Ocean. Mar Biol 37:33–37
13. Jeffrey SW (1980) Algal pigment systems. In: Falkowski PG (ed) Primary productivity in the sea. Plenum, New York, pp 33–58
14. Jeffrey SW (1981) Responses to light in aquatic plants. In: Lange OL, Nobel PS, Osmond CB, Ziegler H (eds) Encyclopaedia of plant physiology, New Ser, vol 12A. Physiological plant ecology I. Springer, Berlin Heidelberg New York, pp 249–276
15. Jeffrey SW, Hallegraeff GM (1980) Studies of phytoplankton species and photosynthetic pigments in a warm core eddy of the East Australian Current. I Summer populations. Mar Ecol Prog Ser 3:285–294
16. Jeffrey SW, Hallegraeff GM (1980) Studies of phytoplankton species and photosynthetic pigments in a warm core eddy of the East Australian Current. II A note on pigment methodology. Mar Ecol Prog Ser 3:295–301
17. Jeffrey SW, Sielicki M, Haxo FT (1975) Chloroplast pigment patterns in dinoflagellates. J Phycol 11:374–384
18. Jeffrey SW, Vesk M (1977) Effect of blue-green light on photosynthetic pigments and chloroplast structure in the marine diatom *Stephanopyxis turris*. J Phycol 13:271–279
19. Jeffrey SW, Vesk M (1981) The phytoplankton: systematics, morphology and ultrastructure. In: Clayton MN, King RJ (eds) Marine botany – an Australasian perspective. Longman Cheshire, Melbourne, pp 138–179
20. Jeffrey SW, Vesk M (1981) Blue-green light effects in marine micro-algae: enhanced thylakoid and chlorophyll synthesis. In: Akoyunoglou G (ed) Proc 5th Int Congr Photosynth, vol VI. Balaban Int Sci Serv, Philadelphia, pp 435–442
21. Jerlov NG (1976) Marine optics. Elsevier Oceanogr Ser 14. Elsevier, Amsterdam
22. Kamiya A, Miyachi S (1984) Blue-green and green light adaptation on photosynthetic activity in some algae collected from the subsurface chlorophyll layer in the western Pacific Ocean. This volume
23. Leith H (1975) Modelling in primary productivity of the world. In: Leith H, Whittaker RH (eds) Primary productivity of the biosphere. Ecol Stud 14. Springer, Berlin Heidelberg New York, pp 237–263
24. Lichtentaler HK, Buschmann C, Rahmsdorf U (1980) The importance of blue light for the development of sun-type chloroplasts. In: Senger H (ed) The blue light syndrome. Springer, Berlin Heidelberg New York, pp 485–494
25. Mohr H (1980) Interaction between blue light and phytochrome in photomorphogenesis. In: Senger H (ed) The blue light syndrome. Springer, Berlin Heidelberg New York, pp 97–109
26. Prézelin BB, Haxo FT (1976) Purification and characterization of peridinin chlorophyll *a* proteins from the marine dinoflagellate *Glenodinium* sp and *Gonyaulax polyedra*. Planta 128: 133–141
27. Round FE (1981) The ecology of algae. Cambridge Univ Press, Cambridge London New York, 653 pp
28. Senger H (ed) (1980) The blue light syndrome. Springer, Berlin Heidelberg New York, 665 pp
29. Shimura S, Fujita Y (1975) Changes in the activity of fucoxanthin-excited photosynthesis in the marine diatom *Phaeodactylum tricornutum* grown under different culture conditions. Mar Biol 33:185–194
30. Taylor FJR (1976) Dinoflagellates from the International Indian Ocean Expedition. Biblio Bot 312:1–234
31. Vesk M, Jeffrey SW (1977) Effect of blue-green light on photosynthetic pigments and chloroplast structure in unicellular marine algae from six classes. J Phycol 13:280–288
32. Wallen DG, Geen GH (1971) Light quality in relation to growth, photosynthetic rates, and carbon metabolism in two species of marine plankton algae. Mar Biol 10:34–43
33. Wallen DG, Geen GH (1971) Light quality and concentration of proteins, RNA, DNA and photosynthetic pigments in two species of marine plankton algae. Mar Biol 10:44–51

34. Waterbury JB, Watson SW, Guillard RRL, Brand LE (1979) Widespread occurrence of a unicellular marine planktonic cyanobacterium. Nature (London) 277:293–294
35. Whittaker RH, Likens GE (1975) The biosphere and man. In: Leith H, Whittaker RH (eds) Primary productivity in the biosphere. Ecol Stud 14. Springer, Berlin Heidelberg New York, pp 305–328

Blue Light Effects in Marine Macroalgae

M. J. Dring[1]

Introduction

The red, brown and green seaweeds, which together make up the marine macroalgae, constitute a taxonomically unorthodox and artificial collection of plants. They are united, however, by the range of habitats which they occupy – marine benthic sites from the upper intertidal to the lower limit of the photic zone in all optical water types [16] – and it is the variety of light climates that they experience in this habitat range [11] that makes their responses to light of particular interest. The diversity of photosynthetic pigments within this group of plants also indicates that the group has evolved different physiological solutions to at least one photobiological problem, and suggests that the photobiology of these plants would repay detailed study. Nevertheless, the investigation of nonphotosynthetic effects of light on marine macroalgae has been slow to develop and, although a wide range of blue light effects has now been reported, this is the first review to concentrate specifically and solely on the responses of seaweeds to blue light. Some of the responses have, however, been covered by recent reviews of photomorphogenesis in marine macroalgae [11, 20], and the blue light effects in one rather exceptional genus, the giant-celled green alga *Acetabularia,* are described elsewhere in this Volume by Schmid. Another macroalga which exhibits blue light effects is the siphonaceous chrysophyte *Vaucheria,* but this genus has been excluded from the present review, partly because most species are found in fresh or brackish waters rather than in strictly marine habitats, and partly because it was featured at the first Blue Light Conference [2].

Many of the responses to blue light reported for marine macroalgae are similar to those observed in other plants, but some responses are of a type unknown in other plant groups. This perhaps provides further evidence for the generalization that blue light effects are more common and more varied among the more "primitive" plants (e.g., [25]). The photobiology of the responses in seaweeds also presents a few surprises and, after a brief survey of the full range of responses reported, this review will concentrate on aspects of particular physiological interest. The ecological significance of these responses and, in particular, the evidence that the blue component of underwater light may control the vertical distribution or the phenological behaviour of marine macroalgae in their natural habitats, has been discussed in detail elsewhere [7, 11].

1 Botany Department, The Queen's University, Belfast BT7 1NN, Northern Ireland

Blue Light Effects in Biological Systems
ed. by H. Senger

Table 1. Nonphotosynthetic effects of blue light on marine macroalgae

Description of response	Genus[a]	Reference
1. Photoorientation responses:		
Induction of polarity in germinating zygotes	*Fucus* (B)	[1]
Negative phototropism of haptera	*Alaria* (B)	[3]
Negative phototropism of rhizoids	*Griffithsia* (R)	[37]
Chromatophore displacement	*Dictyota, Alaria* (B)	[28]
2. Effects on C-metabolism and growth:		
Stimulation of protein synthesis and mobilization of reserves	*Acetabularia* (G)	[4]
	Dictyota (B)	[27]
Stimulation of dark respiration	*Codium* (G)	[23]
3. Effects on vegetative morphology:		
Induction of two-dimensional growth	*Scytosiphon* (B)	[8]
Induction of hair formation	*Scytosiphon* (B)	[8]
	Dictyota (B)	[27]
	Acetabularia (G)	Schmid and Tünnermann this Vol.
4. Effects on reproductive development:		
Stimulation of cap formation	*Acetabularia* (G)	[33]
Induction of egg formation	*Laminaria* (B)	[22]
	Macrocystis (B)	[24]
Stimulation of egg release	*Dictyota* (B)	[18]
Inhibition of egg release	*Laminaria* (B)	[21]
5. Photoperiodic responses:		
Blue light alone effective as night-break	*Scytosiphon* (B)	[9]
Blue and red light effective as night-break	*Ascophyllum* (B)	[34]
	Rhodochorton (R)	[12]
Blue light effective as day extension	*Acrosymphyton* (R)	ten Hoopen 1981 (cited by [30])

[a] B = brown alga; G = green alga, R = red alga

Survey of Responses

Table 1 outlines most of the non-photosynthetic responses to light among marine macroalgae which have been shown to be specifically induced by blue wavelengths. Only a few responses have been included for *Acetabularia* since the wealth of blue light effects in this alga demanded a review to themselves (Schmid this Vol.). The responses have been somewhat arbitrarily grouped under five headings to assist comparison with similar types of response in other plant groups. The photoorientation responses (Group 1) are all typical expressions of the blue light syndrome (e.g., induction of polarity, Weisenseel [38]; phototropism, Dennison [5]; chloroplast movement,

Haupt [14]), and the responses in Group 2, which involve the regulation of carbon metabolism by blue light, are also similar in many respects to effects reported for flowering plants [36] and for microalgae [17]. Blue light initiates the transition from filamentous to two-dimensional growth in sporelings of *Scytosiphon,* and this effect is directly comparable to the action of blue light on fern gametophyte development [15], whereas the induction of hair formation in *Scytosiphon* and *Dictyota* may be indirectly comparable (via speculation about the role of hairs as nutrient antennae, see Dring and Lüning [11]) to blue-light-stimulated nitrate reduction in higher plants, microalgae and fungi ([39], Ninnemann this Vol.). There appear to be no parallels among higher plants or microalgae for the action of blue light on reproductive development in marine macroalgae, but reproduction in fungi is commonly influenced by blue light (e.g. [13, 31], Cerda-Olmeda this Vol.; Horwitz et al. this Vol.). However, the release of eggs from oogonia within 10–20 min of blue light being either switched on (*Dictyota*) or switched off (*Laminaria*) is a type of blue light effect unknown in other organisms, and is of particular interest because of its rapidity. The responses of the final group, in which blue light has been implicated in photoperiodic control, also seem to be unique to marine macroalgae. In all flowering plants and bryophytes investigated, photoperiodic responses are mediated by phytochrome [35], and at least one red seaweed (*Porphyra tenera*) also exhibits a red/far-red reversible inhibition of a short-day response [6, 29]. Phytochrome involvement has been specifically eliminated, however, in the responses of both *Scytosiphon* [9] and *Rhodochorton* [12], and photoperiodism can now be added to the list of symptoms of the blue light syndrome.

Action Spectra

The minimum requirement for a response to be included in Table 1 was the demonstration of a marked contrast in plant performance between broad-band red and broad-band blue light. In some cases, little more than this is known of the photobiology of the response. In others, the response to 4–10 wavelengths through the visible spectrum was measured to give a preliminary action spectrum, but often there was no attempt to determine the dose-response curve at any wavelength, so that highly effective and moderately effective wavelengths cannot always be distinguished. Detailed action or effectiveness spectra are available for only eight responses in five species, four of which are brown algae and the fifth is *Acetabularia.* Species of red algae are conspicuously lacking from this list, and more detailed work on the action spectra for blue light effects in this group deserves a high priority in the immediate future.

The existing action spectra refer to polarity induction in *Fucus* (primary response [1]), stimulation of egg release in *Dictyota* [18], night-break action in *Scytosiphon* [9] and hair-whorl formation in *Acetabularia* (Schmid and Tünnermann this Vol.). Detailed effectiveness spectra have been measured for the induction of both two-dimensional growth and hair formation in *Scytosiphon* [8] and of egg formation in *Laminaria* [22], and for the inhibition of egg release in *Laminaria* [21]. Three of these spectra (night-break action and two-dimensional growth in *Scytosiphon,* and egg formation in *Laminaria*) are reproduced in the review by Dring and Lüning [11]. Most spectra show a main peak at 440–450 nm, with subsidiary peaks at 405–415 nm and at 480–490 nm, although the spectra for *Dictyota* and *Fucus* have a simpler structure in the 400–500 nm region, and peak at 464 and 455 nm, respectively. Wavelengths longer than 500 nm were ineffective in all responses. Not all spectra covered the near-UV region, but most of those that do show a peak at 370–380 nm, and troughs at both 400 and 350 nm. Effectiveness increased further into the UV in the two spectra which included this region, and was highest at the shortest wavelength measured in both *Dictyota* (318 nm) and *Fucus* (254 nm). With the possible exception of these latter two species, all of the responses for which detailed action spectra are available appear to be typical examples of cryptochrome action, as defined by the criteria developed at this conference. The presence of such responses in several unrelated genera of brown algae suggests that cryptochrome is of general occurrence in the Phaeophyta, and the "blue" spotlight should now, perhaps, be directed towards the red algae and the cyanobacteria to establish whether cryptochrome is indeed typical of cryptogams.

There is some recent evidence, however, that not all responses in Table 1 are controlled by cryptochrome, since some of these "blue light effects" appear, in fact, to be "blue-and-red light effects". The short-day response of the red alga *Rhodochorton* was inhibited by 0.6 mmol m^{-2} of blue (448 nm) light in the middle of a long dark period, but the same photon exposure of red (662 nm) light was equally effective [12]. The inhibition by red light could not be reversed by subsequent treatment with a similar photon exposure of far red (731 nm) light, so that phytochrome does not appear to be responsible for the red light effect. There is less complete evidence for a similar response in the brown alga *Ascophyllum*, in which low exposures (1.5–2.0 mmol m^{-2}) of red and blue light were also equally effective as night-breaks, but no dose-response curves were determined and red/far-red reversibility was not tested [34]. Detailed action spectra are clearly needed – but are frustratingly difficult to obtain – for both of these responses, but the evidence already available is sufficient to suggest that some pigment, or combination of pigments, which has not previously been implicated in plant photomorphogenesis, may be operating in these systems. It may be significant that another marine alga, the green flagellate *Dunaliella*, has also been shown to exhibit a blue-and-red light effect (Ruyters et al. this Vol.).

Sensitivity of Responses to Blue Light

The photon exposures that have been shown to elicit a 50% response to blue light in various marine macroalgae are listed in Table 2. The total range covers seven orders of magnitude, but this can be reduced to two orders of magnitude if a group of responses with particularly high photon requirements and a single extremely sensitive response are excluded. The remaining responses, which require between 0.02 and 3.9 mmol m^{-2} for 50% effectiveness, resemble the majority of blue light responses in other plants in their sensitivity. It is the other responses with particularly high or particularly low photon requirements which may prove more interesting, and which are discussed in more detail.

Table 2. Photon exposures required for a 50% response to blue light in various marine macroalgae

Genus	Response	Photon exposure	Reference
a) High photon requirements:			
Laminaria	Induction of egg formation	1.95 mol m^{-2}	[22]
Macrocystis	Induction of egg formation	2.60 mol m^{-2}	[24]
Scytosiphon	Induction of hair formation	1.97 mol m^{-2}	[8]
Scytosiphon	Induction of 2-D growth	2.25 mol m^{-2}	[8]
b) Medium photon requirements:			
Laminaria	Inhibition of egg release	3.8 mmol m^{-2}	[21]
Fucus	Photoinduction of polarity (secondary response)	3.8 mmol m^{-2}	[1]
Rhodochorton	Inhibition of SD response by night-break	0.6 mmol m^{-2}	[12]
Fucus	Photoinduction of polarity (primary response)	0.17 mmol m^{-2}	[1]
Cystoseira	Photoinduction of polarity	0.11 mmol m^{-2}	[26]
Scytosiphon	Inhibition of SD response by night-break	0.02 mmol m^{-2}	[9]
Acetabularia	Induction of hair-whorls	0.02 mmol m^{-2}	Schmid and Tünnermann this Vol.
c) Low photon requirements:			
Dictyota	Stimulation of egg release	0.08 μmol m^{-2}	[18]

The threshold for the egg-release response in *Dictyota* at 464 nm occurs at about 6 nmol m^{-2} [18], which is only one order of magnitude higher than that for the most sensitive blue light response reported in higher plants, the first positive curvature of phototropism in oat coleoptiles (threshold at 458 nm, 0.7 nmol m^{-2} [32]). The phototropic response of sporangiophores in *Phycomyces* is substantially more sensitive, however, with a threshold at about 4 pmol m^{-2} (assuming 10^3 s exposure is required at a threshold irradiance of 10^{-9} W m^{-2}, 450 nm; Lipson this Vol.). The low sensitivity of *Dictyota* relative to *Phycomyces* could possibly be explained by the shading of the cryptochrome by photosynthetic pigments within the oogonium, and by the absence of the light focussing that occurs within the cylindrical sporangiophores of *Phycomyces* [5]. The even lower sensitivity of the algal responses with "medium" photon requirements (Table 2) may also be related to the degree of shading of the cryptochrome within the thalli, but this seems unlikely to provide a complete explanation for such a wide range of sensitivities.

The presence of different blue light effects with medium and high photon requirements within the same species (e.g., *Scytosiphon, Laminaria;* Table 2) indicates that shading of the blue/UV photoreceptor within the cell or thallus cannot explain the extraordinarily high photon requirements of a group of vegetative and reproductive responses in brown algae. In *Scytosiphon,* for example, different responses operate over a range of 10^5 in photon requirements. This is not as great as the range of 10^{10} units of irradiance observed in *Phycomyces* (Lipson this Vol.), but it could similarly be interpreted as evidence that different photoreceptors may be involved in different responses. The high light-requiring responses of these brown algae are also of particular interest because of the extended period over which reciprocity has been shown to operate. In standard reciprocity experiments, the response was proportional to the total photon exposure, irrespective of irradiation times ranging from 24 to 96 h [10], and the fertility of female gametophytes of *Laminaria* after 3 or 4 weeks in various simulated underwater light fields, each of which had a different blue content, was also determined by the total exposure to photons in the active waveband (i.e., 400–512 nm; [19]). Experiments in which the morphogenetical development of *Laminaria* and *Scytosiphon* were followed at different depths in the sea [10] also indicated that these plants are able to count and respond to the total number of photons received over a period of 2–3 weeks. Such long-term reciprocity has not been reported in any other photobiological system, and the biochemical mechanism of the photon-counting system that is implicated deserves detailed investigation.

Conclusions

This review has shown that a full range of symptoms of the blue light syndrome can be found among the seaweeds, and that a few responses in brown and green algae reveal typical cryptochrome action spectra. The marine macroalgae as a group, therefore, clearly deserve a place within the blue light family. There is no question, however, that many of the responses reported and investigated to date are far less convenient as experimental systems than the better known responses in microalgae, fungi and higher plants, and future research with seaweeds needs to concentrate on problems that cannot be tackled in other plants. At present, three main topics deserve attention: (1) the taxonomic distribution of cryptochrome within the group; (2) the mechanism of long-term reciprocity; and (3) the pigments involved in red-and-blue light effects. Hopefully, some progress in these areas will be reported at the next Blue Light Conference.

References

1. Bentrup FW (1963) Vergleichende Untersuchungen zur Polaritätsinduktion durch das Licht an der *Equisetum*spore und der *Fucus*zygote. Planta 59:472–491
2. Briggs WR, Blatt MR (1980) Blue light responses in the siphonaceous alga *Vaucheria*. In: Senger H (ed) The blue light syndrome. Springer, Berlin Heidelberg New York, pp 261–268
3. Buggeln RG (1974) Negative phototropism of the haptera of *Alaria esculenta* (Laminariales). J Phycol 10:80–82
4. Clauss H (1970) Effect of red and blue light on morphogenesis and metabolism of *Acetabularia mediterranea*. In: Brachet J, Bonotto S (eds) Biology of *Acetabularia*. Academic Press, London New York, pp 177–191
5. Dennison DS (1979) Phototropism. In: Haupt W, Feinleib ME (eds) Encyclopedia of plant physiology, New Ser, vol VII. Physiology of movements. Springer, Berlin Heidelberg New York, pp 506–566
6. Dring MJ (1967) Phytochrome in red alga, *Porphyra tenera*. Nature (London) 215:1411–1412

7. Dring MJ (1981) Photosynthesis and development of marine macrophytes in natural light spectra. In: Smith H (ed) Plants and the daylight spectrum. Academic Press, London New York, pp 297–314
8. Dring MJ, Lüning K (1975) Induction of two-dimensional growth and hair formation by blue light in the brown alga *Scytosiphon lomentaria*. Z Pflanzenphysiol 75:107–117
9. Dring MJ, Lüning K (1975) A photoperiodic response mediated by blue light in the brown alga *Scytosiphon lomentaria*. Planta 125:25–32
10. Dring MJ, Lüning K (1981) Photomorphogenesis of brown algae in the laboratory and in the sea. Int Seaweed Symp 8:159–166
11. Dring MJ, Lüning K (1983) Photomorphogenesis of marine macroalgae. In: Shropshire W, Mohr H (eds) Encyclopedia of plant physiology, New Ser, vol XVI. Photomorphogenesis. Springer, Berlin Heidelberg New York, pp 545–568
12. Dring MJ, West JA (1983) Photoperiodic control of tetrasporangium formation in the red alga *Rhodochorton purpureum*. Planta 159:143–150
13. Eger-Hummel G (1980) Blue-light photomorphogenesis in mushrooms (Basidiomycetes). In: Senger H (ed) The blue light syndrome. Springer, Berlin Heidelberg New York, pp 555–562
14. Haupt W (1982) Light-mediated movement of chloroplasts. Annu Rev Plant Physiol 33:205–233
15. Howland GP, Edwards EE (1979) Photomorphogenesis of fern gametophytes. In: Dyer AF (ed) The experimental biology of ferns. Academic Press, London New York, pp 393–434
16. Jerlov NG (1976) Marine optics. Elsevier, Amsterdam, pp 1–231
17. Kowallik W (1982) Blue light effects on respiration. Annu Rev Plant Physiol 33:51–72
18. Kumke J (1973) Beiträge zur Periodizität der Oogon-Entleerung bei *Dictyota dichotoma* (Phaeophyta). Z Pflanzenphysiol 70:191–210
19. Lüning K (1980) Critical levels of light and temperature regulating the gametogenesis of three *Laminaria* spp. (Phaeophyceae). J Phycol 16:1–15
20. Lüning K (1981) Photomorphogenesis of reproduction in marine macroalgae. Ber Dtsch Bot Ges 94:401–417
21. Lüning K (1981) Egg release in gametophytes of *Laminaria saccharina:* induction by darkness and inhibition by blue light and U.V. Br Phycol J 16:379–393
22. Lüning K, Dring MJ (1975) Reproduction, growth and photosynthesis of gametophytes of *Laminaria saccharina* grown in blue and red light. Mar Biol 29:195–200
23. Lüning K, Dring MJ (1984) Action spectra and spectral quantum yield of photosynthesis in marine macroalgae with thin and thick thalli. Mar Biol: in press
24. Lüning K, Neushul M (1978) Light and temperature demands for growth and reproduction of laminarian gametophytes in Southern and Central California. Mar Biol 45:297–309
25. Mohr H (1980) Interaction between blue light and phytochrome in photomorphogenesis. In: Senger H (ed) The blue light syndrome. Springer, Berlin Heidelberg New York, pp 97–109
26. Mosebach G (1938) Über den Einfluss des Lichtes auf die Polarisierung des befruchteten Eies von *Cystoseira barbata*. Ber Dtsch Bot Ges 56:210–225
27. Müller S, Clauss H (1976) Aspects of photomorphogenesis in the brown alga *Dictyota dichotoma*. Z Pflanzenphysiol 78:461–465
28. Nultsch W, Pfau J, Rüffer U (1981) Do correlations exist between chromatophore arrangement and photosynthetic activity in seaweeds? Mar Biol 62:111–117
29. Rentschler HG (1967) Photoperiodische Induktion der Monosporenbildung bei *Porphyra tenera* Kjellm. (Rhodophyta – Bangiophyceae). Planta 76:65–74
30. Rietema H, Breeman AM (1982) The regulation of the life history of *Dumontia contorta* in comparison to that of several other Dumontiaceae (Rhodophyta). Bot Mar 25:569–576
31. Russo VEA, Galland P, Toselli M, Volpi L (1980) Blue light induced differentiation in *Phycomyces blakesleeanus*. In: Senger H (ed) The blue light syndrome. Springer, Berlin Heidelberg New York, pp 563–569
32. Shropshire W (1979) Stimulus perception. In: Haupt W, Feinleib ME (eds) Encyclopedia of plant physiology, New Ser, vol VII. Physiology of movements. Springer, Berlin Heidelberg New York, pp 10–41

33. Terborgh J (1965) Effects of red and blue light on the growth and morphogenesis of *Acetabularia crenulata.* Nature (London) 207:1360–1363
34. Terry LA, Moss BL (1980) The effect of photoperiod on receptacle initiation in *Ascophyllum nodosum* (L.) Le Jol. Br Phycol J 15:291–301
35. Vince-Prue D (1975) Photoperiodism in plants. McGraw-Hill, London, pp 1–444
36. Voskresenskaya NP (1972) Blue light and carbon metabolism. Annu Rev Plant Physiol 23: 219–234
37. Waaland SD, Nehlsen W, Waaland JR (1977) Phototropism in a red alga, *Griffithsia pacifica.* Plant Cell Physiol 18:603–612
38. Weisenseel MH (1979) Induction of polarity. In: Haupt W, Feinleib ME (eds) Encyclopedia of plant physiology, New Ser, vol VII. Physiology of movements. Springer, Berlin Heidelberg New York, pp 485–505
39. Zumft WG, Castillo F, Hartmann KM (1980) Flavin-mediated photoreduction of nitrate by nitrate reductase of higher plants and microorganisms. In: Senger H (ed) The blue light syndrome. Springer, Berlin Heidelberg New York, pp 422–428

Blue-Green and Green Light Adaptations on Photosynthetic Activity in Some Algae Collected from Subsurface Chlorophyll Layer in the Western Pacific Ocean

A. KAMIYA[1] and S. MIYACHI[2]

Introduction

It has shown that the concentration of chlorophyll in the ocean often shows a rather sharp peak in a narrow layer at the depth between 50 to 100 m from the surface (subsurface chlorophyll layer [1]). The light intensity penetrating to this subsurface chlorophyll layer is usually near or below 1% of the surface light, and the main composition of the light at this layer is blue and green, with a maximum at ca. 480–550 nm [8]. It is therefore highly likely that the algae at this layer are metabolically adapted to low intensity of blue and green light. However, surprisingly few studies have been conducted on photosynthetic characteristics of these algae.

Previously, we succeeded in collecting two kinds of algae, *Ochromonas* (Chrysophyceae) and *Cryptomonas* (Cryptophyceae) from the subsurface chlorophyll layer in the western Pacific Ocean and revealed that *Ochromonas* cell is photosynthetically adapted to the short wavelengths and the low intensities of light [9]. In this paper, it was further shown that *Ochromonas* and *Cryptomonas* grown under the short wavelength and the low intensities of light contained much more photosynthetic pigments and exhibited higher photosynthetic activity than those grown under red light.

Materials and Methods

Culture Conditions. Isolation conditions of *Ochromonas* sp. and *Cryptomonas* sp. from the subsurface chlorophyll layer were described previously [9]. These algae were grown in 100 ml Erlenmeyer flasks kept at 20°C under continuous red light (> 580 nm, 3000 mW m^{-2}; Fig. 1a) for 5–12 days. Each flask contained 100 ml of filtered sea water containing 4 ml of Provasoli Enrichment [9].

In preliminary experiments, effects of blue ($340 < \lambda < 510$ nm), green ($500 < \lambda < 620$ nm) and red (> 600 nm) light (see Fig. 1b) on the formation of chlorophyll in *Ochromonas* cells were compared (Table 1). The effects of blue and green light on the

1 Laboratory of Chemistry, Faculty of Pharmaceutical Sciences, Teikyo University, Sagamiko, Kanagawa 199-01, Japan
2 Institute of Applied Microbiology, University of Tokyo, Bunkyo-ku, Tokyo 113, Japan

Blue Light Effects in Biological Systems
ed. by H. Senger

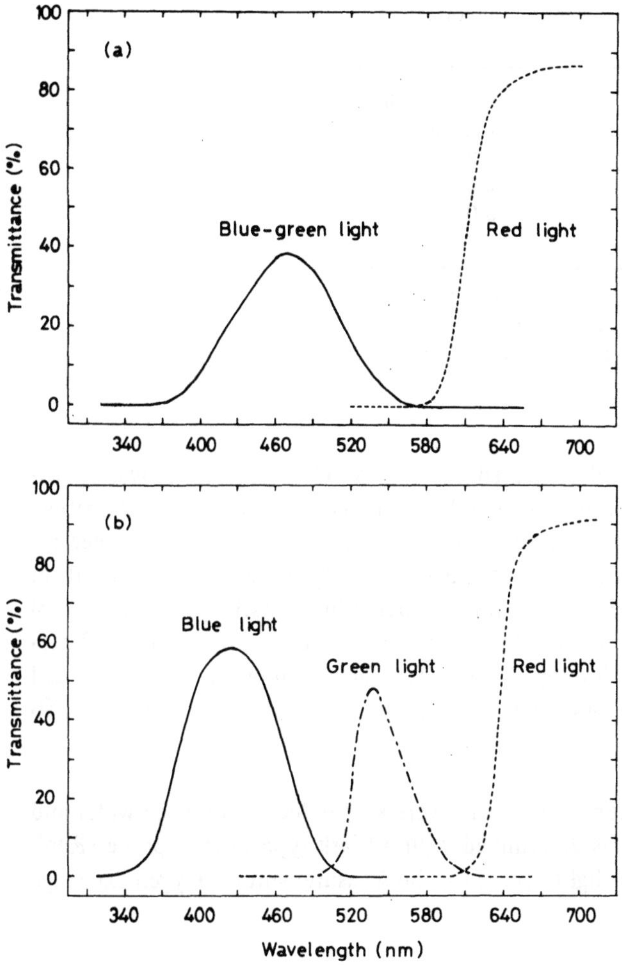

Fig. 1a,b. Spectral transmittances of a blue-green and red light filters used in the experiments for *Ochromonas* **b** blue, green and red light filters used in the experiments for *Cryptomonas.* See Materials and Methods for explanation

formation of chlorophyll were practically equal. Therefore, we used the blue-green light ($370 < \lambda < 570$ nm; obtained with Paraglass 315K, Kyowa Gas Chemical Industry Co. Ltd., Tokyo) and red light (> 580 nm; obtained with Paraglass 102K) (Fig. 1a) for the experiments in this algae. Light source was provided by a bank of 40 W cool-white fluorescence tubes (Fishlux; Toshiba Co. Ltd., Tokyo).

Based on the absorption spectrum of *Cryptomonas* (see Fig. 5b), we used blue, green and red lights (see Fig. 1b) for the experiments in *Cryptomonas.* Blue ($340 < \lambda < 510$ nm) and red (> 600 nm) light was obtained by Corning 5-60 and Corning V-R64, respectively. Green ($500 < \lambda < 620$ nm) light was obtained by the combined use of two filters; Corning 4-96 and VO-52 (Fig. 1b). Light was provided by a bank of 40 W daylight fluorescence tubes (Toshiba Co. Ltd., Tokyo). The light intensity was adjusted with a neutral density filter (TND; Toshiba Kasei Kogyo Co. Ltd., Tokyo).

Analysis of Pigments. Chlorophylls in *Ochromonas* were extracted from the frozen cells in the dark by adding 3–5 ml of 90% acetone. For the determination of the contents of chlorophyll and phycoerythrin, *Cryptomonas* cells were suspended in 0.02 M

Table 1. Effects of blue, green and red light on the formation of chlorophyll in *Ochromonas*

Light condition	Chlorophyll (μg ml^{-1} susp)
Blue light ($340 < \lambda < 510$ nm)	6.2
Green light ($500 < \lambda < 620$ nm)	5.0
Red light (> 600 nm)	3.3

The alga which had been grown under red light (3000 mW m^{-2}) for 7 days was transferred into fresh medium and then illuminated with blue, green and red light for 5 days. Temp, 20°C. Light intensity, 800 mW m^{-2}

K-phosphate buffer (pH 6.5) and sonicated for 2 min. The homogenate was centrifuged at 100,000 *g* for 90 min in a RP65-622 rotor with a Hitachi ultracentrifuge (70P-72). The content of phycoerythrin in the supernatant was determined spectrophotometrically, using the extinction coefficient of cryptomonad phycoerythrin type I [10]. Chlorophylls were extracted from the pellet in the dark by adding 3–5 ml of 100% acetone. Amounts of chlorophyll a and c were determined spectrophotometrically, using an equation developed by Jeffrey and Humphrey [7]. Whole cell spectra of algal cultures were obtained at room temperature with a Hitachi 556 spectrophotometer.

Measurement of Oxygen Exchange. Algal cells were suspended in fresh sea water and the rate of oxygen exchange was determined with a Clark-type oxygen probe (Rank Brothers, London) at 25°C. The light source for the measurement of oxygen exchange was a 500 W Xenon lamp.

Electron Microscopy. Algal cells collected from culture medium were washed once in sea water and fixed with 4% glutaraldehyde and 0.4 M sucrose in 0.07 M sodium phosphate buffer (pH 7.4) at 4°C for a week. Cells were washed three times with the same buffer containing 0.4 M sucrose and then postfixed in 1% osmiumtetroxide in the same buffer at room temperature for 2 h. The sample was dehydrated in an ethanol series and embedded in a low-viscosity epoxy resine. Section was cut with a glass knife on a ultrotome III ultramicrotome and examined with a Hitachi electron microscope.

Results

Blue-Green Light Adaptation on the Activity of Photosynthesis in *Ochromonas* cells

Ochromonas cells grown under red light were transferred to fresh medium and incubated under blue-green and red light, respectively. Then the increases of growth and chlorophyll content were followed for 5 days (Fig. 2). The growth as followed by the increase in packed cell volume (pcv) was more vigorous under blue-green light than that under red light. Chlorophyll formation was also higher under blue-green light than

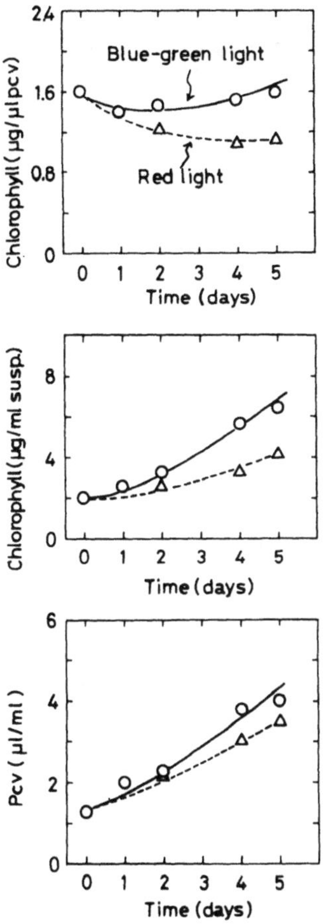

Fig. 2a–c. Time courses of chlorophyll formation under blue-green and red light in *Ochromonas* cells. The alga which had been grown under red light (3000 mW m^{-2}) for 7 days was transferred into fresh medium and then illuminated with blue-green ($370 < \lambda < 570$ nm) and red (> 580 nm) light for 5 days. **a** Chlorophyll content; µg/µl pcv **b** chlorophyll content; µg/ml suspension **c** growth; packed cell volume (µl ml^{-1} suspension). Temp; 20°C. Light intensitiy; 1200 mW m^{-2}

under red light. The enhancing effect of blue-green light on the chlorophyll formation during growth was saturated at intensities as low as 500 mW m^{-2} as reported previously [9].

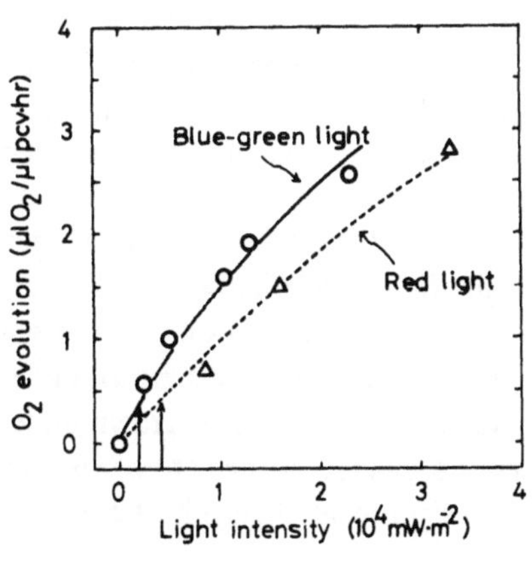

Fig. 3. Oxygen evolution under various intensities of blue-green and red light in *Ochromonas* cells. Oxygen evolution was determined at 25°C. Other conditions were the same as indicated in Fig. 2

Fig. 4a,b. Oxygen exchange under various intensities of blue-green light in *Ochromonas* cells. Algal cells were grown under blue-green (1400 mW m^{-2}) and red (3000 mW m^{-2}) light for 8 days, respectively. a O_2 exchange; μlO_2 μl^{-1} pcv h^{-1}-basis; b O_2 exchange; μlO_2 μg^{-1} chl h^{-1}-basis. Chlorophyll content; the cells grown under blue-green light (3.1 μg μl^{-1} pcv, chl a/c = 6.1), the cells grown under red light (1.7 μg μl^{-1} pcv, chl a/c = 6.5). Temp; 25°C

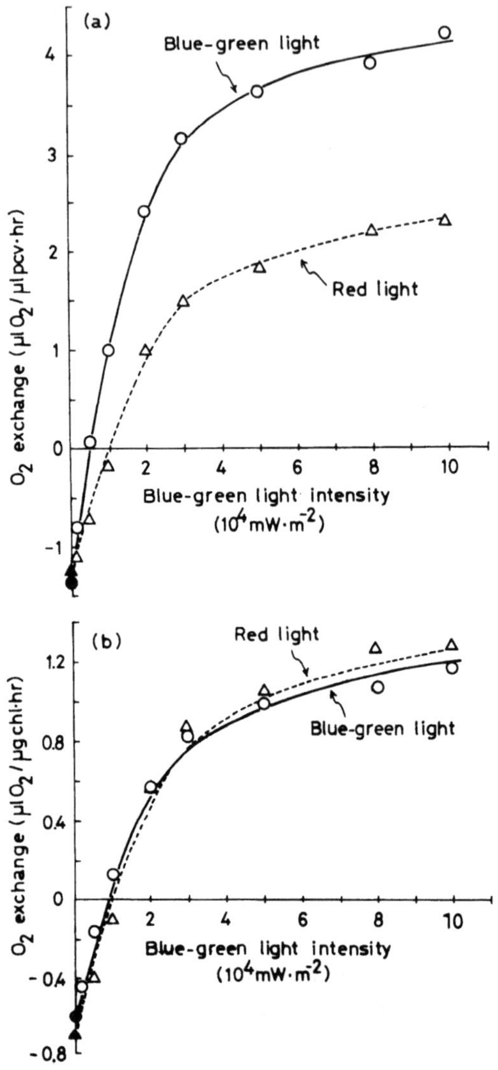

Figure 3 shows the rate of photosynthetic oxygen evolution under several intensities of blue-green and red light in *Ochromonas* cells which had been grown under red light. The rate of oxygen evolution under blue-green light was higher than that under red light. In the next experiment, *Ochromonas* cells were grown under blue-green (1400 mW m^{-2}) and red (3000 mW m^{-2}) light, respectively, which give the same rate of photosynthetic oxygen evolution (see arrows in Fig. 3). Figure 4 shows the oxygen exchange under several intensities of blue-green light in the cells grown under blue-green and red light for 8 days, respectively. The rate of oxygen evolution on pcv-basis in the cells grown under blue-green light was higher than that in the cells grown under red light (Fig. 4a). Since the chlorophyll content was greater in the cells grown under blue-green light than that under red light, there was no change in the rate of oxygen evolution expressed on chlorophyll-basis (Fig. 4b). No significant change in the relative

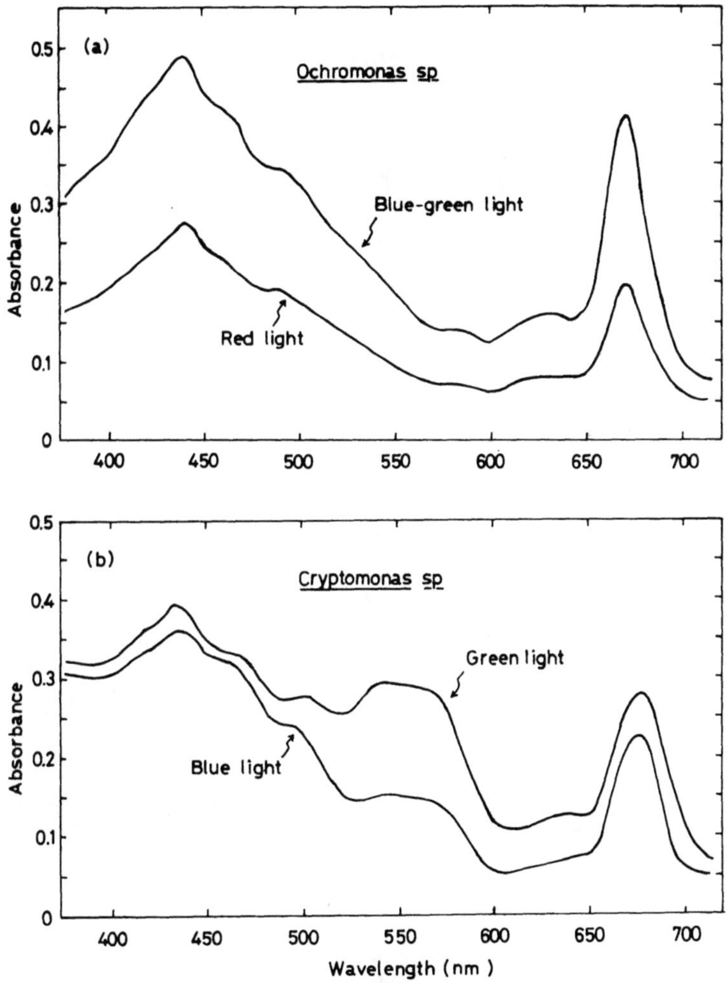

Fig. 5a,b. Absorption spectra of cell suspension. **a** *Ochromonas* cells grown under blue-green ($370 < \lambda < 570$ nm, 1400 mW m^{-2}) and red (> 580 nm, 3000 mW m^{-2}) light for 6 days; **b** *Cryptomonas* cells grown under green ($500 < \lambda < 620$ nm, 800 mW m^{-2}) and blue ($340 < \lambda < 510$ nm, 800 mW m^{-2}) light for 6 days at 20°C. The spectra were compared at the same cell concentration

proportion of the pigment (chl a/c ratio) occurred. Figure 5a shows absorption spectra of cell suspension of *Ochromonas* grown under blue-green and red light, respectively. These results indicate that algal cells grown under blue-green light contained much more chlorophyll, and exhibited higher photosynthetic capacity on pcv-basis than those grown under red light.

Chloroplasts of typical cells grown under red light contained several thylakoid bands which runs parallel (Fig. 6a). On the other hand, chloroplasts of cells grown

---

Fig. 6a,b. Electron micrographs of *Ochromonas* cells grown under (**a**) red light (3000 mW m^{-2}) and (**b**) blue-green light (1400 mW m^{-2}) for 8 days. **a** × 6000, **b** × 4000. Other conditions were the same as indicated in Fig. 2

Fig. 6a

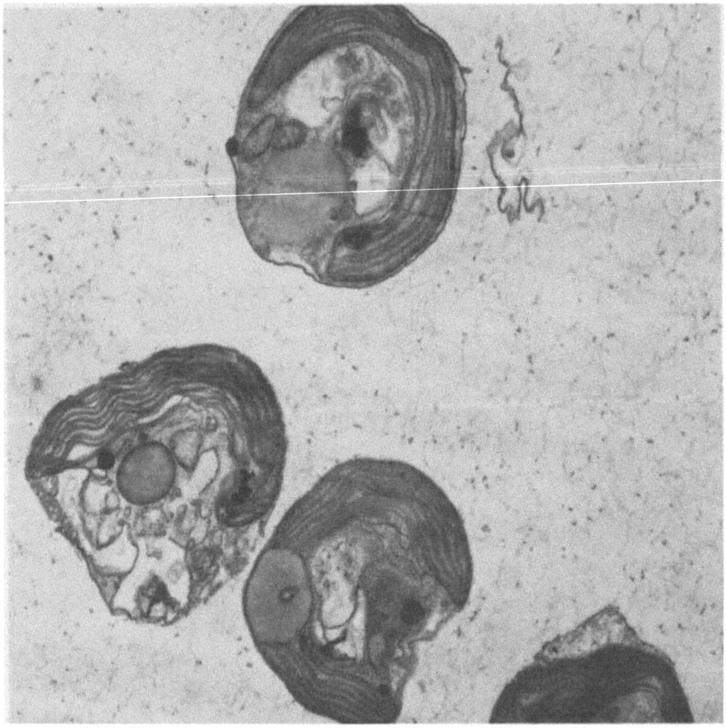

Fig. 6b

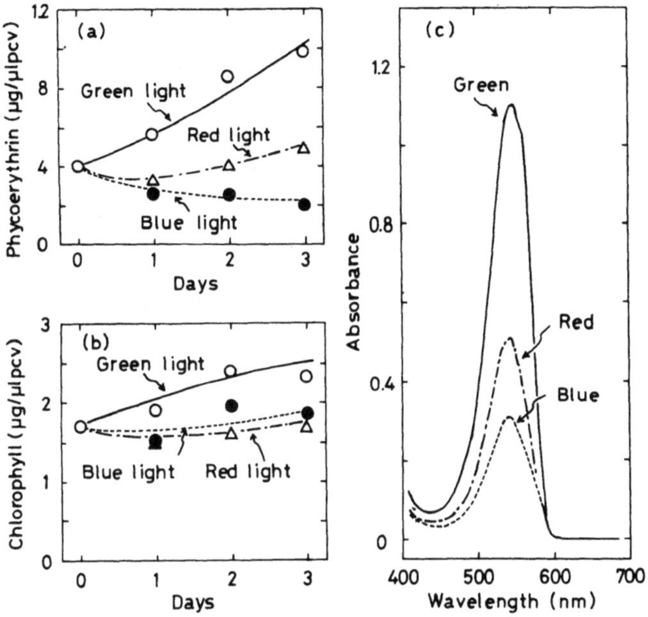

Fig. 7a–c. Time courses of **a** phycoerythrin and **b** chlorophyll formation under blue, green and red light in *Cryptomonas* cells. The alga which had been grown under red light was transferred into fresh medium and then illuminated with blue, green and red light for 3 days. **c** Absorption spectra of 100,000 *g* supernatant obtained from *Cryptomonas* cells grown under blue, green and red light for 3 days, respectively. Light intensity; 800 mW m^{-2}. Temp; 20°C

under blue-green light contained more grana thylakoid bands than those grown under red light (Fig. 6b).

Green Light Adaptation on the Formation of Photosynthetic Pigments in *Cryptomonas* Cells

Figure 7 shows the formation of phycoerythrin and chlorophyll in *Cryptomonas* cells under blue ($340 < \lambda < 510$ nm), green ($500 < \lambda < 620$ nm) and red (> 600 nm) light, respectively. The formation of phycoerythrin and chlorophyll was greatly enhanced by green light, the rate of formation of the former pigment being much higher than that of the latter pigment. Red light slightly enhanced the formation of phycoerythrin. However, blue light did not enhance the formation of both pigments during experimental periods. No significant change in the relative proportion of the pigments (chl a/c ratio; 4.4–4.8) occurred. Figure 7c shows the absorbance spectrum of 100,000 *g* supernatant obtained from the cells illuminated with blue, green and red light for 3 days, respectively. As shown in these spectra, other phycobilin pigment than phycoerythrin was not detected. These results indicate that green light enhances the formation of phycoerythrin and chlorophyll in *Cryptomonas* cells (see Fig. 5b).

Figure 8 shows the enhancing effect of green light on phycoerythrin- and chlorophyll-formation during growth. Green light effects on the formation of phycoerythrin

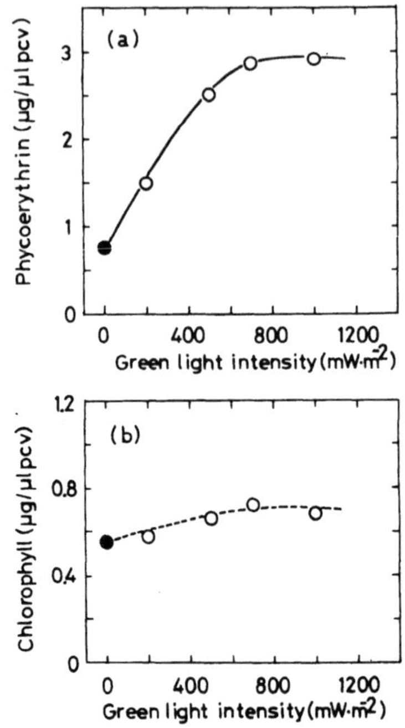

Fig. 8a,b. Green light intensity dependency of **a** phycoerythrin and **b** chlorophyll formation in *Cryptomonas* cells. Algal cells were grown under green light at indicated intensities for 2 days. Temp; 20°C

and chlorophyll were saturated at intensities as low as 600 mW m^{-2}, which were similar to the saturating intensities for the blue-green light effect on the chlorophyll formation in *Ochromonas* cells [9].

Discussion

Attenuation of light with depth in clear water results in the rapid disappearance of red and yellow light, with blue-green light (maximum about 480 nm) penetrating furthest. On the other hand, in coastal waters and upwelling regions, the optical properties of seawater are further modified by increased concentration of phytoplanktons and suspended particles of inorganic or organic origin. These substances shift the under water irradiance from blue-green to green (around 550 nm) light (cf. [8]). The results obtained in the present study clearly showed that the algae grown under low intensity of blue-green or green light (simulated light conditions at the subsurface chlorophyll layer) contained much more light-harvesting pigments and exhibited higher photosynthetic activity. As shown in Fig. 4b, the rate of oxygen evolution on Chl-basis in *Ochromonas* cells grown under blue-green light was equal to that grown under red light. These results indicate that blue-green light seems to enhance not only the formation of light-harvesting pigments but also of the photosynthetic unit.

The formation of chlorophyll and phycoerythrin under blue-green or green light in algae collected from subsurface chlorophyll layer was saturated at intensities as low as 500–600 mW m^{-2} [9] (Fig. 8), which is equivalent to the maximum light intensities in that region [9]. Since the light reaching this region mostly consists of blue and green, it was assumed that the algae used in this experiment are adaptable to the light conditions in the subsurface chlorophyll layer efficiently.

However, there are surprisingly few studies on the photosynthetic adaptability of phytoplankton to the spectral changes associated with the depth of water. Shimura and Ichimura [12] reported that the deep phytoplankton in the Kuroshio off Kii Peninsula possess a higher photosynthetic activity than surface phytoplankton when both were measured at the depth of 60 m. Eppley et al. [3] found evidence for shade adaptation of the deep phytoplankton in the subtropical central gyre of the North Pacific. These results suggest that the phytoplankton in the deep layer can utilize blue-green or green light more efficiently than the phytoplankton in the surface.

On the other hand, laboratory experiments have shown that blue-green light does exert a profound effect on the pigment content. Jeffrey and Vesk [6], and Vesk and Jeffrey [13] reported that blue-green light, compared to white light, caused chlorophyll increases (55–146%) in eight species of unicellular marine algae. However, they reported that there was no significant change in the ratios of chlorophyll a/c and the light-harvesting carotenoid to chlorophyll a in eight species tested. The algae collected from subsurface chlorophyll layer also utilized short-wavelength light and increased the total amount of pigment without changing their proportions. Therefore, deep phytoplankton shows the similar light adaptation to the marine algae used in other laboratory experiments mentioned above. In this connection, using *Chlamydomonas reinhardtii* and *Chlorella pyrenoidosa,* Hess and Tolbert [5] reported that the chlorophyll content per cell volume in the algae grown under blue light was only about 20% higher than that in the algae grown under red light. These results indicate that blue light-enhancing effect on the chlorophyll formation in marine algae is more pronounced than that in freshwater algae.

The phycobiliproteins play a major role as light-harvesting pigments in red, blue-green and cryptomonad algae. The light-harvesting capacity of blue-green algae can be optimized at green and red light, respectively, through complementary chromatic adaptation [2]; green light stimulates the synthesis of green light-absorbing phycoerythrin, whereas red light stimulates the synthesis of red light-absorbing phycocyanin. In our experiments, it was revealed that the formation of phycoerythrin in *Cryptomonas* was higher in green light than in blue or red light (Fig. 7). Thus biosynthesis of biliprotein appears to be controlled by the mechanism of chromatic adaptation. However, no formation of phycocyanin was observed under red light. Usually, only one biliprotein is present in each cryptomonad species [11]. In this respect, biosynthesis of bile pigment in cryptomonad is different from that in blue-green algae.

Haxo and Fork [4] reported that photosynthetic efficiency in cryptomonad cells containing phycoerythrin is highest in the spectral region where the phycoerythrin contributes most prominently to cell absorption. We also observed higher efficiency of photosynthesis in *Cryptomonas* cells under green light (unpublished data). Preferential formation of phycoerythrin and chlorophyll observed under green light would be due, in part, to higher activity of photosynthesis in the spectral region.

Abstract

The cells of *Ochromonas* sp. (Chrysophyceae) and *Cryptomonas* sp. (Cryptophyceae) were isolated from the subsurface chlorophyll layer in the western Pacific Ocean.

1. The effects of blue-green and red light given during the culture on the activity of photosynthesis in *Ochromonas* sp. cells were examined. The intensities of blue-green and red light during the algal culture were adjusted so as to give the same rate of O_2 evolution at the initial stage of the culture. The cells grown under blue-green light contained much more chlorophyll and exhibited higher photosynthetic rate (O_2 evolution measured under blue-green light) than those grown under red light. At the ultrastructural level, more stacking of thylakoid system occurred in the cells grown under blue-green light than in the cells grown under red light.

2. The formation of chlorophyll and phycoerythrin in *Cryptomonas* sp. cells was determined under blue, green and red light. The formation of chlorophyll and phycoerythrin under green light was more active than those under blue and red light and saturated at intensities as low as 600 mW m^{-2}.

These results show two types of light adaptation on the photosynthetic activity in marine algae; one is the blue-green light adaptation on chlorophyll formation in *Ochromonas* sp. and the other is the green light adaptation on chlorophyll and biliprotein formation in *Cryptomonas* sp.

Acknowledgments. We thank Dr. T. Oh-hama of Tokyo University for reading the manuscript and Dr. H. Hirosawa of Tokyo University for his help in electron micrography. Taxonomical identification of *Cryptomonas* was done by Dr. Y. Hara, Tsukuba University. This research was supported by Grant-in-Aid for Special Project Research on Marine Biological Processes (No. 57108011) from the Ministry of Education, Science and Culture, of Japan. Thanks are also due to the Deutsche Forschungsgemeinschaft for the general support for this Congress.

References

1. Anderson GC (1969) Subsurface chlorophyll maximum in the northeast Pacific Ocean. Limnol Oceanogr 14:386–391
2. Bogorad L (1975) Phycobiliproteins and complementary chromatic adaptation. Annu Rev Plant Physiol 26:369–401
3. Eppley RW, Renger EH, Venrick EL, Mullin MM (1973) A study of plankton dynamics and nutrient cycling in the central gyre of the north Pacific Ocean. Limnol Oceanogr 18:534–551
4. Haxo FT, Fork DC (1959) Photosynthetically active accessory pigments of cryptomonads. Nature (London) 184:1051–1052
5. Hess JL, Tolbert NE (1967) Changes in chlorophyll a/b ratio and products of $^{14}CO_2$ fixation by algae grown in blue or red light. Plant Physiol 42:1123–1130
6. Jeffrey SW, Vesk M (1977) Effect of blue-green light on photosynthetic pigments and chloroplast structure in the marine diatom *Stephanopyxis turris*. J Phycol 13:271–279
7. Jeffrey SW, Humphrey GF (1975) New spectrophotometric equations for determining chlorophyll a, b, c_1 and c_2 in higher plants, algae and natural phytoplankton. Biochem Physiol Pflanz 167:191–194
8. Jeffrey SW (1981) Responses to light in aquatic plants. In: Lange OL et al. (eds) Encyclopedia of plant physiology, vol XIIA. Springer, Berlin Heidelberg New York, pp 249–276

9. Kamiya A, Miyachi S (1980) Blue light effects on some algae collected from subsurface chlorophyll layer in the western Pacific Ocean. In: Senger H (ed) The blue light syndrome. Springer, Berlin Heidelberg New York, pp 605–615
10. MacColl R, Berns DS, Gibbons O (1976) Characterization of cryptomonad phycoerythrin and phycocyanin. Arch Biochem Biophys 177:265–275
11. O'Carra P, O'hEocha C (1976) Algal biliproteins and phycobilins. In: Goodwin TW (ed) Chemistry and biochemistry of plant pigments, 2nd edn, vol I. Academic Press, London New York, pp 328–376
12. Shimura S, Ichimura S (1973) Selective transmission of light in the ocean waters and its relation to phytoplankton photosynthesis. J Oceanogr Soc Jpn 29:257–266
13. Vesk M, Jeffrey SW (1977) Effect of blue-green light on photosynthetic pigments and chloroplast structure in unicellular marine algae from six classes. J Phycol 13:280–288

Organism Index

Subject Index

Volume 11
Hormonal Regulation of Development III
Role of Environmental Factors
Editors: A. Pirson, M. H. Zimmermann
With contributions by numerous experts
ISBN 3-540-10197-7
In preparation

Physiological Plant Ecology
Editors of Volume 12A - D: O. L. Lange, P. S. Nobel, C. B. Osmond, H. Ziegler

Volume 12 A
Responses to the Physical Environment
1981. 110 figures. XV, 625 pages
ISBN 3-540-10763-0

Volume 12 B
Water Relations and Carbon Assimilation
1982. 153 figures. XI, 747 pages
ISBN 3-540-10906-4

Volume 12 C
Responses to the Chemical and Biological Environment
1983. 104 figures. XI, 799 pages
ISBN 3-540-10907-2

Volume 12 D
Ecosystem Processes: Mineral Cycling, Productivity and Man's Influence
1983. 61 figures. XI, 644 pages
ISBN 3-540-10908-0

Volume 13 A
Plant Carbohydrates I
Intracellular Carbohydrates
Editors: F. A. Loewus, W. Tanner
1982. 103 figures. XXII, 918 pages
ISBN 3-540-11060-7

Volume 13 B
Plant Carbohydrates II
Extracellular Carbohydrates
Editors: W. Tanner, F. A. Loewus
With contributions by numerous experts
1981. 124 figures. XXII, 769 pages
ISBN 3-540-11007-0

Volume 14 A
Nucleic Acids and Proteins in Plants I
Structure, Biochemistry and Physiology of Proteins
Editors: D. Boulter, B. Parthier
With contributions by numerous experts
1982. 135 figures. XX, 768 pages
ISBN 3-540-11008-9

Volume 14 B
Nucleic Acids and Proteins in Plants II
Structure, Biochemistry and Physiology of Nucleic Acids
Editors: B. Parthier, D. Boulter
With contributions by numerous experts
1982. 173 figures. XVIII, 774 pages
ISBN 3-540-11140-9

Volume 15
Inorganic Plant Nutrition
Editors: A. Läuchli, R. L. Bieleski
With contributions by numerous experts
1983. 131 figures. Approx. 900 pages
(In 2 volumes. Not available separately).
ISBN 3-540-12103-X

Volume 16 (Part A + B)
Photomorphogenesis
Editors: W. Shropshire, Jr., H. Mohr
With contributions by numerous experts
1983. 173 figures. XXX, 832 pages
(In 2 parts, not available separately)
ISBN 3-540-12143-9

Springer-Verlag
Berlin
Heidelberg
New York
Tokyo

MIX
Papier aus verantwortungsvollen Quellen
Paper from responsible sources
FSC® C105338

If you have any concerns about our products,
you can contact us on
ProductSafety@springernature.com

In case Publisher is established outside the EU,
the EU authorized representative is:
Springer Nature Customer Service Center GmbH
Europaplatz 3, 69115 Heidelberg, Germany

Printed by Libri Plureos GmbH
in Hamburg, Germany